# The Routledge Handbook of Medical Anthropology

*The Routledge Handbook of Medical Anthropology* provides a contemporary overview of the key themes in medical anthropology. In this exciting departure from conventional handbooks, compendia and encyclopedias, the three editors have written the core chapters of the volume, and in so doing invite the reader to reflect on the ethnographic richness and theoretical contributions of research on a number of important themes. These key topics include: the clinic and the field, bioscience and medical research, infectious and non-communicable diseases, biomedicine, complementary and alternative modalities, structural violence and vulnerability, gender and aging, reproduction and sexuality. These themes are explored through a rich variety of case histories, presented by over 60 authors from around the world, which reflect on the diverse cultural contexts in which people experience health, illness, and healing. Each chapter and its case studies are introduced by a photograph which illustrates medical and visual anthropological responses to inequality and vulnerability. An indispensable reference book in this fastest growing area of anthropological study, *The Routledge Handbook of Medical Anthropology* is a unique and innovative contribution to the field.

**Lenore Manderson** is Professor of Public Health and Medical Anthropology at the School of Public Health, University of the Witwatersrand, South Africa, and Professor of Anthropology at the Institute at Brown for Environment and Society, Brown University, USA.

**Elizabeth Cartwright** is Professor of Anthropology and Director of the Hispanic Health Projects and the Latino Studies Program at Idaho State University, USA.

**Anita Hardon** is Professor of Health and Social Care at the Faculty of Social and Behavioural Sciences and Dean of the Amsterdam Institute for Social Science Research, at the University of Amsterdam, the Netherlands.

# The Routledge Handbook of Medical Anthropology

*Edited by Lenore Manderson,*
*Elizabeth Cartwright and Anita Hardon*

LONDON AND NEW YORK

First published 2016 by Routledge

2 Park Square, Milton Park, Abingdon, Oxfordshire OX14 4RN
711 Third Avenue, New York, NY 10017

*Routledge is an imprint of the Taylor & Francis Group, an informa business*

First issued in paperback 2018



*British Library Cataloguing in Publication Data*
A catalogue record for this book is available from the British Library

*Library of Congress Cataloging-in-Publication Data*
Names: Manderson, Lenore, editor.
Title: The routledge handbook of medical anthropology / edited by
Lenore Manderson, Elizabeth Cartwright and Anita Hardon.
Description: Milton Park, Abingdon, Oxon ; New York, NY : Routledge,
2016. | Includes index.
Identifiers: LCCN 2015041657 | ISBN 9781138015630 (hardback : alk. paper)
Subjects: LCSH: Medical anthropology.
Classification: LCC GN296 .R68 2016 | DDC 306.4/61—dc23
LC record available at http://lccn.loc.gov/2015041657

ISBN: 978-1-138-01563-0 (hbk)
ISBN: 978-1-138-61287-7 (pbk)

Typeset in Bembo
by Apex CoVantage, LLC

To our mothers, with love—
Mardi Manderson, Agnes Homola Cartwright and Antine Hardon-Baars

# Contents

# Figures

# About the Figures

*Jerome Crowder and Elizabeth Cartwright*

Images do much more than simply illustrate: they convey, evoke and expand our thoughts and feelings about the subject matter very differently than does text. The photographic component of this volume sets it apart from other handbooks in print. These images, counterparts of the written texts of the chapters and their case studies, offer a different kind of commentary on health and the human condition. Lenore, Anita and Liz had discussed how images might work in ways moving beyond illustration and instead adding a richness to our understanding. Jerome Crowder worked with Liz to assist them in collecting and selecting these images to complement the chapters and add another dimension within which to consider contemporary medical anthropology.

We—Jerome and Liz—savored the opportunity because it would allow us to review hundreds of images taken by health researchers from around the globe, as well as make a contribution to an already significant volume. We needed high-quality photographs, and this complicated our task—many anthropologists take photos in the field, but we wanted images that went technically and aesthetically beyond those we routinely collect on our iPads and smartphones. Our call was for photographs taken by researchers and practitioners in the field, about the people they work with, concerning the situations people endure today. The images were to be ethnographic and to demonstrate a connection between the subject and the photographer.

In early 2015, a worldwide call was posted on anthropology listservs, photography sites, Facebook pages and through email, seeking respondents with images from their health-based anthropology fieldwork. We sent the announcement to colleagues and colleagues of colleagues, we racked our brains for ways to spread the word that we needed photographs that explore health and the human condition, the mundanity of life in the street and the suffering of those not so fortunate as ourselves. We received submissions from known and unknown professionals and students, sharing with us their exquisite images, harrowing photographs and their depiction of graphic situations. The call generated an impressive number of images to review and broadened our understanding of the quantity and diversity of work being conducted today by visual anthropologists. As practicing medical anthropologists ourselves, we submitted our own images to the pile, recognizing that ours would only be used for themes that were not well represented from the general call (if the case arose).

Criteria for the final selection required that each image be the strongest of a category, able to stand alone, not as an illustration of text within the chapter but as a prelude to it, capturing some of its points but also complementing the group as a whole. We thank everyone who dared to participate and to take the chance and send in a few field snaps. We are honored to have had the opportunity to review them all and work with those who were finally included in the volume. One image was selected to accompany each chapter of the volume, redolent of its theme, although not necessarily restricted to that one chapter. Whether or not they are viewed in order, the images stand alone as a photo essay exploring the role of medical anthropology; as a group they make a visual commentary on the future of our practice.

## *Cover Image*

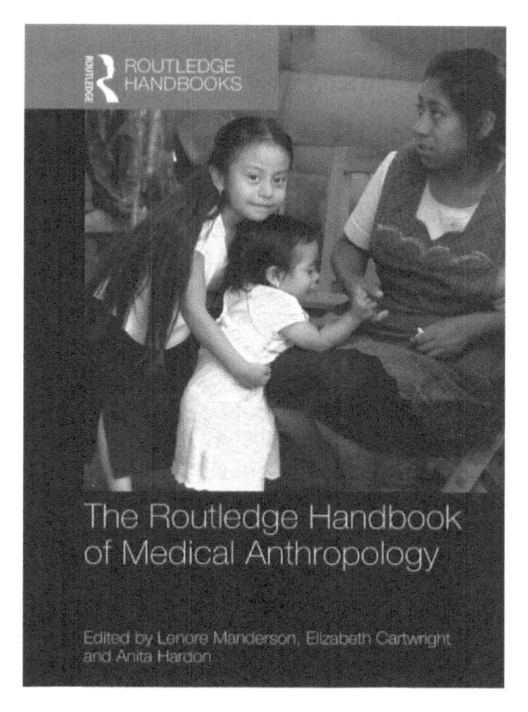


Mother and Daughters, 2006. Oaxaca, Mexico.
Photo by Elizabeth Cartwright.

*This mother brought her young daughter to the local herbalist to be cured for a lingering sore throat. The traditional curing session was one of several that the herbalist did that day. Everyone was dressed up for the occasion, much more than usual, as they knew that I would be there with one other person filming and taking photographs of the healing encounter. The local healer and her patients in this small village in southern Mexico used multiple modalities to cure the common ailments of children, of which one of the most common ailments was* susto, *or fright.*

—Elizabeth Cartwright

# Acknowledgments

Lesley Riddle, at the time Senior Publisher for Religion and Anthropology with Routledge, first discussed this book with Lenore at the 2012 meetings of the American Anthropological Association. The attraction was to create a book which was engaged with issues of contemporary concern to medical anthropologists, to reflect on the theoretical and practical implications of this, and to showcase the extraordinary ethnographic work being undertaken worldwide by younger scholars working within the discipline, and across it to engage also with science and technology studies, cultural studies, public health and clinical research. The three of us met in Tarragona, Spain, in June 2013, at the first international conference of medical anthropologists, co-organized by Liz and Anita, and their colleagues, for the Medical Anthropology Network of the European Association of Social Anthropology and the Society for Medical Anthropology of the American Anthropological Association. The conference, with support from the Wenner-Gren Foundation for Anthropological Research, was a unique global gathering of medical anthropologists, bringing together in exciting and productive ways several hundred scholars from the global North and South, and stimulated our thinking about the book. From this time, the three of us—Lenore, Liz and Anita—worked together and with our case study authors. We are indebted to them for their enthusiasm for the project, their timely submission and gracious acceptance of the editorial excisions that followed. It goes without saying that this volume only exists because of you.

In late 2013, Katherine Ong became the editor of anthropology volumes, and she and Lola Harre, editorial assistant, worked with us as we developed our ideas, commissioned and edited the case studies, selected the photographs and wrote the chapters. We are grateful to both of them, and to Jennifer Bonnar for her careful editing and easy communication with us. This volume was straightforward neither in development nor production, and we are indebted to all at Routledge for trusting our judgment and supporting our unorthodox approach. We are delighted with the result.

We are very grateful to the Social Science and Global Health Program of the University of Amsterdam, which generously supported our retreat at the Netherlands Institute of Advanced Studies in Wassenaar. We were fortunate to spend most of November 2014 as Visiting Research Fellows outside of the regular program of the Institute, and benefited from the opportunity to spend concentrated time together working on the chapters and to interact with other Fellows and so test our ideas about the book with a wider audience. In April, we spent a few days in retreat at Hacienda del Desierto in Tucson, Arizona, as we talked through restructuring and revisions; again we are grateful to the Social Science and Global Health Program of the University of Amsterdam for support.

**Lenore** offers her particular thanks to: Peter Aggleton, who rightly insisted that the project would be worthwhile, and personally and intellectually rewarding, even if he perhaps understated the labor integral to a book this large; Laetitia Rispel and Jane Goudge from the University of the Witwatersrand, who, after my mother's death in late June, insisted that I take time to grieve in Australia, and so gave me the space also to think and write; Victoria Team, for her quiet assurance that the journal *Medical Anthropology* would gently move ahead without my regular input; Ellen Block for running ahead with our shared projects, trusting I would catch up with her and contribute fully when I could; Gwyneira Isaacs for stepping in tandem with me, as she worked through her own pain thousands of miles away; Kimberly Theidon for her wonderful email that simply checked that I was taking a "deep breath in, deep breath out" when I had forgotten to do so; Hannie Rayson and Ailsa Piper, Australian creative writers who tracked my writing and cheered from the side; Amanda Lynch at Brown, for trusting in my silence and my ability to juggle multiple commitments; Wendy Woodson, my longtime collaborator and friend, for always, always being there and ensuring I knew this. In Canberra, my beloved brothers Richard, Roland and Desmond Manderson, and their families, and in Melbourne, Pat Galvin, Kerith Manderson-Galvin and Tobias Manderson-Galvin, for stepping up, caring for me, and surprising me in ways that made my heart sing. Mardi Manderson, my mother, shared with me a passion for medicine, healing, medical research, anthropology and feminism; nurtured the idea that I become a medical researcher when I was only a small child; battled for my right to follow this path; and inspired me. This is in her memory.

**Elizabeth** thanks: Idaho State University for awarding me a sabbatical leave for the Fall term of 2014. This time away from my regular duties was essential and much appreciated. It was a great pleasure being a Visiting Research Fellow at the University of Amsterdam and at the Netherlands Institute for Studies in Humanities and Social Sciences in Wassenaar. Both of these institutions provided me with a wonderful intellectual environment that truly nourished the beginning stages of the writing process. A grant from the Wenner-Gren Foundation for Anthropological Research facilitated the conference "Exchanges and Dialogues: Creating New Agendas for Medical Anthropology," where many of the themes for this book initally emerged from the lively panels, presentations and discussions. To Jerome Crowder for his excellent work on the images in this volume: your contribution has added a wonderful visual dimension to this volume. Many thanks go to Lenore and Anita—two of the best colleagues, co-conspirators and friends that I could ask for. I would also like to thank my partner Mark and my sister Marianne for supporting me back home when I was away working on the book both in Europe and in Tucson. And finally, thanks to my mother, Agnes, for always believing in me—to you, I dedicate this book.

**Anita** thanks: my ChemicalYouth research team members and other colleagues for being patient when I was away from my desk for weeks writing chapters and talking through drafts with Lenore and Liz. Special thanks go to Swasti Mishra, Hayley Murray and Mariana Rios, who read the whole manuscript and gave such good comments. I also thank Erin Martineau whose "other pair of eyes" gave very valuable editorial feedback on my "first author" chapters, and to Hayley Murray who helped with the references. I finally thank my daughter and husband for bearing with me when I spent numerous weekends and late nights on chapters for this book, and my father, who taught me that writing is fun. Special thanks go to my mother, who suffered from dementia in the past years and died peacefully this August, in the final weeks of this work. Caring for and talking with her about quality of life, her confused state of mind, and endings has taught me so much. This book is dedicated to her.

Lenore Manderson, Elizabeth Cartwright and Anita Hardon
September 2015

# Contributors

**Vincanne Adams** is Professor and Vice-Chair in the Department of Anthropology, History and Social Medicine at the University of California, San Francisco. Her research projects are in China, Tibet, Nepal and the US, broadly focused on health, knowledge and society. Her recent books include *Metrics: What Counts in Global Health* (editor, Duke, 2016), *Markets of Sorrow, Labors of Faith: New Orleans in the Wake of Katrina* (Duke, 2013), and *Medicine between Science and Religion: Explorations on Tibetan Grounds* (with M. Schrempf and S. Craig, Berghahn, 2010). She is currently working on two books: *Preoccupied: Living in the Gap in China's Tibet* (Duke) and *Toxic Children: The Search for Integrative Health in America* (with M. Perro).

**Tanja Ahlin** obtained her MA in Health and Society in South Asia from Heidelberg University, Germany, and is presently a PhD candidate at the University of Amsterdam, The Netherlands, and the Institute of Tropical Medicine in Antwerp, Belgium. Her research interests include e-health, m-health and telemedicine; information and communication technologies (ICT); care; global care chains; aging and gender. Her current research, joining medical anthropology and science and technology studies, is on the influence of everyday ICTs on elderly care in Indian transnational families, with multi-sited fieldwork in the southern Indian state of Kerala and in Oman.

**Tim Allen** is Professor of Development Anthropology and Head of the Department of International Development at the London School of Economics and Political Science, UK. He has carried out long-term field research in several African countries, mostly in East Africa. His books and articles have focused on ethnic conflict in Europe, media coverage of wars, links between culture and development issues, mass forced displacement and global health issues. His latest books are *Trial Justice: The International Criminal Court and the Lord's Resistance Army* (Zed Books, 2006), and, edited with Koen Vlassenroot, *The Lord's Resistance Army: Myth and Reality* (Zed Books, 2010). In recent years his research has focused on justice and accountability in African war zones, and social aspects of the control of Neglected Tropical Diseases.

**Adia Benton** is Assistant Professor of Anthropology at Northwestern University, Chicago, Illinois. She is the author of *HIV Exceptionalism: Development through Disease in Sierra Leone* (University of Minnesota Press, 2015). Her recent work addresses the movement to improve access to surgical care and to incorporate this movement into global health equity agendas. She has also written about political economy, race and representation during the 2014–15 West African Ebola outbreak. More broadly, she studies ideological formation in the fields of global health, humanitarianism and biomedicine, and how it relates to efforts to address equity, justice and inequality.

**Ellen Block** is Assistant Professor of Anthropology at the College of St. Benedict/St. John's University in Minnesota. She is completing work on a book manuscript on caregiving practices for AIDS orphans in rural Lesotho, Africa, in which she critically examines how AIDS intersects with kinship, transforming families and reflecting broader concerns about societal change. In 2015, she completed six months of fieldwork on the impact of grandmother aging and death on caregiving networks in Lesotho. This new work highlights the resilience and reconfiguration of families as grandmothers, who are part of the last generation of virtually HIV-free southern Africans, are increasingly unable to shoulder the burden of care for orphans and vulnerable children.

**Pimpawun Boonmongkon** is Professor in the Health Social Sciences Program and Director of the Center for Health Policy Studies in the Faculty of Social Sciences and Humanities, Mahidol University, Salaya, Thailand. She is also a founding member and on the board of the Southeast Asian Consortium on Gender, Sexuality, and Health. Her research areas focus on gender, sexuality, sexual and reproductive health and HIV, health systems and health policy, and infectious diseases. She received the Mahidol Award for Academic Service, 2010; Thailand University Council Award for Distinguished Faculty Staff, 2011; and the Award for Major Advisor Role of an Outstanding Thesis Award, Category: Humanities, Social Sciences Education and Liberal Arts, 2014.

**P. Sean Brotherton** is Associate Professor of Anthropology at the University of Chicago. His research focuses on the anthropology of medicine, the state, psychoanalysis, subjectivity, and the body in Latin America and the Caribbean, drawing upon contemporary social theory and postcolonial studies. He is the author of *Revolutionary Medicine: Health and the Body in Post-Soviet Cuba* (Duke University Press, 2012).

**Elizabeth Cartwright** is Professor in the Department of Anthropology and Director of the Hispanic Health Projects and the Latino Studies Program at Idaho State University, Pocatello, Idaho. She has published widely on structural vulnerability and health among immigrant and ethnic populations, and has worked in Mexico, Peru and Bolivia. She focuses on systematic ethnographic methodologies that use text-based narratives and visual data. She has extensive experience in obstetrical nursing and publishes in the fields of anthropology, nursing and women's health. Cartwright is the co-founder of Crescendos Alliance, a non-profit organization that uses community-based, participatory research to improve the lives of farmworkers in South America. She is an associate editor for *Medical Anthropology*.

**Heide Castañeda** is Associate Professor in the Department of Anthropology at the University of South Florida, Tampa, Florida. Her research activities focus on health inequalities related to unauthorized migration, through the analysis of legal status and constructs of citizenship, informal labor markets, and binational and mixed-status families. She has worked with the Roma and migrants in Germany; her current projects focus on mixed-status families along the US/Mexico border, the effects of health policies and health reform on immigrant communities, and im/migrant youth in Mexico and the US.

**Rachel Carmen Ceasar** is a postdoctoral fellow in the School of Public Health at the University of the Witwatersrand in Johannesburg, South Africa. Trained as a medical anthropologist, Rachel's research and teaching are shaped by feminist science studies and body studies. She is currently writing a book, *Heritage of War, Exhumation of Peace: Death, Disparity, and the Right to Proper Burial in Postwar Spain*, which in analyzing the contemporary politics of exhuming mass graves from the Spanish Civil War and the subsequent dictatorship, explores hierarchies of heritage and knowledge created in conflict and peace. Her ongoing research shifts from inequalities in human rights work in Spain to the racial and postcolonial contexts of forensic practices in post-apartheid South Africa.

**Simonetta Cengarle** has worked in Asia since 1999 in the field of health, human rights, research, sustainability and gender, particularly in relation to managing programs and regional teams in multicultural environments within charities, government and development organizations. From 2009 to 2015 she was based in Singapore, where she conducted anthropological research on umbilical cord blood banking. After a break caring for her three children, she joined the private sector in the UK and is currently Chief Content Officer for a new start-up company concerned with training and career planning, Opporta.

**Chee Heng Leng** is an honorary Research Associate with the Women's Development and Research Centre (KANITA), Universiti Sains Malaysia, Penang, Peninsular Malaysia. She has lectured and conducted research in the field of public health, and was previously attached to the Universiti Putra Malaysia and the Asia Research Institute (NUS). Her current research is on international medical travel in Malaysia.

**Robbie Davis-Floyd,** Senior Research Fellow, Department of Anthropology, University of Texas, Austin, Texas, and Fellow of the Society for Applied Anthropology, is a world-renowned medical anthropologist, international speaker and researcher in transformational models in childbirth, midwifery and obstetrics. She is author of *Birth as an American Rite of Passage* (1992, 2004), coauthor of *From Doctor to Healer: The Transformative Journey* (1998) and *The Power of Ritual* (2015), and lead editor for *Birth Models That Work* (2009), which highlights optimal models of birth care around the world. Robbie serves as Board Member of the International MotherBaby Childbirth Organization (IMBCO) and Senior Advisor to the Council on Anthropology and Reproduction.

**Christine Dedding** is Assistant Professor at the Athena Institute of the Vrije University, Amsterdam, The Netherlands. She worked for 10 years in the field of rehabilitation before undertaking a PhD at the Amsterdam School for Social Science Research. By combining participatory action research and ethnography, she has analyzed the power and dependency relationships between children with diabetes and their medical professionals. Her specific fields of interest are communication and innovation in health care and patient participation.

**Alice Desclaux** is a medical doctor and medical anthropologist. Previously Professor of Anthropology in Aix-Marseille University, France, she is now a full-time researcher in Institut de Recherche pour le Développement in Dakar, Senegal. Her main fields of interest are the study of pharmaceuticals, neotraditional medicines and pharmaceuticalization, as reflected in her book, edited with Marc Egrot, *Anthropologie du médicament au Sud. La pharmaceuticalisation à marges* (L'Harmattan, 2015). She has also worked on HIV prevention and care, gender and health, HIV in children, and social issues related to medical research in West Africa, Southeast Asia and France. She is presently working on the cultural and political framing of response to Ebola in Senegal and Guinea.

**Alexander Edmonds** is Professor of Social and Medical Anthropology at the University of Edinburgh, Scotland. His work explores therapeutic work on the body and self in Latin America and the United States. His book, *Pretty Modern: Beauty, Sex, and Plastic Surgery in Brazil* (Duke University

Press, 2010), awarded the Diana Forsythe Prize, the Eileen Basker Prize and honorable mention for the Sharon Stephens Prize, tracks the emergence of Brazil as a global leader in plastic surgery. He is now conducting comparative research on psychological health and reintegration among soldiers.

**Peter van Eeuwijk** is a social anthropologist and historian and holds a postgraduate degree in development policy studies. Post-doctoral studies were completed in Australia and the Netherlands. He is engaged as senior lecturer and senior researcher in medical anthropology, anthropology of aging, urban anthropology, applied anthropology, political ecology and transformations in Southeast-Asia/East Africa at the Institute of Social Anthropology and the Swiss Tropical and Public Health Institute, University of Basel; the Department of Social and Cultural Anthropology, University of Zurich; and the Institute of Social Anthropology, University of Bern, Switzerland. He has conducted extensive research on aging in Sulawesi, Indonesia, and Tanzania.

**Raphael Frankfurter** studied anthropology and global health and health policy at Princeton University, Princeton, New Jersey. After nearly nine months of anthropology fieldwork in Kono District, Sierra Leone, he assumed the position of Executive Director of Wellbody Alliance, a major health care provider in the country. In this role, he oversaw the organization's response to the 2014–2015 Ebola outbreak and partnership with international NGO Partners in Health to bring high-quality community and facility-based Ebola care to Sierra Leone. He also served as Strategic Advisor for Community Health for Partners in Health—Sierra Leone, managing efforts to find, recruit and accompany Ebola patients to care across the country. He is now an MD/PhD student in medical anthropology at University of California, San Francisco and Berkeley.

**Eugenia Georges** is Chair and Professor of Anthropology at Rice University, Houston, Texas. She has conducted research on medicalization and reproduction in Greece, the movement to humanize childbirth in Brazil, as well as on Dominican transnational migrants in the Dominican Republic and New York City. She is the author of *Bodies of Knowledge: The Medicalization of Reproduction in Greece* (Vanderbilt University Press, 2008) and *The Making of a Transnational Community: Migration, Development and Cultural Change in the Dominican Republic* (Columbia University Press, 1990), and articles in diverse scholarly journals. Her current research focuses on the effects of the Greek economic crisis on maternal health and the provision of maternity care.

**Trudie Gerrits** is Assistant Professor at the Department of Sociology and Anthropology at the University of Amsterdam, after working for five years in the Ministry of Health in Mozambique. Most of her research in the Netherlands and abroad (Mozambique and Ghana) and her publications are related to infertility and assisted reproductive technologies (ARTs). Currently she is involved in a comparative study on the appropriation of assisted reproductive technologies in sub-Saharan Africa, doing fieldwork in Ghanaian private clinics.

**Thomas E. Guadamuz** was trained as an epidemiologist at Johns Hopkins University, Baltimore, Maryland, and is Assistant Professor of Medical and Health Social Sciences in the Department of Society and Health, Faculty of Social Sciences and Humanities, Mahidol University, Salaya, Thailand. His research focuses on the social determinants of health; HIV and AIDS risk contexts; gender, sexuality and health; youth bullying and violence; and community- and structural-level interventions. He has just completed an ethnography of illicit drug use and chem-sex among young gay and bisexual men in Bangkok, Thailand, and a mixed-method national study of LGBT and gender non-conforming bullying among secondary school students in Thailand.

**Marjolein Gysels** is a senior researcher in the Department of Anthropology and Sociology at the University of Amsterdam. She carries out international research on palliative care and long-term care for older people and those affected by dementia, with cross-cutting interests in the global construction and development of the palliative care movement; research synthesis methodology; ethics in end-of life-care research; and the bridging of the research-policy-practice gap. Currently her main research line is participative art for older people with and without dementia.

**Anita Hardon** is Professor of Health and Social Care at the Faculty of Social and Behavioural Sciences (Sociology and Anthropology Department) and Dean of the Amsterdam Institute for Social Science Research, University of Amsterdam. She has conducted a series of studies in Pakistan, the Philippines, Mali and Uganda on how people define, experience and treat their health problems, and their responses to public health programs to promote the appropriate use of medicine. She is convenor of the Medical Anthropology Network, European Association of Social Anthropology, and an associate editor for *Medical Anthropology*.

**Barbara Herr Harthorn** is Professor of Anthropology and Director of the NSF Center for Nanotechnology in Society at University of California, Santa Barbara. Her research examines the social production of health and environmental inequality, technological risk and perception, and gender, culture and mental health. She has conducted field research in East Africa, Polynesia and Melanesia, and rural California, and among diverse experts and publics in the US and UK. Her publications include *The Social Life of Nanotechnology* (Routledge, 2012, with John Mohr) and *Risk, Culture and Health Inequality: Shifting Perceptions of Danger and Blame* (Greenwood/Praeger, 2003, with Laury Oaks).

**Gesine Kuspert Hearn** is Associate Professor of Sociology and the Chair of the Department of Sociology, Social Work and Criminal Justice at Idaho State University, Pocatello, Idaho. She received her PhD in 2006 from the University Erlangen-Nuernberg, Germany. Her research interests include lay and expert perceptions of health and disease, medical uncertainty and patient self-help organizations. Her current research explores the role of sociocultural factors in the experience and management of chronic pain and injury among athletes and veterans. She has published in the areas of medicine, health and illness, gender and family sociology.

**Irene J. Higginson,** OBE, is qualified in medicine from Nottingham University and is dual-trained in palliative medicine and public health medicine. She is Director of the Cicely Saunders Institute, at King's College London, the world's first purpose-built Institute of Palliative Care integrating research, education, clinical services, support and information. Her research interests and publications include the areas of quality of life and outcome measurements, evaluation of palliative care especially of new services and interventions, epidemiology, clinical audit, effectiveness, psychosocial care, symptom assessment, breathlessness and elderly care. She plays an active role in clinical service including on-call. She has developed and validated two outcome measures both freely available: the Support Team Assessment Schedule and the Palliative Care Outcome Scale (see www.pos-pal.org).

**Viola Hörbst,** independent researcher, is interested in biomedical diversification and particularly in mutual engagements of technology and society in different sociocultural settings. From 2004 to 2011 she carried out fieldwork on infertility and assisted reproductive technologies in Mali, Senegal and Togo, and in 2012 and 2013, she conducted fieldwork on the mobility of things and objects in a Ugandan fertility clinic. She lives in Munich, Germany.

**Julia Hornberger** is Senior Lecturer in the Department of Anthropology at the University of the Witwatersrand, Johannesburg, South Africa. She has worked extensively on questions of policing and human rights in Africa, resulting in the book *Human Rights and Policing: The Meaning of Violence and Justice in the Everyday Practice of Policing in Johannesburg* (Routledge, 2011). Her current research interests lie with the policing of health and intellectual property. She is working on a multi-sited ethnography of how health, security and the market intersect around the figure of the pharmaceutical copy, globally and in South Africa.

**Arianna Huhn** is Assistant Professor and Director of the Anthropology Museum at California State University, San Bernardino, California. Her research focuses on foodways, wellbeing and ethnophysiology, in particular perceptions of what food is and what it does to the body. She works primarily with Maravi cluster groups in northern Mozambique. Her most recent project situates dietary choices as both reflecting and shaping projects of social personhood. She has also participated in projects on medical decision-making in alternative parenting circles in the US, and in studies of the cultural context of obesity in rural and urban Africa.

**Susann Huschke** is a postdoctoral fellow at the School of Public Health and the African Centre for Migration and Society (ACMS) at the University of the Witwatersrand, Johannesburg, South Africa. She has conducted research in Germany on undocumented migration and migrants' access to health care (2008–2011), and in Northern Ireland on sex work, policy and moralities (2013–2014). Her current research project (2015–2018) focuses on informal health-seeking practices of sex workers in South Africa. Her study includes female, male and trans sex workers, and explores how migration/mobility shapes the health and illness experiences of sex workers. Among the issues addressed in the study are informal abortions, practices of self-care, spiritual healing and traditional medicine.

**Emmy Kageha Igonya** is currently working on economic empowerment and the political positioning of sex workers, and is involved in the Bridging the Gaps operational research project, with people who use drugs and sex workers, and on lesbian, gay, bisexual, transgender and intersexual (LGBTI) projects in Kenya. She trained at the University of Amsterdam, and her PhD research was on the emergence and transformation of community-based HIV support in Nairobi, Kenya. Her research areas of focus include HIV, men who have sex with men (MSM), MSM and female sex workers, sexual and gender-based violence, reproductive health, HIV and drug abuse, and infant and young child nutrition.

**Noor Johnson** is a Senior Advisor with the Smithsonian's Office of International Relations, Washington, DC, and an Adjunct Assistant Professor of Environment and Society at Brown University, Providence, Rhode Island. Noor's research has focused on how Inuit knowledge is mobilized in climate change science and policy in the Canadian Arctic and in global environmental forums. She has also worked on the development of community-based observing networks that facilitate environmental and health monitoring of contaminants, biodiversity and food security, and environmental change. As a Fulbright Arctic Initiative scholar, Noor's newest research project focuses on community consultation and knowledge practices in decision making about offshore oil and gas development and marine conservation.

**Junko Kitanaka** is a medical anthropologist and Associate Professor in the Department of Human Sciences, Keio University, Tokyo, Japan. Her doctoral dissertation on depression, undertaken at McGill University, received a number of awards including the 2007 Dissertation Award from the American Anthropological Association's Society for Medical Anthropology, and is now published as *Depression in Japan: Psychiatric Cures for a Society in Distress* (Princeton University Press, 2012), which won the American Anthropological Association's Francis Hsu Prize for Best Book in East Asian Anthropology in 2013. She is currently conducting research on dementia and the psychiatrization of the life cycle.

**Alice Larotonda** works on issues surrounding illness experience, care and medical expertise in infant and child health. Her case study addresses the experiences of children affected by rare diseases in Italy, by exploring the alignment and discordance of medical, parental and children's perspectives on being chronically ill. Her more recent research explores the local and global discursive construction of best practices of infant care in the Republic of Cabo Verde. She is particularly interested in investigating the interface between medical authority, lay expertise and everyday care practices like breastfeeding.

**Annette Leibing** first taught medical anthropology in the Department of Psychiatry, Federal University Rio de Janeiro, Brazil, where she founded and directed the Center for Dementia and Aging. She is now full professor at the Nursing Faculty, Université de Montréal, Quebec, Canada, and member of research groups MéOS (social aspects of medications), CREGÉS (social gerontology) and PACTE (politics and territory). Her research focuses on the aging body, diseases such as Alzheimer's and Parkinson's, aging and psychiatry, pharmaceuticals, the situatedness of care and stem cells for a body in decline.

**Stephanie Lloyd** is Assistant Professor of Medical Anthropology at Université Laval (Quebec City, Quebec) who explores the intersections of scientific and medical technologies and identity formation, with a particular interest in genetics and psychiatric knowledge and practices. Her current ethnography examines the production of epigenetic theories of suicide risk, focusing on efforts to construct distinctive biological profiles of 'suicide completers,' shifting temporalities of risk, and the reimagination of the porous nature of human bodies and their interactions with the environment. Her fieldwork is based in Canada and France.

**Lenore Manderson** is Professor of Public Health and Medical Anthropology at the University of the Witwatersrand, Johannesburg, South Africa, and Professor of Anthropology, Institute at Brown for Environment and Society, Brown University, Providence, Rhode Island. She is a medical anthropologist, but she has also contributed to sociology, the social history of medicine and public health, undertaking field research and training across these disciplines primarily in Malaysia, Thailand, the Philippines, China, Ghana and South Africa. Her recent books include *Surface Tensions: Surgery, Bodily Boundaries and the Social Self* (Left Coast Press, 2011), *Technologies of Sexuality, Identity and Sexual Health* (ed., Routledge, 2012), and *Disclosure in Health and Illness* (ed. with Mark Davis, Routledge, 2014). She edits the international journal *Medical Anthropology*.

**Janice McLaughlin** is Professor of Sociology at Newcastle University, UK. Her work focuses on the significance of childhood disability to family life and the influence of new medical approaches to treating and diagnosing disability on both children and their families. Her work draws from embodiment studies, critical disability studies, new kinship studies and medical sociology. Her new book is *Disabled Childhoods: Monitoring Difference and Emerging Identities* (Routledge, 2016).

**Ben McMahan** works at the Institute of the Environment at the University of Arizona, Tucson, Arizona. His dissertation research, on hurricanes and disaster on the US Gulf Coast, focused on human interactions in a dynamic social and environmental context, risk perception and landscape changes during and after disaster, and social network and policy responses. He has also worked on the social, economic and environmental impacts of the US oil and gas industry, and the Deepwater Horizon oil spill. His current research focuses on the acute threats and long-term effects of climate extremes, the effects of climate variability on phenology and temporality of native plants, and how climate information is incorporated into regional decision making in the Southwest US.

**Sebastian Mohr** is an ethnographer and Assistant Professor at the Department of Education at Aarhus University, Denmark. Inspired by medical anthropology and sociology, STS, and feminist and queer theory, his work is concerned with the interplay between (reproductive) biomedicine, gender and sexuality, and the negotiation of intimacy in psychotherapeutic contexts. His doctoral research focused on the experiences of Danish sperm donors and the work conducted at the laboratories of Danish sperm banks.

**Eileen Moyer** is Associate Professor at the University of Amsterdam, specializing in urban and medical anthropology. Her research, which has taken place mainly in eastern and southern Africa, has focused on the entwinement of globalization, health and urban identities, with a special interest in the emergence of cosmopolitan socialities related to HIV. In 2015, she was awarded a prestigious European Research Council consolidator grant to research the relationship between global health gender equality initiatives and transformations in urban African masculinities over the last quarter century.

**Jessica Mulligan** is Associate Professor of Health Policy and Management at Providence College, Providence, Rhode Island, where she teaches courses in public health and medical anthropology. This case study is excerpted from her recent book *Unmanageable Care: An Ethnography of Health Care Privatization in Puerto Rico* (New York University Press, 2014). Her new projects explore health reform and health care finance in the US. She is presently undertaking an ethnographic study of shopping for health insurance on the electronic exchanges that were created by the Affordable Care Act and co-editing a volume on anthropology and health policy.

**Mark Nichter** is Regents Professor with joint appointments in the Department of Family Medicine and College of Public Health, University of Arizona, Tucson, Arizona, and coordinates the Graduate Medical Anthropology Training Program. Mark specializes in the intersections between anthropology and global health, and anthropology and clinical medicine. He has conducted research in Asia, Africa and North America on child survival; infectious, emerging and non-communicable disease; and pharmaceutical practice, and for over 20 years has conducted research on tobacco in the US, India and Indonesia. He is joint Principal Investigator on Project Quit Tobacco International (www.quittobaccointernational.org).

**Mimi Nichter** is Professor of Anthropology at the University of Arizona, Tucson, Arizona. She has conducted ethnographic research in South and Southeast Asia and the US, where her research has focused on topics related to gender and health. She has published extensively on gender and tobacco, and most recently has worked on developing smoke free homes initiatives in India and Indonesia. Her latest book, *Lighting Up: The Rise of Social Smoking on College Campuses* (New York University Press, 2015), focuses on the social utilities of smoking among young adults in the US. She is joint Principal Investigator on Project Quit Tobacco International (www.quittobaccointernational.org).

**Brigit Obrist** is Professor of Anthropology at the University of Basel, Switzerland, with a joint position at the Swiss Tropical and Public Health Institute. Since 1980 she has conducted research in Papua New Guinea, Switzerland, Indonesia and Tanzania, and directed various applied projects.

She currently leads a medical anthropology research group with post-doctoral and PhD projects in Germany, Switzerland, East and West Africa in the fields of malaria, urban health, sexual and reproductive health, aging and health, and on media and health. Her latest project is on Aging, Agency and Health in Urbanizing Tanzania (Swiss National Science Foundation).

**Melissa Parker** is Reader in Medical Anthropology at the London School of Hygiene and Tropical Medicine, UK. Her multi-disciplinary and collaborative research in African and European settings is concerned with global health and international development. Topics investigated include HIV/AIDS in the UK, mental health in war zones, health-related quality of life in Kenya, female circumcision in Sudan, and the control of Neglected Tropical Diseases in Sudan, Uganda and Tanzania. She is a founding member of the Ebola Response Anthropology Platform, and during the Ebola outbreaks in Sierra Leone co-chaired the Anthropology Sub-Group of the UK's Scientific Advisory Group on Emergencies. This group advised the British government about appropriate ways to respond to Ebola.

**Adrienne Pine** teaches anthropology at American University, Washington, DC. Prior to and since the 2009 military coup in Honduras, she has studied how normally invisible forms of violence are embodied as subjectivities by Hondurans, who following the coup formed a nationwide movement (in which Nurses in Resistance played a key role) that named and fought against neoliberal violence and its agents. Since 2004, first in affiliation with California Nurses Association and later from within academia, she has studied the impact of corporate health care technologies on labor practices and patient care in the US and Honduras. Currently, she is conducting a comparative study about nurse strategies for patient advocacy and international solidarity in three neoliberalizing (yet very different) countries: Honduras, Cuba and the US.

**Jeannette Pols** is Socrates Professor in the Department of Sociology and Anthropology, the University of Amsterdam, and Associate Professor and Principal Investigator in Medical Ethics, Academic Medical Centre, Amsterdam. She has a background in philosophy, science and technology studies and clinical psychology. She conducts ethnographic fieldwork in health care settings and in the daily lives of people who live with chronic diseases, particularly in relation to the different forms of normativity in care, the position of patients, the use of new technology, and links between experiential and scientific understandings of care. She is the author of *Care at a Distance: On the Closeness of Technology* (University of Amsterdam Press, 2012). She is working on a project exploring aesthetic values in daily life and care, such as 'quality of life.'

**Robert Pool** is an anthropologist and Professor of Social Science and Global Health at the University of Amsterdam, where he directs the Health, Care and the Body program group and the Long-Term Care Partnership, and co-directs the Centre for Social Science and Global Health. He has carried out individual ethnographic research on food taboos in India, local explanations of illness and misfortune in Cameroon, and euthanasia decisions in the Netherlands, and has led large social science and multidisciplinary research programs relating to HIV and malaria in many countries in Africa. Current research topics include sustainable community health resources in Africa, narratives relating to dementia and euthanasia.

**Panoopat Poompruek** is a Lecturer at the Department of Community Pharmacy, Faculty of Pharmacy, Silapakorn University, Nakhon Pathom, Thailand. Originally trained as a pharmacist, he has a master's degree and doctoral degree in Medical and Health Social Sciences from the Department of Society and Health, Faculty of Social Sciences and Humanities, Mahidol University, Thailand. His areas of interest are in the anthropology of pharmacy, technology and the body.

**Laurent Pordié** is a Senior Researcher with the French National Center for Scientific Research (CNRS) at the CERMES3, a unit focused on medicine, science and society, and a member of the Center for South Asian Studies at the Ecole des Hautes Etudes en Sciences Sociales in Paris, where he teaches. His works concern the social study of science and medicine in South Asia, including the recent volumes *Tibetan Medicine in the Contemporary World* (Routledge 2008—winner of the ICAS Book Prize 2009) and *Les nouveaux guérisseurs* (with E. Simon, Editions de l'EHESS 2013), and *Healing at the Periphery* (Duke University Press, 2007).

**Roberta Raffaetà** is Lecturer in Medical Anthropology at the University of Milan-Bicocca, Italy. In her doctoral work, she explored how different actors in Italy made sense of allergy, and dealt with issues of medical pluralism, medicalization, embodiment and identity. In her post-doctoral work, she has focused on biopolitics, pursuing an ecological understanding of migration on Chinese settlement in the city of Prato and, with a Marie Curie Fellowship (Trento University, Italy), on Moroccan and Ecuadorian parenting in Italy. She has also explored various themes with regard to health governance, pediatric vaccination and transnational medical trajectories, and technology, including e-health and novel treatments for renal diseases.

**Peter Redfield** is Professor of Anthropology at the University of North Carolina, Chapel Hill, North Carolina. Trained as a cultural anthropologist sympathetic to history, he concentrates on circulations of science and technology in colonial and postcolonial contexts. The author of *Life in Crisis: The Ethical Journey of Doctors Without Borders* (University of California Press, 2013) and *Space in the Tropics: From Convicts to Rockets in French Guiana* (University of California Press, 2000), he is also co-editor of *Forces of Compassion: Humanitarianism between Ethics and Politics* (SAR Press, 2011), and a recipient of the Cultural Horizons Prize of the Society for Cultural Anthropology.

**Ria Reis** is Professor of Medical Anthropology at Leiden University Medical Centre, Associate Professor in the Department of Cultural Anthropology, University of Amsterdam, and Fellow of the Leiden African Studies Centre, the Netherlands. She also is Honorary Professor at the Children's Institute, School of Child and Adolescent Health, University of Cape Town, South Africa. With her graduate and doctoral students, her current research focus is on cultural idioms of distress and health perceptions and strategies of children and youth, and the trans-generational transference of vulnerabilities in contexts of inequality and (post)conflict. She is also interested in the articulation of ethnography within multidisciplinary research and interventions, particularly in collaborative projects with partners in policy and practice.

**Emilia Sanabria** holds a lectureship in social anthropology at the Ecole Normale Supérieure of Lyon, France. She has conducted research in Brazil since 2004 on embodiment, menstruation, reproductive health, local conceptions of blood and pharmaceutical cultures. Her monograph *Plastic Bodies: Sex Hormones and Menstrual Suppression in Bahia, Brazil* is forthcoming with Duke University Press. She is interested in questions of evidence making and policy in the field of public health, the circulation of knowledge and ignorance, and theories of embodiment. Her current research examines the multiple relational properties of the category 'chemical' through different modes and localities, from the sensorial dimensions of eating to the forging of new efficacies for different pharmaceutical and herbal substances.

**Margaret Sleeboom-Faulkner** is Professor in Social and Medical Anthropology at the University of Sussex, Brighton, UK. Her work focuses on nationalism and processes of nation-state building in China and Japan, and on biotechnology and society in East Asia. She currently leads the Centre for Bionetworking with two projects on "Bionetworking in Asia" focusing on international science collaboration in advanced stem cell therapies (funded by the ESRC, 2011–2014) and in biobanking in the life sciences and hospitals (financed by the ERC, 2012–2017). Her most recent book is entitled *Global Morality and Life Science Practices in Asia—Assemblages of Life* (Palgrave MacMillan, 2014).

**Carolyn Smith-Morris** is Associate Professor at Southern Methodist University, Dallas, Texas, and directs the SMU Health and Society Program dedicated to interdisciplinary approaches to health, health care and health research. Her ethnographic and biocultural work focuses on chronic disease in cultural and structural contexts. She has conducted ethnographic research among the Gila River Indian Community of Southern Arizona, Mexicans and Mexican immigrants to the US, and veterans with spinal cord injuries. She is currently analyzing neighborhood-based stigma associated with the 2014 Ebola cases in Dallas, and has ongoing projects studying dietary change and healthcare barriers for Spanish-speaking immigrants. She has recently edited *Diagnostic Controversy: Cultural Perspectives on Competing Knowledge in Healthcare* (Routledge, 2015).

**Thomas Stodulka** is a researcher at the Institute of Social and Cultural Anthropology, Freie Universität Berlin, Germany. His work focuses on emotion, stigma, marginality, health and illness. His long-term fieldwork with street-related children, adolescents and young men in Yogyakarta, Indonesia, will be published as *Coming of Age on the Streets of Java* (Routledge, 2016), and led also to the establishment of a shelter for people who are chronically ill at the city's margins. He co-edited *Feelings at the Margins—Dealing with Violence, Stigma and Isolation in Indonesia* (Campus, 2014). He is currently directing an interdisciplinary research project on ethnographic knowledge production entitled "The Researchers' Affects."

**Suli Sui** is Associate Professor of the School of Humanity and Social Science, Peking Union Medical College, China. She holds a PhD in Medical Sociology from the University of Amsterdam and a master's degree in law from Renmin University of China. She was a visiting scholar at Harvard University, Leiden University and London School of Economics. Sui specializes in the interdisciplinary field between medical science and law, sociology and bioethics. She has participated in international research projects on sociogenetic marginalization in Asia and on bionetworking in Asia, with particular attention to bionetworking in advanced stem cell therapies. Her current work focuses on the regulation of life science in China.

**Hilde Thygesen** is an occupational therapist and sociologist. Currently she is working as Associate Professor at the Institute of Nursing and Health at Diakonhjemmet University College in Oslo, Norway. The focus of her research is on care practices, with a particular interest in the relations between care and technology, and in normative issues. Her publications relate to empirical ethics and the use of telecare technologies in the care of elderly persons.

**Ruth E. Toulson** is Assistant Professor of Anthropology in the Department of Humanistic Studies at Maryland Institute College of Art, Baltimore, Maryland. Her research examines transformations in death ritual and in attitudes toward the corpse in Singapore. This research stems from fieldwork in a Singaporean Chinese funeral parlor, where she worked as an embalmer. Broadly, she is interested in the political lives of dead bodies, in the corpse as a distinctive form of material culture and in what it means to die in a Chinese way in the Southeast Asian diaspora.

**Vu Song Ha** is a medical doctor and public health specialist, whose PhD research was on Autism Spectrum Disorder (ASD) in Hanoi, Vietnam. She is a founder of the Center for Creative Initiatives in

Health and Population (CCIHP), an NGO in Vietnam with the mission of promoting health equality, diversity and health for all people in society. For the past 20 years, her research has focused on gender, sexuality and reproductive health, HIV/AIDS and disability. Currently, Ha is working on the research component of a five-year project on HIV, and leading operational research to develop mobile health to promote early identification of and interventions for children with ASD.

**Megan Wainwright** is a medical anthropologist from Quebec, Canada, with a particular interest in chronic illness, ethnography, audiovisual methods/outputs and the medical humanities. While Van Mildert College Trust Research Scholar at Durham University, UK, she carried out PhD fieldwork in Uruguay on the topic of living with and caring for COPD. She is currently a Post-doctoral Research Fellow at the School of Public Health and Family Medicine at the University of Cape Town, and is conducting comparative research in Uruguay and South Africa on the global politics of the medicinal gas industry and the experiential realities of long-term oxygen therapy for people living with chronic lung disease.

**Megan Warin** is Associate Professor and Australian Research Council Future Fellow in the Discipline of Gender Studies and Social Analysis, University of Adelaide, South Australia. Her research interests include theories of embodiment, obesity science (epigenetics and life course), class and gendered dynamics of health and illness, and social change practices. Her current ethnographic work in Australia is investigating the cultural and institutional processes that shape everyday food and eating practices, and how these social practices can be translated into obesity interventions and policy. She is also exploring how developmental perspectives on health and disease and epigenetic understandings of obesity open up new theoretical terrain between the social and life sciences.

**Narelle Warren** is a Lecturer in Anthropology in the School of Social Sciences at Monash University, Melbourne, Australia. Her research is concerned with the relationship between the lived experience of neurological conditions, biomedical representations of the brain, and temporality, from both the perspectives of people living with such conditions and their informal caregivers. Her current research investigates the affective and relational dimensions of stroke and Parkinson's disease in Australia and Malaysia, and how these are shaped by gender, age, geographical location and culture. With Lenore Manderson, she co-edited *Reframing Quality of Life and Physical Disability: A Global Perspective* (Springer, 2013).

**Andrea Whittaker** is Convenor of Anthropology in the School of Social Sciences, Monash University, Melbourne, Australia. She is a medical anthropologist working primarily in the fields of reproductive health, biotechnologies, and medical mobility and travel with a special interest in Thailand and Southeast Asia. She is currently undertaking research on medical travel in Thailand and Malaysia, contraceptive use among CALD women in Victoria, and a longitudinal study of people living with HIV in rural and regional Queensland, and continuing research on reproductive travel, sex selection and surrogacy. Her most recent book is *Thai in Vitro: Gender, Culture and Assisted Reproduction* (Berghahn Books, 2015).

**Meagan Wilson** is a Teaching Associate in the Department of Epidemiology and Preventative Medicine within the Faculty of Medicine, Nursing and Health Sciences at Monash University, Melbourne, Australia. Meagan's research is concerned with forced migration, structural violence and mental health. She is currently completing her doctorate based on field research conducted in northern Thailand, entitled *Survival, Fear and the Future: An Ethnography of Emotional Suffering among 'Burmese' Migrant Women Living in Thailand.* She is also a visual artist.

# Photographers

**Mark Caicedo** has been a shooter for over 30 years, ever since he picked up his first basic Pentax K-1000. In the mid-1980s, he was a Peace Corps volunteer in Honduras. There he learned photography is not about taking pictures, but creating relationships: first, between the subject and photographer; then, the viewer and photographer; ultimately, the viewer and subject. In his work as a photographer and photo editor with the Inter-American Foundation, he has captured and used images of poverty, despair and hopelessness. But as a documentary photographer, he prefers to focus on the positive, the timeless, the beautiful. Ugliness passes, beauty remains.

**Elizabeth Cartwright,** with Jerome Crowder, selected and edited the photographic contributions in this volume. See List of Contributors.

**Arachu Castro** is Samuel Z. Stone Chair of Public Health in Latin America at Tulane University, New Orleans, Louisiana, where she directs the Institute for Health Equity in Latin America. Her major interests are how social inequalities are embodied as differential risk for pathologies common among the poor and how health policies may alter the course of epidemic disease and other pathologies afflicting populations living in poverty. She currently conducts research on maternal mortality in the Dominican Republic and Bolivia and on health equity in the region. Among other awards, Dr. Castro is the recipient of the 2010 Guggenheim Fellowship and the 2005 Rudolf Virchow Award of the Society for Medical Anthropology.

**Jerome Crowder,** with Elizabeth Cartwright, selected and edited the photographic contributions in this volume. A visual and medical anthropologist, his early work, in El Alto, Bolivia, focused on the role of 'community' in primary health care, migrant health decision-making strategies and the use of over-the-counter medications purchased from corner stores for home remedies. His work on the lives of rural-urban migrants in Bolivia, *Sueños Urbanos: Urban Dreams* (2000), has been shown in museums, libraries and universities across the US. He is co-author with Jonathan Marion of *The Visual Research Handbook: A Comprehensive Guide to Issues, Approaches and Methods* (Rowman & Littlefield, 2016). He is an Assistant Professor in the Institute for the Medical Humanities, Galveston, Texas.

**Mary Anne Funk** is a visual ethnographer and documentary photographer, based in Portland, Oregon. Her current ethnographic and photography work aims to redefine our perceptions of homelessness and poverty, and she spends extensive time with individuals and families who are homeless and working poor, interviewing them and photographing their lived experiences. This ongoing project will be used for advocacy, social awareness, program development and policy change, and will be shared online and in local community advocacy settings.

**Athena Madan** is a scholar-practitioner interested in social and political forces in health, specifically in instances of extreme political violence and fragile states. Her research focuses on global mental health, including conceptualizations of mental health and illness; psychosocial legacies of apartheid and genocide; social navigation of child soldiers; sociopolitical contexts of narcotrafficking; and trafficking in recovery from disaster/emergency. Her current projects include reverse innovation in global health, addictions in Afghanistan, and youth insurgency and terrorism. Athena is Indian and half-Filipino, and is currently a postdoctoral research fellow at the University of Toronto, Canada.

**Eileen Moyer,** see List of Contributors.

**Mark Nichter,** see List of Contributors.

**Mimi Nichter,** see List of Contributors.

**Nestor Nuño** is studying for a PhD in medical anthropology and global health at the Universidad Rovira I Virgili, Tarragona, Spain, and the Swiss Institute for Tropical and Public Health, Basel, Switzerland. She is doing fieldwork in the north of Peru, where she is analyzing the sociocultural acceptance of improved cookstoves to reduce air pollution in households. In parallel, her research interests deal with the multilevel impact of development practices, and the long-term sustainability and the construction of health priorities and needs from the perspective of both health centers and rural communities. In 2014, she conducted research in Indonesia on religious experience and the use of 'life-enabling' practices such as silicone injections or hormone intake among *waria*. The photograph in Chapter 3 is from this study.

**Fahim Rahim** is a humanitarian nephrologist, who, like his wife, Beenish Mannan, grew up in Pakistan. They co-founded the JRM (Jamshaid Rahim Mannan) Family Foundation for Humanity, which works to improve the lives of people, in particular the quality of women's and children's health and education, and treatment and research for cystic fibrosis. Dr. Rahim and his partners believe it is a universal right to have access to medical care, and that geographic location should not be a determinant for poor quality of life.

**Amber Urquhart** is a traveler and photographer who has spent years exploring cities and regions around the world. Based in Canada, she maintains an online magazine and travel blog, *A Long Holiday*, and has participated in philanthropic organizations in Mongolia and Guatemala.

**Chelsea Wentworth** is an Assistant Professor of Anthropology at High Point University, High Point, North Carolina. She is a medical and environmental anthropologist with interests in the anthropology of food and nutrition, gender studies, natural resource management, political ecology, visual anthropology and Pacific Island studies. Since 2010, she has conducted research in Port Vila, Vanuatu, and peri-urban areas around the city, working with public health practitioners and families on infant and young child feeding practices, sustainability and childhood malnutrition. Her current research examines the role of urban gardening in child feeding and coping with hunger in the wake of environmental disasters.

Kuna Woman and Child, 2012. Kuna Yala, Panama.

## *About the photograph*

*The Kuna have achieved what very few native peoples have been able to do: adapt to change on their own terms. When I visited Kuna Yala, I saw reminders of this ability all around me: in the way tourism into their territory was regulated, in planning how to confront the effects of global climate change, and in negotiating social change between generations. By and large, young girls choose Western clothes, often wearing shorts or jeans and t-shirts, rather than traditional dress. This image of an older Kuna woman and a young girl reminds me how generational change occurs regardless of our ability to control it. Nonetheless, societies do have the cultural tools to control the rate and extent of how change occurs, and in this control, to maintain a sense of wellbeing. Panama's Kuna are an example for other indigenous peoples adapting to a rapidly changing and modernized world.*

—Mark Caicedo

# 1

# Introduction

## Sign Posts

*Lenore Manderson, Elizabeth Cartwright and Anita Hardon*

---

It is sometimes hard to explain what an anthropologist does; it is even harder to explain 'what *medical* anthropologists do' as we correct assumptions that our work centers on categorizing old bones. Now, our task is more difficult than ever, and a summary descriptor—studying people's experience of sickness and heath, care seeking and care—is banal and inaccurate. Medical anthropology helps make sense of suffering as a social experience, but it does much more than this. It carries us into refugee camps, birthing centers, factories, boardrooms, gaols, rehabilitation centers and schools, across countries and between communities. And, as we describe below and throughout this volume, it is also a field of great privilege; medical anthropology takes us into the most intimate aspects of people's lives, and the most intimate expressions of their joy, anxiety, grief and tenderness.

* * * *

Let us begin by explaining who we are and how we have worked, since this is central to why we have written what we have written. This was a collaborative project, and we brought to our conception of the volume and the task of its construction our experiences as students, researchers and educators in diverse settings.

Lenore was trained in Asian studies and history in Australia, and there, she held positions as a medical anthropologist in public health and medical schools until 2013; she now lives and works in South Africa and the United States, working partly on questions on public health in Africa, partly on environment and climate change. Her earliest field research projects, as an historian and anthropologist, were in Malaysia. For most of her career, she has conducted research and trained others there, elsewhere in Southeast and East Asia, and in Australia with Indigenous, immigrant and settler Australians, on questions of infectious and non-communicable diseases, gender and sexuality, diversity and inequality. Her research students have come from and conducted their research in diverse settings throughout Australia, Asia and Africa. Her work with the Special Programme for Research and Training in Tropical Diseases (TDR) over nearly 30 years has likewise contributed extensively to her understanding of medical anthropology both theoretically and in relation to its practical application in disease control programs and in enhancing people's access to care and improved health.

Liz trained in anthropology in the United States, after initially training and working as a nurse. Her primary work has been in the United States, particularly with Spanish-speaking immigrants, and in Mexico, Central and South America, working on environmental health, immigration, and social justice—her students too are drawn from these areas. Her work among farmworkers in the US resulted in the creation of 'The Hispanic Health Project' that was located in southeast Idaho for over a decade. Through this work she refined her understanding of the value of high-quality data for community-based participatory research methods and community interventions. She now is employing these insights in her work with rural agricultural communities in the Peruvian Andes. Liz conceived and developed the course on Systematic Video Analysis for the Short Course on Research Methods held under the auspices of the National Science Foundation and the direction of Russell Bernard; this course is currently offered in an online format during summer sessions at the University of Florida, Gainesville. She is a trained videographer and has taught ethnographic filmmaking in the US, Mexico, Australia and Vietnam.

Anita was trained in medical anthropology in the Netherlands, after initially training in medical biology. She has conducted extensive fieldwork in the Philippines and elsewhere in Southeast Asia, with her graduate students drawn primarily Europe, Asia and Africa. With her colleagues Sjaak van der Geest and Susan Reynolds Whyte, she spearheaded the anthropological study of pharmaceuticals in the late 1980s. Since then she has been engaged in multi-level and multi-sited ethnographies on immunization, new reproductive technologies and AIDS medicines that have generated important ethnographic insights on the appropriation of these technologies in diverse social-cultural settings, their efficacy in everyday life, the role of social movements in their design, and the dynamics of care and policy making in their provision. She makes it a priority to communicate her research findings to patient advocates, policy makers, and public health researchers and practitioners, through and in collaboration with activist organizations.

This combined geographic and intellectual diversity—of where we studied, where we have taught, where we have undertaken field research, and the diverse backgrounds of our students—inspired us as we thought through the structure and potential uses of this volume, identified the case study authors, and wrote the chapters. Our shared vision, while providing a personal review of medical anthropology in the early twenty-first century, was to keep in focus the directions that contemporary medical anthropology was taking across space. These directions—no single pathway—are shaped by the intellectual traditions of our different continents, despite the travel of ideas between them. They are shaped too by inspirations from various other theoretical, epistemological and disciplinary conventions and concerns across the social sciences, and by medical and public health priorities in our different countries.

Current and emerging economic, political, social and environmental challenges, and the priorities of the people with whom we have worked, have further shaped the content and structure of this volume. The photographs that we have included stand alone, invitations to readers to engage and reflect (see About the Figures, pp. xii–xiii). The 52 case studies, from 66 authors, add polyphony to this volume, offering rich and diverse ethnographic examples of our dominant themes as they play out across geography, theoretical landscape and the political directions of contemporary medical anthropology. Around these case studies, we have woven our own texts in a shared voice.

## Medical Anthropology: A Partial History

We three came to medical anthropology during a period of its vital growth, building on an earlier tradition of ethnographic enquiry into medical practices, ideas about causations of illness and symbolic healing (e.g. Polgar 1962; Rivers 1924; Rubel 1964), and emerging interest in the value of applied anthropology in contributing in practical ways to public and community health. At the time of our training and early careers, other social movements had begun to impact the focus of the discipline; from the 1970s, political, social and epidemiological factors had converged to frame an emerging, engaged medical anthropology. The last wars against colonialism and the violent inequalities that were revealed, the Vietnam War, debates about development, political ecology and environment activism, and trade union, black, Indigenous and gay rights activism, all drew our attention to the political economics of health inequalities within countries and between nations; we gained awareness too of the ways in which race, class, sex and sexuality shaped social exclusion and poor health. French philosophy offered us particular analytic tools in this context, of how knowledge shaped access to institutions, authority, power and practice (Bourdieu 1992; Foucault 1973; 1976; 1980), and of how socioeconomic inequalities are reflected in stratified bodily practices and reproduced across generations (Bourdieu 1977; 1984). Increasingly, anthropologists and others turned to document the impact of local social structures and global relations on the health status of individuals and communities, as in other aspects of social life, including income, education and employment.

Second-wave feminism and its political arms, including the women's health movement, particularly captured our own imagination, supporting our decisions to break from sexist scripts around study and career; we were fortunate because of the number of strong women working in medical anthropology and cognate fields who had worn the path for us (Benedict 1946; Firth 1943; Geertz 1961; Mead 1928; Powdermaker 1966). In the context of health advocacy and feminist politics, many medical anthropologists turned to questions of gender and the silence around women's experiences, and particularly began to write on reproductive health. For feminist anthropologists, this provided a way to trouble questions of nature and culture, the associations of women's social status and biology, and the power of gender relations (e.g. Jordan and Davis-Floyd 1974; MacCormack 1982; MacCormack and Strathern 1980; Rosaldo and Lamphere 1974). It provided an avenue to bring women from the periphery to the center, and in doing so, to interrogate how biomedical regimes have changed women's experiences in different settings and to explore the role of women healers in their reproductive and wider lives. The earliest works on reproductive health were partly inspired correctives to a view of women's lives, bodies and domestic domains as insignificant, and in medical research, to their undisguised omission. Almost half a century later, an impressive corpus of work exists on conception and (in)fertility, contraception and abortion, pregnancy, birth and delivery, the postpartum and breastfeeding, and menstruation and menopause (see, for example, Davis-Floyd 1992; Ginsburg and Rapp 1995; Inhorn 2002; Lock 1993; Martin 1987). These works are centrally about the subjugation of women's bodies and functions to particular discourses of gender and regimes of control. We return to some of these issues at various points in this volume—in Sebastian Mohr's case study of sperm and its donation in Chapter 3; in a number of case studies of genetics and reproduction in Chapter 14; and in Eugenia Georges's and Robbie Davis-Floyd's discussion of midwifery in Chapter 15, for example.

In the late twentieth and early twenty-first centuries, a series of events in international health, involving multilateral agencies primarily, influenced how the field of medical anthropology was to develop. The Declaration of Alma Ata in 1978 (International Conference on Primary Health Care 1978) set out a vision that sought to improve the delivery of health care services, and so improve health in poor countries by redistributing authority from an urban and professional elite to local communities. Ideas about the importance of community participation, identified as critical for effective primary health care to ensure 'health for all,' fed directly into health policy and planning. The Declaration provided a particular role for anthropologists; in operationalizing its principles to ensure effective and sustainable health care, increasing attention was paid to the local acceptability of health programs and services, the involvement of community volunteers to build health infrastructure and extend primary care, and the mechanisms to support populations to use the medical services that were available in their localities. Medical anthropologists had experience with the communities in question.

Arthur Kleinman and his colleagues (1980; Kleinman et al. 1978), in their early work on explanatory models of illness, provided us with a tool that was clear and accessible to medical and public health professionals, enabling us to demonstrate cultural variations in the experiences, diagnosis and treatment of illness. The elegance of this model, and its appeal to health researchers, led a number of medical anthropologists to then develop manuals for rapid assessments, dating from the foundational work of Susan Scrimshaw and Elena Hurtado (1987) and parallel approaches in agriculture and community development (Chambers 1983). Despite criticisms that rapid methods produced superficial findings and that applied anthropology reinforces biomedical hegemony (Manderson and Aaby 1992), these manuals were important for anthropologists and others working in international health programs and in interdisciplinary teams, because the focused data so generated facilitated the translation of anthropology, such that interventions might take account of local understandings of illness, the circumstances that pattern risk, and the community structures that might support prevention (see, among other examples Agyepong et al. 1995; Herman and Bentley 1993; Pelto and Armar-Klemesu 2010; Pelto and Gove 1992).

It was HIV and its lethal consequence as AIDS, however, that most powerfully stimulated medical anthropology. Four years after the Alma Ata Conference, the first case was diagnosed. The subsequent HIV pandemic had singular impact on shaping medical anthropology: the only possible preventions until the successful development and roll-out of anti-retroviral therapy required detailed and nuanced knowledge especially of sexual behavior and injecting drugs—of what people did, in what contexts, and why. Anthropology, as we will demonstrate in Chapter 4, was critical to understanding the pathways of transmission and prevention, and the stigma derived from associating particular marginalized populations with the risks of infection, illness and death (Hardon and Moyer 2014).

## Interplay

Over the past half-century, significant epidemiological changes have precipitated changes in health policies and programs at local and global levels, and in the work that medical anthropologists do. Today, because of early diagnosis and effective pharmaceutical interventions, increasingly infectious as well as non-communicable diseases are of long-term duration, as illustrated by HIV and as we discuss in Chapter 7. At the same time, various chronic conditions are proving to be caused by infectious agents: chronic gastritis and gastric ulcers, and possibly other ulcers, long thought to be due to diet, stress or personality, have been shown to be caused by bacteria; cervical and some other cancers are now known to be caused by human papilloma virus (HPV). However, the vaccines and drugs developed in the wake of this knowledge are not a panacea. Pathogens that cause or contribute to many diseases mutate to resist pharmaceutical interventions, often more quickly than new drugs can be developed. Anopheles mosquitos that transmit malaria rapidly develop resistance to insecticides and change their behavior in response to preventive measures, and the parasite that causes malaria has been equally efficient at developing drug resistance. In addition, as we illustrate in this volume, many people have poor access to health care, and health systems problems limit the effectiveness of interventions for both infectious diseases and chronic non-communicable diseases.

The 'social determinants of health' is one framework used to appreciate how social, cultural and economic environments shape health and wellbeing, disability and disadvantage. The framework, as developed by Richard Wilkinson and Michael Marmot (1999), drew on social, economic and health data from the United Kingdom. In 2005, the WHO established the Commission on the Social Determinants of Health, chaired by Michael Marmot, to examine how social inequalities and injustices compromise people's health, wellbeing and life outcomes (World Health Organization 2008). The original work and the Commission's report both detail the role on health and illness of social, economic, regional and political variables, including food, housing and employment security, education, the reach of health services, early life, childhood development and stress. Discrimination and marginalization from ethnicity, faith, race and gender determine the presence or absence of the preconditions of good health. The absence of social support, various addictions and social isolation create further disadvantages that, if unaddressed, increase the risk of health problems, interfere with treatment, narrow the ability of people to participate in society, and result in premature death. Material conditions, social meanings, and lived experience produce disability-based disadvantage and wellbeing, with the interconnections and intersections of structural and behavioral factors compounding their effects on physical and psychological health. But how do these multiple determinants fit together?

Medical anthropology provides a complex analytic take on the relationships of social structure and context on health. At a broad global level, we have been concerned to illustrate how globalization, industrialization and urbanization reproduce social relations that institutionalize poverty and result in poor health. Contemporary life circumstances in high-, middle- and low-income settings are all powerfully shaped by globalization. Merrill Singer and colleagues (2014; Singer et al. 2011) have inserted into this the interactions of biology and society. The resultant 'syndemic' approach has extended our understanding of disease concentrations and interactions, with social conditions shaping disease clustering and impacting on outcomes concurrent with disease interactions at biological and immunological levels. Examples of syndemics where social, economic and environmental factors interact etiologically and synergistically include HIV and sexual transmitted infections, HIV and diabetes, and HIV and tuberculosis. HIV and food insecurity too, are syndemic, because women living under conditions of food insecurity may be at greater risk of exposure to HIV (through transactional sex, for instance); people with HIV have higher nutritional requirements in terms of both protein and calories; and malnutrition expedites the progression of HIV to AIDS. This understanding of complex recurrent interactions occurs also with diabetes and tuberculosis (Singer et al. 2011: 169–170), for example, and for diabetes and depression among immigrant women (Mendenhall 2012). In such cases, poor health and associated disabilities reduce employment options, and contribute to chronic life difficulties, discrimination, demoralization and social exclusion. In this volume, various case studies explore the synergies of health and social life in different settings, inviting the reader to reflect on comorbidity, syndemics and the persistent interplay of poverty and ill health.

## New Fields

Anthropological interest in local and small-scale communities, and sympathy for those subordinate in relationships of power and authority, has influenced the direction of contemporary medical anthropologists, but various researchers in the field, in the global North and South, have turned their attention from local communities and the individuals and households that comprise them, to their health providers, habitual environments and networks—primary health clinics, marketplaces, and shamans, for instance, as well as surgeons, hospitals, training institutions and boardrooms. Many of us in anthropology, at the same time, have been inspired by critical political theory, cultural studies, and science and technology studies, among other intellectual threads, and reflecting this, our work has extended well beyond the village and its traditional healers, to encompass the body in space (both geographical and aeronautical, see Chapter 15), to nanotechnology and prosthetics, pathology laboratories and tissue banks, investment strategies and medical markets, restorative justice and truth and reconciliation commissions. Where medical anthropology in the past was often criticized by sociocultural anthropologists for being too applied, nowadays it is recognized that medical anthropologists are at the forefront of the discipline, moving anthropology to think beyond dichotomies such as those between ethnomedicine and modern biomedicine, medically defined disease and culturally constructed illness, body and mind, and social and biological anthropology. The case studies in this volume reflect this intellectual and empirical richness.

## Reading the Volume

Although this volume can be read linearly, this need not be so; we have not conceptualized it in structure or use as static or prescriptive. The case studies, for example, have been selected to fit with the themes of the chapters in which they are situated. To a degree we have written around these cases to highlight how anthropologists have engaged with the issues at stake in the past and to explore the particular turns in medical anthropology today. But we also took seriously the challenge to design a handbook as a compendium that might be played with by teachers and students. The volume could be used for a full semester course, although not all semesters run for 16 weeks, and so already, the decision may need to be taken about what to include, what to drop. Further, we have organized the chapters and included particular cases following one logic—from life course and kinship in the earlier chapters, to the globalization and politics of health and illness, with respect to contemporary crises that impact health and the quality of people's lives and in relation to the distribution of technologies, products and people. But this is only one approach, and there are other ways to use the material. It is possible, for instance, to work across chapters, drawing on the case studies of ethnographies from Africa, for instance, or from high-income, high-technology settings, or as they touch on particular health issues (reproduction, for instance, or aging).

We have shied away from encyclopedic coverage; it is neither possible to do this in a field so rapidly expanding, nor to do so in a volume of this length. In selecting the cases, we were interested in illustrating particular concepts and themes, rather than diseases. You, the readers, will find other ways of curating the case studies and will bring other perspectives to the fields that we have included. We now briefly introduce these.

The first substantive chapter, following, is about children, and their sicknesses and suffering. Ironically, many medical anthropologists have attended to reproduction, infertility and delivery practices, and this is reflected in the case studies in a number of chapters. In contrast, medical anthropology has paid far less attention to children, suggesting that, as ethnographers and theorists, we privilege formal communication skills and follow power, even as we resist it (see, for example, Levine and New 2008; Montgomery 2008). Children are especially powerless; in sickness, they are vulnerable to and dependent on their parents and healers to find patterns in frank signs of sickness, pain or discomfort, or in the ambiguities of their behavior. Parents' and children's lives are entwined, but where power and communication are concentrated, as occurs in a medical clinic, parents often speak for and represent their children. With the case studies to guide us, we ask how parents deal with children who are ill, have limited capabilities, or are in other ways suffering. We ask how young people experience illness and disease, negotiate therapeutic demands, and articulate their disapprobation and resistance, opening up the possibility of a stronger research agenda with them.

In Chapter 3, we place medical anthropology's approach to the study of sexuality within a larger context of sexuality studies. Early 'sexology' documented 'normal' sexual behavior; psychiatric definitions classified homosexuality as 'deviant.' Anthropology in contrast has approached sexuality as something both more culture-specific and fluid. Influenced by the scholarship and activism of feminist and queer politics, anthropologists have shown that gender and sexuality intersect with other notions of difference, demonstrating the problem of assuming the universality of Northern biomedical and social concepts of homosexuality and transgender (Jacobs et al. 1997). Societies have unique ways of understanding and accommodating various presentations of gender and diverse sexual practices, and in this context, we consider how people innovate with medical technologies related to sexuality, such as contraception and disease protection, and those related to constructions of masculinity (see Manderson 2012). The refusal or reluctance of some people to use condoms to prevent HIV, for example, has provoked many anthropologists to ask: Who uses contraception, and for what reasons? Understanding the relationship of fertility to status, and contraception to distrust, medical anthropologists have sought to broaden understandings of sexual health.

In Chapter 4, we consider how medical anthropologists have approached the study of HIV. Now that people are living longer as a result of antiretroviral drug regimens and the wider availability of these drugs, HIV infection is less likely to be a near-certain death sentence. Living with HIV brings its own challenges, however, changing how people relate to health care systems, their partners and families, their communities, and others similarly afflicted. Noting the impact of the concept of 'biosocialities' (Rabinow 1996), in this chapter we discuss some of the forms of sociality that have arisen in the 'era of treatment' and consider how treatment protocols that originate in highly industrialized settings travel through global health interventions. Documenting problems of access to treatment, medical anthropologists have been able to show that antiretroviral drug protocols impose a specific form of biopower (Foucault 1976) by mandating that patients follow moralized codes of behavior. Norms surrounding treatment programs originating in the United States and Western Europe—specifically about 'coming out' and 'disclosure'—have settled uneasily in other countries, where secrecy might instead be a way of showing respect and compassion (Manderson et al. 2015). At the same time, the continued real suffering of people infected with HIV has had wider impact on families, particularly as elderly grandparents take on the task of providing care for orphaned children. The persistence of stigma against people living with HIV, and their families, and resistance to disclosure, makes clear that much work needs to be done by both medical anthropologists and activists (Squire 2015).

In Chapter 5 we turn to 'stress,' a term which is used in everyday speech and in psychological frameworks to describe a feeling of being overwhelmed, from being too busy to having too few resources. But medical anthropologists have identified that people understand, enact and speak about stress in quite varied ways, drawing on different symptoms, etiologies and responses for distress. In this chapter, we look at some of the ways that people understand stress, such as the Latin American experience of *nervios* or the Indonesian feeling of *pusing*. People use specific idioms of distress to describe and explain the etiology and physical and emotional dimensions of these experiences (Nichter 2010). Depending on local understandings, effective treatment might involve taking drugs, seeing a spiritual healer, or talking with a counselor or a psychotherapist. In some cases, as in Japan, receiving a diagnosis of depression may help to validate experiences of socioeconomic hardship. Anthropologists relate these particular constructs of illness and stress to larger societal structures that engender social and economic suffering. Alcohol, tobacco and illicit drugs are some of the ways that people use to alleviate stress, and anthropologists have studied how corporate marketing seeks to capitalize on people's desire to escape or to cope. In medical anthropology, studying stress may lead to other insights—about economic precarity, powerlessness, workplace exploitation, inequality, performances of masculinity and femininity and more.

In Chapter 6, we focus on how the body is used as a site of resistance and rebellion against individual, social and cultural norms. Anthropology has a long history of describing resistances of various kinds, and bodily resistances play an important role in that intellectual trajectory. In the most primal of senses, using the body as a site of resistance is one of the strongest acts that an individual can make. To risk one's health and wellbeing, as oftentimes is the case, to make a statement—through a hunger protest, ingesting a dangerous substance, self-harm, or refusing a blood transfusion or other medical treatment, for example—can call attention to injustice, religious difference, personal pain or any number of highly charged issues. Using the cases in this chapter, we think through how acts of resistance vary, how they are often logical on one level yet not on another (Willis 1977), and how resistances are carried out in direct opposition to the established institutions and bureaucracies that reinforce biopower. We explore case studies that describe the promotion and appeal of smoking, or what it is like to starve one's self to the point of near death, and we question the role of bureaucracies in the 'complexification' of daily life (Gupta 2012).

In Chapter 7, we consider the chronicity of both infectious and chronic non-communicable diseases and degenerative conditions, the impact of this on social life, and the variability of outcomes depending on social determinants (Manderson and Smith-Morris 2010). Many health problems are of long duration; in cases such as Chagas disease and onchocerciasis, the full impact of parasitic infection—heart disease and blindness, respectively—may not be apparent for decades. The symptoms, the development of complications, and interactions with other diseases may be able to be controlled through medication or through other interventions, some technical—hip and knee replacements, pacemakers and laser surgery, for example—and some behavioral, as occurs with changes in diet. We note how governments, research scientists and industry have approached the prevention or control of these conditions, and consider the viability of these approaches in relation to different understandings of risk, vulnerability and disability.

We move, in Chapter 8, to the idea of caring. Caring for one another, especially in the transitions of birth, death, serious illness and injury, is an essential part of what we do as humans and what we study as medical anthropologists. We first think through the behavioral acts that surround caring: How is care actually constituted in different settings and who, or what in the new cyborgified world, carries out this care? Our lives are lived with and through technology now more than might have been imagined even only a few decades ago. Humans, ever adaptable and ingenious, quickly integrate friendships and medical consultations into their daily routines via their laptops and mobiles, barely missing a step as they voraciously consume personal communication devices that are ever smaller and ever smarter. In our case studies, we illustrate how medical decisions are made by families spread out across the globe, and we see new forms of sociality established via caregivers, patients and practitioners who are linked through telemedicine and home monitoring systems of 'care.' Patients become their own advocates as they are able to access state-of-the-art medical information; they challenge their practitioners; and practitioners in turn increasingly tell patients to "look it up on the Internet and get back to me if you have questions." In this brave new world of caring, smart houses check in on grandma to see if she's moving around as she should and set off alarms if she's not. Our focus here is on physical care, emotional care and how humans are always "tinkering," as Annemarie Mol and her colleagues (2010) describe it; people find myriad ways of fulfilling this most fundamental and emotionally engaging of life's chores.

In Chapter 9, we turn to the end of life, in a broad sense—to people who are frail and sometimes confused, and to how others interpret and respond to debility, functional impediments and confusion. We consider also the ethics of truth telling and the difficulties that family members face when they must acknowledge that life is finite and so must make decisions that are, for them, ethical and honorable. In the global North, decisions about the care of people in extreme old age, often without the capacity to communicate themselves, are informed by conventions and evolving ideas about the right to know one's diagnosis and prognosis, quality of life, living wills, advanced care directives and palliative care (Kaufman 2005). Globally, people tip toe through this minefield as they try to ensure the best possible care and least suffering for a loved one in advanced age. People do their best, although sometimes to no avail, to protect the dignity of those who are dying and to create a gentle and safe departure for loved ones suffering a terminal illness or the increased dilution of capability and blunting of comprehension. Life outcomes provide empirical examples for people to think through and frame local and global ethics. Death is painful for those left behind, and everywhere, as we explain, its rituals exist to support the bereaved as they make sense of their loss. We conclude this chapter with a case study of undertakers in Singapore, allowing us to reflect on the kinds of work that people do, globally, in order to manage death and the dead.

Anthropologists have long been interested both in the pharmacopea of individual healers and particular medical traditions, and the 'medicalization' of everyday life and the use of products to effect this. Various health concerns have become 'conditions' requiring treatment via surgery or, usually, with medication. In Chapter 10, we examine the movement of medicines. Beyond doctors' prescriptions, synthetic drugs and natural remedies flow in a variety of ways, including through pharmaceutical promotion, small business ventures, off-label use, and informal and illegal trade. By tracing medicines and seeing how they can follow pathways never intended by doctors, regulators or commercial interests, medical anthropologists are able to capture the complexities of the marketplace, its local variations and global influences (Nordstrom 2007). Despite the power of the pharmaceutical industry, people consume medicines according to their own interests and based on their own reasoning. Beyond the global pharmaceutical industry, many smaller actors have an economic interest in the flow of medicines. They include those who mix ingredients for herbal supplements, both home made and manufactured. They include the corner market stall that sells cosmetic products as remedies or that makes small packets of prescription medicine and vitamins; the informal street dealers of drugs with off-label uses; the local pharmacies that compound tailored medications for individuals. Medical anthropologists are engaged in studying the regulations that govern many of these corners of the medicine market, how people circumvent such policies, and for what reasons.

In Chapter 11, we step back and provide a very anthropological account of health and healing within the Anthropocene, a term defined by Webster's dictionary as "the period of time during which human activities have had an environmental impact on the Earth regarded as constituting a distinct geological age." We trouble that definition as we move through the chapter, questioning how best to conceptualize and think about this, our present, age. We use the term *Anthropocene* to highlight the importance of the environment as an integral component of health and wellbeing, and to locate medical anthropology within the larger social science literature surrounding this topic (Latour 1999). We use the term too to engage with the cutting edge, wide-opened thinking that we must have in order for humans and other creatures to survive our current, deeply extractive and destructive ways of being on the planet (Haraway 2015). As a touchstone of sanity, we draw on the early works of anthropologists, starting in the late nineteenth century, who documented Indigenous groups living and thriving in various more or less pristine environments. The old ways of curing using the plants and animals, things inert and things spiritual in the environment, is our starting point as we move into the sullied environment of the present. Recent human history—imperialism and colonization, industrialization and globalization—has been driven by the exploitation of the environment, with little attention to the impact of this appropriation on the health of people immediately or in the long term. Our case studies take us to Indigenous communities dealing with climate change in the Arctic and people suffering from air and other pollutions in Italy. We then move to the ravages of Hurricane Katrina and what Merrill Singer (2009) has referred to as the 'pluralea' of environmental, social and biological threats that come with these mega-disasters. These threats—and the environment itself—are changing in profound ways, shaping the very foundations of our existence on this planet and reshaping the planet itself.

In Chapter 12, we turn to the experiences of growing numbers of people who seek care across international borders, either through volition or necessity. We frame this chapter within the idea of the quest for therapy, as described by Janzen (1978) many years ago. In this chapter, we describe new quests. These include quests for treatments available in the places where people reside, where travel is motivated because people lack health insurance, and so simply cannot meet the costs of basic interventions such as dental surgery, a hearing aid or a pacemaker. Others pursue quests for medications and access to medical personnel across international boundaries, again because of cost at home. Still others live without legal residency papers in countries far from 'home,' and can only access medical care through quests abroad.

People migrate for many reasons: avoiding civil or inter-country war is only one reason. Others flee because of persecution or its threat for personal reasons—sexual preference or religious affiliation for instance, or because their life options are capped because of ethnicity or gender. In Chapter 13, however, we turn specifically to the experiences of war and civil violence, and the scars these leave on soldiers and civilians, however engaged. A century after the 'great war'—World War One—and the trail of violence that stretched from Papua New Guinea to the Middle East and North Africa to Europe—the idiom of 'lest we forget' seems ironic; we have witnessed and still live in a time of unspeakably violent wars; of the continued fall-out from past wars, including on the health of new generations; of nations consistently worrying over further engagement in yet another place; and of the indelible damage of being at war, either as a civilian or as a warrior. The case studies capture the emotional density and political and logistic complexities of these issues: the dilemmas of providing health care in a militarized environment; the psychological scars of soldiers; the indelible traumas experienced by civilians; the search for the remains of people lost in the horror of past conflict; and the confusion of soldiers as they try to reconcile killing with a personal ethics of the preservation of life (Sørensen 2015). Anthropologists have been involved in the past in exhumations and determination of cause of death; in one case study, Rachel Carmen Ceasar returns specifically to this question to address why exhumation matters. The case study authors all also address the impact on the minds of those who fought, were brutalized, fled or were left behind in the shadows of devastation.

Beginning with a history of the gene, Chapter 14 takes us into the future. Here we examine how genetic data and genetic testing have become sites of public concern as well as anthropological study. Medical anthropologists have researched how people come to form new social groupings (i.e. biosocialities, Rabinow 1996), based on their knowledge of having a certain disease or risk of disease. Notions of uncertainty and risk play a significant role in how people conceive of their options and possibilities. Such genomic information also shapes how technology and science seek to tailor treatment to individual circumstances. Anthropologists have shown how notions of risk can make people targets for corporate marketing, as companies promote genetic testing and prey on people's fears (Rapp 2000). Studies have also focused on the understandings of kinship that emerge in relation to genomics, as families incorporate new genetic information into their notions of relatedness and resemblance, to greater and lesser degrees (Finkler 2000; Taussig 2009; Taussig et al. 2001). Looking at the practices of scientists has also been a fruitful avenue for medical anthropologists, not only in genomic medicine but also in relation to synthetic biology, epigenetics and nanotechnology. The avenues of research and experimentation undertaken in these areas open up many social and ethical questions that are important lines of inquiry for medical anthropologists.

The final chapter with case studies focuses on how biomedicine travels across the globe by humanitarian brigades, by teaching and re-teaching techniques and protocols, and by market demand and government-implemented programs. We question what happens to 'biomedicine' when it is dropped, sometimes it seems in a cavalier fashion, onto the least accepting of soils. Without the proper tools and medications, systems of education and patient surveillance, without even the basic administrative technologies to ensure the ability to track patients, biomedicine loses a good deal of its efficacy and power. The lead case study in Chapter 15 demonstrates the ways in which protocols of a particular sub-field of medicine, obstetrics, are challenged by notions of natural birthing—both the original idea and its critique flowing from the global North to the global South, and contrast this with the introduction of IVF technology in Africa. We then critically parse how global surgery is prioritized, or not, within the powerful institutions of academe that produce it in the US and other resource-rich countries, and we chart the course of the ebb and flow of biomedicine into West Africa via the 2014 outbreak of the Ebola virus. We conclude with the imperative of providing care where people are suffering, and so as Peter Redfield illustrates, the necessary continued stop-gap of emergency aid and international humanitarian relief.

We conclude by reflecting on the challenges for medical anthropologists in the decades to come. We revisit a number of critical areas explored in earlier chapters, extending our discussions of involuntary migration and its health effects on adult and children, reflect on zoonotic infections and climate change, and discuss how new biomedical technologies and new discoveries in medical research are providing us with new areas for new research. We encourage medical anthropologists to rethink the complex connections between body and mind, and people and the societies in which they live. We draw attention also to the need for anthropologists to engage in reflexive dialogues with biomedical and other scientists who are increasingly acknowledging the importance of local biologies and the role of environment in people's heath.

## References


Agyepong, Irene A., Bertha Aryee, Helen Dzikunu, and Lenore Manderson 1995 *The Malaria Manual. Guidelines for the Rapid Assessment of Social, Economic and Cultural Aspects of Malaria. Methods for Social and Economic Research in Tropical Diseases No. 2.* Geneva, Switzerland: Special Programme for Research and Training in Tropical Diseases.

Benedict, Ruth 1946 *The Chrysanthemum and the Sword: Patterns of Japanese Culture.* Boston, MA: Houghton Mifflin Harcourt.

Bourdieu, Pierre 1977 *Outline of a Theory of Practice.* Cambridge, UK: Cambridge University Press.

——— 1984 *Distinction: A Social Critique of the Judgement of Taste.* Cambridge, MA: Harvard University Press.

——— 1992 *The Logic of Practice.* Stanford, CA: Stanford University Press.

Chambers, Robert 1983 *Rural Development—Putting the Last First.* Harlow, UK: Longmans Scientific and Technical Publishers.

Davis-Floyd, Robbie 1992 *Birth as an American Rite of Passage.* Berkeley, CA: University of California Press.

Finkler, Kaja 2000 *Experiencing the New Genetics: Family and Kinship on the Medical Frontier.* Philadelphia, PA: University of Pennsylvania Press.

Firth, Rosemary 1943 *Housekeeping among Malay Peasants.* London: London School of Economics.

Foucault, Michel 1973 *The Birth of the Clinic: An Archaeology of Medical Perception.* A.M. Sheridan, transl. London: Tavistock Publications Limited.

——— 1976 *The History of Sexuality*, Vol.1. London: Penguin Books.

——— 1980 *Power/Knowledge: Selected Interviews and Other Writings, 1972–1977.* New York: Pantheon.

Geertz, Hildred 1961 *The Javanese Family—A Study of Kinship and Socialization.* New York: The Free Press of Glencoe.

Ginsburg, Faye, and Rayna Rapp, eds. 1995 *Conceiving the New World Order: The Global Politics of Reproduction.* Berkeley, CA: University of California Press.

Gupta, Akhil 2012 *Red Tape: Bureaucracy, Structural Violence, and Poverty in India.* Durham, NC: Duke University Press.

Haraway, Donna 2015 Anthropocene, capitalocene, plantationocene, chthulucene: Making kin. *Environmental Humanities* 6:159–165.

Hardon, Anita, and Eileen Moyer 2014 Medical technologies: Flows, frictions and new socialities. *Anthropology & Medicine* 21(2):107–112.

Herman, Elizabeth, and Margaret Bentley 1993 *Rapid Assessment Procedures (RAP) to Improve the Household Management of Diarrhea.* Boston, MA: International Nutrition Foundation for Developing Countries.

Inhorn, Marcia, ed. 2002 *Infertility around the Globe: New Thinking on Childlessness, Gender, and Reproductive Technologies.* Berkeley and Los Angeles, CA: University of California Press.

International Conference on Primary Health Care, Alma-Ata, USSR 1978 *Declaration of Alma-Ata, 6–12 September.* Geneva, Switzerland: World Health Organization. Accessed from http://www.who.int/publications/almaata_declaration_en.pdf

Jacobs, Sue-Ellen, Wesley Thomas, and Sabine Lang, eds. 1997 *Two-Spirit People: Native American Gender Identity, Sexuality and Spirituality.* Champaign, IL: University of Illinois Press.

Janzen, John M. 1978 *The Quest for Therapy: Medical Pluarlism in Lower Zaire.* Berkeley, CA: University of California Press.

Jordan, Brigitte, and Robbie E. Davis-Floyd 1974 *Birth in Four Cultures: A Crosscultural Investigation of Childbirth in Yucatan, Holland, Sweden, and the United States.* Montreal, Canada: Eden Press.

Kaufman, Sharon R. 2005 *And a Time to Die: How American Hospitals Shape the End of Life.* Chicago, IL: University of Chicago Press.

Kleinman, Arthur 1980 *Patients and Healers in the Context of Culture: An Exploration of the Borderland Between Anthropology, Medicine, and Psychiatry.* Berkeley, CA: University of California Press.

Kleinman, Arthur, Leon Eisenberg, and Byron Good 1978 Culture, illness, and care: Clinical lessons from anthropological and cross-cultural research. *Annals of Internal Medicine* 88(2):251–258.

Latour, Bruno 1999 *Pandora's Hope: Essays on the Reality of Science Studies.* Cambridge, MA: Harvard University Press.

Levine, Robert, and S. Rebecca New, eds. 2008 *Anthropology and Child Development: A Cross-Cultural Reader.* Oxford, UK: Blackwell.

Lock, Margaret 1993 *Encounters with Aging: Mythologies of Menopause in Japan and North America.* Berkeley, CA: University of California Press.

MacCormack, Carol P., ed. 1982 *Ethnography of Fertility and Birth.* Prospect Heights, IL: Waveland Press.

MacCormack, Carol P., and Marilyn Strathern, eds. 1980 *Nature, Culture, Gender.* Cambridge, UK: Cambridge University Press.

Manderson, Lenore, ed. 2012 *Technologies of Sexuality, Identity and Sexual Health.* London: Routledge.

Manderson, Lenore, and Peter Aaby 1992 An epidemic in the field? Rapid assessment procedures and health research. *Social Science & Medicine* 35(7):839–850.

Manderson, Lenore, and Carolyn Smith-Morris 2010 *Chronic Conditions, Fluid States: Chronicity and the Anthropology of Illness.* New Brunswick, NJ: Rutgers University Press.

Manderson, Lenore, Mark Davis, Chip Colwell, and Tanja Ahlin 2015 On secrecy, disclosure, the public and the private in anthropology. *Current Anthropology* 56(S12):S183–S190.

Martin, Emily 1987 *The Woman in the Body: A Cultural Analysis of Reproduction.* Boston, MA: Beacon Press.

Mead, Margaret 1928 *Coming of Age in Samoa.* New York: William Morrow Paperbacks.

Mendenhall, Emily 2012 *Syndemic Suffering: Social Distress, Depression, and Diabetes among Mexican Immigrant Women.* Walnut Creek, CA: Left Coast Press.

Mol, Annemarie, Ingunn Moser, and Jeannette Pols, eds. 2010 *Care in Practice: On Tinkering in Clinics, Homes and Farms.* Piscataway, NJ: Transaction Publishers.

Montgomery, Heather 2008 *An Introduction to Childhood: Anthropological Perspectives on Children's Lives.* Oxford, UK: Blackwell.

Nichter, Mark 2010 Idioms of distress revisited. *Culture, Medicine and Psychiatry* 34(2):401–416.

Nordstrom, Carolyn 2007 *Global Outlaws: Crime, Money, and Power in the Contemporary World.* Berkeley and Los Angeles, CA: University of California Press.

Pelto, Gretel H., and Margaret Armar-Klemesu 2010 *Assessing the Behavioral and Local Market Environment for a Commercial Complementary Food: A Focused Ethnographic Study Manual.* Geneva, Switzerland: Global Alliance for Improved Nutrition.

Pelto, Gretel H., and Sandy Gove 1992 Developing a focused ethnographic study for the WHO Acute Respiratory Infection Control Programme. In *Rapid Assessment Procedures: Qualitative Methodologies forPlanning and Evaluation of Health Related Programmes.* N.S. Scrimshaw and G.R. Gleason, eds. Pp. 215–226. Boston, MA: International Nutrition Foundation for Developing Countries.

Polgar, Steven 1962 Health and human behavior: Areas of interest common to the social and medical sciences. *Current Anthropology* 3:159–205.

Powdermaker, Hortense 1966 *Stranger and Friend: The Way of an Anthropologist.* New York: W.W. Norton & Company.

Rabinow, Paul 1996 Artificiality and enlightenment: From sociobiology to biosociality. *Essays on the Anthropology of Reason.* Pp. 91–111. Princeton, NJ: Princeton University Press.

Rapp, Rayna 2000 *Testing Women, Testing the Fetus: The Social Impact of Amniocentesis in America.* New York: Routledge.

Rivers, William H.R. 1924 *Medicine, Magic, and Religion.* New York: Harcourt Brace.

Rosaldo, Michelle, and Louise Lamphere, eds. 1974 *Woman, Culture, and Society.* Stanford, CA: Stanford University Press.

Rubel, Arthur J. 1964 The epidemiology of a folk illness—Susto in Hispanic America. *Ethnology* 3(3): 268–283.

Scrimshaw, Susan C.M., and Elena Hurtado 1987 *Rapid Assessment Procedures for Nutrition and Primary Health Care: Anthropological Approaches to Improving Programme Effectiveness.* Tokyo, Japan: United Nations University.

Singer, Merrill 2009 Beyond global warming: Interacting ecocrises and the critical anthropology of health. *Anthropological Quarterly* 82(3):795–819.

——— 2014 Syndemics. In *The Wiley Blackwell Encyclopedia of Health, Illness, Behavior, and Society.* W. Cockerham, R. Dingwall, and S.R. Quah, eds. Pp. 2419–2423. Hoboken, NJ: Wiley-Blackwell.

Singer, Merrill, D. Ann Herring, Judith Littleton, and Melanie Rock 2011 Syndemics in global health. In *A Companion to Medical Anthropology.* M. Singer and P.I. Erickson, eds. Pp. 159–179. Malden, MA: Wiley-Blackwell.

Sørensen, Birgitte Refslund 2015 Veterans' homecomings: Secrecy and post-deployment social becoming. *Current Anthropology* 56(S12):S231–S240.

Squire, Corinne 2015 Partial secrets. *Current Anthropology* 56(S12):S201–S210.

Taussig, Karen Sue 2009 *Ordinary Genomes: Science, Citizenship, and Genetic Identities.* Durham, NC: Duke University Press.

Taussig, Karen Sue, Rayna Rapp, and Deborah Heath 2001 Flexible eugenics: Technologies of the self in the age of genetics. In *Genetic Nature/Culture: Anthropology and Science Beyond the Two Culture Divide.* A. Goodman, D. Heath, and S. Lindee, eds. Pp. 58–76. Berkeley, CA: University of California Press.

Wilkinson, Richard, and Michael E. Marmot 1999 *Social Determinants of Health.* Oxford, UK: Oxford University Press.

Willis, Paul 1977 *Learning to Labor: How Working Class Kids Get Working Class Jobs.* New York: Columbia University Press.

World Health Organization 2008 *Closing the Gap in a Generation: Health Equity through Action on the Social Determinants of Health.* Commission on Social Determinants of Health Final Report. Geneva, Switzerland: World Health Organization.

Generations of Village Life, 2012. Efate, Vanuatu.

## About the photograph

*The burdens of bus fare, arranging for travel, and the requirements of a full day away from work and subsistence agriculture are significant impediments to accessing health care, illustrating a multi-generational battle with poverty where women and young children are most vulnerable. Without monthly visits from a mobile clinic, many children living in rural Vanuatu would not have regular access to biomedical care. Everywhere, intersections of poverty, health, illness, and medicine require new approaches and culturally diverse solutions, informed by ethnographies of children, to improve health care and so health outcomes for infants and children.*

—Chelsea Wentworth

# 2
# Changing Childhoods

*Lenore Manderson, Elizabeth Cartwright and Anita Hardon*

Anthropology has a long history of neglecting children, both ethnographically and theoretically. Only a few anthropologists have made the study of children central to their work (Abadia-Barrero 2011; Gottlieb 2004; Hirschfeld 2002), and in general, perhaps unwittingly, we have taken on an understanding—from our own settings and from the people with whom we have worked—of children as not yet 'human,' that is, as not yet enculturated or socialized (Geertz 1960; Herdt 1981). Such understandings of what children are, and what they can do, are informed by ideas about age, cognitive development and ideation. Accordingly, children are the subjects of patterns of socialization, formal and informal learning, and ritual occasions that 'make' them human (Gould and Glowacka 2004), but they are rarely represented as social actors who might contribute to our knowledge of social processes, structures and institutions or to our imagination of social life. In this chapter, we concentrate on children and adolescents, and their experiences of sickness, discomfort, disability, vulnerability and marginality, so to gain insight into how they experience disease and distress, and how these unfold in different family and cultural contexts. In the case studies, we highlight children's points of view to the extent that we have access to it. Children have their own perspectives on health, disease and agency. Those who work with children argue cogently that adults need to listen to them and grant them power to care for themselves. Children participate in their illnesses, acute, curable, chronic and fatal, with insight and fear, humor and anger, and, ultimately, quite often, with grace.

Anticipating the material we explore here, a number of scholars set the groundwork for contemporary research with young people as key informants. Margaret Mead's monograph *Coming of Age in Samoa* (1928), based on her doctoral dissertation, was particularly influential. Her descriptions of the relaxed attitudes towards life and the apparent sexual freedom of the young Samoan women with whom she worked was discrepant with American normative behavior at the time of the publication of the book—it was shocking and unsettling to conservative sectors of the public. In many respects, Mead paved the way both for an anthropology of children and an anthropology of sexuality. Much later, both John and Beatrice Whiting (1975) and Sandra Burman and Pamela Reynolds (1986) carried out comparative fieldwork, highlighting the diverse experiences of childhood in different social systems. As children became a more popular topic of research, ethicists and review boards concomitantly began to question whether and how children might ethically participate in research, how such research might be conducted, what children might reveal, and the obligations and boundaries that need to be honored by researchers. Increasingly, medical and other anthropologists have been asked to reflect on the capacity of children to exercise informed consent, to make their own decisions about participating in research, and thus, to contribute to our understandings of human relationships, social processes and social organization from their points of view. The case studies that we present in this chapter take up this challenge, offering us fresh ways of thinking about children's health and wellbeing, social engagement and resistance.

In *Rites of Passage* (1960), Arnold Van Gennep drew attention to social elaborations of transitions from birth to puberty and beyond, depending on cultural context and local tradition: hair-cutting, tattooing, naming, circumcision and religious ceremonies have been and continue to be held to link children to particular families and faith, lineage and community. Such practices mark sexual and economic maturation, marriageability and civic responsibility. These ideas are not immutable, however. Family structure and composition have changed substantially everywhere with changes in economic production, monetization, industrialization, and urbanization, influencing what children are and what they able to do (Ariès 1962; Reynolds 1991). In all of this, children have increasingly been the subject not authors of their lives, that is, their childhood and the activities in which they might be involved are defined for them. Everywhere children's lives are circumscribed and protected through regulations that follow a consensus of adult views and global norms, based on ideas about age-appropriate development goals, activities, age at marriage, the right to education (and its content, often still gender-biased), and the limits to their employment until the age of assumed agency and autonomy. As Allison James and Adrian James (2008) note, no-one writes of 'the adult' to refer to all adults, yet 'the child' is routinely the subject of international conventions, national policies, and medical, educational, and other social institutions that directly impact on the health and wellbeing of individual children. Compulsory school attendance and strict controls over work are mandated by international conventions and nation states. Multinational agencies advise and governments determine when children should attend health clinics for monitoring, surveillance, treatment and disease prevention. Yet even so, children work as unpaid family labor in and around the home, and formally or informally as waged labor. And as illustrated in the case studies in this chapter, they manage their own and others' sickness, disease, disability and difference; they provide care both to other younger children and adults; and they question the extent to which others govern their lives.

## Childhood and the Positioning of Children

The most successful public health interventions of the past two centuries have focused on children. In the last hundred years especially, maternal and child health care, immunization programs to prevent infectious disease, the provision of potable water and sanitation, and nutrition programs have driven down the incidence, prevalence and severity of infections in infants and small children, dramatically reducing morbidity and mortality rates. Although vaccination safety and risk have been part of a sustained critique of the medicalization of childhood, immunization remains a key strategy to control infections that, in children especially, can be fatal or cause long-term disability; it is standard in global infant and child health protocols even in settings where general access to health care is poor.

Clinical metrics feed into parental concerns about infants who might be at one end or the other of growth percentiles, or about a child's attainment of milestones—teething, walking, talking, weaning, sleeping, toileting. These metrics of the 'normal' child and normal development influence the decisions that women make in the care of their infants and small children. Slow weight gain may be interpreted as a sign of insufficient milk supply (van Esterik 1988), for instance, leading to early weaning to bottled milk or the introduction or increase of other foodstuffs. Slow speech development might signal problems in intellectual capacity, leading parents and children to a circuit of diagnostics, medical and behavioral interventions, and lay advice (Sobo 2010). Although these metrics vary with local norms of infant behavior—ideas about 'healthy appetite' or 'healthy sleeping,' for instance—global consensus about infant and child health, and global programs that follow from this, are universally applied despite local norms and other unique positionalities (Justice 2000). At the same time, a parent or another caregiver may see no link between clinical advice proffered and their own child's health, or for financial and other practical reasons, they may be unable to follow the advice. Problems in the procurement of vaccines, the organization of health services, or the capacity of health staff may also lead to the uneven delivery of child health programs. While 'the child' is the subject of state and global interventions, advice and monitoring, the health of children is inevitably shaped by local exigencies.

## Sickness and Subjugation

Research conducted in the mid-twentieth century, particularly by demographers, highlighted the instrumental reasons that gave children value to families and, in various settings, explained large family size: to maintain lineage and property, contribute to family prestige, provide labor, and ensure care in old age, among other reasons (Nieuwenhuys 1996). Yet, regardless of how the number of children was rationalized, and how and why a child of one sex might be preferred over another, deep affection still binds parents, and other family members and caregivers, to their children. This emotional investment in children—the affective underpinning of reproduction—and the commitment to having a family, influences why people turn to services and their various technologies to ensure conception, safe pregnancy, fetal health, safe childbirth, and secure life outcomes.

Historically, reproduction has been very much a concern contained within families, despite some informal surrogacy, wet-nursing and adoption that extended its boundaries. But from the late twentieth and into the twenty-first century, conception, gestation and reproduction have increasingly been medicalized and commercialized, highlighting inequalities in in-vitro fertilization, non-familial surrogacy, and pre-implantation screening. For example, women involved in commercial surrogacy sequestered in Ghana and India have a very different relationship to their pregnancies, medical services, and the children they bear compared with the wealthier men and women for whom they are incubating (see also Chapters 3 and 15).

In resource-rich environments, the reach of technology into pregnancy includes amniocentesis and chorionic villus sampling (CVS), with other new tests always rolling out into the market. Rayna Rapp (2000) has described how these tests and their results influence decisions about whether or not to proceed with a pregnancy. Where such facilities are available, technologies are used to monitor women throughout pregnancy, capturing changes in health that might indicate compromised maternal or fetal outcomes, using such information to determine or advise on mode of birth, and identifying the need for emergency prenatal or just postnatal surgery or other high case interventions. Highly sophisticated technologies in neonatal wards may ensure that preterm and very low birth weight infants survive, even though many of these infants will experience lifelong medical complications and impairments as a result of these interventions. For most women worldwide, these technologies of surveillance, safety and survival are elusive; for them, even basic emergency obstetric care may be an ideal rather than a right.

Likewise, advanced knowledge of the pathophysiology and management of functional impairments and progressive disease, and developments in surgery, medication, and physiotherapy, have significantly extended the lives and ensured the capabilities of children, from wealthy households and wealthier nations, who are born with a range of medical conditions—hemophilia, cystic fibrosis, osteogenesis imperfecta, severe asthma, spina bifida, epilepsy and cerebral palsy, among others. These children's lives, depending on the severity of their condition, are disrupted by illness and disability, and they depend on others to ensure and provide appropriate care. Such care is generally provided by women, most often their mothers. Here, there is scope for much medical anthropological work. As an example, Elisa Sobo (2010) has illustrated the complex pathways that parents of children with congenital talipes (club foot) and Down syndrome must navigate between family doctors, specialist pediatricians, clinics and other therapies and rehabilitative services in search of effective treatment and care, evoking foundational medical anthropological work on hierarchies of resort (Holten 2013; Romanucci Schwartz 1969). The visibility of these conditions, Sobo suggests, attracts discussion about interventions and possible action; the clinician is the expert. Children experiencing these conditions are positioned as therapeutic subjects, defined by their conditions as 'disabled,' with possible lifelong social disadvantage depending on household income and access to services.

In clinical encounters, children, parents and health providers interact in particular, scripted ways. Parents or guardians sift out and represent what children feel and experience, including symptoms that they cannot know in certainty—pain, for example. Parental power is, even so, subordinate to medical authority even when that authority is couched in the most banal terms, as Mara Buchbinder (2012) describes in a paediatric clinic. In her example, patients aged from 5 to 23 years, the majority adolescents, had a range of development disorders, and communication and social difficulties. Clinicians explained persistent pain to children and their parents by talking about 'sticky brains' as a metaphor of the difficulty a child might have in disassociating from pain. Given the access of parents and older children to information on the Internet, we might question the effectiveness of this apparent talking down, and in other contexts, both children and parents are expected to be experts.

Even when children are subject both to parental and clinical control, they are not invisible (Christensen and James 2008). Children can and want to participate in decisions concerning their health. In her book *The Private Worlds of Dying Children*, Bluebond-Langner (1980) illustrates the capacity of very young (3–6 years of age) children with leukemia who know that they are dying, yet who do not reveal this in order to protect their parents. Alderson (1993) similarly argues in her book *Children's Consent to Surgery* that very young children want to be informed about their condition, and are able to absorb the information and make sense of it.

As they age, children are expected to take increasing responsibility for their own health, and in doing so, they gain competence in reading their own bodily signs of illness, and are able to anticipate, for example, an asthma attack or hypoglycemia. While participating in ongoing therapeutic routines and surveillance in addition to school attendance and other activities, children with chronic conditions need to manage the everyday constraints to which they are subject, negotiate the meanings of their illnesses with peers and others, and manage their own and others' anxiety. The routines around illness and self-care, as well as time out from usual childhood activities as a result of episodic sickness, result in a strong sense of 'difference' for some children. This is evident when we consider how this difference manifests in diabetes. In resource-rich environments, homecare technologies allow for blood glucose measurement, and insulin pens and pumps enable parents and children to frame the condition as controllable. Despite this, young people as well as adults (as we discuss in Chapters 6 and 7) may find it difficult to control their health on a day-to-day basis, and may be frustrated by the social, psychological and physical impediments to living a 'normal life.' As Christine Dedding illustrates in her case study below, children may learn to manipulate prescriptions to fit in with their everyday schedules, disregarding certain restrictions, and resenting and rejecting certain aspects of these regimens of care. Like adults, children develop ways to minimize the extent to which illness controls their lives, and often stealthily resist parents' and clinicians' prescriptions, finding their own ways to manage their health and so their lives.

Well children, of course, also find ways to resist the authority of their parents and others: that is part of what being a child is about.

## 2.1 Children with Diabetes

*Christine Dedding*

"Must have fallen next to my bag, when I left the house," ten-year-old Nassar, who had come alone to the clinic, calmly replied to the doctor's question of why he hadn't brought his blood sugar device with him to the hospital *again*. Visibly frustrated, the doctor continued the conversation by asking him how things had gone with his diabetes in the last couple of months.

In the past, doctors in the Netherlands asked children with diabetes to keep a diary of their blood sugar levels during the day. Nowadays, this is not necessary, since the blood sugar device keeps a record. The 'only' thing children need to do is to measure their blood sugar level several times a day and to bring the device with them to the clinical encounter. Why didn't Nassar do this, again? This was not the first time that he had 'forgotten.' Perhaps he was in a hurry and so forgot. Or perhaps he hadn't measured his blood sugar levels as often as the doctors had advised him, or he expected that the measurements would not be received well by the doctor if there were too many high, too many low, or too precarious blood sugar readings. Perhaps he didn't want to be bothered by difficult, boring and unwelcome questions.

The agency of children in clinical encounters is easily overlooked, due to sociocultural images of them as innocent, vulnerable, and not yet competent, and associated with this, the idea that childhood should be a time without concerns (Panter-Brick 2002). There is also the dominant presence of doctors and parents in the consultation room; an assumption that patients of all ages have no real power; and, with the exception of young children, the often subtle expression of agency by children. Young children, contrary to older children and adults, can scream, yell, hide behind their mothers back, fly under the table, or fight if they don't want things to happen to them or they wish things to be different.

To gain insight into the power and dependency relationships of children, parents and medical professionals, I combined the approaches of participatory action research and ethnography. I invited children with diabetes Type I, between the ages of 8–14, to become co-researchers. My aims were to acknowledge them as social actors, to develop interventions that fit their reality and needs, and to learn what happens if we hand over power to children in a medically dominated domain. In the ethnography, I looked at the interaction between children, parents and medical professionals in clinical settings. I conducted participant observation, interviewed children, parents and medical professionals, organized focus group discussions, and wrote detailed notes of all our formal and informal meetings. As the children could not travel on their own, I picked them up and took them back home, and this offered us many occasions for reflection. With help of professionals, we developed two interventions, a book and a rap, both meeting the needs of young people to educate those around them, including medical professionals, about their perspectives on and experiences with their diabetes. The book is available at bookstores and libraries. The rap has been distributed on a DVD and found its way successfully onto YouTube. More importantly, the children, including the ones who usually didn't want to talk about their disease, were eager to show their products at their schools, to their relatives and friends.

One of the things I learned during this co-creation process and working closely together is that what is seen as 'one touch' of a button of a blood sugar device, from the point of view of medical professionals, means a lot of work for a person with diabetes. It means that you have to have your device with you all the time; to remember to use it; to decide when and where to do it; to think about the social consequences—they will look at me, I will miss parts of the conversation, they might leave without me—and to anticipate welcome and, more often, unwelcome outcomes. An unwelcome outcome means that you have to ask yourself why your blood sugar level is too high or too low; you might need to inject yourself; you either cannot eat or must eat. Even when you have finished this, you are not done. You have to stay alert to whether the action taken was appropriate. If not, the circle starts again. This work, invisible since it mainly takes place outside the walls of the hospital and in the heads of children, disturbs the flow of a normal life:

As usual, I brought cookies and drinks to share when we were working on the book. The cookies were well received; obviously I had made a good choice. Suddenly I saw that Malika, who had stepped aside, didn't have a cookie yet. I asked in a friendly way, 'Don't you want one?'

*Malika (gruffly):* No.
*Christine:* Don't you like them?
*Malika:* I do, but then I have to measure first.
*Christine:* Well, why don't you then?
*Malika (angry):* No, because then I might not be allowed.
*Christine (still ignorant):* Then you measure first, and then we'll see together what to do with the results.

The moment I said it, I knew I had made a mistake: neither she nor I could wave aside the outcome. The device, and all that it represents, determines how the cookie tastes. With a high blood sugar level, the cookie won't taste good; with a low outcome, the cookie and its sweetness would be welcomed. A bad measurement doesn't only have consequences for food intake, but it is also an infringement of an ideal—of control, of coping well with the disease, and of being a good patient. After all, since the discovery of insulin in 1921 and with the help of homecare technologies such as blood sugar measurements, insulin pens and pumps, diabetes is increasingly depicted as a controllable disease (cf. Feudtner 2003). Not being able to reach that control, day in, day out, raises various questions for young people: What do I do wrong? What do I have to change? Why do other people manage well? And 'what will my mother or doctor say when they find out?' In anticipation of an adverse judgement, Malika decided to avoid confrontation by not measuring her blood sugar level, and so resisting the temptation of the cookie. While the other children continued working, it took a while before Malika had settled her feelings of anger and frustration. This example of one little cookie shows that 'the work' starts before the device is even taken out of its case.

Children referred to the tasks they have to do in relation to their diabetes as *work*. Furthermore, they referred to this work as analogous to that of a doctor: they observe their body, measure their blood sugar, inject medicines, and make interpretations. Children make interpretations of what the disease means to them, and on how to treat it appropriately, especially in relation to their daily lives. Children understand what medical professionals consider a good blood sugar reading, but this doesn't mean that they don't have their own ideas about it. Based on years of lived bodily experience, they have their own meanings too. Bart and Fatita, for example, are aware that medical professionals strive for a blood sugar level between 4 and 10 (mg/dL), but they themselves consider a result of 4 far too low: "Then you still feel shaky. And then you have to eat something." Ghalid, on the other hand, is happy when his level is below 4: "Yeah I like that, then I can eat extra." Yet he considers 16 good too: "Normally that is high, but I consider it good, (because) then I say that I won't need an extra injection."

The blood sugar levels that medical professionals target are generally determined by the risk of long-term complications (e.g. blindness, amputations, renal failure), but children mainly determine levels on the basis of how they feel and the implications of levels for what they can do and eat. By taking such an approach, children withstand the strong disciplinary power of the health education that they receive routinely, and the classification of symptoms based on medically defined facts and relations. This example of the meaning of measuring blood sugars in daily life and the work it demands gives some context, however limited, to why Nassar might have 'forgotten' to bring his blood sugar device to the consultation room.

Observations, confirmed by children in the interviews, demonstrate that children hardly ever openly question medical discourse and treatment rules in the clinical encounter. The unequal power relationship, dependency, and the controlling presence of a parent in the consultation room, restrict direct and open confrontations with medical professionals. Instead, children choose silent resistance and non-compliance. Non-verbal signs of resistance, such as demonstratively sitting in a slouched position, watching the clock, closing the zipper of their jacket, and answering questions with a jaded 'Yes,' 'No,' or 'OK,' can be read as silent protest. As many children explained, after years of experience, they are fully aware of the questions to expect and of the 'appropriate' answers they are expected to give.

Outside the clinic, children have far more scope to enact their agency and this is much easier to recognize. Children are often out of sight of their parents, playing, attending school or sports clubs, or hanging out at friends' houses while their parents are involved in other activities such as working, and taking care of the household and other children:

> Ghalid is visibly at ease when he explains to me: "I can't eat exactly on time anyway, and I can also inject later. I do that often, I eat first before I inject. You don't notice the difference." When asked why he ignores medical advice to inject before dinner, he explains that his family starts eating without him if he injects before dinner. By the time he is ready to eat, they have finished and his food is cold. When I suggest that he could start injecting earlier, he answers that he does not like to do that. It would mean that he needs to inject quite some time before dinner which implies he needs to watch the clock, needs to know what his mother will cook and at what time she will be ready. Taking all this into consideration would make it impossible to fully enjoy playing football with his friends.

This example shows the agency of children, but it shows also the practical logic behind their acts and the work they have to do. The medical project is just one of many projects in which children with diabetes are involved. An easily overlooked project is how children look after their parents. Most children are well aware of their parents' fears and concerns about their health and happiness now and in the future. This leads them to not always or not explicitly share negative thoughts and experiences about their disease.

Another overlooked aspect, defined by children as a major problem, is general misunderstanding about the nature of diabetes. Children suffer from erroneous ideas about what is in the Netherlands still often called 'sugar disease.' Many people still think that someone with diabetes is not allowed to eat sugar or that they have diabetes because of eating too much sugar in the past. Neither is true, but the image is more persistent than the truth. These misperceptions are difficult to refute for children, especially in confrontation with adults. Adults are verbally stronger and also think they know better. The educating that children have to do is not limited to adults (teachers, trainers, other parents); they also need to explain to their friends what diabetes entails. When I asked Teun how he explains to his friends what diabetes is, his answer was, "well just . . . ," followed by a long and complicated medical speech. This, he confirmed, was not understood by his peers. As Zayna explained, "then children don't understand and you get tired. Then you have to tell it again, and then they still don't understand."

To explain a malfunctioning pancreas and the working of insulin is a difficult task, which few adults can achieve. However, the children who do manage—and many do—still experience little benefit. The abstract medical story works within the walls of the hospital, but not outside. Outside children with diabetes need a different story, a story that explains what they can and can't do, what they have to do, and what might happen. They spend a lot of time educating others, finding practical translations for a very abstract medical story, and in legitimizing what they have to do and why. The common misunderstandings about diabetes, and the lack of respect for their own tacit knowledge, were things that they wanted to address in their intervention. The book and rap not only provided clear explanations of what diabetes entails in daily life, but also explicitly stressed the importance of being listened to. In the book, they included a chapter about their idea of what being a good doctor entails. In the rap, they sang 'doctors open your ears, then we will explain how we really feel.' Implicitly, in developing the book and the rap, they sought answers to their own concerns—about sexual relationships, for example, and whether they will be able to have children; on what happens with their blood samples in the laboratory; on how to deal with (over)concerned parents.

* * * *

This case study is not a critique of medical professionals. On the contrary, all the professionals I met were, without exception, very dedicated. They worked hard keeping up to date, collaborating with other professionals, being available during evenings and weekends, and balancing friendliness and being strict in order to both prevent complications in the future and to offer their young patients optimal quality of life in the present. All tried hard to involve children in conversations in the clinic. However, conversations are not only shaped by medical professionals, but also by children, their parents, spaces, technology, managers, policy makers and insurance companies. Also, acknowledging children as knowledgeable social actors makes consultations much more complex: a 'trialogue' is much more difficult to accomplish than a dialogue, especially in the short time period usually available for a consultation.

If we want to understand how children deal with diseases and its treatments, it is helpful to step outside the walls of the hospital. This allows more focus on the daily lives of children. It makes it easier to recognize them as knowledgeable social actors, rather than passive patients who are treated and controlled by adults. Moreover, it raises questions about current approaches to diabetes: the idea of diabetes as a controllable disease; the changes in tasks and responsibilities due to the use of home technologies; the difficulty of incorporating tacit knowledge and of making this available in medical books and patient education material; and questions about how well we are able to support children to deal with this demanding disease and its treatment in their daily lives. In this situation, children have a lot to offer to each other. They can teach their peers about how to deal with diabetes in different settings and find support from knowing that other children suffer from the same problems. If we create a cultural shift in our thinking about children, we might be able to create circumstances in which children can show their competences, practical needs and critiques more openly, so enhancing our understanding of what children are capable of, and of their needs and how these can be met.

### *References*

Feudtner, C. 2003 *Bittersweet: Diabetes, Insulin, and the Transformation of Illness.* Chapel Hill, NC: University of North Carolina Press.

Panter-Brick, C. 2002 Street children, human rights, and public health: A critique and future directions. *Annual Review of Anthropology* 31:147–171.

## Facing Extreme Illness

The diagnosis of childhood cancer and other serious and often fatal illnesses creates terrible ruptures in the fabric of a family. Diagnosis and disease produce constant breaks in domestic life and in interactions among household members, and between the household and the clinic. Filmmaker Julia Reichert describes her experience of being present in the hospital while filming children living with cancer when she was working on the documentary film, *A Lion in the House.*

> Kids are vagabonds, pirates and natural troublemakers, whether they have cancer or not. It was a joy to hang out with these kids, to give them the camera so they could run around filming, to see them play practical jokes on their parents, nurses and doctors. Meanwhile, down the hall, to see these same nurses and doctors grappling with life and death decisions around a family, and to struggle to understand what was going on in each family's emotional life—this was amazing and very moving. A children's hospital is an incredibly vital environment, where every moment matters.
>
> *(Reichert 2006)*

Like the children living with cancer who are so beautifully portrayed in this documentary film, the children in the following case studies deal with each day, each round of vile medications, and each invasive procedure as best they can—sometimes with stoicism and sometimes with tears. Complex health problems impact at multiple points in children's lives, requiring that they develop an understanding of their health condition; manage pain, treatments and their side effects; deal with stigmatization from others; and come to terms with the deaths of friends made through clinical communities with similar health problems. Illness shapes children's attitudes to and decisions about their education and employment, friendships, intimate relationships, reproductive and sexual health, and their own possible families (Drew 2007).

The children of whom Alice Larotonda writes are in this group. They have uncommon conditions, and regardless of the sophistication of health services in Italy, where her study was conducted, and their access to quality health care, they live their lives with serious impairment and probable reduced life expectancy. The embodiment of their conditions, such as atypical short stature, distinctive morphology or skin depigmentation results in their social exclusion, even though the health problems that impact their daily life may be unrelated to the most visible signs of disease and although they may have similar energy levels and intellectual capacity as other children.

Children born with such conditions survive largely when they live in high-income settings where their families can access and afford high-quality care. Throughout their lives, they may have to interact with multiple clinical specialists to manage their conditions and head off complications. In clinics, as Christine Dedding (above) and Alice Larotonda (below) illustrate, children may resist the intrusions of doctors through their silence and quiet rebellion. While small children may fight and scream, older children are expected by parents and health professionals to show control, to cope with the disease and its impositions on their lives, and to deflect the anxiety of others with whom they share its unpredictability. Sick children are socialized to take responsibility and so to be responsible for the 'social life' of their health, and to take account of their parents' fears and concerns about their health and happiness. Children witness the impacts of their health conditions on others, and so often manage their health as a way of caring for and protecting their parents and siblings as they care for themselves.

Serious diseases are rarely stable. While their unfolding may be familiar to clinicians, this is not so for affected families. Like the children with Type 1 diabetes in Dedding's case study, in Larotonda's example following, both Mattia and Ettore must consistently make adjustments in their everyday lives, manage invasive procedures while they learn in more mundane ways to work with and around their conditions, and negotiate their interactions with other children, family members and clinicians.

---

## 2.2 Experiencing Rare Diseases

*Alice Larotonda*[1]

*Alice:* Tell me a bit about yourself, Ettore.

*Ettore:* Well . . . I do well at school. I will go into fourth grade next year. I can play soccer but only a little, five minutes max, very little actually. Since I have this problem, I get tired very easily, so my parents said I have to adjust to it. We come here (to the hospital) for some check-ups, because I have a disease, one of the rare diseases, called McCune-Albright, so I have to do check-ups. This McCune-Albright (syndrome) makes my bones—how can I say?—more fragile than those of other children. I was born with this disease.

Syndromes such as McCune-Albright are part of a list of pathologies also known as rare diseases. These are defined by the European Union (Fregonese and Aymé 2008: 2) as conditions that affect less than five out of 10,000 individuals. Comprising 5,000 to 8,000 distinct pathologies of different origins (genetic, endocrine, immune, etc.), the category includes a wide group of diseases that are often difficult to diagnose; the majority remain without a cure. Rare diseases are chronic conditions that disrupt the lives of those affected in profound ways. They plunge individuals into a life of complete unpredictability, condemning them to an ever-present threat for their health, while binding them to a constant need for medications and medical screening.

Diagnosed prenatally or during childhood, these pathologies affect social, cultural, and emotional aspects of the experience of children and their families. The difficulty of diagnosis, the lack of treatment, and the challenges of carrying a chronic and debilitating disease generate experiences of disruption and uncertainty, of individual and social suffering entailed in living at the margins of the categories of health and disease, of 'normalcy' and 'difference.'

I conducted fieldwork in an Italian pediatric hospital looking at what having a 'rare' pathology meant to children, their families and their doctors. I participated in medical rounds and examinations in outpatient and inpatient clinics, observing the activity of physicians and other health professionals, and their clinical encounters with chronically ill children and their families. In this way, I met many parents and children, whom I asked to share their stories and experiences of illness.

For children affected by rare diseases, illness seems to be a chronic condition coming in, on, and through the body, entailing physical suffering, impairment or pain. Illness becomes incarnated—personified and tangible—on the child, whose physical appearance can be marked by particular phenotypic traits or malformations. This influences the child's self-perception and the ways in which he or she is considered by others, so shaping individual and social experiences. In my participants' narrations, however, chronic illness was also an experience of negotiation, in which the traditional definitions of 'suffering,' 'difference' and 'normalcy' were called into question. These active negotiations and re-definitions of illness, uncertainty, and difference were more or less successfully elaborated in the everyday social experience of children with rare diseases and their families.

## *Ettore*

I meet Ettore and his parents at the hospital, where he is regularly admitted to undergo routine tests and check-ups on the status of his symptoms. Ettore is affected by McCune-Albright syndrome, which entails "the clinical triad of fibrous dysplasia of bone (FD), café-au-lait skin spots, and precocious puberty (PP)" (Dumitrescu and Collins 2008); his bones are weak and need to be supported, his skin spots require screening, and his endocrine system demands monitoring to prevent PP.

He comes in the examination room shyly, limping, and looking quite unenthusiastic at the prospect of the medical visit. He tries to jump on the examination table, without success. His mother intervenes, but he resolutely refuses help. After a couple of failed tries, he finally succeeds, proudly. Sullen and quiet during the whole visit, Ettore seems quite relieved when the examination and doctor's questioning are over.

After the visit, the physician introduces me to the family and Ettore's parents willingly agree to participate in an interview. The doctor leaves and we all sit in one of the empty, bare rooms of the clinic, where I explain the purposes of my project. Ettore promptly offers to be the one talking about his own experience, "so that [he] can make [him]self helpful." In contrast with his previous shyness, now he is talkative, quite lighthearted in his experience, with a sharp sense of irony.

*Alice:* So, tell me about this disease you said you have. Do you know what it is about?

*Ettore:* Yes, my mom and dad told me. At first I did not understand very well, but then they explained it to me. In our DNA there are different patterns, for example: A-B, C-D; basically what happened was A-D and B-C, so to speak. This is why I have this problem.

*Alice:* I understand. So what does it mean for you to have this problem, in your everyday life? You told me, for example, that you cannot play soccer a lot, because you have to be careful not to get too tired. But is there anything else?

*Ettore:* I don't play soccer at all. I really cannot play soccer, because . . . well, I *can* kick the ball. The problem is with fouls, because I can get hurt . . . So I don't have to play, because other kids may bang into me and make me fall. I don't play soccer, I can play very few sports. . . .

*Alice:* I see. Are there other things? What about your leg, for example: I hear you had an operation.

*Ettore:* Both legs! They operated on both my legs several times, about six times. I had the first surgery when I was two, and then until when I was eight, last year.

*Alice:* Do you remember these surgeries?

*Ettore:* About the first ones . . . not really. But I remember from when I was six, until now: I had three surgeries.

*Alice:* What did they do in the surgery?

*Ettore:* They fixed my bone. That is, they gave me a plaster, so callus could form again, and become bone. In this case, inside the bone they put a metal bar, inside. So even if it breaks, the bone does not bend, it stays like this and you would have to stay in bed twenty days, if it ever broke.

*Alice:* So you have to be very careful!

*Ettore:* Yes, even at school, but not all that much . . . because everybody knows about the disease.

*Alice:* Really? Who told them about it? Was it you?

*Ettore:* The teacher and I. Well, some of my friends, the ones who were in preschool with me, knew already. The new ones, the 22 new classmates, didn't know, so the teacher and I explained it to them.

*Alice:* So, what exactly did you tell them?

*Ettore:* I explained how come I have this disease, and that they had to be careful not to push me, make me fall down, and so on.

*Alice:* And what did they say? Did they ask any questions?

*Ettore:* Yes, more or less . . . for example: "How did you get it?" "I was born with this disease." Like this.

*Alice:* And did you know the answers?

*Ettore:* Not all of them. Because for example, "what is the cause?" At that time I didn't know about the DNA thing.

*Alice:* Well, that is a hard one! So you told them in first grade, what about now? Are they still curious?

*Ettore:* Sometimes, they ask: "How many times did you break your bone?" and so on. . . .

*Alice:* How did your mom and dad find out about this disease, do you know?

*Ettore:* Because once—I think that's the answer—I fell from my bed, when I was very young, and I broke my bone. So we went to have surgery, and that's how they found out.

*Mother:* Well, that's not exactly it . . . he started limping when he was two . . . so we did some check-ups. And finally, it turned out (to be this disease).

*Alice:* So you have quite some experience with doctors! Do you mind the medical visits?

*Ettore:* No, no. Because the surgeries, in the end, are useful . . . I mean, what would you have to fear? Well, yes, the pain. But then, they are helpful for improving. I mean, they are useful when the bone is broken, to set it back in place, stick it back.

*Alice:* So, thinking it is useful helps you, somehow.

*Ettore:* Yes. But I also have a high . . . what's the name? Pain threshold . . .

*Alice:* Do you still have to undergo many surgeries?

*Ettore:* Yes, I do. Because, as I said earlier, they close my thighbone: when the bone grows it becomes bigger, and they have to change the metal bar. They take it out and put a new one.

*Mother:* It will be stabilized at the end of bone development, around the age of 17. (Turning to Ettore) You still have to be patient. . . .

*Alice:* It seems to me that you are quite at ease with this experience, Ettore!

*Ettore:* Yes, yes! Once the surgery is gone, you don't have that concern any more. Especially when they change the pins, because then you can say: "I have done this, now the next (surgery) is in two or three years."

*Mother:* He really seems quite relaxed. . . .

*Father:* Anyway all of us have problems, even those people who seem not to have any, isn't that so, Ettore? . . . We have this, but other people have other problems.

*Ettore:* Yes, just like this child who is in my class: he has glasses. Even though he says he puts them on because he sees too much! (i.e. he is far sighted).

Ettore's narrative gives insight into different aspects of the experience of children with rare diseases. The troubles of everyday life, the practical and material impairments that the disease causes, the experience of physical pain are contrasted with the protection and care that surrounds the child. Ettore's account is exemplary of how children with rare diseases experience illness *in* the body through physical pain, *on* the body, which can become remarkably 'visible' and 'different' because of the specific features or impairments that are caused by the disease, and *through* the body, as a sensory means for experiencing the world. His narrative, however, also offers some clues about how these experiences of uncertainty and difference can be actively challenged and negotiated.

With the help of his parents, Ettore constructs and negotiates meaning through narrative, so making sense of his predicament. By the age of nine, he has already had six surgeries, but he argues that these procedures are useful to him, and his high pain threshold helps him cope with them. He cannot play soccer, but drawing on his parents' recommendation that he adjust what he does to his own possibilities, he compromises: he can play, but just for a few minutes. He stresses how the problem is not that he is unable to play—even though he limps quite visibly—but that it can be dangerous for him since soccer involves fouls: pushing, falling down, physical contact with other players, are a threat for his fragile bones.

His parents support Ettore in embracing a particular interpretation of his experience: there is a problem, but problems affect everybody. The exceptionality of Ettore's disease is placed in a framework of normalcy. Ettore limps noticeably and his appearance stands out because of the café-au-lait skin spots that cover his body and parts of his face; additionally, long scars from his numerous surgeries cover his thighs. Many children with rare diseases have pathologies that entail tangible bodily marks, specific phenotypic traits or malformation, making the child's physical appearance remarkably 'visible' and 'different.' Ettore's father, however, questions this 'visibility,' maintaining that what one can see from the outside is not always relevant: "even those who *seem* not to have any (problems)" may have some. By attributing normalcy to an experience that might be lived as unique or isolating by Ettore, the exclusivity and marginalization that he might otherwise feel is reduced. Ettore embraces this discourse, and actively re-shapes it with a personal interpretation taken from his own experience: a boy from his class has a problem too, and not because of impairment, but because his sight is 'too good.' Too good or too bad can both range outside of normalcy and thus may be considered problematic. Therefore, indirectly, shaping meaning also consists in molding the body and its characteristics, and in establishing what is relevant and what is not, and what the meaning of these differences is. And as the clinic's psychologist remarked, after all, 'visible' is not what is objectively there; 'visible' is what *one* sees.

## *Mattia*

Mattia is a cheerful 10-year-old, with beautiful olive skin and almond-shaped eyes. Extremely skinny and a bit short for his age, his appearance and features are very different from those of his parents and sister, with whom he shows no resemblance. After a suspected hydrocephalus at birth and an unfounded diagnosis of Silver-Russell syndrome, Mattia was eventually pronounced to be suffering from a serious growth hormone deficiency, and has been under treatment since the age of two. Despite falling sick quite frequently, Mattia goes to school and has plenty of friends, and—in defiance of some isolated episodes of schoolyard mockery—his classmates stand by his side and support him. As the years go by, however, the anomaly of his growth patterns has become increasingly noticeable; his peers are generally taller and more robust. In a pre-teen age when body size comes to gain unprecedented relevance, this is becoming particularly important to Mattia.

*Mattia:* I wanted to say something . . . I got quite scared when the gastroenterologist told me: "You go see the geneticist, and there are two things: either a problem for which you are normal and you grow up and reach 1.60 meters height, or it is a disease." Well, there I got scared . . . because even 1.60 . . .
*Mother:* Listen, Mattia . . . Look, I am going to tell you something: if you get to 1.60, you will crow with victory!
*Sister:* Just like your sister! Then you buy Nike sneakers—the tallest ones—and you gain five more centimeters! (Laughing)
*Mattia:* No, you know how tall I wish I would become? 1.90 or 1.80 meters!
*Mother:* Well, that's not possible, c'est pas possible!
*Father:* Yeah, one meter and a lot of willingness to grow up! (A typical ironic expression in Italian, we all laugh)
*Mattia:* 1.70 at least!
*Mother:* Well, I don't know Mattia. . . .

What does being 'different' or 'visible' mean? What does a person's face, body, height or weight tell about him or her? What is at stake in inhabiting different bodies?

*Alice:* What would happen if you remain small? What would it mean for you?
*Mattia:* . . . Life! Ok, let me explain this to you. If I remain short, I mean, it's not like I stay there thinking and thinking . . . I stay there a bit, but not much. It's life, and I go on, because anyway . . . I am not sick! I am healthy, only I am short, but other people can also be short. Being ill is different, if you are ill you should really be crying, but if you are just short, you are healthy, basically! . . . 90 percent healthy!

But what does 'short,' 'tall,' 'healthy' or 'ill' mean? The construction of these meanings is negotiable. Mattia's family takes the effort to try and co-negotiate these categories, showing him how definitions can be malleable.

*Alice:* Mattia, listen, but why do you say you want to be so tall, 1.90 meters? What does it mean to be so tall?

*Mattia:* Because if you are tall . . . Not too tall . . . Not two meters, because that is not good either!

*Mother:* Then you cannot find a girlfriend, you find a short girlfriend! At that point, it's better to be smaller! Don't you think? You, very, very tall, with a short girlfriend? (Laughing)

*Mattia:* Well, what about the opposite, then: me short, and a tall girlfriend?

No matter how unsuccessful, his mother's intervention teaches Mattia something: meanings can shift to shed light on a different outlook.

* * * *

In social interaction, the criteria that determine the positioning of the different actors are not fixed nor objective. On the contrary, they can be negotiated depending on what specific characteristics those interacting decide to attribute relevance to. For children affected by rare diseases, the opportunity of redefinition of these criteria can be fundamental in shaping social relations and granting empowerment. Shifting attention from their 'visible' difference to other empowering characteristics, they get to negotiate their social positioning in interactions, and consequently to challenge—insofar as possible—the uncertainty and precariousness of their whole social experience despite illness. It is arguable, as a consequence, that 'difference' and its perception are subjective and, above all, negotiable, just as 'normalcy' is.

Ettore and Mattia are but two isolated instances of how rare disease can affect children's everyday individual and social experience. Even though the personal, physical and social suffering in pediatric rare diseases cannot be underestimated, what we can take from these cases is the complexity of experience of children with rare diseases, never definitively positive or negative, and always controversial. Children might have an experience of 'difference,' but they learn to manipulate and negotiate its meanings by incorporating and interweaving empowering narratives. Ultimately, their lives stand in ambiguous spaces at the margins between stigma and empowerment, the existential experience of pain and its compensation with the strength and serenity of childhood.

## Note

1. The data presented here were collected during fieldwork in a pediatric hospital in Italy. I warmly thank the health professionals, parents and children who made this research possible by accepting to participate and share their experiences. I am especially grateful to Professor Robert Pool for his enlightening advice and support, and to the rest of the Medical Anthropology Faculty at University of Amsterdam, NL. Further references to this work are Larotonda (2012; 2014).

## References

Dumitrescu, C.E., and M.T. Collins 2008 McCune-Albright syndrome. Accessed from http://www.ojrd.com/content/3/1/12

Fregonese, L., and S. Aymé 2008 Health indicators for rare diseases: State of the art and future directions, Rare Diseases Task Force. Accessed from http://www.eucerd.eu/?post_type=document&p=1207

Larotonda, A. 2012 In, on and through the body: Experiencing and negotiating uncertainty and difference in paediatric rare disease. Master's thesis in Medical Anthropology, Department of Social and Behavioural Sciences, University of Amsterdam, The Netherlands.

——— 2014 The 'imperfect child': Parents' expectations disrupted by having an infant with a rare disease. In *Medical Anthropology: Essays and Reflections from an Amsterdam Graduate Program*. S. van der Geest, T. Gerrits, and J. Challinor, eds. Pp. 43–54. Diemen, The Netherlands: AMB Publishers.

## Cognitive Development

Few anthropological studies have examined intellectual, developmental or mental health problems in children, despite the substantial numbers of children diagnosed with these conditions worldwide. In resource poor settings, families adapt to a child's capabilities as best they can, in some cases with great care, in other cases, with few ways in which they can address the child's needs, by physical constraining or sequestering the child. However, when a child is noticed by or brought to the attention of primary health care workers, doctors or specialist medical authorities, in industrialized countries and in urban settings everywhere, parents and the child may be catapulted into multiple early interventions to support the child's motor, cognitive and social development. In the case study below, Vu Song Ha describes the circumstances of Vietnamese children identified as being on the autism spectrum, with or without formal diagnosis, because their social development, interactions, communication, concentration and behaviors are atypical. Vu is concerned with children living in Hanoi, the capital city of Vietnam, where the facilities and services for education, health care and therapy are far beyond those available in rural areas. Depending on parents' perceptions of the child's capacity and school policy, in Hanoi a child with learning and communication difficulties may be able to attend a mainstream school, but some parents have established special schools for their children to better address their needs. However, understandings of developmental delays and mental illnesses are confused whether or not the child is 'mainstreamed,' and parents may isolate themselves and their children to avoid stigma, blame and discrimination. This is especially so when community members believe that limitations to a child's capabilities are the fault of the parents, the consequences of parental or familial misdeeds, inheritance or 'genetics,' however understood (see also Chapter 14).

---

### 2.3 Autism Spectrum Disorder in Vietnam

*Vu Song Ha*[1]

I first met Long, an eight-year-old boy, at a group established by parents for children with autism spectrum disorder (*tự kỷ* in Vietnamese). Like the other six children in the group, Long studied in a mainstream school in the morning, and came to the group for lunch and afternoon activities. In this group, he had individual sessions, in which a private teacher helped him review the morning lessons and do homework. He also had physical, speech and sensory therapies, either in group or individual sessions. After a few days spent with him at this group, I joined him in his mainstream school. My notes follow:

> Long went to his school with Miss Loan, his private teaching aide. Long and Loan sat together at a table in the far corner of the classroom. Loan repeated softly what the teacher said, and guided Long to follow the teacher's instructions. In the two lessons of mathematics and Vietnamese I attended, the teacher ignored Long. She walked around the classroom but did not come to his table, invited other students to answer questions or do some math at the blackboard, but neither asked Long nor looked at him. Long talked to himself about advertisements, and Loan kept reminding him to not do so. During the break, Long did not play with other children, but sat at his table and talked with Loan. After the break, Long and Loan moved to the music class. The teacher was cheerful, and Long was excited. He could sing a little with his peers. The teacher asked some students to sing solo, and then asked students to sing as a group. Long sang with them. Then I moved to his physical education class. Again, the male teacher seemed to ignore Long. He called various students to go to the front to perform. Long's classmates in his row went up to perform, but Long stayed there, playing by himself, disregarded by the teacher. I questioned Loan about this, and she said she had told the teacher on the first day of school about Long's condition. He said nothing, but never requested Long to do anything. When Loan thought that Long was tired, she asked the teacher permission for him to skip the lesson, and Long was driven to his parent group.

Among the children with ASD in Hanoi, Long was one of lucky ones, able to go to a mainstream school, even though his inclusion is questionable. His family can afford private school aides and private therapies. Most children going to mainstream schools do not have school aides, and others do not attend mainstream schools either because the school refuses to take them or because their parents are concerned that their children are not capable. Other options for children with ASD of school age are limited—a few special schools for children with disabilities, private centers and parent-run groups. Educational exclusion is only one problem for children with ASD in Vietnam. They are also excluded from health care and social participation.

## *Educational Exclusion*

The Government of Vietnam has made significant efforts in creating a favorable legal framework to support people with disabilities. The Disability Law, enacted in January 2011, ensures the rights of people with disabilities in all aspects of life, including equal access to health care, education, vocational training, employment, residential and government buildings, transportation, cultural activities and entertainment (National Assembly 2010). However, there are significant gaps between law and practices.

Inclusive education programs, based on the principle of integrating children with disabilities into mainstream schools, were initiated in the early 1990s and have had some success in encouraging children to attend schools (Rydstrom 2010; Villa et al. 2003). However, there is no specific policy to support children with ASD in education, and this remains a challenge because of fixed curriculum, emphasis on academic knowledge and achievement, crowding in classrooms (with 40–60 children in classes in public primary schools), and lack of human resources with the capacity and motivation to support them.

ASD was first used as a diagnostic label at major children's hospitals in Vietnam in 2000, and it is still very new for both lay people and professionals working with such children. Many parents and service providers are concerned that ASD is a form of schizophrenia, or that children will become schizophrenic without intervention. Sensory seeking and stereotypic behaviors such as hand waving, rocking, moving and talking constantly are often misinterpreted. Due to the stigma toward mental illness in Vietnam, children with ASD are seen to be threatening and bring shame to the family, and like other children with disability, they are seen as worthless and burdensome. These opinions influence the responses of professionals, as two teachers explained:

> These children are very slow. Linh (a child with ASD in her class) cannot take notes. I remind him, but he fails to do so. He also cannot study . . . Teaching these children is very tiring, and frustrating. Linh is good because he doesn't have behavioral problems—he's just slow. Other kids have lots of problems, and don't want to study.
>
> *(Secondary school teacher, mainstream school)*

> These children would be called *thần kinh* (mental, crazy) in a rural area. It's partly correct. Twice I saw them take food out of the waste bin. It was pitiful . . . Sometimes I wish I only had to teach children with hearing difficulties. They have normal intellectual ability, so they can learn what they are taught. Teaching these children with ASD is frustrating. It takes lot of time to teach them a little, and then they forget. At the end of the day, they are still dependent on their families. They couldn't live independently.
>
> *(Primary school teacher, class for children with special needs)*

The social and cultural values toward childhood, parenthood and disability also disadvantage parents of children with ASD. In patriarchal Vietnam, a couple is expected to have numerous children, including at least one healthy son to continue the male line. Parents are expected to sacrifice their own needs to raise their children properly. In turn, children are expected to please their parents because of the moral debt and gratitude they owe them and their ancestors (Hunt 2005; McLeod and Nguyen 2001), and to provide financial support and care for their parents when they are old (Malarney 2002). Vietnamese culture emphasizes the importance of *gia tộc* (family) and its good reputation. Women are blamed for childlessness, disease, misfortunes in the family, and disabling conditions of their children. Furthermore, beliefs of karma and rebirth construct disability as a punishment for the sins or immorality of previous generations. Although many parents stated that they did not believe in karma, they confirmed that the belief existed and made them uncomfortable. For example, Hà explained that "some people might say that I had done some (bad) things, so my son suffered. We can't avoid other people thinking badly about us. We can't prevent their thoughts. I think someone may think this about me. Sometimes it distresses me."

Due to stigma, a number of parents feel inferior—they have failed to produce a healthy child. Thus even though parents recognize the difficulties (being excluded, bullied) their children encounter, they are often reluctant to ask for support beyond securing a place for their child in school:

> As parents of these children, we are disadvantaged in all aspects. We sacrifice a lot. When our child is of school age, we have to run around, beg for a place in school for him/her. There is no support, no sympathy from society . . . I know that they let my son just sit in the classroom. The class is too crowded, 56 children. My son can't follow everything [class activities]; however, I think it is good enough. He knows how to function in a group . . . They accept these children. That's already an improvement.
>
> *(Bách)*

## Health Care Exclusion

There are no routine child development check-ups in Vietnam. Screening, assessment and diagnosis services for children with ASD are extremely limited in both quantity and quality. The government provides no guidance on assessment procedures, and service providers have limited capacity. Health care insurance law indicates that children under six are fully covered for the examination fee and treatment at health care centers (National Assembly 2008), but only a few public facilities provide intervention services for children with ASD. Almost all parents with whom I worked sought support from private early intervention centers or private providers, not covered by medical insurance, and so they paid all costs themselves. As a result, many children with ASD in Vietnam remain undetected and untreated. In addition, other medical specialists (for example, dentists) have no knowledge of working with children with ASD, and so many children with ASD do not get proper health care. One mother recalled:

> I am so scared whenever I have to bring my son to hospitals. These children are afraid of crowds; whenever they go to crowded places, they scream and cry. They also are afraid of strangers, and do not stay still and follow the instructions of doctors. Doctors, who have knowledge and experience [in this field], might sympathize with us. However, hospitals [in Hanoi] often are overloaded, so nobody has time to be patient and gentle with these children. So, it is really a nightmare if I need to bring my son to hospital. I never use medical insurance; I often go to private doctors. We do not have rights to [access services], we have to go to doctors who do not have many clients, and who are most easy-going . . . [My son] has a lot of trouble with his teeth, decay and malpositions. We took him to a dentist many times, but had to take him home as the dentist could not do anything with him. So I am the person who takes care of his teeth.
>
> *(Nga)*

## Social Isolation

When I worked with nine children with ASD in a photo-voice project, I found that most of the children's photos were taken within the home environment. Of 2,142 photos, almost three quarters were taken at home, 6 percent at their grandparents' homes. Some children studying at parent-run schools took photos there, but few children took photos at their schools even if they had cameras with them. They did take photos at special excursions organized for them, and a few photos were taken on streets or at public places. In general, children with ASD tended to take photographs of themselves and of things, and of people they knew well. At the same time, they had few opportunities to engage in other environments and perhaps minimal opportunities to be comfortable outside of their homes.

Going out with children with ASD is often challenging. Some parents reported difficulties in managing their children's behaviors, especially in new environments and with strangers. Fear of stigma also prevents some parents from bringing their children into social gatherings. Mai explained:

> I only bring my child with me when I am with close friends. You see, this is my feeling of inferiority. A couple of times, I brought my child to attend parties at work. However, I regretted

> this, as my daughter has some behaviors not seen as normal by others. These made me feel uncomfortable, inferior . . . It is hard to describe [these feelings] . . . .

In addition, in Hanoi, traffic can be extremely busy. People often travel by motorbikes, and public transportation is crowded. Children with ASD are often unable to enter the world outside their family by themselves. Quỳnh's parents recalled that once Quỳnh, a 15-year-old girl, had crossed the street herself on the way home from school. Her entire family was terrified as she just walked onto the road without looking out for traffic. Thereafter, they would not let her onto the street by herself. In the photo-voice project, Quỳnh took photographs from her favorite spot at home, in a glassed-in balcony from where she would watch the street below, crowded with shops, food stalls and people.

## *Parents' Efforts*

With limited support from government and society, parents are the key actors taking care of their children and changing knowledge and social responses on ASD. They are almost solely responsible in identifying, searching and making decisions on assessment, interventions, and education for their children; paying for interventions, health care and education; and protecting their children. In response to stigma and discrimination, and the lack of services available for their children, they employ defensive strategies (such as keeping the child's condition secret and restricting the child's activities and interactions). But they also employ proactive strategies, including home schooling, capacity building for other parents and service providers, initiating various services, and providing public education to address and change social attitudes and policies. Parents build on biosociality (Rose 2006) around ASD, as they work together to demand knowledge and recognition, shape identity, and claim their expertise to fill in service gaps for their children. A number of groups have been established, for example, for parents of newly diagnosed children, parents with teenagers with ASD, and parents using biomedicine. In Hanoi, besides intervention centers established by professionals, three parent-run groups provide interventions for their own children.

The Hanoi Club for parents of children with ASD is a good example of a biosocial grouping. It has organized various training workshops for parents and professionals, held dialogues with school leaders and policy makers in education, disseminated information through its own website, and engaged with the media and initiated public awareness raising events on ASD, for example, walks for autism in 2010 and 2011, in which thousands of people participated.

## *Conclusion*

Although ASD has been recognized as a major public health concern globally, the stories of Long, Quỳnh, and many other children with ASD and their parents, illustrate their neglect in both policy and services, not only because of unfamiliarity with the condition, but also because of stigma and discrimination toward mental illness and disability. The social exclusion of people with ASD reflects the structural violence that operates toward people with disabilities, and those with mental illness, who lack social and economic status and lack power to access to knowledge and services. The biosocial groupings around ADS are helping to make changes, however—negotiating for better services, working for educational inclusion, reframing ASD as a developmental disorder not a mental illness, and countering stigma and discrimination.

## *Note*

1. This case study draws on work conducted on autism spectrum disorder (ASD) in Hanoi, Vietnam from July 2011 to May 2012, as part of my PhD program. I employed in-depth interviews with 27 parents of children with ASD and 17 key informants; participant observation of the daily life of children with ASD at home, in schools, at events, and routine practices of clinics and intervention centers; an on-line survey with 125 parents of children with ASD; and photo-voice with nine teenagers with ASD and six parents. I am grateful to the University of Queensland and the Organization for Autism Research for supporting this work. I thank my advisors Andrea Whittaker, Sylvia Rodger and Maxine Whittaker, and colleagues at CCIHP for their support during my study. I am indebted to the children with ASD, their parents and service providers in Hanoi for sharing their experiences and views with me.

### *References*

Hunt, P.C. 2005 An introduction to Vietnamese culture for rehabilitation service providers in the United States. In *Culture and Disability: Providing Culturally Competent Services*. J.H. Stone, ed. Pp. 203–223. Thousand Oaks, CA: Sage.
Malarney, S.K. 2002 *Culture, Ritual and Revolution in Vietnam*. New York: RoutledgeCurzon.
McLeod, M.W., and T.D. Nguyen 2001 *Culture and Customs of Vietnam*. Westport, CT: Greenwood Press.
National Assembly 2008 *Luật bảo hiểm y tế (Law on medical insurance)*. Hanoi: National Assembly.
National Assembly 2010 *Luật người khuyết tật (Disability Law)*. Hanoi: National Assembly.
Rose, N. 2006 *The Politics of Life Itself: Biomedicine, Power, and Subjectivity in the Twenty-First Century*. Princeton, NJ: Princeton University Press.
Rydstrom, H. 2010 Having 'learning difficulties': The inclusive education of disabled girls and boys in Vietnam. *Improving Schools* 13:81–98.
Villa, R.A., L.V. Tac, P.M. Muc, S. Ryan, and J.S. Thousand 2003 Inclusion in Viet Nam: More than a decade of implementation. *Research and Practice for Persons with Severe Disabilities* 28:23–32.

## Coming of Age

Local ideas about sexual practice and desire influence when and with whom young men and young women have sex; these practices vary across time and space (Manderson and Liamputtong 2001; Simpson 2009). Institutional and social factors inform debates about age at first sex and pregnancy. Early marriage and pregnancy (before 18 years of age) and especially very early pregnancy (before 15 years of age) impact young women's own health and physical development for physiological reasons, as well as disrupting schooling and employment. Yet young women may chose to conceive or continue with an unintended pregnancy while still very young, explicitly to strengthen kinship ties, as Nolwazi Mkhwanazi (2010) has illustrated. In the townships and informal settlements near Cape Town, South Africa, young women live in an environment with some of the highest rates worldwide of HIV infection, and equally concerning high rates of sexual assault, property crime, and violence. Each newborn is a symbol of hope and connectedness, a tangible body that maintains social and familial connections—a compelling symbol for young women.

Early parenting is only one way in which young women and men take on responsibilities for their families and kinship networks. Young people contribute tangibly to economic production, domestically and in the informal labor market, acting in ways that reflect local values of the reciprocity and co-responsibility that exist between parents and children. Such engagement is also often vital. Worldwide, there are unknown numbers of children-headed households; in Sub-Saharan Africa, by 2005 already an estimated 12 million children were singly or doubly orphaned, not only due to HIV/AIDS, but because of war, environmental pressure, poverty and violence, matters that we return to later in this volume. Children may be de facto heads of household too where parents have migrated for work, or when their parents are present but too ill or incapable of caring for others. In her research in Dhaka, Bangladesh, for example, Susan Bissell (2000) followed the lives of a number of children whose waged labor in the garment industry ensured they could keep their families together, providing care for parents or grandparents unable to work and supporting their young siblings to stay in school. But in face of a possible boycott against these industries by the US government, these children lost their jobs. The majority ended up not back in school, but earning less money in more dangerous ways—selling sex, selling snacks, begging, crushing stone to make gravel, carrying heavy loads on the streets and through markets, and working in bicycle factories and as domestic servants. These children were often the sole income earners for their families and while they had some choice about how to make money, they had no real choice that they needed to do so.

Worldwide, many children take on the everyday practical work of caring for others, as determined by family circumstances and individual family and gender-based values. Young girls particularly take on the physical and emotional labor involved in caring for those who are ill. Sometimes the care work is limited—collecting drugs from a pharmacist or local herbalist, for instance (Craig 2002; Granado et al. 2009), but often care is ongoing, with children cooking for, feeding, administering medications, and bathing, dressing and toileting their parents (Jennaway et al. 2015). Young people often hide the extent of their caregiving from others for fear of state intervention and separation.

While many children live their lives in inventive and resilient ways, the pressures they experience can lead to behaviors that might be regarded as inappropriate, pathological or 'risky.' Recreational drug taking and alcohol use, eating disorders, cigarette smoking, and interpersonal violence are all areas where young people inhabit an ambiguous terrain between childhood and adulthood. These can also be sites of active resistance and contest, as we explore in Chapter 6. With illness and its prevention, children are the subject of multiple gazes—a clinical gaze, a public health gaze, and a parental gaze. In everyday life, they often have little power over their own lives; rather, they are witness to disruptions that occur around them. This helps make sense of some problematic behavior, including dangerous acts. In situations of poverty and despair, in particular, often very young children drink alcohol; smoke cigarettes; sniff petrol, glue, household aerosols and other substances; and in doing so, risk immediate and long-term health problems including organ failure, brain damage and death. In the same communities, other socially disruptive behaviors and mental health problems, violence and suicide, are present. These children often live in environments where adult mental and physical health is poor, and gender-based violence, sexual abuse and child abuse, suicide, gambling, drug and alcohol abuse, and high rates of incarceration are all endemic (Currie et al. 2013; Gone 2013; Stevens and Bailie 2012). These pathologies are often the long-term outcomes of colonization, persistent structural violence, social exclusion and state neglect. Children's responses are idioms of endemic social and cultural distress and community suffering.

In the final case study below, Ria Reis illustrates in Suriname and Swaziland how some children respond to and express distress, reflecting the strategies available to them to elicit the concern, care and interventions of adults. Children's outbreaks of violence, trance, and other unusual behaviors are examples of resort to cultural idioms of distress (Nichter 1981; 2010). Embodied distress is typically linked to powerlessness and resistance, and to a sense of helplessness when individuals are unable to meet social expectations. In the examples below, children are not accountable for their behavior and may feel unable to speak of their distress, and so they use other means to draw attention to it. Away from the school or community, with the medical anthropologist, children in both settings were able to speak of endemic poverty, conflict in the home, abuse and gossip, conflicting identities and feelings of loneliness, distrust, anger and depression. But they had few ways of resolving these problems. Reminiscent of classical discussions of *susto* and *nervios,* their public enactment of distress provided them with a culturally legitimate way to express their feelings, and an effective means of gaining attention and care.

## 2.4 Children's Idioms of Distress

*Ria Reis*[1]

It was a normal day in a recently renovated high school in the capital city of Suriname. Children were quietly working on a task the teacher had given them, when Robert's friends noticed he had stopped working and looked very pale. They asked him what was wrong, but he did not respond, staring past them. A classmate called the teacher who found the boy slumped over his desk as if fainted. She shook him by the shoulder, calling his name. Robert reacted as if he was stung, wildly jumping up, his chair crashing to the floor. He screamed and growled, his face distorted in an angry snarl. While the teacher backed off alarmed, his friends tried to calm him but he fought them off, grasped a chair and violently threw it to the wall. The teacher called out to his classmates to leave the room, made a dash for the door herself, and ran to call the school director. By the time the director came running, Robert was on the floor, sliding across it like a snake. Now there was noise from another classroom, where Suzie had been complaining of stomach pain and feeling dizzy just an hour before. The normally soft-spoken girl began to scream, violently tearing at her clothes and lashing out and shouting abuse at anyone trying to come near her. In a matter of a few hours, four children in different classrooms had fallen prey. Robert was now out in the school yard, where teachers and children were herding in groups, some of the latter crying, others taking pictures of Robert with their cell phones. With the help of some colleagues and older boys, the director took matters in hand efficiently, attending to the affected children and summoning their parents to come and fetch them home, not to be returned before they were well. All other children were also sent home, and the school was temporarily closed while the director called the school board to report the event and seek advice.

## *How Trance Becomes Epidemic*

Epidemics of seemingly contagious dissociative behavior are known to occur worldwide in bounded communities of peers, particularly younger people, in institutions such as schools, orphanages and factories. They have been reported on since the sixteenth century, mostly anecdotally or in the format of medical case studies (cf. Bartholomew and Rickard 2014). However, research remains unsystematic and opportunistic. It mostly takes place in answer to a specific outbreak, involving professionals or academics who happened to be around or were called in to help solve a particular crisis. Different approaches lead to findings that are difficult to compare. Whereas outbreaks come at great costs for individual children, their families and schools, many questions remain unanswered. What motivates children living in vastly different sociocultural contexts to exhibit the same behavior? Do similar experiences or problems underlie epidemics of possession trance? How can insight into children's own experiences, perspectives and strategies contribute to effective preventive and crisis interventions? These questions guided an ongoing multiple case study into children's cultural idioms of distress, in which I collaborated, with local doctoral researchers and small multidisciplinary teams; here I discuss two cases of outbreaks in schools, in Suriname in South America and Swaziland in southern Africa. In each country three recently affected schools were selected and from these we gathered information about past episodes. In Swaziland we also observed children's trance behavior and teachers' responses during an outbreak.

In Suriname, outbreaks in schools of what is locally called 'trance phenomena' have taken place almost on a yearly basis over the last decade, the first report dating from 2006, but there may have been earlier, unreported incidents (Nannan Panday-Jhingoeri et al. 2014). In Swaziland, yearly episodes of *lihabiya*, a cultural syndrome among high school girls, have been reported since the 1980s (Reis 2000: 65). Over the last decade no year has passed without a Swazi high school closing because of what is now called 'demons.'

The description above is typical of outbreaks in both countries. Predominantly girls are affected, but boys can be affected as well, and outbreaks take place at primary as well as secondary level. It always begins with one child feeling physically unwell, having a headache or stomach pain, feeling dizzy, having difficulty breathing, or experiencing strange (temperature) sensations in the body. Some but not all children see or hear things or beings (for instance, animals or dead people) that others can't see or hear. The child becomes unresponsive and observers, both classmates and teachers, undergo a rapid process from feeling concerned and bewildered to realizing something out of the ordinary is going on. It is when a normally nice and well-behaved child bursts out in explosive behavior—e.g. fighting off with seemingly non-human strength peers who want to prevent them running wildly into the bush, damaging furniture or window shutters—that realization seeps in that 'the children are not themselves,' and that some 'thing' may be possessing them. Certain clues in children's behavior confirm this. Some clues are the same for both settings. People report how the child feels uncannily heavy, his or her voice changes, their eyes become red. Both in Suriname and Swaziland, teachers were predominantly of mainstream or Evangelical Christian faith, and likened what they saw to what they knew of exorcist rituals in their own churches or from television.

Children also displayed behavior reminiscent of culture-specific ritual settings. In Suriname people referred to Winti-prey or Jaran Kepang, ritual dances during which descendants of escaped African slaves (Maroon) or of Javanese migrant laborers become possessed by the spirits particular to their religion. In Swaziland some children clasped and moved their stomach as if something was stirring in there (reflective of *nyoka*, the belief in a snake in the stomach that reacts to impurity), or frantically rubbed their arm where they believed they had been marked by the demons as accomplices in their evil work. Swazi eye witnesses described how children often shouted names and this was considered proof of spirit possession: children recognized people they had met while they carried out chores for demons in secret underwater worlds. Imagery in Swaziland was most elaborated, with children relating how 'pastors' whom only they could see had enticed them with promises of wealth, and their narratives were full of references to witchcraft for an informed listener (e.g. having drunk red water and been fed meat that did not taste like animal meat).

In Suriname outbreaks would last for more than a week, in Swaziland some outbreaks lasted for months. Each outbreak caused debates about accountability amongst the adults. Teachers frequently pointed fingers at the families and communities of the children, accusing them of involvement in the occult. At the schools, dormant tensions between staff or between staff and higher levels of management were exposed. Individual staff members were at risk of being singled out as culprit. In Suriname, outbreaks were contained by school management; in Swaziland, such people were sometimes 'chased away.'

In both settings *personalistic* explanations dominate: an evil supernatural force manifests itself, outside of the control of the child it possesses. All adults we spoke to, also in our local teams, shared the strong belief that intentional spiritual evil is present in the world. Those who had witnessed a child in trance act on these notions and on the fear they induce; teachers showed their suspicion or conviction of negative spiritual agency by their involuntary bodily expressions (e.g. fearful expressions, crying, taking physical distance) but also by purposive acts (such as praying or exorcist rituals). The enfolding panic (children running out of class crying, teachers gathering frightened in a corner of the school yard or locking themselves in a room, and only some teachers and students approaching the 'victims') feeds back into the varying levels of distress already experienced, and leads to emotional expressions by children and teachers to different degrees. In the context of this heightened and generalized anxiety, other children's mild dissociative experiences may then also develop into trance and an outbreak is born.

## *Sources of Distress*

That reactions to illness and misfortune may open windows to tensions in social relations is well known, but what was at stake for the affected children? Unlike other reports on epidemics in schools, pressures that specifically related to the school environment, such as the curriculum, the exam period, unfair treatment by teachers, or being bullied by peers, could not be identified. We learned about the drivers of outbreaks from focus groups with children. Both boys and girls pointed at problems in children's daily lives. They suggested children who suffered from trance or demon possession might be an orphan, live in a stepfamily, or have caretakers who drank or quarrelled a lot. They discussed maltreatment, neglect or sexual abuse, poverty, and gossip or conflicts with peers.

> I don't think they came from a happy family . . . Because the demons attacked mostly those who came from under privileged families so that it helps them out with their needs . . . I think the ones that are attacked by demons are the ones that stay with guardians and not their real parents and you find that they abuse them . . . When you don't stay with your parents, the people you live with sometimes beat you for nothing. . . .
>
> *(Girl, focus group, Swaziland, translated from siSwati)*

Problems such as these were widely shared among children in general, but they were acute in the life histories of children like Robert and Suzie, with whom an outbreak had started. In the outbreak we witnessed in Swaziland, all possessed children were single or double orphans. Both in Swaziland and Suriname, some individual children spoke about identity problems, about being torn between a parent's traditional belief and the child's wish to be a Christian, a few hinted at homosexual feelings that were condemned by the majority of the community. All children with whom outbreaks started had experienced trance in the home environment, sometimes long before the outbreak at school.

Teachers, caretakers and key informants all recognized that children in these communities grow up in dire circumstances, but a range of sources of distress on all ecological levels was mentioned.

In Swaziland, the devastating HIV epidemic has deeply affected communities and families. Almost one third of the population of 18–49 years old is infected, and 10 percent of the population under 18 years of age is a (single or double) orphan. Lack of social protection against livelihood and lifecycle risks, high unemployment, food price inflation, and outdated farming methods have caused widespread poverty and increasing inequality. Some 60 percent of the population lives under the poverty line, 90 percent in rural areas.

In Suriname, ethnic tensions fed into seven years of civil war (1986–1992) that caused disruption and displacement in many regions, particularly among Maroon living in the interior South, and among the Indigenous American population of the Savannah. Many families fled to the city or abroad. Some who returned home after the peace treaty found their villages destroyed and their traditional authority structures undermined; they were no longer able to live in a self-sustainable way. Many now live a marginalized life in the cities. Suriname's economy is heavily dependent on fluctuations in mining profits and although poverty is worst in the country's interior, more than half of the urban population also lives under the poverty line.

In the daily lives of children, such stressors at macro-levels translate into stressors in their daily lives. At an individual level, poverty, health care crises, inequity, and a damaged social fabric translate into concrete experiences of going hungry to school, loss of loved ones, and neglect or maltreatment by caretakers who are distressed themselves. Teachers, key informants and children, particularly the girls, also reflected on the fact that children are generally not allowed to express their emotions in the presence of adults because it amounts to disrespect.

> Children are not being recognized as human beings. Whatever happens at home, even if one of the parents dies . . . the elders talk alone (while) the child is being left outside. You see, this is still burning in its heart: "What is happening? Why did my mother die? Why did my father die? Why do they not call me? Am I the cause of my father's death, since the family are not talking anything to me about the death of my parents?" These are all the things that pile in the child . . . and cause this.
>
> *(Key informant, Swaziland).*

Although all actors were well aware of the stressors in children's lives, few children or adults directly connected the social suffering in children's lives with an outbreak. Predominantly occult forces were blamed. This shared understanding allows children to take resort to a cultural idiom of distress. In principle children are not accountable for their behavior, they are not themselves: demons are at work in them. However, accountability in possession trance is not straightforward. Occult forces are everywhere, but their ability to take hold of a person also depends on that person's spiritual strength. Children are considered spiritually weaker than adults, girls weaker than boys, and spiritual weakness allows for evil forces to enter a child. Whereas adults mostly considered children's weakness as a given and looked no further, children's own perspectives helped us understand how, from an emic viewpoint, social suffering causes vulnerability for possession by evil. Boys and girls, but particularly the latter, eloquently described how poverty, conflict, abuse and gossip would lead to feelings of loneliness, distrust, anger, and somberness. These feelings in themselves, and the fact that children are not allowed to express their emotions, contributed to vulnerability to something evil entering them.

*Janet:* (maybe) that person gossiped this about me and I know it is not like that.
*Aminah:* . . . and maybe someone comes to you to tell you what is being said and then it becomes . . .
*Rosa:* then you hear it all the time . . .
*Janet:* you are not able to get it out of your head, it just stays in your head.
*Mariam:* you take everything into yourself and if you cannot keep it anymore, then . . .
*Iris:* then, then they think about senseless things, for instance, Little Busch bridge . . .
*Ella:* they think of jumping . . . think of suicide . . .
*Janet:* you just frustrate away.
*Ella:* plus if you can't discuss it with your parents.
*Mariam:* you keep everything to yourself, you can't discuss it with anyone, nobody understands you

| | |
|---|---|
| *Janet:* | or you're afraid to trust someone |
| *Rosa:* | especially your parents |
| *Ella:* | if your parents are strict, you . . . are scared to tell them |
| *Mariam:* | and you do not want to let the face of your parents fall, so you have no one. |

*(Girls' focus group, Suriname, translated from Dutch)*

## *A Few Words on Interventions*

Swaziland and Suriname are on opposite sides of the globe, and the political and societal challenges they face are very different. Yet at the micro-level, children's social suffering is remarkably similar, as are the structural obstacles for children to express the negative emotions related to this suffering. Possession by spirits or demons offer children culturally shared ways of experiencing and expressing their distress (cf. Nichter 1981; cf. Reis 2013). From an emic viewpoint, responses to outbreaks need to aim to exorcise these forces and enhance children's spiritual strength. Whether or not people accept the presence of demons and evil supernatural force in the world, it is important to intervene in factors that cause children's vulnerabilities and contribute to and compound their suffering. Our findings point to the possibility of preventing or responding to outbreaks of dissociative behavior in schools, by listening to what children can tell us about the challenges in their lives and by providing them with a safe environment to express their (negative) emotions and share their experiences of suffering with actors who may help them. Schools are the perfect arena outside of children's homes to engage adult actors relevant to them.

## *Note*

1. Data were collected through focus groups with children and teachers, and in-depth interviews were conducted with key informants and children and adults who had been part of or witnessed an outbreak. To help elicit children's opinions, a brief vignette was used of a child with dissociative symptoms. I thank Kamla Nannan Panday-Jhingoeri and her team in Suriname; Fortunate and Jabulani Shabalala and their team in Swaziland; Joop T. de Jong for our inspiring collaboration; and the participating children, caretakers and teachers for sharing their perspectives.

## *References*

Bartholomew, R.E., and B. Rickard 2014 *Mass Hysteria in Schools: A Worldwide History Since 1566.* Jefferson, NC: McFarland.

Nannan Panday-Jhingoeri, K., B. Sabajo-Cederboom, J. de Jong, R. Reis, and J. Menke. 2014 *Trance bij schoolkinderen. Rapport van een onderzoek op drie scholen in Suriname.* Paramaribo, Suriname: Antom de Kom Unversiteit van Suriname.

Nichter, M. 1981 Idioms of distress: Alternatives in the expression of psychosocial distress: A case study from South India. *Culture, Medicine, and Psychiatry* 5(4):379–408.

Reis, R. 2000 Kinderen en conversieverschijnselen: Over de gevolgen van de stress hypothese. *Medische Antropologie* 12(1):57–70.

——— 2013 Child idioms of distress as a response to trauma: Therapeutically beneficial, and for whom? *Transcultural Psychiatry* 50(5):623–644.

---

Reis's ethnographic examples of idioms of distress, like Dedding's research with children with diabetes, highlight how anthropologists have worked with children, not only as subjects of enquiry but also as the co-producers of ethnographic accounts. Children's lives are a constant in ethnographic accounts of the family, and individual anthropologists interact with children regularly in clinics, villages, schools and in families, their own and those of their friends and research participants. But few have made children central to their enquiry, and few have questioned how changes in the constitution of the family, with other economic, social and epidemiological changes, have impacted on children's wellbeing. Across societies, children play multiple roles within their families, in relation to their own health and healing, and in other aspects of social life. These roles disrupt easy connections between youth and powerlessness, age and power. Children create their own ways of being; they should play a central role in medical anthropology in the twenty-first century. In attending to child health, anthropologists need to acknowledge the diverse ways in which children exert agency over their health conditions and disabilities while growing up in different societies. The case studies presented in this chapter suggest that they should not only be informants. They can also be co-investigators, helping anthropologists produce ethnographic insights into the ways in which they live with and confront their illnesses and disabilities, and suggesting ways in which care modalities can help them improve the quality of their lives.

## References

Abadia-Barrero, César Ernesto 2011 *"I have AIDS but I am Happy." Children's Subjectivities, AIDS and Social Responses in Brazil.* Bogotà, Colombia: Universidad Nacional de Colombia.

Alderson, Priscilla 1993 *Children's Consent to Surgery.* Milton Keynes, UK: Open University Press.

Ariès, Philippe 1962 *Centuries of Childhood: A Social History of Family Life.* London: Jonathan Cape Ltd.

Bissell, Susan 2000 Child labour in Bangladesh: The "best interests" of children. PhD disseration, School of Public Health, The University of Melbourne.

Bluebond-Langner, Myra 1980 *The Private Worlds of Dying Children.* Princeton, NJ: Princeton University Press.

Buchbinder, Mara 2012 "Sticky" brains and sticky encounters in a U.S. pediatric pain clinic. *Culture Medicine and Psychiatry* 36(1):102–123.

Burman, Sandra, and Pamela Reynolds, eds. 1986 *Growing Up in a Divided Society: The Contexts of Childhood in South Africa.* Johannesburg, South Africa: Ravan Press.

Christensen, Pia, and Allison James, eds. 2008 *Research With Children: Perspectives and Practices.* London: Routledge.

Craig, David 2002 *Familiar Medicine: Everyday Health Knowledge and Practice in Today's Vietnam.* Honolulu, HI: University of Hawaii Press.

Currie, Cheryl L., T. Cameron Wild, Donald P. Schopflocher, Lory Laing, and Paul Veugelers 2013 Illicit and prescription drug problems among urban Aboriginal adults in Canada: The role of traditional culture in protection and resilience. *Social Science & Medicine* 88:1–9.

Drew, Sarah 2007 'Having cancer changed my life, and changed my life forever': Survival, illness legacy and service provision following cancer in childhood. *Chronic Illness* 3(4):278–295.

Geertz, Clifford 1960 *The Religion of Java.* Chicago, IL: The University of Chicago Press.

Gone, Joseph P. 2013 Redressing First Nations historical trauma: Theorizing mechanisms for indigenous culture as mental health treatment. *Transcultural Psychiatry* 50(5):683–706.

Gottlieb, Alma 2004 *The Afterlife Is Where We Come From: The Culture of Infancy in West Africa.* Chicago, IL: University of Chicago Press.

Gould, Drusilla, and Maria Glowacka 2004 Nagatooh(gahni): The bonding between mother and child in Shoshoni tradition. *Ethnology* 43(2):185–191.

Granado, Stephanie, Lenore Manderson, Brigit Obrist, and Marcel Tanner 2009 The moment of sale: Treating malaria in Abidjan, Côte d'Ivoire. *Anthropology and Medicine* 16(3):319–331.

Herdt, Gilbert 1981 *Guardians of the Flutes: Idioms of Masculinity: A. Study of Ritualized Homosexual Behavior.* New York: McGraw-Hill Book Co.

Hirschfeld, Lawrence A. 2002 Why don't anthropologists like children? *American Anthropologist* 104(2): 611–627.

Holten, Lianne 2013 *Mothers, Medicine and Morality in Rural Mali: An Ethnographic Study of Therapy Management of Pregnancy and Children's Illness Episodes.* Berlin: LIT Verlag.

James, Allison, and Adrian James 2008 *Key Concepts in Childhood Studies.* London: Sage.

Jennaway, Megan, Alexandra Gartrell, Lenore Manderson, Judith Fangalasu'u, and Simon Dolaiano 2015 Disability and constellations of care in the Solomon Islands. Unpublished manuscript.

Justice, Judith 2000 The politics of child survival. In *Global Health Policy, Local Realities: The Fallacy of the Level Playing Field.* Linda M. Whiteford and Lenore Manderson, eds. Pp. 23–38. Boulder, CO: Lynne Reiner Publishers.

Manderson, Lenore, and Pranee Liamputtong, eds. 2001 *Coming of Age in South and Southeast Asia*. London: NIAS/Curzon Press.
Mead, Margaret 1928 *Coming of Age in Samoa*. New York: William Morrow and Company.
Mkhwanazi, Nolwazi 2010 Understanding teenage pregnancy in a post-apartheid South African township. *Culture, Health and Sexuality* 12(4):347–358.
Nichter, Mark 1981 Idioms of distress—alternatives in the expression of psycho-social distress—a case study from South India. *Culture, Medicine and Psychiatry* 5(4):379–408.
——— 2010 Idioms of distress revisited. *Culture, Medicine and Psychiatry* 34(2):401–416.
Nieuwenhuys, Olga 1996 The paradox of child labor and anthropology. *Annual Review of Anthropology* 25:237–251.
Rapp, Rayna 2000 *Testing Women, Testing the Fetus: The Social Impact of Amniocentesis in America*. New York: Routledge.
Reichert, Julia 2006 *A Lion in the House, Independent Lens Filmmaker Q & A*. Arlington, VA: PBS.org. Accessed from http://www.pbs.org/independentlens/lioninthehouse/02_02_b.htm.
Reynolds, Pamela 1991 *Dance, Civet Cat: Child Labour in the Zambesi Valley*. Athens, OH: Ohio University Press.
Romanucci Schwartz, Lola 1969 The hierarchy of resort in curative practices: The Admiralty Islands, Melanesia. *Journal of Health and Social Behavior* 10(3):201–209.
Simpson, Anthony 2009 *Boys to Men in the Shadow of AIDS. Masculinity and HIV Risk in Zambia*. London: Palgrave Macmillan.
Sobo, Elisa J. 2010 Caring for children with special healthcare needs: "Once we got there, it was fine". In *Chronic Conditions, Fluid States: Chronicity and the Anthropology of Illness*. Lenore Manderson and Carolyn Smith-Morris, eds. Pp. 212–229. New Brunswick, NJ: Rutgers University Press.
Stevens, Matthew, and Ross Bailie 2012 *Gambling, housing conditions, community contexts and child health in remote indigenous communities in the Northern Territory, Australia*. BMC Public Health 12.
van Esterik, Penny 1988 The insufficient milk syndrome: Biological epidemic or cultural construction. In *Women and Health*. Patricia E. Whelehan, ed. Pp. 97–109. Granby, MA: Bergin & Garvey.
van Gennep, Arnold 1960 *The Rites of Passage*. Chicago, IL: The University of Chicago Press.
Whiting, John, and Beatrice Whiting 1975 *Children of Six Cultures: A Psychocultural Analysis*. Cambridge, MA: Harvard University Press.
Wizemann, T.M, and M.L. Pardue, eds. 2001 *Exploring the Biological Contributions to Human Health: Does Sex Matter?* Washington, DC: National Academy Press.

The Last Farewell, 2014. Jogjakarta, Indonesia.

## *About the photograph*

*Ibu Maryani founded the Pondok Pesantren Khusus Waria Senin-Kamis. (S)he was an important figure in the LGTB movement of Jogjakarta, known nation-wide for her kindness and efforts to support and promote the integration of waria within Indonesian society. (S)he was considered wealthy among other waria, and they would often receive from her money, protection, shelter and employment. Her death, after several months of sickness, shocked waria and the LGBT community. At her funeral, depicted here, more than 250 people, including waria, Ibu Maryani's relatives, NGO staff, LGBT activists, journalists, neighbors, and other people turned out to pay their respects, cramming into the narrow street where Ibu Maryani lived. The farewell was visceral, her mourners wailing the loss of such an influential leader and loved friend.*

—Nestor Nuño

# 3

# Sexuality and Technology

*Anita Hardon, Lenore Manderson and Elizabeth Cartwright*

'Gender' and 'sexuality' are elusive terms, commonly used as if self-evident yet as problematic in their definitions as in their politics. The simplest distinctions are between sex and gender, in order to differentiate biology and social and behavioral traits, but these distinctions are not made and do not make sense in all cultures (Rubin 1975); anthropologists studying sexuality, as well as queer theorists and activists, have made the insensitivity of this demarcation clear (Jackson 2011). 'Gender' references broader configurations of social life, encompassing ideologies and practices of kinship; sex, in contrast, is generally seen to be a biological classification of living things as male or female according to their external and internal genitalia, their chromosomes, endocrine systems and reproductive organs (Karkazis 2008; Wizemann and Pardue 2001). But as controversies illustrate, such as that over the gold medal for the 800-meter race of the South African woman, Caster Semenya, definitions of male and female are not straightforward. Semenya was found to have an intersex condition, leaving her with no uterus or ovaries and high levels of androgens. Defining one's sex is complex—there is not one biological marker that allows for a simple categorization of people as male or female. Moreover, gendered life experiences impact endocrinological processes, and thereby affect biologically defined sex differences between men and women (Karkazis 2008).

Together sex/gender as an intertwined concept contributes to the meanings given to sexuality, and to how it is understood in social, political, and cultural life. 'Sexuality' can refer to sexual feelings, sexual desire, and pleasure; to identity and its implications; and to social arenas where moral discourses on 'good' sexual behavior are played out. In this chapter, we use the term 'sexuality' as a relational concept within a medical anthropological framework, and reflect on the ways that different strands of sexuality research, involving historians, sexologists, and queer and feminist scholars, have contributed to our understanding of these issues. We focus on how sexuality is affected by and how it shapes medical technologies. We approach sexual behavior not as a biologically determined drive, but as a socially and culturally constructed practice shaped by power relations (see, for example, Spronk 2009). Using four case studies, we show how sexuality has been shaped by access to medical technology, and how medical technologies can be experienced and analyzed as liberating and oppressive, reflecting and shaping cultural notions of appropriate sexual identities, practices, and gender roles.

## Taming Non-normative Sexualities

In the late nineteenth and early twentieth centuries, in Europe and North America, medical researchers began to study variations in sexual behavior and desire, identifying people considered 'deviant' from the heterosexual norm of given times and places (Irvine 1990). Michel Foucault (1976) described how in the nineteenth century scientists started defining sex as something core to our identity, and so began to establish sexual standards and measure individuals and groups. Sexology as a discipline is conventionally traced to the publication of Krafft-Ebbing's book *Psychopathia Sexualis* (2011 [1886]), but this is simply a place marker in the development of the 'science' of sex in the Western academy.

Krafft-Ebbing's work built on and reflected popular interest in sexual difference and its embodiment throughout the nineteenth century (Foucault 1976; Roberts 2011). The story of Saartjie (Sarah) Baartman, born in 1879, is a particularly stark example of this early fascination with the embodiment of sexuality. Baartman, with at least one other Khoi woman, was taken from Cape Town, South Africa, to London in 1810, and then to Paris in 1814, and was paraded through the streets and displayed, examined, and ridiculed because of her large buttocks. Following her premature death in December 1815, she was dissected, and her skeleton and a body cast remained on display at the Muséum d'histoire naturelle d'Angers in France until 1974; her remains were returned for burial in South Africa only in 2002. Her treatment has been the subject of numerous scholarly articles, books, plays, and poems for many reasons, not only to expose ideas about sex and gender, bodies and sexuality, but also to illuminate slavery, racism and colonialism, stereotypes of race and sexuality (libido, desire, and so on), and the role of science in perpetrating racism (Gilman 1985; Willis 2010).

The interests of medical scientists and curious publics in Baartman's physicality were only part of the oppressive view about sexuality and bodies at the time. Historians of sexuality, in emphasizing the disciplining role of medical technology, have documented the recourse of surgeons in Europe and the United States to sexual surgeries to tame sexual desires and practices, such as masturbation, that did not conform to prevailing moralities on gender relations and sex. Sexual surgery like clitoridectomy continued until the 1940s in the United States to treat women's 'neuroses,' hysteria, and other behavior regarded as aberrant, even though at this same time other doctors were experimenting in stimulating sexual response (Kapsalis 1997; Maines 1999). In the United States, too, sexologists and other medical professionals used sex hormones to try to change the sexual desires and drives of men identified as homosexual, who were thought to have an excess of female hormones. Later in the mid-twentieth century, psychiatrists and other doctors started using electric shocks and lobotomy surgery with this aim. Not surprisingly, such pathologization of non-normative sexual identities and practices reinforced high levels of stigma, preventing people from openly expressing their sexual desires (Roberts 2011). At the same time, the sharpening of psychological interventions and the use of ways of measuring sexual desire contributed to the production of stereotypes about and laws targeting gay men and women, in contrast to the diversity that was permitted in certain other cultural settings. Reflecting this attitude, until 1974, homosexuality was considered to be a mental illness by the American Psychiatric Association, and this repressive regime had profound negative effects on the mental health of gay men, lesbians, and others whose sexualities were non-normative.

## Celebrating Sexual Diversity in Queer Studies

In the 1970s and 1980s, gay and lesbian activists played a key role in challenging the pathologization of non-normative sexual identities in medicine. Stimulated by this activism, scholarship on sexuality and gender in humanities departments in the United States and Europe has revealed the heteronormativity underlying much of the earlier research and writing on sexuality, and fundamentally changed our ways of thinking about sexuality. Queer scholars urged us not to assume sexual identities in binary terms (masculine versus feminine), but instead to examine the plurality and fluidity of desire and identity. Women's studies, feminist studies, and gay, lesbian, and queer studies have further critically recast theories that previously used reproduction to link gender and sexuality in order to explain women's subordination. Sexual and gender identities are not fixed, but draw on multiple axes of difference, style, sexual practice, and performance (Butler 1990). At the same time, anthropologists and other scholars have described the material aspects of sexual practice, so illustrating the ways in which local notions of desirability and various protocols, pharmaceuticals, surgeries, and personal practices affect how people present their bodies. Various chemicals, for instance, may be used to induce sexual desire in the global North (Race 2009); similarly, women in southern Africa use vaginal products to dry their vaginas, and stretch their labia to enhance pleasure (Hilber et al. 2010). In his ethnography of Brazilian *travesti*, Don Kulick (1998) described how his interlocutors, born biologically as men, adopted female clothes, names, and hairstyles, and used female hormones to achieve more female body forms, including growing breasts. His interlocutors, however, did not want to *be* women. They defined themselves as homosexuals, as men who desired men. They dressed as women to make themselves attractive to men, in response to local ideas of same-sex relations between men, independent of sexual identity.

In Southeast Asia, too, differences in gender, sexual identity, and same-sex desire are understood through a variety of categories that do not map neatly onto conventional biomedical classification systems. *Waria* in Indonesia, *bakla* in the Philippines, and kathoei in Thailand are all culturally structured categories of sexuality, which provide ways of dealing with people who might be categorized in the global North as gay, transgender, or intersex. Peter Jackson (1997; 2011) has described how sexual and gender difference is accommodated in Thailand, provided that there is conformity to the alternatives: as an effeminate man, a 'ladyboy' who identifies as female, or as a transvestite. Here, sexual identity is reinforced through practices such as dress style, habitus (speech, walk, gestures), and occupation; many people who identify as kathoei work in stereotypically female occupations. These Southeast Asian transgender/transsexual identities have been represented as premodern ritualized variants of contemporary sexual fluidity. More recent research emphasizes the differences between these forms of sexuality and identity and contemporary modes of being gay and queer that are consistent with ideas of sex and gender found in the global North (Jackson 2011).

In the first case study below, located in Thailand, we meet three young male-to-female transgender kathoei. Like travesti, kathoei use a wide variety of techniques to perform their desired gender identity, and in doing so, they relate to dominant notions of femininity, associated with conventions of beauty, such as having soft skin and breasts. Kathoei use hormonal pills and injections to grow breasts and otherwise feminize their bodies. Older kathoei are unsympathetic and warn the youth that they risk their health through such practices.

## 3.1 Feminizing the Body

*Panoopat Poompruek, Pimpawun Boonmongkon and Thomas E. Guadamuz*

At the Dok Mai Studio, a group of Thai *kathoei*, aspiring to live as women, meet regularly to inject hormones. Here, they provide each other with advice about various oral medications and injectables, and about managing negative side effects to achieve the desired results—breasts, glowing skin, femininity. They provide support too, as they seek to negotiate a life in Bangkok that is at the same time accepted and transgressive.

### *Sai's Story*

Among the other Dok Mai Studio crowd members, Sai was the last to embark upon her journey of medicine taking. Some of her friends had taken medicines when she was in her fourth year of high school, and she had tried it herself, but she did not continue because an older student warned her that if she took a lot, she would become confused, would not have a clue about her studies, and would not be able to pass university entrance exams. So, Sai first resisted the urge. The first kind she took was Diane, an oral contraceptive brand. An older friend told her it would make her beautiful and her skin would become white. She took one tablet per day, and it was effective—she got breasts:

> My senior at school told me it'd make you pretty. Her skin was very white. So I bought and took Diane and my breasts grew. In the past, I was in a boys' school. So, when my breasts grew a bit, oh, it was a rage. I became interesting to the boys. And so I bought more, taking a pill a day. I just took them in any order, didn't follow the arrow on the pack or go against it. Just pressed them off the blister pack and took them.

Sai began to use medicine in earnest in her fifth year of high school. By this time, she was permitted to wear a *rong song* hairstyle (short in the back but longer on the top of the head) as she was now a senior in her school. The pills gave her headaches so badly she could not bear it anymore. A friend advised her to switch to another brand, Preme. Sai was in luck. She did not get side effects from this brand, and so she has used it ever since. But when she talked with some older kathoei students,[1] they told her that if she wanted to become more beautiful, she should take a combination of different medicines, like they did. Preme alone, they thought, was insufficient for becoming beautiful. So, she added another kind, Premarin. Her expectation was that by taking this second drug, her breasts would grow bigger. When two kinds of pills were insufficient for the job, Sai decided to add hormone injections to the mix. A clinic near her school provided Progynon injections for a promotion price. She had them once a month. The injections did help her breasts to grow, but she felt weak and devoid of energy. She herself called these side effects 'laziness symptoms.' Because of them, she could not prepare for her entrance exams. So Sai stopped taking these injections for a while, and just took her pills.

## *When She Hit It Hard*

When Sai finished her final year of high school, her life had become freer—she did not continue her studies. She started working in a large cosmetics company, and there she could already dress as a woman. During this time, there was a shortage of Premarin on the market, so she could not take it any longer. Instead, she found so-called Lao birth control pills (*ya khum Lao*). She took these each day with Preme, combined with weekly injections of Progynon. She was able to reduce the side effects by adjusting the timing of intake, so ensuring that none affected her too much when she took them:

> I'd mostly heard about these ways from others, and now I tried them out by myself. Preme, for example, I took three pills a day. I might split that into one in the morning and two in the evening, together with one Lao birth control pill. In other words, at the time when I was hitting it hard, I took them mostly in the evening, and so they would not have too much effect—I'd fall asleep and not feel anything. Because with these drugs, when one takes a lot of them. . . . I also felt at one point that when I went out to work, did makeup for the clients, if I faced really bright sunlight, I felt dizzy, didn't have much energy.

After working for a while, Sai increased her dosage again when she met her first boyfriend and felt it necessary to make her body as womanlike as possible: "To make me beautiful, to not make him feel embarrassed about having me as his girlfriend," Sai explained. She now took hormone injections, Proluton, twice a week, every Monday and Thursday. But Sai experienced numerous side effects, and after three months, she had to reduce her dosage:

> It lasted some three months. During these three months it was like this, non-stop. My body really changed. My emotions changed so much that I felt like I was someone with violent mood swings, like someone who exploded with anger for no reason. And I ate so much. I was at my fattest at the time. At the time, I was some 53 kilos. I normally can eat whatever and not get fat, but during that time I ate so much. Phi Meng, I could eat over ten plates of rice in a day. Fifty-three kilos was a lot. I looked radiant, was fat, and had good skin. I was getting a lot of injections, had a new boyfriend, and worked as well. I felt that problems were circling me—my emotions were so violent and I spoke so arrogantly that my closest friend told me that if I didn't cut down on

my hormones, they would not want to come near me, because I'd complain and explode with anger just like that. I also had no sexual desire. I felt like, er, I was a woman and just wanted to do things related to beauty. Like this. But no sexual desire. My sexual organ didn't have any feeling—it was as good as dead.

Although Sai could not tolerate the side effects any longer, she did not stop taking hormones entirely, because she still wanted her body to appear feminine to her boyfriend. Instead, she cut down her hormone injections to once weekly or once every two weeks. She also reduced her doses of Preme and Premarin to one tablet of each per day. The Lao pills she stopped taking altogether. For four years, whenever she observed that her body looked less feminine, or her boyfriend mentioned it to her, she lost confidence and increased her dosage again, both oral and injected hormones, to her previous levels. When she had regained her self-confidence, she would slowly cut down her dosage.

## *Sai, Revisited*

Sai's self-medication had a lot to do with having a boyfriend. She wanted him to feel that she was as much like a woman as possible. But he told her: "Sai, you've got to accept that in any case, you're not a woman." He was also unfaithful many times, and Sai finally decided to end the relationship. She moved out of Bangkok to "flee her love." Her new life constituted only work. Having lost her motivation, medicines no longer seemed necessary to her, and she stopped taking them. When she began to hang out with her friends in Bangkok again, however, she resumed taking hormones. This time, she did not take such heavy doses, but used the same brands as before—Preme and Premarin, one each per day. Sai surgically augmented her breasts and had to adjust her dosage once more, to fit with the changes in her body. She was now taking hormones to ensure she would not develop bulging muscles, and to take care of her skin. She took two pills of Progynova each day and half a tablet of Androcur every two days to control the production of male hormones, as she had not had sexual reassignment surgery.

## *Twins Nam Daeng and Som*

*Nam Daeng:* I didn't know about it before I visited Corner, [a bar for] kathoeis who sell their cunt on Patpong. They were really in the top league of beauty. They said to me: "Why don't you take birth control pills? When I was kid, I wasn't as pretty as you. Go ahead, take them, in secret. If your mum does not let you take them, then hide it." They advised me that people usually took one pill a day. I wanted to be more beautiful, so I took a lot. Took a lot first and then slowly tapered it down, because this can increase your hormones. So I did take them secretly. When I first took them my spots disappeared. When I got spots again, I took them again. After a while, I had no muscles, and began to feel pretty. I began to get moody, my moustache wouldn't grow, and I had no spots. My moods were more feminine—whatever I did, I was so emotional. I didn't have much energy, but I was ever so moody. I didn't have energy but I had a fighter's heart.

Nam Daeng's and Som's medicine use began when they were 17 years old, but they did not start using hormones at the same time. Nam Daeng began first, because at the time she already felt confident enough to dress as a woman. Som, at the time, was still choosing between gay and kathoei identities because she was not quite sure if she would pass as a woman. She tried taking small doses of hormones from time to time, then stopped, because she had decided to be gay. Nam Daeng started with the oral contraceptive Diane, but it gave her migraines and she switched to another brand, Preme. In her first month of taking Preme, she took three pills a day—one in the morning, one at midday, and one in the evening. Based on advice she had acquired from the kathoeis at Corner—her role models in medicine taking—she later tapered down her use to one or two pills a day. She believed that by beginning with a strong dose, the effects of the medicine would stay with her longer. When she noticed the effects wearing off, she went back to taking heavier doses, and so on.

## *Hitting It Hard*

When Som began to feel that her experiments with dressing as a woman were successful—people were complimenting her on her beauty and men had begun to take an interest in her—she abandoned her gay persona and started to dress as a woman full-time. What was needed though, were hormones, just as they had been for her sister Nam Daeng. Som began taking hormones five years later than Nam Daeng, at the age of 22, which kathoeis consider a late start. Som felt that she had to hurry her feminization, and she did so by taking large doses of hormones. She began with six tablets of the oral contraceptive Preme per day—three times per day, two tablets each time. Now Som and Nam Daeng were both experimenting with medicines—they tried whichever kind their friends said were effective, including the Lao birth control pills, which are considered the strongest available. But both had to discontinue these, as, while they made their breasts grow, they had heavy side effects:

*Nam Daeng:* The Lao birth control pills are bought from Burma, by the ten or by the dozen. Some come in from China, too. These are really weird, scary birth control pills. They say that women only take one a month, but kathoeis take one a day. I tried taking one a day. The first time I took them was with this kathoei on Silom (Road), who told me to take them because they'd be quite something, two tablets. I took them and became beautiful in an instant. After the first blister pack, oh, my breasts grew straight away. My breasts hurt. But I couldn't take the second blister pack because I puked really hard. But just that one pack made my breasts stand out. If something touched my breasts, they hurt so much. Oh, no. Like, they were really something. I put on a shirt and when the shirt touched my breasts they hurt really bad. They became real humps I could push up with a bra. If I did that, I looked like a *chani* (derogatory term for women) who's had kids, and the bra got soiled with, like, white, thick milk that came out of my breasts.

After the Laotian pills, the twins switched to injectable hormones twice a week. Then they added two tablets of Androcur per week to the mix to reduce their male hormones. They also took their previous dose of Preme. In combination, these medicines constituted what the twins considered their full option. But they also kept on trying new medicines, so eagerly that they became the ones to whom others would turn to if they wanted to know which medicines worked well. If the twins said a certain medicine worked, others would start taking it.

Nam Daeng and Som took many kinds of medicines and so had to take several tablets at a time, several times a day. This complicated their lives. Older kathoeis also told them: "Don't take too much, you'll ruin your kidneys." So both quit all hormones except the injectables, which impressed them because they were fast acting and convenient; they did not have to carry around pills every day. However, these injections could still be considered hitting it hard because they had them once weekly (or four times per month):

*Som:* We had so many injections. So many that Nam Daeng and I stopped taking pills altogether. At that time, we just got the injections. Like, if I got the injection today, next week I'd already start to feel something. But with pills, I think you can see results in two months' time, like in the second month. And though you've seen me complain, they're good. Better than pills.

## *Quitting Hormones*

Having supercharged herself with hormones, both oral and injectable, for some two years, Som quit. She had tried to adapt the medication regimen many times, changing the dose or the timing, but to no avail. The side effects were numerous. She felt nauseated and vomited, but was willing to put up with these side effects for beauty's sake. But she also felt very weak and devoid of energy, because her muscles had shrunk. Both she and Nam Daeng had to work hard to earn money, and lack of energy made this difficult. In addition, she became depressed, then suicidal. She was lucky to have her mother warn her:

*Som:* When I was taking a lot of birth control pills or injections, I knew that I wasn't in my normal shape. Like, some people said that I'd cross the street without realizing it. Like I've told you,

> I'm someone who likes to observe things. I knew that eight injections in two months was a lot for my body. There was a time I just sat still and wondered why I was feeling so terribly sad and empty . . . I began to realize that this was an effect of the drug. I warned myself about it. One day, I sat in front of a mirror, and suddenly, I was talking with the mirror, just sitting there, combing my hair, as if this was nothing out of the usual. But my mother asked me, 'child, are you all right?' My mother had seen me sitting there, talking to the mirror for a while, talking to myself, complaining about this and that—it's hard to explain, like saying 'I'm so fed up—why is my life so tedious?' to the point I wanted to kill myself, that bad. And that wasn't the only time or the first time . . . I realized that if I'd keep on taking a lot of birth control pills, or have my cunt done, and then kill myself, what would happen? I'd accumulated this for my entire life, just wanting to be beautiful. So, I chose to quit.

Som was not the only one to suffer from mood swings. It happened to everyone who used hormones—Nam Daeng, Sai, and their friends. Although they were not so badly affected, and hormones did not make them suicidal, it was not that much better, either:

*Som:* I asked a friend: 'What's wrong with you, sitting there, playing with that lipstick?' She said: 'My big sister gave it to me. I see this lipstick and I miss my sister.' So I asked her: 'These past three months, have you still been taking two in the morning, two at noon, and two in the evening?' She said yes, she had. So I said: 'It's high time for you to cut down to one a day.' Phi Meng, [she said] 'I got a haircut and didn't like it. Shave my head, if I'm not beautiful, shave it all off.' So when they didn't shave her hair, she cried. [So I said,] 'when you've finished shaving your head, go become a nun.' [But in a little while] she [started to] like [her hairdo] and cried again. Really scary, these birth control pills.

Like Som, Nam Daeng quit taking hormones because she did not have enough energy to work. She also had mood problems, but she said she could control these symptoms. Som found quitting the hormones difficult because she was depressed, but she persisted. Her *kathoei* friends were stunned: she had been a pioneer of hormone use and had more knowledge about hormones than anyone else in the group. Many of her friends tried to persuade her to restart, asking her: "Why did you quit? You wanna be a gay or what?" She replied: "I don't wanna die."

### Note

1. Kathoei refers to sexually diverse people. See Jackson (2010) and Käng (2012).

### References

Jackson, P. 2010 Thai (trans)genders and (homo)sexualities in a global context. In *Routledge Handbook of Sexuality, Health and Rights*. P. Aggleton and R. Parker, eds. Pp. 88–96. Abingdon, UK and New York: Routledge.

Käng, D.B. 2012 Kathoey "in trend": Emergent genderscapes, national anxieties and the re-signification of male-bodied effeminacy in Thailand. *Asian Studies Review* 36:475–494.

---

## Taming Reproductive Women through Contraceptives

Contraceptive technologies can have both liberating and disempowering effects. Among *kathoey*, contraceptive hormones are used to achieve desired sexual identities; among heterosexual women in the West, beginning with the introduction of 'the pill' in the 1960s, they caused a sexual revolution. And, from the 1960s, population planners promoted contraception as an instrument to control fertility in the global South, as high rates of fertility were seen to be a threat to global food security and the environment.

In this context, sexuality studies were conducted both with a focus on family planning and within a health and development framework. Anthropologists were called on to study 'barriers' to contraceptive use in order to increase the uptake and sustained use of family planning methods

(Bledsoe et al. 1994; Hardon 1997; Richey 2008). While population planners expected that the low acceptance and uneven use of contraceptives were associated with people's lack of knowledge, anthropologists found that women strategically navigated their fertility (Bledsoe et al. 2000). Fertility is strongly valued in many societies: pregnancy proves fertility, and is thus cherished; women are valued by virtue of their fecundity; and infertility may be used as a reason for divorce or marriage to a second wife. Men may prohibit their wives from using contraception, but at the same time, anthropologists have found that women in many settings dislike contraceptive pills because of headaches, nausea, weight gain, and other perceived understandings of the technology and its effects—such as pills piling up in the uterus and causing cancer (Hardon 1997).

Rather than preventing pregnancy by taking a pill every day, women may resort to massage, herbs, or modern pharmaceuticals to induce menstruation when they fear that they are pregnant. Erica van der Sijpt (2013), in an intriguing study of reproductive behavior in East Cameroon, has documented how young women keep their pregnancy secret for as long as they can, while they try to catch a 'big fish' as the father-to-be. Only when women cannot negotiate a suitable father do they resort to herbs or pharmaceuticals to self-induce abortion, or seek the services of a traditional abortionist.

## Towards a Broad Definition of Sexual Health

At the 1994 International Conference on Population and Development in Cairo held by the United Nations, in response to a decade of campaigning by women's health advocates on the violations of women's reproductive rights in family planning programs that emphasized population control, the global community committed itself to programs to ensure that men and women might enjoy safe sex lives. Reproductive and sexual health was defined in the conference's Programme of Action (section 7.2) as:

> a state of complete physical, mental and social well-being and not merely the absence of disease or infirmity, in all matters relating to the reproductive system and to its functions and processes. Reproductive health therefore implies that people are able to have a satisfying and safe sex life. . . . It also includes sexual health, the purpose of which is the enhancement of life and personal relations, and not merely counselling and care related to reproduction and sexually transmitted diseases.
>
> *(ICPD [International Conference for Population and Development] 1994: 59)*

The Programme of Action called for reproductive and sexual health programs that would empower women, and technologies that would provide them with protection against unwanted pregnancy and sexually transmitted diseases. The program signalled a radical change in global population policies. It acknowledged women's agency in fertility and sexuality, and granted individuals, including unmarried women (ignored in the United Nation's earlier population policy statements), the right to sexual health. As a result, the conference changed both international agencies and national family planning programs worldwide, as they started to redefine their aims in terms of reproductive health rather than population control. It also changed sexuality research in diverse sociocultural settings, with the goal of informing sexual health policies. While research in the 1980s was aimed at developing long-acting contraceptives to help reduce population, by the late 1990s, contraceptive development increasingly focused on reproductive choice (Hardon 2006).

Advocacy for female-controlled contraceptive methods also gained momentum after the Cairo conference. Studies supported by the Population Council pointed to the widespread reluctance of men to use condoms; there was also a consistent message in studies of condom use in the context of HIV prevention (Manderson et al. 1997). Non-use partly again related to lack of knowledge and lack of access to condoms, but many studies also pointed to cultural barriers. Many qualitative studies had found that male condoms had a negative influence on sexual desire; as one study conducted in Uganda, using condoms was compared to eating sweets with the wrapping still on: they prevent the "real taste of sex" (Obbo 1995: 80).The non-use of condoms was also associated with trust; conversely, the use of condoms within a relationship implied lack of trust. Linked to this, in certain settings, condoms were associated with promiscuity and were thus considered inappropriate for use in stable relationships. Attempts to negotiate condom use within marriage invariably met resistance.

To elaborate on this, it is helpful to consider a study conducted in Zambia by Anthony Simpson (2009), who followed students from an earlier study into adulthood. Men, he found, wanted skin-to-skin contact; they argued that women needed to feel the semen entering their bodies. Many of Simpson's respondents said that they felt uneasy ejaculating into a condom; echoing Mary Douglas (1966), this was seen as 'matter out of place.' Some men even feared that their semen would be captured and used by women for love potions. Other men feared that condom use would prevent speedy 'rounds,' a sign of sexual strength (Simpson 2009: 140), which was important given their struggle to live up to expectations of manliness by demonstrating sexual potency. And while they acknowledged the need to protect themselves and their sexual partners from HIV, they could not imagine how this could be done in real life. Those who did use condoms reportedly came to enjoy taking a longer time in sex, acknowledging that women had the right to expect their partners to attend to their needs and desires, but even in this case, condoms were still seen to reduce male potency and to result in a 'waste' of semen. Following a 'hydraulic' model of sexuality, they wanted their semen to flow.

## Bodily Fluids for Sexual Health

Arianna Huhn in the next case study points to the importance of the flow of bodily fluids, not only for sexual health but also for health more broadly. Her interlocutors, she explains, believe that both foods consumed with the mouth and sexual fluids absorbed by the sexual organs influenced whether an individual gained or lost weight. Men and women believe that sexual fluids are essential for vitality, just like vitamins. They are seen to enter the body through women's vaginas and men's urinary tracts. A woman, her interlocutors asserted, would not become fat—a desired state—without such sexual vitamins. For men, the challenge is one of balance: too little sex makes their bodies big and soft, due to the unreleased sperm, while too much sex makes them frail. Huhn points to the implications of her findings for safe sex programs. If women need sperm to grow fat and healthy, they are not likely to want to use condoms.

---

## 3.2 Body, Sex and Diet in Mozambique

*Arianna Huhn*[1]

It was a warm afternoon in early June 2010, and I was among a group of women waiting for a meeting of the local chapter of Organização das Mulheres Moçambicanas (Organization of Mozambican Women) to commence. We sat outside of the group secretary's home, legs outstretched and covered in the multi-functional and colorful *capulana* cloth ubiquitous in East Africa. I wore my standard all-terrain sandals. The other women's feet were either bare or sporting flimsy, colorful flip-flops adorned with images like the Portuguese flag or Mickey Mouse. The shoes were cheap in price and in quality, but they were one of few affordable footware options for a population living far below the poverty line. The winds blew lightly, stirring up barely audible waves in the nearest bays of Lake Niassa.

The women chatted and joked quietly, the conversation eventually turning to my recent weight gain. "Arianna," Maria began cautiously, "You have really grown fat." I am a smallish-framed woman, and had gained enough weight over the previous six months to elevate my Body Mass Index (BMI) from around 19 to 20—not a dramatic difference, and certainly not qualifying me as 'fat' by biomedical standards, but still a noticeable change.[2] Several other women turned to join the conversation, adding their own observations of my increased breast size, the girth and number of lines around my neck, and my emerging love handles. The women laughed boisterously, clapped their hands together, and slapped their thighs in response to the litany. They jockeyed for space near to me so as to point out the specific features being discussed.

It wasn't that gaining weight was something in itself funny or negatively viewed—in fact, becoming larger in bodily dimension is usually evaluated quite positively in much of Africa (Fogelman 2009). The humor lay in the novelty of the conversation topic, and the perceived source of my enlargement. Fatima finally breached this topic openly, commenting between fits of laughter that, "Arianna's husband must be feeding her *very* well." While I agreed that the observations about my body were on target, I had to quibble with Fatima's explanation—my husband primarily occupied himself for our year and a half in residence by writing emails and sleeping. I responded to Fatima by explaining that my research grant was our family's only source of income, and because these funds were purchasing our food, it was logically *I* who was feeding my *husband* well, not the other way around.

The women burst into a fit of laughter at this literal and preposterous statement. Fatima, more sympathetic to my despondency, went about explaining that by 'feeding' she was euphemistically referring to sexual intercourse. Through further conversations that day and in the ensuing months, I came to understand that the transference of sexual fluids is locally reckoned to be an essential determinant of body size. In this case study I explain these principles of local ethnophysiology and consider their potential ramifications.[3]

## *Diet and Body Size*

To begin to understand sex as equivalent to food in contributing to bodily dimension, we must first consider local conceptions of nutrition. In brief, people in the Lake Niassa region see food as consumed primarily to ensure vitality (*thanzi*). They categorize foods by whether or not they provide strength and power for the body to do its daily tasks. Deceptively similar to biomedical understandings, 'vitality' is distinct from life-giving properties, such that food is consumed specifically for *energy*, rather than to ensure the continuation of life (which is dependent on spiritual forces). Vitality enables individuals to care for themselves and for others, which is a fundamental moral duty and a necessity for achieving social personhood, or status as a community member via interconnectivity and interdependence. The Portuguese word *vitamina* (vitamin) was ubiquitous in the discussions that I engaged in and overheard about food in relation to these effects. Acquaintances explained that foods with vitamina become blood, which provided the consumer with vigor and health to farm, cook, care for children, and so forth. Alternately, foods deficient in vitamina left or rendered the consumer's body weak, tired, and unable to engage in productive work (Huhn 2013).

Where a person eats plenty of *vitamina* foods, his or her body becomes plump to its full capacity. For some this may be a BMI of 19, for others a BMI of 26. Eating more vitamina is a sure way to become fat*ter*, but not necessarily *fat*, and a person who is not regularly consuming vitamina will become thin*ner*, but not necessarily thin. Fat and thin, in local formulation, are relative states dependent on each individual's body, circumstance, and predispositions, rather than objective categorizations based on a calculation of weight and height. This explains why I received smirks when I first arrived in Metangula and explained my research agenda as related to body size and diet. "It is fine that you have an interest in our traditions," one of my research assistants eventually said to me. "But, what a person eats does not change his or her body." Words from Verónica, a middle-aged obese woman, typified others' comments: "Some people say that diet makes a person thin or fat. But this isn't true." She reasoned, as did many, that she ate the same thing each day as others in her family who were of drastically different bodily dimensions. Verónica pointed to her adult son, who had a healthy BMI of 21. "A person's body size," she explained, "just comes out that way." No specific foods can aid a person to achieve *absolute* fatness or thinness, as it is only the *relative* states of becoming fat*ter* or thin*ner* that correlate with diet.

Further complicating matters, foods do not possess stable properties for provisioning vitamina. Social circumstances also impact vitamina: an individual suffering from depression, for example, will only taste bitterness when consuming food. The unpleasant experience impedes the transformation of foods into blood and vitality, resulting in smaller bodily dimensions rather than largeness. It was thus not unusual for me to hear a clinically thin person told enthusiastically that he or she was 'fat' when regaining strength after a recent bout of poor health or after experiencing a financial windfall. Similarly, 'thin' was a relative evaluation, and it was not oxymoronic to hear a person who was quite rotund declare his or her thinness if he or she was ill or in emotional distress. The ability of a situation to affect vitamina properties was driven home for me one afternoon when conversing with a man at a Metangula bar. Discussing his recent bicycle accident, Paolo suggested that unpleasant things diminish our energy and will to live, and so our sanguinity. "Blood," he said, "is things that taste good." He added, "If there are problems in the house, people in that home do not eat. They have no appetite. Nothing tastes good to them." Helena joined the conversation and added that even sugar would taste bitter to someone who had problems. "The tongue only works when a person is without worries," she said. An unhappy person, in other words, will veer toward their thinner form, regardless of diet.

In addition to diet and circumstance, people seemed to be working out the idea that an individual's body could also artificially swell when suffering from *bipi*, or high blood pressure (B.P.), referencing a diagnosis that some larger individuals are given at the local hospital. Rather than being understood as a *result* of largeness, bipi seemed to be emerging in local formulation as an illness itself, with obesity a key *side effect* and symptom. The blood of a person suffering from bipi, informants explained, would be foamy, causing the *appearance* of largeness, despite lack of health. Several of the bigger-bodied women I knew justified their size with bipi, seemingly as a way to deflect their better-off status, which can be an invitation for family petitions for assistance, or accusations of sorcery.

## *Sexual* Vitamina *and Gender*

In addition to diet, social circumstance, and bipi, another factor influencing body size in Metangula is sexual activity, and specifically the exchange of *ubazi* (sexual fluid). The sexual fluids of a man enter into a woman's body through her vagina, and the sexual fluids of a woman enter into a man's body through his urethra.[4] From there, the substance becomes part of the recipient's bloodstream, making sexual 'vitamina' just as necessary as alimentary vitamina for the production of blood, and so vitality. Sexual vitamina is, however, distinct from and complementary to the dietary variety, rather than a substitute for it. As Lúcia said when I asked her why a sexually inactive person couldn't just, say, eat a lot of tomatoes to ensure his or her strength, "Can a woman eat tomatoes with her vagina?" Nope.[5]

The effects of sexual vitamina on the body, however, unlike alimentary vitamina, seem to be distinctly gendered. For a woman, acquaintances explained to me, the vitamina of a man is necessary to reach the larger dimensions of her body size potential. A woman cannot fatten (though she can be fat), or otherwise reach her predisposed girth extreme, which is a sign of health and social interconnectedness, without regular sexual intercourse. While estimates varied, 'regular' seemed to mean at least roughly once per month. If a woman does not have sex at this interval, I was told, she deviates toward her predisposed thinner form. At the Health Center and around Metangula, I encountered many women who said they felt weak and complained that they were becoming thin from being unmarried or otherwise lacking sexual encounters. Those who had been seen by a medical professional often reported that they were diagnosed as suffering from anemia. This signaled to both the medical technician and the women being treated that they were deficient in blood. For the women, this indicated that their blood was 'moving alone'—without a male intimate partner. When I expressed skepticism, multiple women added that medical professionals they met with had asked if they 'had a husband' (culturally implying a regular sexual partner) as part of the diagnostic process, suggesting concordance with their own interpretations. And rather than being given medicine, several said, they were instructed to have sex to address their health problems. I could not find any member of the current medical staff who would confirm these rumors. Similar perceptions, however, were so strong in the provincial capital Lichinga that the Anglican diocese put together a brochure extolling the value of semen for sexual reproduction *only*, not female health (Rebecca Vander Meulen, pers. comm., 11 February 2011).

The import of men receiving female ubazi through sexual intercourse is not as clear. Acquaintances did not conceive of a woman's sexual fluids as necessary for a man to fatten. In fact, *lack* of sexual activity makes a man softer and larger as his body fills with unreleased sperm and, like bipi, makes his blood foamy and his body artificially big. This happens, I was told, if a man does not have sex often—at least once every six months by some accounts, at least once every week by others. Acquaintances pointed to the round bellies of male government officials whose families had not traveled with them to Metangula as evidence that male abstinence leads to rotundity. When I observed that some of these men were also thin, I was reminded that everyone's body has its own predispositions. Men having 'too much' sex (regular intercourse three or more times a day) are in equal danger of frailty, I was told. Again, however, this is unrelated to female ubazi. Rather, such men run out of their own ubazi, and new supplies have to be created from vitamina siphoned from their blood, negatively prejudicing vitality and so leading to bodily thinning.

Abstinence is one way that such men could regain their strength. Lourenço, an acquaintance in his 30s, for example, explained to me that he refrained from sex with his wife whenever he found that he was lacking the strength to collect bamboo or make bricks. Being poor, he said, he did not have the luxury to simply eat well to regain strength when sanguine vitamina was redirected for sexual activity. Another strategy for enhancing sexual fluid production without risking blood loss was to eat specific foods. Men commonly snacked on sugarcane in the marketplace, soaked rice was one of few foods I witnessed men preparing for themselves, and acquaintances often sent me home with raw cassava "for my husband." Additional ubazi-enhancing (but *not* blood-building) foods include peanuts and coconut meat.

While women can also eat ubazi-enhancing foods to increase their sexual fluid output, they are less likely to do so for three reasons. First, any excessive ubazi just leaks out of a woman's body; it cannot be stored internally as it can in a man's body. Informants pointed to regular vaginal secretions as evidence. Second, both women and men suggested that a woman's sexual appetite is not as voracious as that of a man's. A woman can thus have sex once and be satisfied for long enough, before having sex again, for a new supply of ubazi to be created without haste or depleting her blood supplies. Finally, female ubazi is not an especially prized part of sexual experience in Metangula. As elsewhere in Africa, informants indicated a male preference for dry sex, and women took measures to both tighten their vaginas and reduce their sexual secretions through use of astringent substances inserted into the vaginal canal. Adult woman I knew candidly pointed to plants like kobwe beans and tomato vines maintained in their yards, from which they harvested leaves from for such purposes. The preference for vaginal dryness is also perhaps why the importance of female ubazi for male health is culturally unelaborated.

## Conclusions

The correlation between diet and body size in Western biomedicine is complicated in Metangula, where the energy-provisioning properties of foods are malleable, evaluations of largeness in bodily dimension are based on individual predisposition rather than standardized scales, and sexual intercourse is conceived as a literally embodied experience integral to female health. While Metangula's ethnophysiology may be unique to the locale, or to the region, the logic complements broader African philosophies that envision the merger of bodies through sexual union as creating a metaphysical bond between regular sexual partners, through which earthly life continues and wrongdoing can summon the wrath of spiritual forces (Heald 1995), and interpretations of larger bodily dimensions indicate both good health and community engagement. The necessity of another person's presence in order to be personally complete further speaks to the central importance of interconnectivity in broad African conceptions of both personhood and wellbeing. This is not in the sense of selfless collectivism effacing individuality, but in persons being literally composed and created through their encounters with and dependencies on other human beings (Geissler and Prince 2010).

Considering such manifestations and integrations of local ideologies about personhood in relation to the body can be useful when implementing programming related to health, sex, and diet anywhere. In African settings, and perhaps elsewhere, local ideas about sexual fluids have important implications for the prevention and spread of sexually transmitted diseases (STDs). If receipt of sexual fluids within the body is necessary for female wellbeing, for example, women are likely to be reluctant to use a condom during sex, and condoms were not positively evaluated by my acquaintances in Metangula. While many condoms were sold in the market and health professionals gave away many more at the Health Center, the end use was often children filling the condoms with air and using them as balloons, or wrapping them in plastic bags and string to form homemade soccer balls. Few people regularly purchased condoms and used them for health or contraceptive purposes, and even local NGO–organized volunteers were at a loss to explain to onlookers at a STD-themed lecture I attended how a woman was supposed to get her vitamina if she used a condom. Similar public health campaigns may fall flat if condom use is the only pathway for health presented, as it is directly at odds with local perceptions of what leads to wellbeing. Similarly, programming focused on extolling the nutritional properties of foods and the importance of a balanced diet may present information that is not easily compatible with local notions of nutrition in relation to broader social circumstance, with conceivable impact on the effectiveness of programming related to both under-nutrition and over-nutrition.

## Notes

1. This case study is based on 16 months of ethnographic research conducted between 2009 and 2011 in the town of Metangula, Mozambique. The study was made possible through generous funding from a Fulbright-Hays Doctoral Dissertation Research Abroad Fellowship and the Department of Anthropology and Office of the Dean at Boston University.
2. BMI is calculated by dividing an individual's weight in kilograms by the square of his or her height in meters, to roughly determine if a person is under- or over-nourished. BMI less than or equal to 18.5 is categorized as 'malnourished,' between 18.5 and 25 is categorized as 'normal,' greater than 25 is 'overweight,' and greater than 30 is 'obese.'
3. During the period of research, Metangula was home to approximately 10,000 residents, most subsistence-level farmers or petty traders. The vast majority of these individuals ethnically identified as Nyanja, an ethnicity that scholars class as part of the group of Bantu-speaking matrilineal peoples called the 'Maravi Cluster.' My data come primarily from unstructured conversations held in Chinyanja with a diverse group of 20 principal informants. All names have been replaced with pseudonyms to protect the identities of informants. The body of water on which Metangula is located is known as 'Lake Niassa' in Mozambique, 'Lake Malawi' in Malawi, and 'Lake Nyasa' in Tanzania, all of which border the body of water.
4. The heterosexual focus of cultural renderings of ubazi exchange does not, of course, rule out the possibility of same-sex attraction or intimate encounters. It simply speaks to the gender complementarity embedded in Metangula and many sub-Saharan African societies, whereby male and female are considered incomplete entities made whole only through their union.
5. Tomatoes are one of four foods locally categorized as vitamina-enhancing substances. The others three are onions, oil, and sugar. See Huhn (2013) for additional discussion of this point.

## References

Fogelman, A. 2009 *The Changing Shape of Malnutrition: Obesity in Sub-Saharan Africa. Issues in Brief, No 11.* Boston, MA: Pardee Center for the Study of the Longer Range Future.

Geissler, W., and R. Prince 2010 *The Land Is Dying: Contingency, Creativity and Conflict in Western Kenya.* New York and Oxford, UK: Berghahn Books.

Heald, S. 1995 The power of sex: Some reflections on the Caldwells' 'African sexuality' thesis. *Journal of the International African Institute* 65(4):489–505.

Huhn, A. 2013 The tongue only works without worries: Sentiment and sustenance in a Mozambican town. *Food and Foodways* 21(3):186–210.

---

The highly dynamic field of family planning and reproductive health illustrates how medical technologies are intertwined with societal notions of appropriate sexuality, and how such technologies can oscillate between being oppressive and being liberating. Anthropological research in this field has not only described the appropriation and use of reproductive health technologies, but also the way in which understandings of sexuality and health more broadly can inform the design of reproductive health programs, and how societal influences affect their development. Such links and influences are also visible in HIV prevention and AIDS care programs.

Early in the AIDS epidemic, anthropologists collaborated with AIDS activists and policy makers in setting up HIV prevention programs. Many such programs, however, have had a hard time addressing the lack of willingness among men in many settings to use condoms to prevent HIV transmission. Anita Hardon (2010) described how a group of feminist researchers based at the Population Council in New York, in consultation with women's health advocates from around the world, sought to resolve this problem by developing microbicides that can be used vaginally, to help women protect themselves. Microbicides are chemical agents that are being developed to prevent and treat vaginal infections and STDs, including HIV. Developing these new female-controlled barrier methods required creating new alliances between researchers, donors, health-related manufacturers, clinics, health workers, patients, and women's health advocates. Like the contraceptive pill, microbicides provide women with more power over their sexuality and sexual health than did the use of male condoms, while giving them more responsibility. Hardon (2010), one of the authors of this chapter, participated in the Women's Health Advocacy Committee on Microbicides, which was set up to advise researchers at the Population Council on engaging with women's sexuality in developing these technologies. She was asked by the Population Council to prepare a report for a meeting on the acceptability of microbicides from a women's health perspective. She did so by reviewing the scientific literature on women's use of vaginal products, and by asking friends working in different fields to explore vaginal products and their use. They found that women used many different substances vaginally for diverse reasons, ranging from sexual pleasure and the cleansing of sperm to the prevention and treatment of reproductive tract infections. In her report, Hardon argued that vaginal products were acceptable to women, and she suggested that protecting against reproductive tract infections, enhancing sexual pleasure, and contraception should all be key goals if microbicidal products were to be developed.

The envisioned users of such microbicides were women at risk because their sexual partners were unwilling to use male condoms and/or women who wanted to protect themselves from HIV without their husbands' knowledge. Sexual enhancement and broader protection against reproductive tract infections were therefore not included in the terms of reference provided by the Population Council. How do female-controlled barrier methods intersect with sexuality, thus defined, in the various settings where they are introduced? In the following case study, Robert Pool describes how in clinical trial settings women have appropriated microbicides to fulfill their own sexual desires. While developers expected that women would keep the use of the gels secret, social scientists found that women often told their partners about the gels, and that these became a means to increase sexual pleasure.

## 3.3 Empowerment and the Use of Vaginal Microbicides

*Robert Pool*

Vaginal microbicides were originally conceptualized by Northern researchers and advocates as a female-controlled means of HIV prevention—something that women, mainly in Southern countries, particularly in sub-Saharan Africa, could use to protect themselves in situations in which male partners were reluctant to use condoms and in which they were unable to negotiate condom use. The assumption was that women would be able to use these products secretly, without informing partners. This assumption was supported in early hypothetical acceptability studies, in which women were asked whether they would inform their partners if they were to use a microbicide. The empowerment of women was central to the development of vaginal microbicides, thus bringing together feminist ideals and biomedical research. However, when products started to be tested in trials in Africa, women's response was not always what researchers and advocates had anticipated. In what follows, I describe how individual women in South Africa redefined empowerment to fit their particular situation in the context of a microbicide clinical trial.

The trial, called MDP 301, was a randomized, double-blind, placebo controlled trial that aimed to determine the efficacy and safety of PRO-2000 gel in preventing vaginally acquired HIV infection. It was conducted at three sites in South Africa and one each in Zambia, Uganda, and Tanzania. The enrollment of 9,385 women was completed in August 2008, and follow-up was completed in August 2009. PRO-2000 turned out to be safe but not to prevent HIV transmission.

The trial had an integrated social science component, which focused on a random sample of almost 8 percent of trial participants. The aims of this component were to assess the accuracy of the data on sexual behavior, adherence, and condom use, to investigate acceptability of the product, and to assess participants' comprehension of the trial and the informed consent procedures. Women were also asked whether they had informed their partner about the trial and the gel, and they were asked about partner involvement in gel use. Where possible male partners were also interviewed. Below, I draw on data from the South African sites.[1]

Before enrollment women received detailed information about the trial and the product. As part of the informed consent procedure, they were taught about HIV and risk behavior, and repeatedly they were told that the efficacy of the product was unknown and that it was therefore essential for them to use a condom to protect themselves whenever they had sex. Women were tested on this knowledge before they were enrolled, and their comprehension was continually monitored throughout the trial. They also received detailed information on the various clinical procedures and they were taught, using anatomical models, how to insert the gel using the pre-filled applicator.

Women who participated in the trial attended the trial clinic every month, where these messages were repeated and where they underwent clinical tests and examinations and received any necessary treatment. At the clinic they could also ask questions related to the trial and their health. As a result, they not only learnt about the trial and the gel, but were also educated on their own anatomy and on sexual and reproductive health more generally. The trial also formed a context in which women could gather and communicate about 'women's issues' and discuss partners and relationships. They encouraged friends and neighbors to join, and met other women during clinic visits. As a result, a feeling of community and ownership developed among participants, who often spoke of 'their' trial.

### *The Appropriation of Empowerment*

Contrary to the notion that microbicides would be used secretly, almost all women in the trial informed their partners about the gel. The most important reason was that they felt that men had a right to know, given that they were also being exposed to the gel.

> I think it is not right keep it a secret from your partner, because if you come across a problem, if the gel affects him, he won't know what affected him, whereas you knew but you kept it a secret. I think it's the right thing to tell your partner about the gel.

How, women asked, would they explain to their partners if the gel turned out to have side effects, and especially a negative effect on their (the men's) health? Even women who did not inform their partners still thought that it was right that they knew.

> I think it is the right thing to do, but I made a decision not to inform him about using the gel [because] men ask too many questions. He would have asked me: why are you using the gel, and are you sure that it is going to protect you, and so forth. So I was avoiding questions like that when I decided not to tell him.

A second reason was that men would probably notice the gel anyway: they would feel a difference in lubrication during sex, or discover the boxes of applicators, or catch their partner inserting it before sex. Women said that men thought of themselves as being in charge and would be angry if 'their' women ignored their authority. Both men and women said that if a man noticed that his partner was inserting something in her vagina before sex, he would be likely to suspect that she was using *muti* (magical substances) to bewitch him.

As a result, almost all women did inform their partners about the gel. Most women (75 percent) told their partners about the trial when they first heard about it, and before they enrolled. Almost all the others told their partners soon after enrollment. A few men discovered the gel early on while the women were still procrastinating about informing them, and a few others discovered it later, after their partners had decided not to inform them.

This 'informing' was sometimes a gradual and diffuse process. For example, one woman said she had 'told' her partner straight after enrollment. But she had already been discussing the trial with him for a long time in relation to a cousin who was already enrolled, and she had told him she wanted to go to the same clinic and get tested. So she had been grooming him for a while before she actually told him that she had enrolled. It was common for women to first tell their partners that they were going to the clinic to find out about a study that was taking place, or to get themselves tested, and then coming home with the information sheets and the gel and explaining further. Confronted with a *fait accompli*, men found it difficult to refuse.

Women with 'difficult' partners were often patient and creative in the way they ended up getting their partners to accept gel use.

| | |
|---|---|
| *Interviewer:* | Did your partner know that you were using the gel? |
| *Woman:* | He did not like it. I told him after I enrolled in the study and he refused to allow me to use it. I left it like that and I said to him: it's fine . . . [but] I used the gel without telling him and we had sex all the time. |
| *Interviewer:* | Did he feel that you were using it? |
| *Woman:* | No, he did not feel it. And after two months using it in secret I talked to him again about it. But I did not tell him I was using it in secret. Then he said that if I used the gel he would not have sex with me. I asked him why. He said that he did not know if sex was going to be the same or not. Then I asked him how he felt about our sex and he said that it was okay. I then told him I had been using the gel, and he said that I could continue to use it. |
| *Interviewer:* | How did he feel when you told him that you used the gel secretly? |
| *Woman:* | He was angry and we had a little argument. But he realized there was no difference when I used the gel. |

Sometimes women made use of the formal gender power relationship of male dominance and female obedience to neutralize suspicion. For example, one man refused to allow his partner to continue using the gel, but she ignored him and continued to use it anyway, and he never suspected anything. She could do this because it was inconceivable to him that she would disobey him.

> Because he knows that when he says he does not like something I am doing, I don't argue with him: I stop straight away. He is always proud that I respect him by obeying his instructions. He does not know that I am continuing using the gel.

Other men forbade their partner to use the gel but, when the women continued nonetheless, appeared to be aware of this but pretended not to notice, possibly in order to avoid confrontation and open questioning of their 'authority,' thus avoiding potential loss of face. Some men confirmed this, and in some of the focus groups, men explicitly made a distinction between actually having power in the relationship and only *appearing* to the outside world to have power. So while being dominant in the relationship was important for them, it was more important that *other men* saw them as dominant. As one man put it:

> There are some men that meet and talk. It's just that I'm going to be proud. I cannot meet with my friends and tell them how my girlfriend and I really live at home. I make out as if I do not take orders from a woman. They don't know that when I get home I dance to her tune.

Finally, as the quotes from two women below show, the use of the gel in the context of the trial gave women the pretext and the confidence to address the issue of condom use, which is generally considered to be unacceptable in regular relationships. Indeed, this was one of the main reasons for the development of vaginal microbicides in the first place.

> I had to sit him down, show him the gel, and tell him: each time we have sex I will use one gel and you have to use a condom. I explained that if I use the gel for one round of sex, then we have to use another gel for the next round. And he also has to use another condom.
>
> *(Woman A)*

> I told him that I was participating in a study that involved using gel. I told him I had to use the gel and he had to use a condom. And that when the study finished, it didn't mean that he could stop using condoms, oh no!
>
> *(Woman B)*

### *Conclusion*

The empowerment of women envisioned with the development of vaginal microbicides entailed women using such products secretly. No one took into account the possibility that women would consider it wrong to use such products secretly, or that it might be impractical to do so in regular relationships. What was also not taken into account is the importance of the precise nature of the relationship between the woman and her partner. So whereas some men immediately were open and communicative in their relationship, accepted the gel easily, and integrated the gel into their foreplay, others were uncommunicative about the relationship and sex (except to say when they wanted it), and were suspicious of the gel.

Advocates and researchers also underestimated the power that women had, because they had tended to be too focused on the public surface of power relations and ignored the underlying dynamics of relationships in practice, and as a result they assumed that secrecy was the solution. So for example, when women discussed power and authority in relationships, they rarely said that men *were* the decision makers, but that they *liked to think* that they were, suggesting that they—the women—had more power than they appeared to have on the surface, but that they had to exercise it in more subtle and indirect ways.

Arguably, the sort of empowerment discussed here, of course, is not 'really' empowerment because it operates within the existing hierarchical structures and relationships without changing them. But the same applies to secret use. The difference is that the former is emically directed, by the women themselves, and grounded in their everyday practice, whereas the latter is etically directed, by the research and advocacy community, without taking the diverse practices on the ground into account.

This 'power from below' was stimulated and supported by the trial and clinic setting. This was reinforced through contact with other women with whom they could exchange experiences relating to the use of the gel and the response of partners. This knowledge, and the feeling of being part of the trial community, empowered women participants to take on partners and persuade them to accept gel use, and if they were unable to convince reluctant partners, it gave them the confidence to defy them and use the gel secretly anyway. It also gave them the tools and the confidence to go home and initiate a broader discussion about HIV prevention and condom use.

### *Note*

1. This case study is based on data from 154 in-depth interviews with 45 couples (women who shared information about the trial and involved their partners from early on, and their partners, all interviewed at least once), 60 interviews with 30 women who did not immediately inform their partners, 31 focus group discussions with trial participants, 18 focus groups with non-trial community women, and 18 focus groups with community men. The women in this case study were all in regular relationships.

---

As Pool has illustrated, the clinical trials in which the microbicides were tested thus became sites where new sexualities were constructed. While the designers did not imagine that the products would enhance sexual pleasure, women and men appropriated them for this purpose. Facing fear of HIV transmission, the participants in the trial appropriated microbicide gels into their sexual relations. The trials gave women the pretext and confidence to address the issue of HIV prevention and condom use, which, as we have seen above, is generally considered to be unacceptable in intimate relationships.

## Sexual Enhancement and Confidence

Recent studies of sexuality and gender in Africa point out that economic constraints and the absence of educational and job opportunities disempower men, and make it difficult for them to live up to the expected ideal of the man as the provider of the family (Silberschmidt 2004).

Young men try to overcome feelings of unmanliness by exerting a kind of masculinity that is based on sexuality (Aboim 2009; Cornwall 2003). For example, Christian Groes-Green (2011) has illustrated that in Maputo, Mozambique, young working-class men lack the means to offer their girlfriends gifts or financial support, and therefore become preoccupied with becoming skilled lovers. As he described it, these young men invest in learning new sexual tricks and positions, and consume certain drinks, drugs, and food to enhance their performance. Mozambican men's 'sexualized masculinity' was 'based on the man's ability to perform sexually, give erotic pleasure and become respected due to his sexual satisfaction of the female partner' (Groes-Green 2011: 289).

Similarly, Jennifer Cole (2005) writes about unemployed young men in Madagascar, the *jaombilo*, who use their good looks and sex appeal to gain support from young women working in the sexual economy. Although being financially supported by a woman could potentially lead to feelings of diminished manhood among *jaombilo*, their masculinity was based on cultivating their appearance and sexual desirability. Another study shows how young men in Addis Ababa resort to using Viagra to gain sexual confidence (Both and Pool Forthcoming; Both 2015). Over time they become dependent on the drug, fearing that they will not be able to perform sexually if they do not take it. From a young age, boys in many parts of Ethiopia learn that they can only be proud if they carry out manly tasks, like being good at sports or doing well in school, and later in life they are expected to achieve financial or social success. The absence of promising socioeconomic prospects has been noticed as a constraining factor for young men in fulfilling these normative tasks of adult men. To overcome their frustrations and feelings of hopelessness these men turn to chewing *khat*, watching movies, and taking Viagra.

## Rethinking Masculinities

The final case study contributed by Sebastian Mohr takes us to a specific clinical setting, a sperm bank in Denmark, where healthy potent men come to donate their sperm. Mohr illustrates how female technicians handle the sperm with much gentle humor, through which they subtly acknowledge the sexual feelings of the men who donate sperm, and, in this case, also the male ethnographer. In his analysis of this case, Mohr describes how gender and sexuality are made meaningful in the sperm bank, and how the technicians overcome their own feelings of disgust in handling men's sperm. Men boost their sense of their own masculinity by providing the bank with 'good quality' sperm, which Mohr analyzes as a performance of masculinity. While repeated performances make it seem as though sexual identities are natural, men are created as individuals through their employment and appropriation of society's ideas about gender.

---

## 3.4 Donating Semen in Denmark

*Sebastian Mohr*

> It is just before nine when I arrive at Andersen Sperm Bank on a cold Monday morning in February. Lise and Signe, the two technicians whom I have gotten to know since I first started my fieldwork a year and a half ago, are already busy in the laboratory. So is Martin, the leading lab technician in charge of coordinating daily activities. After I greet all three of them, Lise and Signe tell me that I have missed one of the best weeks at Andersen Sperm Bank. A number of young soldiers had been here the week before, Lise says and adds while smiling: "It was wonderful to have all those young guys in here." "Why were there so many soldiers?" I inquire. Signe explains, that, apparently, the Danish Army had developed new underwear for combat situations, but this underwear also exposed men's testicles to more heat than is deemed healthy for sperm

development: "They wanted to know whether there were side effects and how they would go about leaving a deposit before taking off to combat," Signe explains. I was aware that some soldiers deposit their semen before being deployed, but that concerns about underwear would drive groups of soldiers into a sperm bank to seek advice was surprising. I mention that I had thought that the Danish Army would have specialists for these instances. Lise and Signe agree. "But too bad you missed all those well-trained guys," Lise jokes.

Lise and Signe joke a lot as part of their work. Humor is one way in which they frame their daily encounters with sperm donors and their work with semen samples at the lab. Yet, the fieldnote is remarkable because of the subtle ways sexuality and gender are invoked through the use of humor when Lise and Signe describe their encounters with men who donate their semen. Without directly referring to either sexual desire or practices, or concrete interactions with sperm donors, they were nevertheless able to create a meaningful context that invoked sexuality and masculinity. They also created a situation in which my own possible desires as an anthropologist were addressed. Lise's comment, that it is too bad that I had missed all the well-trained guys, marks us both as individuals who desire these kinds of encounters and bodies, while simultaneously also making the young soldiers—representing a particular kind of masculinity—into desirable men with desirable bodies. Thus, Lise and Signe remade the sperm bank as a 'sexualized' and gendered space. But there is also a more obvious observation to draw from the fieldnote: young men are concerned about their fertility and make use of possibilities to conserve their potential to procreate. Their bodies are exposed to certain technological innovations and interventions that directly influence whether they will be able to father children or not. Biomedical knowledge about male bodies and semen, its assessments, and technological intervention in male bodies are thus changing the ways that men think about themselves and their bodies. Attending to instances at Danish sperm banks in which gender and sexuality were invoked, as well as interviews with sperm donors about their experiences of donating semen, I want to explore how and when gender and sexuality are made meaningful in the context of sperm donation in Denmark, and how biomedicine and biotechnology make their ways into Danish men's gendered self-conceptions. I draw on fieldwork at three sperm banks—Andersen, Jensen, and Miller—and interviews with 26 sperm donors. Whereas the Andersen and Jensen sperm banks are located in Denmark, Miller Sperm Bank is a subsidiary location of Jensen Sperm Bank in the USA.

* * * *

Alfred and I had agreed to meet in my office for the formal interview. He had contacted me via my research project's homepage. Jensen Sperm Bank, where Alfred was a donor, had sent out an email to all its current sperm donors explaining my research project. The email also contained a link to the research project's homepage for those interested in participating. In his mid-20s, Alfred spent much of his professional and spare time on activities connected to physical exercise. His body was built accordingly. When I met him, he had been donating semen for about a year and would usually visit Jensen Sperm Bank twice a week. Alfred had no children of his own, self-identified as heterosexual, and was not in a relationship at the time. He had participated in my research project out of curiosity, he explained, but also because he thought that he had an obligation to critically reflect on what he had committed himself to. It became clear early in the interview that having to masturbate at the sperm bank was of central concern for him. We were talking about the organizational and legal frameworks of sperm donation in Denmark when I asked him if he had been aware of how sperm donation was organized before becoming a donor:

*Alfred:* As I said, I was curious and my roommate at the time was a sperm donor before I started. I was curious to hear how that was, and I was maybe also a little bit outraged about, maybe not outraged, but also curious about, do you just stand next to each other, do you just close a curtain, do you go into a room, because I thought that that would be a very strange situation. I mean, this is not, usually you think of your sperm, that is something that you only have with yourself or with your girlfriend, something intimate, and just leaving it at some random place, that is, well that was a strange thought somehow. If you don't think that is strange, then you're probably missing some kind of empathy, if you're really completely indifferent about that. So, yes, of course one asks about how that works. And my mind was put at ease that once you're done, then you don't have to hand it over to someone that is just waiting for it, but you just set it [on the counter] and leave.

Anticipating a transgression of personal space, Alfred was concerned about encountering other men masturbating, and he felt ashamed about having to approach people working at the sperm bank with his semen sample in his hand. His feeling of shame and embarrassment stems from having to negotiate his sexual desires in the semi-public environment of a sperm bank, for him a *transgressive act* (Donnan and Magowan 2009) which marked the boundaries of his concepts of (appropriate) male heterosexual conduct. Donating semen in a sperm bank breaches normative frameworks that enforce compliant behavior in relation to gender and sexuality (Mohr 2010; 2014a).

* * * *

These kinds of transgressive acts are also at play when semen samples are handled at the sperm bank laboratory. As already noted, sperm bank staff use humor to handle these kinds of transgressions, and thus playfully negotiate the ambiguous context of handling donor semen. Lise and Signe, the two lab technicians at Andersen Sperm Bank, both had experiences working with bodily fluids and substances (animal semen, human feces, urine) before joining Andersen Sperm Bank. This kind of educational and experiential background was very common among the lab technicians I met during my fieldwork. However, working with human semen provided for a context marked by potential transgressions of normative claims to gender and sexuality, something that is not at stake in other contexts and that relates to semen's symbolic meaning as a sexual substance in Danish culture.

Whereas semen at the sperm bank laboratory is made into a procreative compound, a substance that has been called *techno-semen* (Moore and Schmidt 1999) in order to enter a global reproductive supply chain, semen is usually encountered as part of sexual practices and therewith embedded in symbolisms of lust, desire, and disgust, a point that also becomes clear in Alfred's quote above. Handling semen's ambiguity thus becomes the focal point of work in the sperm bank laboratory, something that I call *containment*: material-semiotic practices which involve semen, its various containers (specimen cups, test tubes, vials, and straws), as well as computers, centrifuges, and cryopreservation tanks in order to make donor semen into a governable reproductive substance by managing its lust and disgust potential (Mohr 2014a). The containment of semen ensures that work at the laboratory is successful. In situations in which the containment of semen is not successful, on the other hand, semen is re-embedded into its symbolic meaning as a lustful and/or disgusting substance.

During participant observation, I paid special attention to instances in which lab technicians were talking about getting semen on their hands and equipment or about encountering blood or pubic hair in semen samples. These encounters provoked disgust since they reminded technicians that semen samples actually came from real men, something that they normally did not think about when processing semen samples. For Julia, a lab technician at Miller Sperm Bank, a subsidiary location of Jensen Sperm Bank in the USA, it was the smell of semen that was nauseating:

> It is Thursday morning in early December, and Julia and I are in the lab. It is not busy today, so Julia has time to talk about working in the lab. I ask her if there are special challenges when working with semen: "The smell is just horrible. I can't get used to it," says Julia. Having noticed a particular smell at sperm bank labs myself, I want to know how she would describe that smell: "It is a musky smell, almost like dark mustard but just a little bit sharper," Julia replies. When opening specimen cups, she would deliberately hold samples away from her body in order to avoid encountering the smell: "But I also don't like it when the sample is clumpy," Julia adds. She then tells me that there once had been a donor whose samples would always be clotted and she had found it hard to work with them. "Another thing that is gross is when you have a hard time getting a sample into the pipette, and then it just drips on your hands and stuff, disgusting." Julia then reflects on how her work at the laboratory is perceived by outsiders: "For a really long time, my boyfriend would not say anything about my work to his colleagues. And when he did, they made fun of him as if this were some kind of sex work or something." I am astonished by this story. Julia then says while smiling: "But I look at this from the bright side. I keep saying to him [her boyfriend] that I am his dirty little secret," and we both break out in laughter.

Work at sperm bank laboratories is centered around encounters with semen. Aimed at containing semen as a reproductive substance, these encounters involve technological equipment and biomedical substances like dilution and freezing solutions that help to remake semen into techno-semen, an exchangeable substance used for third-party reproduction. Nevertheless, despite all this containment work, semen's symbolic meaning as a lustful and/or disgusting substance breaks its way into the context of lab work at all times. As Julia's experience shows, semen's physical characteristics such as smell and consistency can provoke disgust and serve as a reminder of semen's symbolic meaning as a sexual substance in an otherwise hyper-technologized environment. Julia's story points to how gender and sexuality operate in the context of work with semen samples. Semen is perceived as the result of gendered sexual practices. It is made meaningful as a substance emitted by male bodies due to sexual excitement, and working with donor semen is, as a result, understood as a transgressive act that breaches established scripts for gender and sexuality, "some kind of sex work," as Julia put it. Staff at sperm bank labs are aware of how gender and sexuality converge in semen, and they thus handle this particularity of their work accordingly through humor—in Julia's case framing her own work as "a dirty little secret."

★ ★ ★ ★

This convergence of gender and sexuality in semen is a cultural phenomenon particular to contemporary Danish and Western European contexts. As Gilbert Herdt (1987) demonstrated in the case of the Sambia in Papua New Guinea, the repeated insemination of boys with semen from older men is understood as securing the successful transition from boy to manhood and not necessarily as a homosexual or even pedophile act, as would be the case in Denmark or most other contexts. Yet as will become clear, beyond this particular convergence of gender and sexuality in semen, the context of sperm donation in Denmark also makes for a coming together of biomedical practices and men's gendered self-conceptions (Mohr 2014b). As men were narrating their experiences of being a sperm donor to me, it struck me how their understandings of masculinity were tied to the biomedical practices involved in sperm donation.

I met William, in his mid-20s and a sperm donor for about a year, at his apartment for the formal interview. Like Alfred, William had replied to the email sent to donors by Jensen Sperm Bank. Training to become an engineer, he had heard about being a sperm donor from some of his classmates. William was proud of his semen quality. For him it was a sign that he was a healthy man with valuable genetic traits, even though the assessment of semen quality says nothing about genetic health. When I asked him in which way he felt proud about having good semen quality, he said:

*William:* Well, just that feeling, you know, that you are healthy and well, just as when you are at the doctor's and he says to you: you are a healthy young man. Then you feel like: ahhh, this is beautiful, wouldn't you agree, to know that you don't have any illnesses. The worst thing is to be sick, and it was just wonderful to know that one has good genes.

For William, having good semen quality made him into a "healthy young man," whereby he understood good semen quality to be the result of his healthy male body. Being a sperm donor enabled William to see and experience himself as that young healthy man that he refers to in his quote, and this specific self-experience is only possible because of the biomedical assessments of semen which are part of sperm donation in Denmark. But besides connecting knowledge about his semen to his embodied gendered self-conception, being a sperm donor also had implications for William's experience of sexuality and especially for his sexual relations with his girlfriend:

*William:* I want to stop at some point [with donating semen] because the biggest price you pay is, and as a guy I can tell you this, you are really restricted by the fact that you have to go down there and have to ejaculate [*at have udløsning*] and that they are supposed to go 48 hours before. I don't live together with my girlfriend, but that doesn't mean that you don't miss this freedom, you know, you can't fall asleep at night or whatever, and then this becomes the biggest sacrifice. Sometimes, I also think about this when I am together with my girlfriend. It is just really hard to avoid those kinds of thoughts, you know, thinking about the money and then wondering if it is worth it. But as soon as I get these thoughts, I try to suppress them. Sometimes though I can't help it; you can't always stop yourself from keeping an eye on this. But in a strange way, somehow I think all of this has also been

> good for our sex life. I mean, we have been together for almost 4 years and somehow, just as if, when I have to wait a little longer, in that way that has been good for us. But it is nevertheless a small sacrifice. Sometimes, I have money on my mind, you're thinking: ups, there you go, 300 crowns.

Abstaining from sexual activity in order to be a sperm donor is a sacrifice William is willing to make. Yet in his narrative, thinking of this as a sacrifice seems to be a result of valuing his semen through the amount of money that he receives for his semen samples. The organizational logic of sperm donation in Denmark—orgasmic control, monetary compensation—thus impacts how he experiences his sexuality and how he thinks about his intimate loving relationships. Donating semen changes the ways that men experience themselves. It makes for enactments and embodiments of masculinity particular to the context of contemporary reproductive biomedicine in which the control of masturbation and the assessment of semen regulate what it means to be a man.

### *References*

Donnan, H., and F. Magowan 2009 Sexual transgression, social order and the self. In *Transgressive Sex: Subversion and Control in Erotic Encounters*. H. Donnan and F. Magowan, eds. Pp. 1–24. New York: Berghahn Books.

Herdt, G. 1987 *The Sambia. Ritual and Gender in New Guinea*. New York: CBS College Publishing.

Mohr, Sebastian 2010 What does one wear to a sperm bank? Negotiations of Sexuality in Sperm Donation. *Kuckuck.notizen zur alltagskultur* 25(2):36–42.

——— 2014a Containing sperm—managing legitimacy. Lust, disgust, and hybridity at Danish sperm banks. *Journal of Contemporary Ethnography*. November 21. DOI: 10.1177/0891241614558517.

——— 2014b Beyond motivation: On what it means to be a sperm donor in Denmark. *Anthropology & Medicine* 21(2):162–173.

Moore, L.J., and M.A. Schmidt 1999 On the construction of male differences: Marketing variations in technosemen. *Men and Masculinities* 1(4):331–351.

---

The technologies of sex, sexuality, and gender include, as we have demonstrated, surgeries and chemicals, practices, rules and regulations, and the body itself. Access to these technologies varies by time and place. Some people, globally, by happenstance and access to resources, have relative freedom in defining who they are, what they can do, and what they look like. Other people live in constrained environments as a result of access to resources, personal income, and prevailing social values. International and national legislation, conventions, and programs further frame the control people have over their bodies, and how their bodies influence their positioning in society. These issues—the choices that people make; the divisions in society of 'men,' 'women,' and other; and the role of sex in forging interpersonal relations and maintaining kinship systems—and in local economies—are key issues for medical anthropologists, and an enduring field of study.

## References

Aboim, Sofia 2009 Men between worlds: Changing masculinites in urban Maputo. *Men and Masculinities* 12(2):201–224.

Bledsoe, Caroline H., Allan G. Hill, Umberto d'Alessandro, and Patricia Langerock 1994 Constructing natural fertility: The use of western contraceptive technologies in rural Gambia. *Population and Development Review* 20:81–113.

Bledsoe, Caroline, Susana Lerner, and Jane Guyer, eds. 2000 *Fertility and the Male Life Cycle in the Era of Fertility Decline*. Clarendon, UK: Oxford University Press.

Both, Rosalijn 2015 A matter of sexual confidence: Young men's non-prescription use of Viagra in Addis Ababa, Ethiopia. *Culture, Health & Sexuality*. DOI: 10.1080/13691058.2015.1101489.

Both, Rosalijn, and Robert Pool 2016 A secret practice: Young men and sildenafil citrate (Viagra) in Addis Ababa, Ethiopia. Unpublished manuscript.

Butler, Judith 1990 *Gender Trouble and the Subversion of Identity*. New York, NY: Routledge.

Cole, Jennifer 2005 *Sex and Salvation: Imagining the Future in Madagascar*. Chicago, IL: The University of Chicago Press.

Cornwall, Andrea A. 2003 To be a man is more than a day's work: Shifting ideals of masculinity in Ado-Odo, Southwestern Nigeria. In *Men and Masculinities in Modern Africa*. Lisa A. Lindsay and Stephan F. Miescher, eds. Pp. 231–248. Portsmouth, NH: Heineman.

Douglas, Mary 1966 *Purity and Danger: An Analysis of Concepts of Pollution and Taboo*. London: Routledge.

Foucault, Michel 1976 *The History of Sexuality*, Vol.1. London: Penguin Books.

Gilman, Sander L. 1985 Black bodies, white bodies: Toward an iconography of female sexuality in late nineteenth-century art, medicine, and literature. In *Race, Writing and Difference*. Henry Gates, ed. Pp. 223–261. Chicago, IL: University of Chicago Press.

Groes-Green, Christian 2011 Hegemonic and subordinated masculinites: Class, violence and sexual performances among young Mozambican men. *Nordic Journal of African Studies* 18(4):286–304.

Hardon, Anita 1997 Women's views and experiences of hormonal contraceptives: What we know and what we need to find out. In *Beyond Acceptability: Users' Perspectives on Contraception*. Marge Berer, T.K. Sundari Ravindran, and Jane Cottingham, eds. Pp. 68–78. Geneva, Switzerland: Reproductive Health Matters for the World Health Organization.

——— 2006 Contraceptive innovation: Reinventing the script. *Social Science & Medicine* 62(3):614–627.

——— 2010 From subaltern alignment to constructive mediation: Modes of feminist engagement in the design of reproductive technologies. In *Feminist Technology*. Linda Layne, Sharra Vostral, and Kate Boyer, eds. Pp. 154–178. Urbana, IL: University of Illinois Press.

Hilber, Adriane Martin, Terence H. Hull, Eleanor Preston-Whyte, Brigitte Bagnol, Jenni Smit, Chintana Wacharasin, Ninuk Widyantoro, for the WHO GSVP Study Group 2010 A cross cultural study of vaginal practices and sexuality: Implications for sexual health. *Social Science & Medicine* 70:392–400.

ICPD (International Conference for Population and Development) 1994 *Programme of Action*. New York, NY: United Nations.

Irvine, J. 1990 *Disorders of Desire: Sex and Gender in Modern American Sexology*. Philadelphia, PA: Temple University Press.

Jackson, Peter A. 1997 Kathoey <>Gay <> Man: The historical emergence of gay male identity in Thailand. In *Sites of Desire/Economies of Pleasure: Sexualities in Asia and the Pacific*. Lenore Manderson and Margaret Jolly, eds. Pp. 166–190. Chicago, IL: University of Chicago Press.

——— 2011 *Queer Bangkok: 21st Century Markets, Media, and Rights*. Hong Kong: Hong Kong University Press.

Kapsalis, Terri 1997 *Public Privates: Performing Gynecology from Both Ends of the Speculum*. Durham, NC: Duke University Press.

Karkazis, Katrina 2008 *Fixing Sex: Intersex, Medical Authority, and Lived Experience*. Durham, NC: Duke University Press.

Krafft-Ebbing, Richard von 2011 [1886] *Psychopathia Sexualis: The Classic Study of Deviant Sex*. New York: Arcade Publishing.

Kulick, Don 1998 *Travesti: Sex, Gender, and Culture among Brazilian Transgendered Prostitutes*. Chicago, IL: The University of Chicago Press.

Maines, Rachel P. 1999 *The Technology of the Orgasm: "Hysteria," the Vibrator and Women's Sexual Satisfaction*. Baltimore, MD: The Johns Hopkins Unversity Press.

Manderson, Lenore, Lee Chang Tye, and Kirubi Rajanayagam 1997 Condom use in heterosexual sex: A review of research, 1985–1994. In *The Impact of AIDS: Psychological and Social Aspects of HIV Infection*. Jose Catalan, Lorraine Sherr, and Barbara Hedge, eds. Pp. 1–26. Chur, Switzerland: Harwood Academic Press.

Obbo, Christine 1995 Gender, age and class: Discourses on HIV transmission and control in Uganda. In *Culture and Sexual Risk: Anthropological Perspectives on AIDS*. H. ten Brummelhuis and G. Herdt, eds. Pp. 78–95. Amsterdam, The Netherlands: Gordan and Breach.

Race, Kaine 2009 *Pleasure Consuming Medicine: The Queer Politics of Drugs*. Durham, NC: Duke University Press.

Richey, Lisa Ann 2008 *Population Politics and Development: From the Policies to the Clinics*. New York: Palgrave Macmillan.

Roberts, Celia 2011 Medicine and the making of a sexual body. In *Introducing the New Sexuality Studies.* Steven Seidman, Nancy Fischer, and Chet Meeks, eds. Pp. 65–74. New York: Routledge.
Rubin, Gayle 1975 The traffic in women: Notes on the "political economy" of sex. In *Toward an Anthropology of Women.* Rayna R. Reiter, ed. Pp. 157–210. New York: Monthly Review Press.
Silberschmidt, Margrethe 2004 Men, male sexuality and HIV/AIDS: Reflections from studies in rural and urban East Africa. *Transformation: Critical Perspectives on Southern Africa* 53:42–58.
Simpson, Anthony 2009 *Boys to Men in the Shadows of AIDS: Masculinites and HIV Risk in Zambia.* London: Palgrave Macmillan.
Spronk, Rachel 2009 Sex, sexuality and negotiating Africanness in Nairobi. *Africa* 79(4):500–519.
van der Sijpt, Erica 2013 Hiding or hospitalizing: On dilemmas of pregnancy management in East Cameroon. *Anthropology and Medicine* 20(3):288–298.
Willis, Deborah ed. 2010 *Black Venus 2010: They Called Her 'Hottentot.'* Philadelphia, PA: Temple University Press.
Wizemann, T.M., and M.L. Pardue, eds. 2001 *Exploring the Biological Contributions to Human Health: Does Sex Matter?* Washington, DC: National Academy Press.

Where Is the Money for HIV? 2009. Addis Ababa, Ethiopia.

## *About the photograph*

*Global funding for HIV treatment in Africa began to stagnate in 2008. Although increasing numbers of people entered HIV treatment programs across the continent, international donors began to decrease or cut spending, claiming that African states should increase their financial contributions to national programs and find ways to become more efficient. The bottom line was that African countries were going to have to treat more people with less money.*

*The woman in the photograph holds a sign and wears a t-shirt asking African governments, 'Where's the money for HIV?" The protesters were mobilized primarily by the Global Network for People living with HIV (GNP+) to disrupt a morning plenary session featuring spokespeople from African ministries of health and international organizations. The 'disruption' turned out to be quite playful, with many protesters taking selfies with politicians. There was little dissent regarding the idea that African governments could do more.*

—Eileen Moyer

# 4
# The Socialities of HIV

*Anita Hardon, Lenore Manderson and Elizabeth Cartwright*

Viruses cause all kinds of illnesses, some potentially fatal or with severe side effects, others less virulent—seasonal flus, measles, polio, herpes, hepatitis, shingles, HIV/AIDS, Ebola. Their infectiousness makes them a particular concern to communities, public health programs, and policy makers. Being infected by a deadly virus affects social relations between people, sometimes leading to extreme stigma or silence, other times giving rise to health activism and global health campaigns. Viruses are hard to treat, many antiviral medications are expensive, and these medications often suppress but do not cure infections. Given this, health programs tend to emphasize prevention, an approach that requires people to engage with health programs when they are not sick. Vaccination is a preferred option; inoculation against smallpox beginning in the early nineteenth century, for example, led to the worldwide eradication of this disease. This public health success continues to inspire global immunization programs for children and adults, but shifting attitudes and variable adherence to childhood immunization have made it less easy to approach the goal of eradication for other infections. And for some deadly diseases, including HIV and Ebola, we do not have yet effective vaccines.

In this chapter, through the lens of HIV, we explore how people deal with infectious agents for which there is no cure and, in some cases, no sure treatment. We consider how people engage in prevention efforts that require lifestyle changes, and participate in treatments that suppress but do not cure their condition. We use HIV as a case that helps us understand how people confront blame and stigma associated with infection, and how they confront the fear of contagion.

As the global landscape of HIV changed from near-certain death to a chronic condition, so too did the ways that medical anthropologists approached its study. Anthropological research was critical to understanding how the disease was transmitted through needle use, both in clinics and on the streets, and through sex between known partners and strangers. It was critical to understanding the stigma derived from associating particular marginalized populations, including men who have sex with men, sex workers, injection drug users, people with hemophilia, poor immigrants, and other poor people. These are intersecting and overlapping categories in many cases. Medical anthropological research was important in providing a deep understanding of the risks of HIV infection and illness, and in helping to develop strategies for prevention, early diagnosis and care, and, more recently, timely treatment. HIV expanded the subjects that medical anthropologists studied, and an appreciation of this work, because of the necessity of understanding the conditions in which people were infected. In particular, HIV drew attention to social structure and disadvantage, causing us to rethink our language, as reflected in the evolution in terminology from 'prostitution' to 'sex work' to 'survival sex'; HIV community advocates and researchers emphasized the need to engage with affected communities, policy makers, and health services. Along with this expectation and the urgency of applied research, this focus on sociality led us to interrogate the notion of risk, the parameters of vulnerability, and the meaning of agency. These insights from HIV have informed medical anthropological research in all domains.

In the twenty-first century, the focus in relation to HIV is increasingly on people living with HIV as a chronic disease, and so we have turned to explore the new social forms, including expert patients, that shape contemporary AIDS care. Even so, despite some three decades of education, advocacy, awareness raising, and experience, stigma and discrimination continue to affect people who are HIV-positive because of their status, their assumed sexuality and sexual identities, and other marginalities. The continued challenges around HIV, the difficulties of intervening in intimate life, social exclusion, and difficulties related to the affordable and regular supply of medications, continue to make HIV an important field of anthropological research.

Recently, a new phase in the global fight against AIDS has commenced. The new aim is to eliminate the disease by expanding HIV testing and starting people on treatment earlier (including through pre-exposure prophylactic treatment). This drive to eradicate HIV is being promoted by global health planners, who have calculated that early treatment, which results in suppressed viral loads, can be used as an effective prevention tool. The next generation of AIDS ethnographers is conducting studies in sites where elimination is the goal, and they are finding that there is significant resistance both to expanded testing and to beginning treatment when one is not yet ill. Such resistance appears to undermine the overall effort, but the basis for this resistance merits reflection and further enquiry (on resistance, see Chapter 6).

## Beyond Biosocialities

HIV and its lethal consequences as it progressed to AIDS engaged and inspired a generation of social scientists. Until antiretroviral treatments (ART) were introduced, the only interventions that were possible required detailed and nuanced knowledge of what people did, in what contexts, and why. The history of HIV—emerging first in gay communities in the United States and Europe, and then tied to countries and populations who were marginalized and racialized—has framed how various communities have responded to the infection. Because it is, in part, sexually transmitted, HIV is associated with promiscuity and irresponsible behavior. Where its transmission occurs through intravenous drug use, the coincidence of the two reinforces the idea of HIV as an infection that affects people who are already marginalized, while the association with drug use marginalizes those who are infected. Practices such as the safe handling of blood products, increased care in standard clinical procedures, and interventions to reduce high-risk needle use—through the use of disposable needles in clinics, for instance, as well as through the introduction of safe injecting practices to people using injectable drugs for recreational purposes—have had an impact on the transmission of HIV through this pathway. On the other hand, preventing HIV transmission through changing sexual practices is a complex task. Awareness of the challenges of behavioral changes has sharpened our understanding of the social determinants of infection and the social life of disease. We now readily talk about the 'global health assemblage' of research, trials, action, and patient support programs, much of which was founded on the massive amount of HIV-related research that entered the global scene in the 1990s.

The activism that emerged around HIV seems a perfect example of what Paul Rabinow (1996) referred to as 'biosocialities,' capturing how biological understandings and diagnostic practices shape the formation of new groups and identities. Rabinow's research highlights how diagnosis and treatment are criteria for inclusion in patient support groups and advocacy groups, in the way that biology becomes a basis for the formulation of people's rights and claims (Nguyen 2006). Rabinow focused on people who shared an identity based on genetic testing for particular conditions such as neurofibromatosis. However, such commonality cannot be taken for granted. In resource-poor settings, as Susan Reynolds Whyte (2009) has argued, if clinics are too far away or wait times are too long, people may not even engage with biomedicine, and their identities may not be defined in biomedical terms regardless of their access to clinical care. Further, even if people do identify as HIV-positive, they do not relate only to fellow HIV-positive individuals. They work, have families, and are neighbors; other kinds of sociality are implicated.

In the past decades, patient activism has been ascendant around the world and specific instances of it have attracted much scholarly attention (Brown and Zavestoski 2004; Landzelius and Dumit 2006). This interest comes at a time of transition for health care, from a classic public health model in which governments are expected to oversee public wellbeing, to a neoliberal model in which patients are expected to be informed, self-mobilizing, biological citizens (Petersen and Lupton 1996). Below, we reflect on the dynamics of activism and social support surrounding AIDS in the course of the pandemic. We consider how different strands of the AIDS movement represent and mobilize the voices of people living with HIV, including in relation to the medical establishment and the pharmaceutical industry. In so doing, we highlight how anthropologists have engaged with AIDS activists, and how they have contributed to a reflexive understanding of HIV socialities. What stands out in early AIDS activism is the close collaboration between people living with the virus and those pharmaceutical companies and clinical researchers involved in designing new technologies. They were jointly engaged in what Rose and Novas (2005: 5) called "the political economy of hope," a new mode of governance in which life becomes wedded to the generation of wealth. In the context of HIV, this political economy involved alliances between medical practitioners, patients hoping for effective treatments, and pharmaceutical and biotechnology firms seeking greater profits and shareholder value.

## Involving People Living with HIV/AIDS

From the late 1980s, international and local meetings of HIV researchers and policy makers became venues for activists to present their concerns to scientists and to assert their right to participate as equal partners in response to the epidemic. While their calls were initially ignored, 42 countries at the 1994 Paris AIDS Summit agreed that the 'Greater Involvement of People Living with HIV and AIDS' was crucial for responses to the virus to be effective and ethical, and thereafter referred to this by its acronym as the GIPA Principle. Five years later, UNAIDS (1999: 4) suggested that the notion of 'greater involvement' could mean many things, from sitting on the boards of international funding agencies to involvement in national-level programs and policy making, to membership in community-based organizations participating in prevention outreach and care. Others have emphasized involvement in the design, governance, and delivery of such programs as "a necessary reminder of the importance of taking active control of our lives and our health" (Morolake et al. 2009).[1] In participating in the design of HIV-prevention programs, and thus co-constructing the treatment for the virus that was affecting their health, people living with HIV and AIDS forced health care providers and policy makers to take a different approach than they otherwise would have.

Transnational activism, alongside the arrival of drugs that could control viral load (highly active antiretroviral therapy, or HAART), brought a dramatic change in the provision of AIDS care globally. As they first came onto the market in the 1990s, AIDS medicines were unaffordable in Africa and for poor people elsewhere. While development donors were willing to invest in prevention programs and palliative care, they were unwilling to invest in antiretroviral drugs

(ARVs). American and European AIDS activists therefore sent limited supplies of ARVs to their counterparts in Africa in another mode of activism—international solidarity. However, as Vinh-Kim Nguyen (2010) has described, only a fraction of people living with HIV in Africa could gain access to ARVs. They did so by positioning themselves as model activists—'living positively,' attending discussion groups, and caring for others through humanitarian networks and AIDS activist organizations. Most people in Africa did not even know ARVs existed.

At the turn of this century, activists in the global North increased their collaborations with their counterparts in countries such as Thailand, Brazil, and South Africa, and formed alliances with health and development organizations such as the Geneva-based Médecins Sans Frontières (MSF, see Chapter 15, this volume) and Partners In Health at Harvard University. In Brazil, activism led to legislation ensuring universal access to AIDS medicines, which João Biehl (2007) has noted saved the government money it would otherwise have spent on care for AIDS patients. The Thai government linked its AIDS program to a universal access insurance scheme. In Uganda, the Ugandan AIDS Service Organization played a key role by providing counseling and social support, and, as AIDS medicines became available, by training 'expert clients' to introduce ARV treatment in that country. Thus, different strands of AIDS treatment activism, government action, and international programs have had a major impact on mobilizing resources for health and in designing care programs based on the principle of patient engagement (Hardon and Dilger 2011).

Below, we present three case studies that provide detailed insight into this new therapeutic landscape. In the first case study, Emmy Kageha Igonya and Eileen Moyer discuss the dynamics of an informal support group of men who identify as gay and have sex with men. The men meet in a corner of a bar in Nairobi, Kenya, to drink beer, support each other morally and economically, and confront the devastating effects of AIDS in their community in an era when, in theory, AIDS treatments are available in local clinics and hospitals. The men, who no longer have relationships with their biological families, refer to each other as *kuchus* (brothers). They are confronted with multiple layers of stigma—being gay and poor, doing sex work, lacking secure housing, and being infected by HIV.

## 4.1 Freedom Corner

*Emmy Kageha Igonya and Eileen Moyer*

In Kenya, public debates about homosexuality and sex work, shaped by public health, human rights, and moral discourses, have becoming increasingly prominent. Religious and political leaders have condemned homosexuality and sex work as 'immoral,' 'corruption,' or 'pollution of the nation.' Homosexuals, sex workers, and their supporters have, on the other hand, begun to make public demands for health and human rights through protests and speaking engagements. Given the current political and legal climate around homosexuality in East Africa, we jumped at the opportunity to include a case study of male sex workers in our ethnographic project on the new forms of social support that have emerged in Nairobi over the last several decades in the wake of the HIV epidemic.

In most instances, the 'gay community' of Nairobi is a few articulate and educated homosexual men. Because homosexuality remains illegal, if somewhat tolerated in Nairobi, these gatekeepers, quite rightly, tested us for several months before introducing us to the underworld of male sex work in the city. Most likely, we could have used our clout as researchers to engage with these same men through one of the public health interventions currently targeting them. Hoping to distance ourselves from such interventions, however, we decided to go through community gatekeepers, building rapport and trust slowly over several months. We did attend and observe public health events and activities focusing on HIV treatment and prevention among 'men who have sex with men,' or MSM, but only in the company of our research participants. Most of our research took place in informal settings, relying heavily on observation, conversations, and 'deep hanging out' in a bar in downtown Nairobi where male sex workers gather in the early evening before hitting the streets.

Here, we offer a glimpse into a space of unexpected social support, where young men living in extremely precarious circumstances become their 'brothers' keepers.' In this space, known as Freedom Corner, men collectively engage in impromptu performances of 'gayness,' assigning one another female nicknames and sashaying about in tight clothes as they drink cheap beer, flat and warm. In between the sometimes forced frivolity, they check up on one another and offer advice on ways to be safe from gay bashing and other forms of gendered violence, from HIV, and from the police. As a social group, these men also support one another when attending NGO-sponsored activities, sometimes taking over those spaces to voice their frustrations with one another's behavior and to seek allies among NGO staff to pressure their peers. At other times, they demonstrably disengage, showing up only to collect 'transport allowances' and offering little to the forced discussions.

We refer to these men as 'men' while using the female names they prefer to use when they are together. This is purposefully done to bring a degree of gendered tension to the description, replicating the gendered tension that characterizes their day-to-day existence. When performing their particular version of 'gayness,' they refer to themselves as *shogas*, a Kiswahili word referring to effeminate homosexual men; when performing for the public health world, usually as people living with or at high risk for HIV, they refer to themselves as MSM; when alone together, they define themselves as male sex workers, appropriating the term *kuchu*. They use this term because it is unfamiliar to most Kiswahili speakers, but widely known among homosexuals; it suggests their sense of shared oppression. In our text, we use all these terms consciously, depending on the context.

## *Freedom Corner*

The name Freedom Corner is taken from a hand-painted sign hanging above the table in the back of the bar, where the men gathered before and in between their engagements with clients. When we first began working with them, they did not even define themselves as a group. Our identifying them as a group and giving them a name served as an act of interpellation, increasing a sense of group identity, facilitating the selection of leaders, and lending language to their growing political consciousness. We were immediately pulled into their political project, and asked to participate actively—making referrals to health care, offering advice, pressuring non-compliant peers, and making financial contributions when money was collected to aid a fellow in need.

The name Freedom Corner derives from the famous historical spot of the same name located in Nairobi's Uhuru Park. The late Nobel Laureate, Professor Wangari Mathaai, named Freedom Corner when she and other activists held a hunger strike there to pressure the government to release political prisoners. The kuchus identified with Mathaai's political activism; for them Freedom Corner symbolized their longing for freedom and safety as citizens of Kenya and their struggle against the triple discrimination they faced as HIV-positive, homosexual, sex workers.

The run-down bar that housed Freedom Corner was located in Nairobi's downtown River Road area, sandwiched between two shops. The owners of the bar tolerated the kuchus, who were accorded the back corner of the establishment. Freedom Corner was spatially marked out by four long benches arranged in square, at the center of which was a low table, usually filled with plastic beer mugs topped with drinking straws. The space could comfortably accommodate 20 people and often overflowed as the evening picked up. Most notable about the space was the overpowering smell emanating from the toilets, less than two meters away. The first doorless toilet was used by men who stood in the corridor when urinating. When visitors expressed concern with the deplorable and unhygienic state of Freedom Corner, they were quickly told that compared to most places sex workers congregate, Freedom Corner was 'high class,' and that the cheap beer on offer was comparatively expensive. The kuchus felt at home at Freedom Corner; there they could relax, tell stories and jokes, and not worry about harassment from police or homophobes.

Participation in the group was purely voluntary. There were 24 core group members from diverse ethnic backgrounds; all except one came from relatively poor families. Nineteen were primary school dropouts, three secondary school dropouts, and two had attained tertiary education. Poorly paying sex work was their main source of income, although eight occasionally also worked as peer educators in various programs targeting MSM sex workers, for which they received a monthly allowance. HIV training allowances supplemented their income. Occasionally, a few engaged in petty business.

Everything at Freedom Corner was impromptu: no group facilitator and no agenda. With a fluid quorum, those who hung around Freedom Corner updated others on what had transpired in their absence. Text messages were shared, in part because the messages concerned everyone, and because the majority did not own phones. The kuchus of Freedom Corner were welcoming to visitors, including male transgenders, kuchus from Mombasa, and, less often, current and ex-boyfriends, lesbians, and female sex workers. Kuchus socialized with one another and visitors while drinking, smoking cigarettes, and chewing khat; they laughed, gossiped, quarreled, cried, and engaged in various discussions collectively or in smaller groups; they networked for group sex, shared security alerts, and organized participation for HIV training workshops and street advocacy; and mobilized support for those who were ill. Discussions covered ailing colleagues, abuse by clients and police, security, sexual practices, and experiences with homophobia, health issues, and nutrition.

## *My Brothers' Keeper*

The majority of kuchus had cut ties with family and kin, often referring to the Freedom Corner group as their family. Most were homeless. On a 'good' day, when they had clients, they might sleep in cheap lodgings (about US$5 per night). On bad days, they would either sleep on the streets or seek temporary shelter from colleagues who had managed to rent a place for the night or week. Some exchanged sex for a place to sleep. They drank and ate cheaply, sponsoring one another when able.

Only two members received occasional support from their mothers. A few who had requested support from family had been denied or told they would only be helped if they first denounced their sexual identity. Kuchus consciously invested in the group, providing social support to colleagues and peers in anticipation of reciprocity. They worked together to care and support sick members of their community. Although many suffered from minor health ailments as a result of poor nutrition and sleeping on the streets, periodically, some required extensive care when they suffered from HIV-related opportunistic infections. In such instances, the sick were taken in by colleagues who would do their best to nurse them back to health—bathing, feeding, ensuring access to medication and treatment, taking them to the clinic, and updating the others who were expected to provide economic support. This monetary support was meant for the sick person, but also those who missed work to care for the sick. Group members also offered emotional support and encouragement to sick members and those providing care. Those who were very sick were hospitalized, in which case group members would visit, bringing gifts of food and paying hospital bills, either out of pocket or by mobilizing support from HIV intervention projects (and willing anthropologists).

Because of the heavy burden of HIV-related care resulting from poor treatment adherence, it was common for group members to pressure those on antiretroviral treatment to take their medication, to eat well, and to refrain from excessive alcohol consumption and smoking. Members also often shared medications. Because most were homeless and rarely slept in the same place from night to night, they were often caught without their daily dose. Not everyone in the group had HIV; several were probably HIV-positive but were not open about their status until they were too sick to hide it. For as much as there was support in the group, there was also stigma. Being known to be HIV-positive affected one's ability to find work, as other, presumably HIV-negative, sex workers were not above stealing clients from those known to be positive, disclosing their status to gain the upper hand.

## *Disappointing Development*

Although the Freedom Corner kuchus were heavily targeted by NGOs and had come to depend on them for medicines and occasional income-earning opportunities through trainings and workshops, the most important source of support for most of them were the friends they had made among their colleagues at Freedom Corner. In fact, most felt discouraged by the NGOs that tried to help them. As the example below demonstrates, the training seminars offered by NGOs rarely addressed the underlying economic, political, and emotional concerns of kuchus. Workshops intended to empower more often contributed to feelings of disillusionment.

In recent years, Nairobi has been the site of various public health interventions aimed at reducing HIV risk among the 'key populations,' including sex workers and MSM. The kuchus of Freedom Corner have high 'biovalue' in this global health context, in which aid organizations and researchers often have trouble recruiting participants for planned intervention and research activities. Members of high-risk communities are usually by definition hard to reach, resulting in a paradoxical scarcity that means aid organizations often share clients, and the same MSM repeatedly serve as key informants for diverse research projects. The kuchus of Freedom Corner were aware of their value in this context and did not hesitate to capitalize on it when possible.

As 'targets' of global health interventions, they are invited to many trainings and workshops, where they are given advice on subjects intended to educate and empower them, with the assumption that such interventions will motivate them to protect themselves, their clients, and their lovers from HIV infection by using condoms and taking ARVs, both made freely available to them. Their lives, however, are characterized by violence, uncertainty, fear, powerlessness, and anxiety. Kuchus are more concerned with day-to-day survival, homophobia and identity issues than they are with negotiating condom use with paying clients. Hence the disconnect between such trainings and their MSM targets, as one of us—Emmy—experienced.

On the morning of 23 August 2011, twenty-five MSM aged between 25 and 45 gather in a small room on the eighth floor of one of downtown Nairobi's three-star hotels for a harm-reduction training on HIV, Alcohol and Substance Abuse, facilitated by the USAID-funded AIDS Population and Health Integrated Assistance (APHIAplus) project. A middle-aged Kenyan woman, I (Emmy) sit in the back row and observe the proceedings. To my left is a young man I know as Maggie; dressed in an oversized jacket he sleeps, slumped over his desk throughout the three-hour session. All is calm as the facilitator reviews the previous session. When he asks the participants to evaluate their 'self-concept,' however, Carol, another young man, begins weeping uncontrollably and the room falls silent. A participant places a hand on Carol to console him, while others look to the facilitator to intervene. After a few uncomfortable moments, the facilitator asks Carol what the problem is. Looking at the ground, he responds: "My life is wasted . . . I did not know I would end up as a sex worker and also infected with HIV (he sobbed) . . . I was a bright boy but dropped out of school for lack of school fees. . . . I feel so useless and wasted."

Carol's outburst transforms the previously boring training into a support group session. His friends perk up and gather around, taking turns to offer support, but also to direct remarks to others in the group. Jemima begins with a statement to build unity by offering a critique of donor-sponsored events like the one they are attending. He follows this with the message that they must take care of one another, but also declaring that each person is responsible for his own health.

> We are our brothers' keeper. We are shogas . . . kuchus and sex workers. People do not want us . . . we have to rely on each other. Most of us are not wanted by our family members because we are gay. It is difficult to explain our work to our family members . . . they will despise us. I had been living with my sister but I think she suspected my orientation and has put me out of her house.
>
> We are also fed up with gay groups who keep using us to get donor funding . . . most of us did not go far in school. I dropped out of school in class two. We have to care for each other. I know everyone here knows Jeremiah. He refused to take ARVs, became very sick and was admitted at Kenyatta National Hospital. He died on Saturday and he did not have his people . . . we have to make arrangements to bury him . . . if you know you have been told to take ARVs, you better start . . . we do not want to bury any more of our colleagues.

Georgina follows Jemima, mentioning friends they have lost who refused to take treatment despite group support and ends with the declaration: "We do not want to lose others when we have medicines available!" Marian then turns to Kate, whom they accuse of throwing away his medicines: "Our brother here started taking ARVs but he has stopped taking them . . . now it has been a year." Kate tries to defend himself, prompting responses and advice from the others.

*Kate:* I was on ARVs but I kept missing to take them for days because of being drunk and working odd hours. Also the drugs were too many. I thought the drugs were not helping me. I spoke to the counselor and I was given Septrin and a multi-vitamin and advised to take ARVs when ready.

*Jane:* See how he is thinking about it. We have lost Chidi because he refused to take drugs. Go to Liverpool VCT, talk to the counselor and you will be given ARVs. You can also go to SWOP (Sex Workers Organization Programme).

*Latifa:* When you work late you can bring an extra dose with you for emergency.

*Marian:* I do not want to be called to take someone to hospital or contribute toward funeral expenses when one has refused to take drugs. We have free ARVs and we should take

> them . . . if you do not have food you can ask us to support you. Some come to Freedom Corner and ask for alcohol when they have not eaten yet and are on drugs . . . then they say they did not take drugs because they do not have food. It is better you ask for a plate of food first and later alcohol. If you come to me and you have not eaten I will not buy alcohol for you.

Sofia is then asked about another colleague who is sick and homeless. Sofia reports that he had washed 'his patient,' and given him his ARVs, and that he was sleeping when he left. The group quickly appoints three representatives to visit the patient and report back to the group, but Sofia objects, fearing neighbors will ostracize him if other MSM are seen visiting him. The group members decide to collect money to support Sofia's efforts instead.

Once this is settled, Georgina, Sofia, and Marion take turns trying to convince the meeting facilitator to give them time to 'share experiences,' claiming that they are 'wounded' and need to 'be open' so that they can know their lives are not over. The facilitator promises to reserve time for 'experience sharing' the following morning, and returns to the alcohol and substance abuse training agenda that had been interrupted by Carol's outburst. Those who had become animated during the discussions about Carol, Kate, and Sofia's patient withdraw again. Maggie continues sleeping next to me.

The following day, like other participants, I was eagerly waiting for experience sharing. When the training facilitator arrives, however, she highlights the main issues covered during the training session and then distributes evaluation forms. This does not go down well with the participants. One interrupts, telling the facilitator he is very disappointed that she overlooked their request. "We are very disappointed because most of us are wounded, we go through a lot, and we need to talk about it." The facilitator, seemingly eager to conclude the training, responds that no one had responded when she had asked if there were remaining issues to be discussed. Soon the APHIAplus coordinator arrives and pays allowances to the participants before concluding the training. I leave together with the young men who invite me to join them at Freedom Corner, where they 'share' their most recent experience of feeling used by development groups.

### *Extraordinary Support*

Two months after the APHIAplus training, Norah arrived at Freedom Corner carrying a handbag. She sat down and took out a book with a list of members who had made contributions to help Maggie. As she went through the list, she reminded those who had not yet contributed. When she finished I asked her about Maggie, the same young man who had been sleeping next to me during the training. Norah told me that they had forced Maggie to go to the hospital. When he was discharged he chose to return to his mother in Mombasa to convalesce. The Freedom Corner group members had paid his hospital bill and continued to collect food money to send to Maggie's mother while she cared for him. Every week, each member contributed 20 Kenya Shillings (less than five cents) or more, which was sent by M-PESA (a mobile banking system) every month. This continued until Maggie had regained his health and returned to Nairobi. Fifteen months later, a healthy-looking Maggie served as the Master of Ceremonies at the 2012 Mr. Red Ribbon Gala, a camp beauty pageant for HIV-positive men.

During the period we worked at Freedom Corner, we witnessed the revival of three other group members, in addition to Maggie and Sofia's patient. Similar care was also extended to other kuchus who were not Freedom Corner regulars, when their needs came to the attention of the group. In contrast, support systems across the city have collapsed, as once well-funded community-based initiatives have gone dormant in the wake of a global economic crisis that has led to a widespread divestment in HIV care. Whereas funding for antiretroviral treatment, condom distribution, and formal support mechanisms are readily available to MSM sex workers, existing interventions have failed so far to recognize the everyday life challenges faced by these multiply marginalized men, resulting in their being alienated and excluded from the very programs that target them.

At Freedom Corner, male sex workers gather every evening as they prepare for the night in town, offering one another camaraderie, advice, and a degree of security in the midst of a hostile working world. They also encourage one another to adhere to their antiretroviral treatment and mobilize social support. In this case study, we have attempted to provide insight into the challenges these young men face and ways in which public health interventions unintentionally exclude those who are most at risk.

The men who join these groups seek support in an adverse environment. They do not see themselves as shaping the response to the epidemic—they just hope that collectively they can live a better life and perhaps gain access to some of the AIDS resources that circulate in Nairobi. Although they are not in the center of power of the Kenyan AIDS movement, and although they do not see other parts of Nairobi activism as relevant, the idea of forming support groups has reached them.

Igonya and Moyer echo the analysis of Beckman and Bujra (2010), who observed that in Tanzania, donor funds for programs based on the GIPA Principle mainly reached better-educated and middle-class people for whom AIDS work provided an attractive source of income. Such studies of HIV socialities show how activist dynamics are different in resource-poor settings from those in the global North. In East Africa, GIPA programs are heavily funded by donors, and AIDS organizations mobilize to get a share of the funds and jobs, sometimes at the expense of those living with HIV who have less power.

When empowering technologies such as programs designed according to the GIPA Principle flowed to Africa, they became embedded in national health systems that did not necessarily support the active role of patients. The social practices of peer support and community involvement were empowering in the United States, and were key in trials that provided 'proof of concept' that ARV treatment could be used effectively in Africa. But in remote local settings, they also discipline people's everyday lives. ARV treatment, as is often the case in the treatment of chronic conditions, comes with a moral code about how to behave responsibly as a patient; similar moral codes are inflicted on patients with many diseases—diabetes, obesity, mental illness, and substance dependencies. In the case of HIV, the code demands that before treatment can begin, individuals must disclose their HIV status to 'treatment buddies'—a family member or a friend who are expected to help them adhere to the ARV medication regime (Hardon and Posel 2012; Mattes 2011). They must also disclose their status as HIV-positive to clinic staff, eat well and abstain from alcohol, and forgo traditional treatments. The ARV regime thus subjects people living with HIV to a particular form of biopower. Nurses and other medical professionals decide who is in and who is out of such programs. People who do not adhere to biomedical regimes, and do not or cannot follow social and lifestyle recommendations, risk being excluded from future care.

## HIV, AIDS, and Families

UNAIDS estimated in July 2015 that around 70 percent of adults and children living with HIV (25.8 million), and 70 percent of new cases, were in sub-Saharan Africa (www.unaids.org). The magnitude of the illness and death in eastern and southern Africa, the regions hit hardest by the HIV/AIDS epidemic, has overwhelmed both nuclear and extended family systems. As Whyte (2009) asserts, people diagnosed with HIV are not only clients of clinics, members of support groups, and mediators of AIDS technologies and programs. They are also sons, daughters, husbands, and wives, all living in communities affected by the virus, and they depend on their families and communities to care for them. Echoing the importance of families ties, Abadia-Barrero, who conducted ethnographic research among children and adolescents living with HIV in a Brazilian support house, points out that for them 'biological belonging,' that is their connection with family, is a permanent reference point for the self (2011: 185). In the case study that follows, Ellen Block provides us with an ethnographic window into the impact of the virus at the level of kin-based networks. In her case study, located in Lesotho where according to the 2013 HIV Sentinel Surveillance (Ministry of Health, Lesotho 2015) one out of four adults is infected with HIV, Block shows how Lebo, a child orphaned in infancy after both his parents had died of the infection, survives with the help of AIDS medications. Block emphasizes the role of his grandmothers from both maternal and paternal families, who, while not infected with HIV, are frail themselves, and who, although reluctant to do so because of their poverty, take turns in caring for the child (see also Williams 2003).

## 4.2 Chronicle of a Mosotho Boy

*Ellen Block*

Shortly after his first birthday, Lebo Mabotse was admitted to Mokhotlong Hospital in the rural mountains of Lesotho, a small country nestled completely within South Africa. His mother had died only a few months prior and he was living with his 77-year-old maternal grandmother, 'M'e Masello. He had been sick at the time of his mother's death, and since then, his illness had worsened. For a while, his grandmother thought his appetite might return once he adjusted to the rapid weaning forced by his mother's death, but instead his health deteriorated. With chronic diarrhea, cough, fatigue, and poor appetite, she took him to the hospital. She recalled, "He was very sick. He nearly died."

At first, 'M'e Masello did not want to care for Lebo, who had been living with his father's family at the time of his mother's death. She explained to me how at his mother's funeral she implored the paternal family to take him: "How will I take the child yet I'm sick? And the thieves have taken all my animals. At least here, he will have milk to eat! And I refused to take him." In fact, neither side of the family was anxious to take Lebo, whose young age and illness meant intensive care labor for whoever took the job. But, after his mother's funeral, the paternal family brought Lebo to 'M'e Masello's house and left him there, forcing her hand.

For the two weeks that Lebo was admitted to the hospital, 'M'e Masello dutifully stayed by his side, washing him, holding him, feeding him, and sharing his small cot in the crowded children's ward, while his siblings stayed home alone. Lebo's chest X-ray showed active tuberculosis (TB). After two weeks of intensive TB treatment, Lebo's health had improved marginally, but he was still severely underweight, and it would have been difficult to care for him in a remote village setting, especially for 'M'e Masello who suffered from her own chronic health problems. Fortunately, during his stay at the hospital, Lebo came to the attention of a small locally run NGO that serves AIDS orphans, called Mokhotlong Children's Services (MCS). MCS decided to admit Lebo to its temporary place of safety, or safe home, to oversee the completion of his six-month TB treatment and make sure that he was well enough before returning to his grandmother's house. MCS paid the hospital bill and 'M'e Masello agreed to let him stay there for a while.

Lebo moved to the MCS safe home the day after Christmas. Though 13 months old, he weighed only 12 pounds, the average weight of a five-month-old. During the first six weeks he barely ate, so a nasogastric tube was inserted into his stomach through his nose, allowing liquefied food and medicine to be administered with a syringe. But Lebo's condition did not improve. 'M'e Nthabeleng, the managing director of MCS, was surprised to find that the hospital had not tested him for HIV when he was there. She immediately tested him and, unsurprisingly, his test was positive.

Lesotho has a 23.6 percent adult HIV-prevalence rate, the second-highest globally (UNAIDS 2012). High adult mortality leads to both the transmission of HIV from mother to child and high rates of orphaning. There are around 120,000 orphans in Lesotho, and 12,000 of them, like Lebo, are HIV-positive (UNICEF 2010). Once diagnosed, Lebo was put on antiretroviral treatment (ART), a three-drug cocktail that can have unpleasant side effects, but that has been instrumental in turning the tide of the African AIDS epidemic. When I met Lebo in 2007, during my first fieldwork trip to Lesotho, I would never have known he had once been frail and malnourished, underweight and listless. He was the boss of the safe home, a solid two-year-old whose enthusiastic and physical style of play bowled over many of the younger children living there.

After almost a year in the MCS safe home, Lebo returned home to live with 'M'e Masello. He was two years old. Like many orphans in Lesotho and across southern Africa Lebo depended on his grandmother who cared for him and four older siblings as well as another orphaned grandchild. Grandmothers like 'M'e Masello have come to care for so many orphans that they have helped cushion the blow of the AIDS pandemic, enabling children to remain with their extended families. Lebo became quickly attached to his grandmother, who said he didn't like any of his siblings to help him or feed him; he only wanted his grandmother. Laughing, she told me: "This one I'm living with, he doesn't want his brothers and sisters to touch him. He only says, '*nkhono, nkhono*' [grandmother, grandmother]." I often found them sitting out together in the sun under a blanket, eating some biscuits or sharing a piece of fruit.

In February 2008, a few months after returning home from my first trip to Lesotho, I was chatting with 'M'e Nthabeleng on the phone, when she revealed the news: "Guess what? Lebo is HIV-negative." I was shocked and more than a little skeptical. Apparently, he had gotten a routine follow-up HIV test and it was negative. The doctors were confused because he had had several positive tests, both of his blood and his DNA, which confirmed his status back in early 2007. So they tested and retested him, and could not come up with a positive test. Finally, after consulting with a visiting American doctor, they decided to take him off ART. Shortly afterwards, he stopped receiving services from MCS.

A few months after returning to Lesotho in 2009, I rode my motorcycle out to Lebo's village to visit him and his grandmother. Lebo was now three years old and it had been a year and a half since I had last seen him at the MCS safe home. When I arrived, 'M'e Masello greeted me warmly and welcomed me into her home. She struggled up the two small steps, stopping to breathe heavily.

Once we were seated, her on the bed, me and my research assistant on a small bench, I told her I had heard the good news about Lebo's HIV status. She said, "Yes, if you could see inside me you would see that I am very happy because of the way my heart is moving." When I inquired about Lebo, she said he was doing well but had had a poor appetite and stomach problems for over a month. She said her chronic asthma was also acting up, and indeed, she was wheezing heavily, and had to stop speaking with us for several minutes at a time to regain her breath. She had struggled with asthma since adulthood, and had been hospitalized for it several times. But the doctor's solution—for her to move to the lowlands where it is dry and warm—did not take into account the challenges of leaving the home she had shared with her late husband for half a century and taking her six grandchildren with her. Like many of the recommendations for sick people in Lesotho, this one was out of reach for 'M'e Masello. Instead she was prescribed Tylenol, a multivitamin, and two different types of inhalers, which she kept by her side.

It was a few months before I returned to visit 'M'e Masello. When I entered her house, she was lying on the single bed she shared with Lebo. The room was so dim and crowded it took a moment to realize that Lebo was lying in bed next to her. 'M'e Masello told me that they were both feeling unwell. Since I had last visited, Lebo's stomach problems and poor appetite had worsened and 'M'e Masello's wheezing was persistent. She was too tired for a lengthy visit, but told me that as soon as she was feeling better, she would take him to the doctor.

Two weeks later, on a Friday afternoon, I ran into 'M'e Masello and Lebo at the district hospital. I told them to stop by MCS on their way back to the taxi rank in order to say hello to the staff. When they arrived, 'M'e Nthabeleng was immediately worried. She had seen so many sick children, knew what signs to look for, and decided to test Lebo again just for good measure. The test was positive. 'M'e Nthabeleng gave them money for their taxi and told them to come back for a blood test first thing Monday. Lebo's test showed that his immune response was dangerously low, so he started on ART immediately.

'M'e Masello had mixed feelings about his complicated series of diagnoses. She was of course thrilled when she found out he was negative, but also confused. "Firstly they told me that there was no cure for that disease," she said, "and I was so surprised when they said it has gone away. How has it gone away, yet they say there is no cure for that? . . . I was just asking myself and answering." When he tested positive again, she was angry at first. But shortly after getting back on his medication, 'M'e Masello reported that Lebo was back to his usual self; he regained his appetite and was eager to play. His re-diagnosis also came with access to services, both from MCS and the hospital—support that was much needed by this chronically ill grandmother whose caregiving duties went well beyond her physical and material capacity. When I asked her a few months later if she was still angry, she replied, "No, I'm very happy now. If I was a dog I would be wagging my tail. Yes. I'm very happy, *kanete* [I swear]."

For a while, 'M'e Masello and Lebo's life returned to a sort of normalcy. Then, in the winter of 2009, near the end of my fieldwork, 'M'e Masello was hospitalized several times. The second time, she could not even walk to the road to catch a taxi, so her children carried her down the hill. Her doctor put her on medication for her asthma and for a weak heart. During her extended hospital stays, the children stayed alone in their house. I went to the hospital most days to visit her, and to bring her tea and bread. She lamented that she was feeling overwhelmed by her caregiving burden.

Yet the children were immensely helpful to 'M'e Masello, likely a large part of why she wanted them to stay with her. She told me: "Sometimes I'm sick, not able to do anything, but they will know how to gather the wood, and they cook. I'm coming from the hospital now, that's why I'm looking like this [pointing to her skinny body]. And I was not able to do anything, and they were making *leshelshele* [sour porridge] and other foods for me . . . They are so helpful."

Like many grandmothers, 'M'e Masello worried about what would happen to the children if she died. Grandmothers make good caregivers both because of the cultural expectation that they will provide care, but also because they have not been directly affected by HIV to the same extent as the so-called missing generation of younger adults. They are part of the last generation of virtually HIV-free southern Africans, as the disease spread after their sexually active days had subsided. Yet they are starting to age and die in greater numbers, leaving orphaned children behind. Younger generations of caregivers are fewer in number, due to AIDS mortality among their ranks. The next cohort of caregivers will have to cope with aging, chronic disease, and HIV—an emerging and complex cluster of factors that have yet to be dealt with effectively in a resource-poor context like Lesotho.

Around Christmas of 2009, when Lebo was four years old, 'M'e Masello died. His grandmother was the only caregiver he had ever known. Several members of the paternal side, the Mabotse family, came to the funeral, and one of 'M'e Masello's in-laws advised them, saying: "Here is your luggage," so they took Lebo and his siblings back to their village. This was no small burden. The children went with very few belongings, and Lebo's complicated drug regimen was new to his maternal grandmother, who was chiefly responsible for his care despite her own health problems.

During my most recent trip to Lesotho, in 2013, I went to visit Lebo in his new home. I was concerned because 'M'e Masello had told me that the paternal family had not wanted to care for the children at the time of their mother's death. However, after speaking with several members of the Mabotse family, I learned that 'M'e Masello had not shared the entire story with me.

Lebo's mother and father had been separated for some time prior to their deaths in 2005 and 2006. Lebo's mother, during a period when she was fighting with her husband, had moved back to 'M'e Masello's village and brought her children with her. While at her mother's village, and in her husband's absence, she became pregnant with Lebo. When her husband found out, he was livid. However, near the end of her life, her husband's family fulfilled their responsibility as her caregiver. Since bridewealth had been paid at the time of her marriage and the paternal family honored her and her children as kin, they took her back to their village to attend to her during her final days. Because Lebo was very young he accompanied her, and they accepted him as their kin, emphasizing the importance of social roles and maintaining kin connections, despite that he had been conceived outside of marriage. Lebo's aunt told me about his mother's last moments: "The terrible thing is that he was suckling his mother . . . She died in my hands. He was suckling his mother there." After Lebo's mother died, neither side of the family wanted to care for him, perhaps because he was young, perhaps because of his questionable paternity. But 'M'e Masello's characterization of the paternal family as reluctant and not fulfilling their obligations was misleading. The paternal family had made it clear that they wanted to care for the older children, as they had cared for their mother, and at one time had taken them back to their village. But 'M'e Masello encouraged them to return to her home, which they did in secret, without the knowledge or permission of the paternal grandparents.

The convoluted way in which this story unfolded revealed numerous fictions, embellishments, and hidden truths that speak more to the complexity of people's needs, social obligations, and fears than the original truth would have done. 'M'e Masello wanted to live with her older grandchildren because she needed them, but also because they gave meaning to her life. Despite having seven pregnancies, she had four miscarriages and only gave birth to three live babies. She was outlived by only one child, a son. Her grandchildren, even if they were maternal grandchildren, were essential to her connections during life and to ensuring her ancestral connections after her death. Her initial refusal to take Lebo likely had as much to do with the difficulties of taking care of a small sick child as with her desire to secure Lebo's place within his paternal family, despite his questionable paternity. It also had to do with protecting the memory of her daughter. As 'M'e Masello's niece, herself an old woman, told me: "So, you know that we as parents, we protect our children. We don't know how to speak truthfully." Then, referring to Lebo's mother's infidelity, she added: "We protect them, saying that the child is for that family, but we know what type of child we have here in our household."

AIDS clearly played a large role in Lebo's life. It killed both of his parents, as well as 'M'e Masello's other son, and it brought Lebo near death twice, creating significant challenges for his caregivers. It would be easy to pin this entire case on AIDS. Yet, as Lebo's story shows there is much more to be learned from following the stories of AIDS orphans. They reveal children's many needs in a network of kin, they highlight the disruption of kin as a result of disease, and they demonstrate the challenges of chronic diseases, which add to the burden of many aging caregivers. Lebo's story also underscores the potential for massive social disruption as grandmothers like 'M'e Masello, who are relied upon so heavily to care for AIDS orphans, are aging and dying. Yet, as we have seen, kin-based networks of care are continuing to adapt in order to keep children close to home, reinforcing the strength and resilience of kinship networks.

As of my last visit, in May 2013, seven-year-old Lebo was doing well. He has continued to adhere well to his ART regimen and was attending school in his new village. 'M'e Masello's niece keeps her eye on the children and says they are thriving. She said the Mabotse family willingly took the children after their maternal grandmother's death: "We showed the Mabotse family that, 'here are your children.' So, they didn't have a problem. They said, 'We know that they are our children.'"

### *References*

UNAIDS 2012 Lesotho global AIDS response country progress report. Electronic document. Accessed January 26, 2014 from http://www.unaids.org/en/dataanalysis/knowyourresponse/countryprogressreports/2012countries/ce_LS_Narrative_Report%5B1%5D.pdf

UNICEF 2010 Children and AIDS: Fifth stocktaking report. Electronic document. Accessed February 5, 2013 from http://www.unicef.org/publications/files/Children_and_AIDS-Fifth_Stocktaking_Report_2010_EN.pdf

The shared expectation that family members, more so women than men, should care for orphaned children is part of the morality surrounding HIV. Like Ellen Block, Josien de Klerk (2013) examines the morality surrounding HIV in relation to family dynamics. Studying grandmothers who care for their adult children dying of AIDS in their family homes in Tanzania, de Klerk highlights the intimacy between caregivers and patients with AIDS as their health declines. She traces how both concealment and compassion enable older caregivers to maintain the humanity of their dying children. In the moral world in which caregiving takes place, when confronted with a disease associated with sexual promiscuity, the act of concealment is an intrinsic part of proper care. The community's compassionate silence serves to maintain the dignity and position of caregivers and their families, and to shield the family's affairs from public scrutiny: the public secret protects family honor (Iliffe 2005).

## Secrecy and the Closet

Secrecy is also a theme in the next case study, situated in Indonesia, where HIV prevalence is much lower in the general population than in Africa (less than 1 percent), although relatively high (between 25 and 50 percent) in the sub-populations of sex workers and intravenous drug users (IBBS 2007). Thomas Stodulka describes the life of Monchi, a young man who left his family due to violence and socioeconomic hardship, and who lived on the streets of Yogyakarta until he moved in with his girlfriend's family. He connects to an NGO that conducts AIDS programs in these marginal communities, and finds out that he is HIV-positive. Soon after, his girlfriend learns she is pregnant. Being a good *bapak* (father), one who provides for his family, is an important Indonesian moral value, and Monchi takes on the challenge by marrying his girlfriend, without disclosing his HIV status to the mother of his child.

## 4.3 Coming of Age on the Streets

*Thomas Stodulka*[1]

Monchi (pseudonym) started his street career at the age of 13. I met him five years later, when he was already the leader of a street community of almost 60 children, adolescents, and adults. Like others in this community, he had transformed his lack of self-esteem and negative emotions as a society drop-out into a street pride that he stressed when interacting with his community members, with friends from other street communities, and with NGO activists, street-related women, students, expatriates, and researchers (Stodulka 2009; 2014). Even so, stigma and marginality can outlive the deaths of former 'street children' and are passed on to the next generation.

### *Leaving Home*

Monchi broke with his family after many years of economic and social hardships. His father was a truck driver who often left the family for extended periods, without news of his whereabouts. His mother cleaned houses and worked in the market to make ends meet for Monchi and his three siblings. He started working at a bus terminal after school to help out his mother, and eventually left school to work full-time. "It was better that my sister and brothers went to school and not me. I always wanted to finish school but I was just too tired in the evenings after washing the buses at the terminal." He started to enjoy spending time at the terminal even at night and grew closer to the other children working there. One day a group of friends took him to the train station where he saw even more children selling newspapers, sweets, and soft drinks. Some of them collected waste plastic bottles and sold them to a local liquor store.

A violent episode with his father turned Monchi to the streets and to a different city, where we met some years later. His father had come back from a trip, claiming to be the head of the household. He was drunk mostly during the day and brought other women to the house. Monchi was furious and had had enough. He decided to leave his family for good.

> The last time I met my father he was drunk and I hit him with a big glass . . . I'd had enough. I almost killed him, because I hit him with a heavy glass many times, on the head, on the chest, in the face. I hardly remember it, and when I ran out of the house, and stopped when I was far enough from our house, I couldn't remember what happened. I hoped, but somehow was afraid, that I killed him, so my mother could live in peace. That was the last time I saw him.

### *Twelve Years Later*

After the shacks on the waste land where Monchi and his community lived had been burned down by local authorities, he moved to Bambangrejo, a neighborhood of scavengers, street musicians, and street vendors. The 'marginal neighborhood' (*kampung marjinal*), as he called it, was located in north Yogyakarta, just behind the superhighway that connected the Sultanate city with the Northern Javanese seaport Semarang. The parents of his 15-year-old girlfriend, Dyah, had rented a small hut there. Although Monchi was used to sleeping on the streets, he decided to move in with his 'new family.' He slept in one of the two rooms, together with Dyah and her seven siblings. His new neighborhood consisted of 10 huts and was located at the fringes of a prospering kampung. The inhabitants of the latter called the area where Monchi lived *belakang*, an Indonesian term that means 'behind' or 'backward.' None of the belakang houses had access to clean water, sanitation, or sewage disposal, and only few of them had cement floors. The drinking and cooking water was taken from the little creek behind the huts, which was also used as the communal lavatory. Electricity was provided almost 24 hours a day, however, and the residents participated and contributed to social activities of the adjacent kampung, such as *arisan* (community meetings) and *gotong royong* (mutual help). While Monchi was grateful to have a roof over his head, the new location also imposed prevailing social norms expected from a Javanese kampung.

Monchi and his girlfriend Dyah met when she was 10 years old. Her family also worked on the streets and was part of Monchi's street community. Dyah's mother would dress in rags and beg for money, while her father, Pak Tanto, occasionally busked and worked as a hired laborer at construction sites. He would more often, however, sit in the shade, drinking tea, smoking cigarettes, and playing cards while 'monitoring' his children working on the streets. After they moved to Bambangrejo, he accompanied his children to the street junction in the mornings, and in the evenings he sometimes drove his wife and the youngest of their children back home on a motorbike, which was purchased on installment. The payment of these installments were frequent, and created diffuse networks of indebtedness and economic dependence (*utang-piutang*); money was borrowed from a creditor (*rent-enir*) at high interest rates (up to 50 percent of the actual loan) in order to pay back loans from other creditors. For Pak Tanto and his family, such practices led to high debts, frustration, distress and suffering among all family members.

Dyah's brothers Tirto (14), Dian (12), and Heri (11) continued working at the street junction as buskers (*pengamen*). Their sisters Ratna (10) and Anita (8) both worked on the streets as beggars (*pengemis*) in the late afternoons after they returned home from school. Their little brother Monti (5) accompanied them whenever he was bored sitting around with his father and his company.

Economic pressures and scarce material and kinship resources were not surprising in a family that lacked culturally esteemed family wealth (*bibit*), honorable descent (*bebet*), and professional background (*bobot*). But despite the children's hard work, the economic pressures mostly affected them not their father. Malnutrition, the denial of a basic school education for most of the siblings, frequent diarrhea, fever, and heavy coughs without proper treatment affected everyone in the family but Pak Tanto. In contrast, he was rarely ill, or without food, cigarettes, or alcohol, and neither busked nor begged in the rainy season. Those who suffered most worked the hardest: 10-year-old Ratna, 11-year-old Heri, 12-year-old Dian, their mother, and Monchi, who worked on the streets every day.

After the huts had burned down, Dyah found out that she was pregnant. Monchi wanted to take responsibility and become a caring husband and father, better than his own, so they married three months later. However, Monchi had been recently diagnosed HIV-positive. He had no idea of how to cope with this, and so kept it to himself; Dyah did not know that her husband was HIV-positive Their daughter Ary was born, six weeks premature, a few weeks after Dyah and Monchi's wedding. Feeling guilty (*merasa bersalah*) for not using prevention to protect his wife and daughter from HIV, Monchi tried to make it up to his young family by literally living for them. He reduced his work on the streets except when essential, and instead started cycling the streets with a big rear basket rack on his heavy hand-me-down bicycle looking for scrap iron to sell. Besides providing for his daughter, his wife, her siblings, and his mother-in-law as much as he could, he volunteered for an NGO that he knew since his 'heyday' on the streets. The local organization appointed him as a contact person in belakang Bambangrejo on a 'voluntary' basis.

## *Entering a New Home*

Although the rules of conduct were not especially rigid in Bambangrejo, kampung life still consisted of many obligations to ensure integration and harmony (*rukun*). When asked about his new life in the kampung, Monchi shook his head and answered, "Meetings, meetings, meetings, rules, rules, rules." To attain integration in the kampung, Monchi added more and more communal activities to his everyday 'schedule' of preventing his brothers- and sisters-in-law from working on the streets. By means of his contacts with hospitals and NGOs, which he had developed as the boss of his now-dispersed street community, he started to provide his new neighbors' children with access to appointments with health practitioners in the city's hospitals and cost-free medication. Once the head of the kampung found out that Monchi was volunteering for an NGO, he assigned him the honorable duty to set up an extra curriculum program for all neighborhood children (not only those at the margins) in order to teach them English, and to make Bambangrejo a fixed stop for the NGO's mobile clinic, to provide medical treatment free of charge. The NGO equally built on Monchi's marginality to achieve its targets indicated by its international sponsors, by getting better access to Bambangrejo's alleged 'children at risk,' and implementing their projects there. Monchi's moral dilemma and bad conscience towards his wife and daughter, his marginal status in the kampung community, and his stigmatized social identity as a *jalanan* (street kid) turned him into an important broker between the NGO, the kampung authorities, the neighborhood children, their parents, and his extended family.

Monchi tried to convince me that 'being busy' (*sibuk*) was an adequate strategy to cope with his HIV status and stay healthy, but his many responsibilities in Bambangrejo, his increasingly ailing body, and the guilt he felt towards his wife and daughter made him visibly stressed and tired. He had chronic diarrhea and increasing opportunistic infections, and at times his weight dropped to 43 kg (he was only 48 kg when he was diagnosed as HIV-positive in the year before). And without feeling ashamed (*malu*), he could not find a way to refuse the ever-increasing tasks assigned by his parents-in-law, his neighbors, the head of the kampung, and NGO workers.

What made Monchi increasingly furious (*emosi*), confused (*bingung*), and tired (*pusing*) were the same issues that drove him to the streets almost 15 years ago when he was about the age of his brothers- and sisters-in-law: economic pressures and exploitative family structures that were based on seniority and patriarchy and manifested in the all-encompassing 'Javanese way' (*cara Jawa*) of avoiding and overlooking (potential) conflict and publicly demonstrating the family's and neighborhood's social harmony (*rukun*). Again, the consequences of an exaggerated 'machismo,' this time by his father-in-law and the heads of the new kampung, painfully affected the most vulnerable. Even worse, this time he played a crucial part in jeopardizing the lives of his wife and daughter. Although Monchi frantically contributed to community life and complied with the rules of the kampung, so that he and his family were tolerated there, he became more and more frustrated because he did not have the courage to confide his HIV status to his wife and have his daughter tested. At that time, he repeatedly concluded our conversations under the big banyan tree with tears in his eyes and the words "I think I'm going crazy, and I will burn in hell!"

At the back door of his new home, I asked Monchi one night whether his wife didn't actually assume that he was HIV-positive.

> I don't know. Sometimes I think so. I talk about it a lot with her. Sometimes I also bring leaflets and little booklets from NGOs and the HIV-AIDS clinic home. And I talk about the basic information. But I think she does not really understand yet. Because the rumors are so strong, I think. The fear of HIV is huge. Everybody thinks you die straight away when you have HIV. And people are very afraid of you once you have it. And because I still look healthy, I think she would not believe me anyway. But if the neighbors found out, they would chase me out of Bambangrejo.

Monchi started talking about HIV more often, whenever Dyah and I were both in their house. I also started to insinuate and provoke unpleasant questions, because as the months passed, my lethal conspiracy became more and more unbearable. We spread brochures from various NGOs on the old refrigerator that they used as a closet, where everyone could see them. Contrary to his otherwise very proficient skills of circumnavigating potential harm, Monchi started provoking unpredictable situations inside his house, in which anyone could have asked him about his HIV status: he talked on the phone with doctors and made appointments for CD4-check ups while Dyah and her mother were sitting next to him watching TV; he insisted that Dyah join him during our street art performance on World AIDS Day; he started wearing a T-shirt that said, "HIV-AIDS—Take care of yourself!" He continuously exposed himself, but nobody ever asked.

This situation of provoking and insinuating became particularly awkward after Monchi came out publicly in a video documentary, produced by a local filmmaker whom he had known since his early days on the streets. We decided to screen the video at an event in a cultural center that included talks, workshops, concerts, and theatre performances related to HIV. But Monchi neither appeared at the public screening nor told his wife that he was HIV-positive until a year later. By then, his daughter had reached the age of three.

## Conclusion

The marginalization of street-related young people and their difficulties in blending in with local society does not stop after they exit the streets. In the kampung, tattoos, piercings, 'tattered' clothes, and recurring illnesses all unfolded as social inscriptions of marginality and a stigmatized cultural identity on Monchi's and his family's conspicuous bodies. Monchi, who died of AIDS in late 2013, did not jeopardize the health and lives of his wife and daughter, who continue to live on the streets together with her cousins, because he was a morally degenerate 'street kid.' Rather, the cultural ideal of securing social harmony and silencing 'aberrant' matters (in this case, HIV) conspired with aggravated social pressure in the face of marginality and stigmatized identity as a *jalanan*—a 'street kid'—regardless of the protagonists' chronological age. Despite the remarkable and creative coping styles of young men and women on the streets (Stodulka 2014; 2015), marginality, stigma, and inequity were persistent.

A long-term perspective highlights the need of a wider ethnographic lens when designing 'intervention strategies' in order to alleviate the psychological and physical suffering of street-related children, adolescents, and young adults. It is not necessarily the 'immorality' of 'deviant' street communities that undermines local moralities of mutual respect and compassion. Sometimes it is the sociocultural fabric of the family and the surrounding local society—the purported haven of care, support, and security itself—that stirs up and fosters painful transgenerational spirals of shame and deference and puts the lives of infants and young children at great risk.

## Note

1. The case study is based on applied anthropological research with street-related children, youths and young adults in Yogyakarta, Indonesia, between 2001 and 2013. The longitudinal study has yielded a shelter and ongoing support network for chronically ill street-related people as well as a doctorate thesis, titled "Coming of Age on the Streets of Java" (Institute of Social and Cultural Anthropology, Freie Universität Berlin, 2013).

## References

Stodulka, T. 2009 'Beggars' and 'kings': Emotional regulation of shame among street youths in a Javanese city in Indonesia. In *Emotions as Bio-cultural Processes.* B. Röttger-Rössler and H.J. Markowitsch, eds. Pp. 329–349. New York: Springer.

——— 2014 'Playing it right': Empathy and emotional economies on the streets of Java. In *Feelings at the Margins—Dealing with Violence, Stigma and Isolation in Indonesia.* T. Stodulka and B. Röttger-Rössler, eds. Pp. 103–127. Frankfurt, Germany/New York: Campus.

——— 2015 Emotion work, ethnography and survival strategies on the streets of Yogyakarta. *Medical Anthropology* 34(1):84–97.

---

In the early years of AIDS advocacy, activist discourses and identity politics emphasized 'coming out'; secrecy and silence were seen as obstacles. In addition to assumptions about the health benefits of disclosure, these repertoires of judgment and intervention were often informed by gay and lesbian metaphors of the 'closet' and assumptions about the liberating effects of coming out (Persson and Richards 2008). Secrecy was framed not only as leading to physical death, but as complicit in the stigmatization and shaming that inflicts social death on people living with HIV (Manderson 2014; Robins 2004). The empowerment-speak of HIV-status disclosure—a kind of coming out—dovetails with the identity politics outlined above, in which people with HIV can demand their right to life-saving treatment, but only once they have 'confessed' (Nguyen 2010; Robins 2004). But, as the case study of Monchi shows, the act of withholding information from others is a relational practice, embedded in a social milieu, with particular repertoires of truth telling, histories of power, and social consequences. Secrets have audiences and contexts, and logics of making and unmaking particular to these contexts (Hardon and Posel 2012; Manderson 2014). In fact, one could argue that sociality is in part constituted by secrecy, in varying ways and degrees. For Monchi, secrecy facilitates morality: by withholding information on his HIV status, he can fulfill his role as a good bapak for his newly acquired family.

In Monchi's case, making and keeping secrets is a defensive response to avoid social stigma, and a way to live a life worth living. At the same time, concealment tactics need to be understood in relation to the cultural politics of 'telling' through indirect speech and 'active not-knowing' of "that which is generally known but cannot be spoken" (Taussig 1999: 50). Monchi expects that his family does know that he is HIV-positive because of the active role that he plays in an AIDS-related NGO and the many AIDS-related leaflets that he brings home. Likewise, Mark Davis and Paul Flowers (2014) write of men in London who resist speaking to new partners of their status as HIV-positive, but show it by the open display of their ARV medications.

Gender is an important factor that shapes responses to HIV and AIDS. Sakhumize Mfecane (2011), in a study that echoes Monchi's struggles, reveals how young men in Limpopo Province, South Africa, deny that they have the virus for as long as possible. Young men drink and brag about their latest sexual conquests in community-based groups, and their involvement in such groups is an essential part of their identity as men. As HIV treatment requires abstaining from drinking and unprotected sex, men delay visiting the clinic. When they finally seek out the clinic, they must disclose to other HIV-positive individuals in support groups. Mfecane (2011) finds that these young men disclose their HIV status to their families and neighbors only after their medication has produced physical improvement, although paradoxically, in regaining health, they are better able to keep their HIV status concealed. Their disclosure wins them great admiration from their families and friends, who view such openness as unusual. By withholding the disclosure until their bodies are on the mend, a particular version of embodied masculinity remains intact. In the absence of achieving other traditional markers of masculinity, such as providing for their family through a well-paid job, disclosure becomes a marker of masculine prowess and status (see also Chapter 3 on masculinities).

★ ★ ★ ★

In this chapter we have examined various social forms that have emerged in response to the HIV epidemic, showing how activists engaged in identity politics, demanding that AIDS be taken seriously and fighting stigma by creating an 'out' HIV-positive identity. Some strands of activism have had tremendous impact on new technologies and care programs, with activists working in close collaboration with clinicians and pharmaceutical companies. AIDS activists have achieved remarkable success in making medicine accessible in places where people initially could not afford ARVs.

The case studies show, however, that despite activism, stigma associated with HIV and the secrecy that is sustained as a consequence continues to shape how people relate to each other and seek care. Further, despite the proliferation and global reach of AIDS care programs and support structures, families are still the primary place where health conditions are confronted.

In the past decade, various discursive, legal, health policy, and institutional efforts have attempted to reframe HIV as a chronic disease that should be treated 'like any other.' Despite these efforts to 'normalize' HIV, anthropologists have shown that this normalization has not taken place (Moyer and Hardon 2014). In many parts of the world, people living with and affected by HIV struggle with various social and economic stressors in their daily lives. Stigma and discrimination continue to affect HIV-positive people, not only because of their status but also at times because of their sexual identities, even when HIV-related stigma has been challenged through education, awareness raising, and experience. The persistence of stigma seems to confound the expectations of public health planners and medical practitioners, as well as the early AIDS activists, many of whom seemed to believe that the very process of biomedicalization through treatment would depoliticize the disease (Moyer and Hardon 2014). As the case studies show, when faced with the social and economic challenges of HIV, people seek support from each other through reconfigured family relations, constellations in which they assert their identities as good grandmothers, brothers, and fathers.

## Note

1. This definition of GIPA echoes the 1978 declaration on primary health care adopted by the WHO in Alma Ata, which encouraged community participation at every stage from needs assessment to implementation. Like GIPA, the Alma Ata Declaration defined community participation not only as a right, but as a process whereby individuals come to view health as a responsibility.

## References

Abadia-Barrero, César Ernesto 2011 *"I Have AIDS but I Am Happy." Children's Subjectivities, AIDS and Social Responses in Brazil.* Bogotà, Colombia: Universidad Nacional de Colombia.

Beckman, Nadine, and Janet Bujra 2010 The politics of the queue. *Development and Change* 41(6):1041–1064.

Biehl, João 2007 *Will to Live: AIDS Therapies and the Politics of Survival.* Princeton, NJ: Princeton University Press.

Brown, Phil and Stephen Zavestoski 2004 Social movements in health: An introduction. *Sociology of Health & Illness* 26(6):679–694.

Davis, Mark, and Paul Flowers 2014 HIV/STI prevention technologies and 'strategic (in)visibilities.' In *Disclosure in Health and Illness.* Mark Davis and Lenore Manderson, eds. Pp. 72–88. Abingdon, UK: Routledge.

de Klerk, Josien 2013 Being tough, being healthy: Local forms of counselling in response to adult death in northwest Tanzania. *Culture, Health & Sexuality* 15(Supp 4):S482–S494.

Hardon, Anita, and Hansjoerg Dilger 2011 Global AIDS medicines in East African health institutions. *Medical Anthropology* 30(2):136–157.

Hardon, Anita, and Deborah Posel 2012 Secrecy as embodied practice: Beyond the confessional imperative. *Culture, Health & Sexuality* 14(Supp 1):S1–S13.

IBBS 2007 *Integrated Biological-Behavorial Surveillance among Most At Risk Groups (MARG) in Indonesia. Surveillance Highlights.* Jakarta, Indonesia: IBBS.

Iliffe, John 2005 *Honour in African History.* Cambridge, UK: Cambridge University Press.

Landzelius, Kyra, and Joseph Dumit 2006 Introduction: Patient organization movements and new metamorphoses in patienthood. *Social Science & Medicine* 62(3):529–537.

Manderson, Lenore 2014 Telling points. In *Disclosure in Health and Illness.* Mark Davis and Lenore Manderson, eds. Pp. 1–15. London: Routledge.

Mattes, Dominik 2011 "We are just supposed to be quiet": Patient adherence to antiretroviral treatment in urban Tanzania. *Medical Anthropology* 30(2):158–183.

Mfecane, Sakhumize 2011 Negotiating therapeutic citizenship and notions of masculinity in a South African village. *African Journal of AIDS Research* 10(2):129–138.

Ministry of Health, Lesotho 2015 Global AIDS Response Progress Report, 2015. Follow-up to the 2011 Political Declaration on HIV/AIDS: Intensifying Efforts to Eliminate HIV/AIDS. Lesotho Country Report. Accessed from http://www.unaids.org/sites/default/files/country/documents/LSO_narrative_report_2015.pdf.

Morolake, Odetoyinbo, David Stephens, and Alice Welbourn 2009 Greater involvement of people living with HIV in health care. *Journal of the International AIDS Society* 12:4. DOI: 10.1186/1758-2652-12-4.

Moyer, Eileen, and Anita Hardon 2014 A disease unlike any other? Why HIV remains exceptional in the age of treatment. *Medical Anthropology* 33(4):263–269.

Nguyen, Vinh-Kim 2006 Attivismo, farmaci antiretrovirali e riplasmazione del sé come forme di cittadinanza biopolitica. *Antropologia* 6(8):71–92.

——— 2010 *The Republic of Therapy: Triage and Sovereignty in West Africa's Time of AIDS.* Durham, NC: Duke University Press.

Persson, Asha, and Wendy Richards 2008 From closet to heterotopia: A conceptual exploration of disclosure and 'passing' among heterosexuals living with HIV. *Culture, Health & Sexuality* 10(1):73–86.

Petersen, Alan, and Deborah Lupton 1996 *The New Public Health: Health and Self in the Age of Risk.* London: Sage.

Rabinow, Paul 1996 Artificiality and enlightenment: From sociobiology to biosociality. *Essays on the Anthropology of Reason.* Pp. 91–111. Princeton, NJ: Princeton University Press.

Robins, S. 2004 'Long live Zackie, long live': AIDS activism, science and citizenship after Apartheid. *Journal of South African Studies* 30(3):651–671.

Rose, Nikolas, and Carlos Novas 2005 Biological citizenship. In *Global Assemblages: Technology, Politics, and Ethics as Anthropological Problems.* Aihwa Ong and Stephen J. Collier, eds. Pp. 439–463. Malden, MA: Blackwell.

Taussig, Michael T. 1999 *Defacement: Public Secrecy and the Labour of the Negative.* Stanford, CA: Stanford University Press.

UNAIDS 1999 *From Principle to Practice: Greater Involvement of People Living with or Affected by HIV/AIDS (GIPA). UNAIDS/99.43E.* Geneva, Switzerland: Joint United Nations Programme on HIV/AIDS (UNAIDS).

Whyte, Susan Reynolds 2009 Health identities and subjectivities: The ethnographic challenge. *Medical Anthropology Quarterly* 23(1):6–15.

Williams, Alun 2003 *Ageing and Poverty in Africa: Ugandan Livelihoods in a Time of HIV/AIDS.* Farnham, UK: Ashgate.

Girl Working in Silk Factory, 2006. Hanoi, Vietnam.

## *About the photograph*

*Silk is still an important part of the economy in small villages outside of Hanoi. Like many places around the world the working conditions are terrible. Young women made up the majority of the workers in this particular factory. The machines, many of which were over 100 years old, made an earsplitting noise, and only the most minimal ear protection was being used. Exposure to high noise levels, suffocating dust, and long hours standing over the machines makes this a difficult job, especially for young women like this one pictured, who continue to work late into their pregnancies.*

—Elizabeth Cartwright

# 5

# Stress in Everyday Life

*Anita Hardon, Elizabeth Cartwright and Lenore Manderson*

People express, experience, and manage the stresses of everyday life in many ways. Stress can be defined as the response of humans to environmental demands, also referred to as a biological 'fight or flight response.' Some stress is good: it increases alertness and performance, enabling people to adapt to the needs of a given situation. But when people are subjected to a continuing assault of stressors, they can become exhausted and sick. Social epidemiologists assert that stress is exacerbated when people experience ongoing unpredictable circumstance and a lack of control (Brunner and Marmot 1999).

Stress has become a popular topic in talking about health and wellbeing among relatively well-off people in global cities, who feel economic and social pressures to perform and are exposed to the many different stimuli of the current digital age. Popular magazines and websites encourage people to test their stress levels by answering questions such as "Are you irritable?" "Do you sleep well?" and "When did you last have a good laugh?" They advise people to stop fretting, concentrate on breathing, undertake mindfulness training, and go for walks. Urban yoga studios, massage parlors, and nail spas all invite people to take time out in order to de-stress, not only in cities in the global North, but also in most cosmopolitan centers worldwide.

Expressions of stress are not new, and they are not associated only with the upwardly mobile middle classes and the well-to-do; the discourses of stress have all over the world penetrated everyday discussions of fluctuations in health, wellbeing and mood. In this chapter, we present anthropological accounts of the ways in which people live with and talk about stress in a variety of sociocultural and economic settings, and we reflect on the diverse ways that people strive to manage or cope with stress. Worldwide, cultures have ways of diagnosing and treating stress that have been handed down over generations, that have specific local meanings, and that are often effective in ways that may be illuminating for biomedical practitioners who so often rely solely on psychiatric manuals and pharmacopeia.

## Diverse Idioms of Distress

What are the common ways of talking about, or otherwise conveying, stress, tension, and negative emotions in a given society? How do people typically respond to, handle, or attempt to reduce such stresses? We can think through these questions with the notion of 'idioms of distress,' in reference to the ways that people make sense of and express such feelings, and the ways that they work to alleviate them. For example, although the term *stress* exists in Bahasa, the official language of Indonesia, when doing fieldwork in Makassar, Anita Hardon observed that her female friends more often used the term *pusing* (literally translated as 'dizzy') to express a general feeling of malaise associated with all kinds of stress, in the family, at work, or otherwise. When one declares that one is pusing, it is understood that one should take a good rest. Friends and colleagues will send you home, urging you to get a traditional massage. After a while Hardon started describing herself as pusing too, even if she was only slightly tired. Doing so changed her lifestyle, as her Makassar friends encouraged her to do such things as take some time off and to have regular massages at a local women's wellness center; she had effectively entered into the local logic of maintaining her health and wellbeing.

More than 30 years ago, Mark Nichter (1981: 379; see also Nichter 2010) called on anthropologists to examine such idioms of distress as ethnopsychiatric phenomena, arguing these idioms are "underscored by symbolic and affective associations which take on contextual meaning in relation to particular stressors, the availability and social ramifications of engaging alternative expressive modes, and the communicative power of these modes." He illustrated this by presenting alternative idioms of distress among Havik Brahmin women in South India, who face tensions when they marry and go to live with their husbands and mothers-in-law. Family tensions were expressed through purity discourses, and these tensions were exacerbated as more young women pursued higher education. Nichter showed how young, well-educated wives often purposefully neglected purity rules regarding food preparation and serving when they were unhappy with their husbands, and how mothers-in-law criticized their behavior and its impact on the prestige of Brahmins. These forms of resistance bear interesting similarities to some of those we discuss in Chapter 6.

Elsewhere, anthropologists have observed that social tensions and distress caused by family and community conflicts are expressed through accusations of witchcraft. Peter Geschiere, for example, observed in Cameroon that local witchcraft narratives offer a "seductive discourse to address the riddles of modern development: the rapid emergence of shocking new inequalities, the enigmatic enrichment of a happy few, and the ongoing poverty of many" (2013: 5). Jeanne Favret-Saada (1980), in a seminal study of sorcery in a rural farming community in Normandy in the late 1970s, describes how accusations of witchcraft are related to struggles between households stretching over years. She noted that generally neighbors are identified as witches, not relatives or people living under the same roof. Favret-Saada (1980) was able to explore the world of sorcery in rural France by taking on a role of being bewitched, and enrolling in a therapeutic process with a local healer.

Thus, idioms of distress may appear very differently from place to place. They may be encapsulated in a word or simple phrase directly related to physical feelings, like being 'pusing'; they may be ways of defining or responding to social relations or events, such as the purity discourses mentioned above; they may be invoked in connection to other key frameworks and systems of belief; or they may be expressed through accusations of witchcraft.

Byron Good has argued that "a semantic network analysis conceives the meaning of illness categories to be constituted not primarily as an ostensive relationship between signs and natural disease entities but as a 'syndrome' of symbols and experiences which typically 'run together' for the members of a society" (1977: 25). He suggested that such an analysis could draw attention to configurations of social stress that are embedded in such a syndrome. During his fieldwork in Iran in the early 1970s, he found through such semantic network analysis that stress was located in the heart, with many different kinds of expression: 'my heart is pounding,' 'my heart is trembling' (*qalbim tittirir*) or 'fluttering,' or 'beating rapidly,' and 'my heart feels pressed' or 'squeezed,' 'bored,' or 'lonely.' Focusing on semantics and meanings, Good does not relate these embodied experiences to biological stress responses. Rather, he sticks to the analysis that his respondents make, pointing out that women related heart distress to their use of the contraceptive pill, and that because it affected their menstrual cycle, they worried about future fertility. Men related heart distress to tense interpersonal situations that were also often closely associated with complaints of weak nerves (Good 1977).

A more recent example of this kind of research is that undertaken by Mysyk and colleagues (2008) in Canada. They describe the occurrence of 'nerves' (*nervios*) among Mexican agricultural workers who enter Canada as seasonal agricultural laborers, under a bilateral agreement that is seen, at least by the governments involved, as solving both the problem of unemployment in Mexico and the shortage of agricultural laborers in Canada. Mysyk and colleagues describe how Mexican men feel uprooted from their families, fear becoming ill, and suffer from language barriers when trying to interact with angry bosses. As one young worker explained, "Sometimes with the boss, you can't understand very well. . . . They don't understand Spanish either. And sometimes, maybe you don't do things like he wants and sometimes he gets angry, he gets annoyed, and sometimes your morale falls because you'd like to do things well" (Mysyk et al. 2008: 390). Another interlocutor explained: "Nerves betray you, no? Because I've seen, in the four seasons I've come [to Canada], I've seen cases of co-workers who sometimes start to cry. . . . 'I want to go back because my son is sick,' 'My wife is sick,' or 'I can't [stay] here any longer.' And nerves betray them and they start to cry" (2008: 397). The participants in the migratory worker program have to receive a positive evaluation in order to be allowed to work in Canada the following year, making it especially hard for newcomers who have less experience and lower English-language skills. Mysyk and colleagues argue that, in this context, nervios "embodies a general lack of control over their lives" (2008: 396). Other anthropologists have described nervios and *susto* (fright) as illness terms that express stress and depression (Baer et al. 2003; Cartwright 2007; Guarnaccia et al. 2003; Weller et al. 2002).

Kaiser and colleagues (2014) also discuss how a lack of control relates to feeling distressed in the context of work, as they unravel a Haitian idiom of distress described as *reflechi twop* ('thinking too much'). This syndrome is associated with sadness, suicidal ideation, and social and structural hardships such as unemployment and poverty. As one of the community leaders explains, "There is no work! It is the impossibility; it's poverty that puts everyone in all these things because people are sitting down, only sitting down, eating. And the food, they don't know where it will come from, and they are thinking about how to get food. . . . You can't think of anything else" (2014: 458).

Kirmayer and Young (1988) argue that cultural models supply individuals with a 'vocabulary' of symptoms and provide explanations for these symptoms and associated suffering. One could add to this line of thought that new patterns of stress lead to new forms of somatization, such as those experienced by factory workers in Malaysia in the 1970s when jobs became more routinized and sedentary, as described by Aihwa Ong (1987). Similar examples of workplace 'hysteria,' or in the case study by Ria Reis in Chapter 2, of hysterical responses in schools, further illustrate the ways in which people enact and express feelings of stress, distress, and loss of control over their everyday lives. While transcultural psychiatrists have conducted epidemiological studies to find associations between cultural expressions of stress and psychiatric conditions, Kirmayer and Young emphasize that it is important for medical practitioners to listen to how patients describe their illness, in their own terms, in order to understand the way idioms of distress are linked to "social predicaments, moral sentiments, and otherwise unexpressed emotions" (1988: 424). Listening to the stories of somatized stress can provide clues to non-medical solutions that could prevent—or at least acknowledge—the underlying causes of the physical stresses individuals are experiencing.

## Managing Stress

Anthropologists have long observed that spiritual healers, diviners, and shamans mediate tensions of everyday life through rituals and collective healing sessions. They do so, for example, by going into trance states, speaking in a special vocabulary and tone, and answering questions about misfortunes (Beattie 1967). A key characteristic of such healing sessions is that they are social events, involving patients, friends, and relatives. Geschiere (2013: 527) describes how in Cameroon, spiritual healers called *nganga* can solve problems, because they can 'see' the witches, 'fall upon them,' and force them to deliver their victims to safety. Through diagnoses of distress, such healers are able to prescribe appropriate healing acts: a gift to an ancestor, the payment of reparations, a special feast. In Tim Asch, Linda Connor, and Patsy Asch's classic films, *Balinese Trance Séance* and *Jero on Jero* (2015 [1980–1981]), and associated text (Connor et al. 1996 [1986]), the filmmakers richly illustrate how Jero, a Balinese trance healer, manages family distress and helps family members come to terms with a child's unexpected death.

Critical medical anthropologists worry that such social healing approaches may be replaced by biopsychiatric models of healing, which are increasingly dominant all over the world, in part due to the flow of these therapeutic regimes through aid efforts in areas affected by conflicts and natural disasters (Good 2010; Watters 2010). Anthropologists are critical of the global applications of the American *Diagnostic and Statistical Manual of Mental Disorders* (*DSM*), the current version of which is called the *DSM-5*. These manuals provide psychiatrists and other biomedical practitioners with diagnostic tools that are assumed to be applicable and valid in any cultural setting. Watters shows how American definitions of depression, anorexia, and post-traumatic stress disorder, among other states, are spreading worldwide; that local expressions of distress are being discounted; and that, as a result, "indigenous forms of mental illness and healing are being bulldozed" (2010: 3).

The attention to mental health is part of a more general recognition among global health planners of the importance of non-communicable diseases—a domain that was neglected in the past when programs were mainly directed at the control of acute infectious diseases (see Chapter 7). Up to 30 percent of the global population experiences a mental health condition every year (Prince et al. 2007), and mental health problems are increasingly prevalent compared with other conditions. Some anthropologists fear that this recognition will further fuel the use of psychopharmaceuticals in the treatment of all kinds of symptoms of distress, thereby medicalizing social suffering (Ecks 2014; Watters 2010). Others have pointed out that it is important that mental needs enter the global health agenda (Summerfield 2012). Good takes this latter view, and argues that in settings like Indonesia, "the salient issues seem to be the scarcity of mental health resources, including access to pharmaceuticals, the poor quality of care for those seeking treatment, and the utter indifference to mental health services among many ministries of health" (2010: 121). Elsewhere, however, Good also draws attention to the different ways in which mental health states are understood, and hence the different expectations of the chronicity of conditions and the relationship of this to treatment, recovery, and social inclusion (Nichter 2010).

The first case study in this chapter outlines how in Japan, a country hit recently by a severe economic recession, psychiatric diagnosis, treatment, and the prescription of antidepressants have played an empowering role in helping people deal with the pressures of work and the stresses of unemployment. Junko Kitanaka describes how the biomedical diagnosis of depression as a work hazard gives men (more so than women) recognition that they have suffered injustice. Psychiatrists have played a key role in creating a socializing discourse, Kitanaka maintains, by providing powerful testimonies for depressed workers, showing that depression is not only a pathology of the brain but is also rooted in the stress of losing one's job or of having to work excessively long hours. Thus psychiatrists in Japan articulate depression as an affliction of hardworking people, who, unable to sense their own fatigue, end up being completely exhausted. While this definition of depression has resulted in large increases in antidepressant sales in Japan, Kitanaka states that depression is not seen in the local setting as the medicalization of mood, but as the result of precarious economic conditions.

---

## 5.1 A Cold of the Soul

*Junko Kitanaka*

In 2001, at a psychiatric hospital in Tokyo, I met Kobayashi-san, a civil servant in his fifties, working for the central government. After we exchanged our business cards, he began sketching out the timeline of his illness. He told me how, on a Sunday when he was in the office, his mind suddenly went blank. He had been involved in an intense budget negotiation with the Ministry of Finance and had been sleeping only three hours every night, continuously working on weekends. At the time, it was not unusual for people in his ministry to work more than 100 hours of overtime per month—"If someone dropped dead, then superiors would talk about giving the family a workers' compensation but that was it." Even when he felt exhausted, he blamed his own 'weak' personality, feeling that he was doing a job that was beyond his capabilities. He had heard about cases of other employees throwing themselves out of an office window in the high-rise government headquarters. That day, a thought occurred to him that he might just do the same . . . he just wanted to escape from it all.

Such was a typical story I heard from the people with depression whom I met in psychiatric hospitals and clinics in and around Tokyo in the early 2000s. Many of these were men in their prime, hard hit by the recession that had triggered high unemployment rates and record-high national suicide rates. As banks went bankrupt and companies that had prided themselves on their lifetime employment system began to resort to massive layoffs, Japanese people's belief in their society's stability was fundamentally shaken. In this context, Japanese people began to talk about depression as an illness of stress and overwork, a theme that consistently appeared in almost all the narratives I heard from patients with depression. Takashima-san, a businessman in his forties, said he was regularly working from 7 AM to 2 AM when he developed this ailment. Enduring an abusive boss who scolded him in front of his colleagues, Takashima-san began to lose confidence and did not know what he was doing anymore. His own children discovered him, at home, when he was trying to hang himself with a rope. Machida-san, a former executive director of a construction company in his sixties, bitterly described how, after the burst of the bubble economy, his parent company began to dump unprofitable projects on his firm. While striving to survive by scrambling for funds day and night, ensuring that his employees were paid every month, he said he didn't even realize he had become depressed. He recalled how humiliating it was to be in the bankruptcy court that had confiscated his house and every other asset, while being condescendingly lectured by the judge about the meaning of 'collective responsibility.' He criticized the government for promoting rhetoric like 'sharing the pain' to protect big businesses in recession while doing little to save little guys like himself. For these men, the diagnosis of depression as a work hazard seemed to give some recognition of the structural injustice they had endured, and allowed them to recover their impaired sense of self through the care of a psychiatrist, who often served the role of a benevolent superior by not only prescribing medication but also ordering ample rest.

The global rise of depression since the 1990s, spearheaded by the advent of new antidepressants, has often been discussed as a new era of biological management of everyday distress. In the way that the medicalization of depression has prompted workers to pharmacologically control their negative affect, some critics argue that psychiatry might be serving as an insidious tool for manufacturing productive and delusionary happy workers, oblivious to the social roots of their distress. This view, however, fails to capture the complexity of what has been happening in Japan, where the rapid medicalization of depression has coincided with—in fact has been spurred on by—a new type of workers' movement. Through a series of litigations over what Japanese now call 'overwork suicide' and 'overwork depression,' depression has emerged as a much-politicized biomedical category used by burned-out workers and their family members as a basis for economic compensation for their suffering and sometimes death. Japanese psychiatrists have played a pivotal role in creating this 'socializing' discourse by providing powerful testimonies for depressed workers, showing how depression is not only a pathology of the individual brain but is also rooted in workplace conditions. Their arguments have prompted the Japanese government to change its traditional stance that mental health is a private matter, and to officially adopt a stress-diathesis model that clearly defines mental illness as a product of interactions between the individual and the environment. Psychiatrists have also aided the government in establishing Stress Evaluation Tables, which list 62 stress events as potential causes for psychiatric breakdown; these tables are now routinely used by medical experts for determining if a worker's mental illness has been caused by work stress. As a result of these changes, almost all forms of psychopathology—including schizophrenia—can now be administratively examined in terms of social causes and are subject to workers' compensation (Kitanaka 2012). By depicting depression as both a neurobiological and social pathology, psychiatrists, industrial leaders, and the government have seemed as if they are trying to lift the nation out of economic depression by turning its attention to individuals' clinical depression.

This particular rendering of depression in Japan as an illness of stress—even evocatively called a 'cold of the soul'[1] that one catches in a time of hardship—is based upon a century-old theorization about the relationship between overwork and psychopathology. Japanese psychiatrists have long noted how the depressed tend to be hardworking people who seem unable to sense their own fatigue and end up breaking down at the height of their exhaustion. Understanding depression as a somewhat adaptive mechanism for self-preservation, psychiatrists have also discussed this ailment as a product of a particular personality (called *Typus melancholicus*), shaped and reinforced by a Japanese work ethic. Psychiatrists I observed in their everyday clinical practice in the early 2000s would often assure patients that their overwork—marked by their own diligence, strong sense of responsibility, and consideration for others—had driven them to a psychiatric breakdown. This therapeutic narrative seemed to be embraced by all the patients I met, many of whom talked about depression as not simply a biological defect or a distortion in cognition but a bodily insight that is calling for change in the way they live. Some saw this ailment as an opportunity to retreat from excessive social obligations and even contemplated the nature of their own self-subjugation. Their critical awareness became reflected in revised labor laws in the 2000s, which seemed to provide structural possibilities for Japanese to distance themselves from a cult of overwork that had bound them.

Reconceptualizing depression as an illness of stress has thus brought about a new sensibility in Japan about the burden of affective labor. It has also created 'gender equality' in the medical and popular discourses about psychiatric suffering. Curiously, female depression had long been under-recognized in Japan. Although the theory of *Typus melancholicus* and its emphasis on work were not necessarily gender-bound, given men's prominence in workplaces (and the social visibility of the 'public' work in which they are often engaged), it seems to have contributed to foregrounding men's suffering. Some of the veteran psychiatrists I interviewed pointed out that their model of depression had traditionally developed around male depression and women's depression defied easy classifications. Psychiatrists thus used to be much less eloquent about the plight of depressed women, as if the doctors themselves could not quite grasp or explicate the nature of these women's suffering. This gender bias was reflected in the narratives of the depressed women I met, some of who discussed how their complaints had been dismissively treated as 'psychosomatic' both by family members and doctors. Such stories were quickly becoming a thing of the past in the early 2000s, however, when the media began to emphasize that 'even women become depressed,' and depression was becoming a top reason for both women and men taking sick leave.

Yet, using a biomedical category to mediate social injustice and to claim a status of victimhood—especially to secure one's place in the rapidly changing labor force—is fraught with moral and political ambiguities, as it often leaves those people vulnerable to the instability of biomedical (re)definition of their suffering (Petryna 2002; Young 1995). This was the case in Japan through the 2000s, as contradictions in the simplified notion of depression began to be exposed. First, psychiatric campaigns depicting depression as an illness of stress helped prompt a sudden increase in the number of people seeking and receiving this diagnosis, resulting in an unprecedented number of people with depression. Sales of antidepressants rose from approximately 17 billion yen in the 1990s to reach 90 billion yen by 2007, and the number of depressed patients more than doubled over a ten-year period, exceeding one million by 2008 (Yomiuri 2010). (This was remarkable given that psychiatrists themselves had assumed until the mid-1990s that depression was a rare occurrence.) The patients I met told me that, prompted by psychiatrists, media, the pharmaceutical industry, and the government to become aware of their mental state, they began to realize how 'depressed' they had been. This change in people's self-awareness set in motion what Ian Hacking (1999) calls a (bio)looping effect; the rise of an illness category, as it becomes embraced and internalized by those afflicted, begins to change the manner in which the illness itself is expressed and experienced. In Japan, as people voluntarily sought medical care, psychiatrists and internists began to see people with milder forms of depression, for which traditional psychiatric treatment proved ineffective, even detrimental. As some patients began to suffer severe side effects of medication, developing chronic and protracted forms of depression, people's initial hope for their linear, straightforward path of recovery collapsed, unsettling the public's expectations for psychiatry's efficacy.

The patients I interviewed in 2008–2009 lamented how they had casually begun taking antidepressants and thought little about the consequences of carelessly adopting a psychiatric identity and the effects of long-term psychotropic use. Kawai-san, a nurse in her thirties, told me how she was surprised to receive a depression diagnosis (particularly because she suspected that her problem stemmed from years of marital conflict), but nonetheless embraced it as a way of obtaining sick leave. A year after treatment, however, she found herself uncured, divorced, and unemployed, while suffering severely from the side effects of heavy medication. After three years of struggle, she met a psychiatrist who gradually took her off all medications, only to find that she was 'hardly depressed at all.' As this case and many others like hers attest, contrary to the then-popularized image of depression as an illness of stress, depression is far more complicated in its etiology and often leaves a mark on the afflicted, at times fundamentally altering their sense of identity. This is especially the case with chronic patients, for whom 'depression' often remains irregular, irrational, and unpredictable, and for whom a certain period of social disengagement may be crucial for recovery. Yet, because of the generally simplified and optimistic understanding of depression, people who long remain depressed may appear to others (particularly personnel staff at the workplace) as puzzling and unpredictable. Thus, the overly normalized understanding of depression is beginning to generate new forms of disability and stigmatization against those who remain uncured. By remaining disengaged for too long, they risk being stigmatized as suffering a 'new type depression'—a sign of personal immaturity—and reclassified as a moral threat to the labor-obsessed society.

Partly in response to such confusion, corporations and the government have begun to approach depression as a form of risk to which every worker is subject, which should come under rational management rather than strictly traditional psychiatric practice. Increasingly, the care of depressed workers is being outsourced to specialized clinics that provide not only medical care but also occupational training and group cognitive therapy designed to rehabilitate—even improve—the work skills of the depressed. As depression becomes an object of corporate risk management, companies are beginning to scrutinize workers' mental health and monitor their recovery in order to promptly restore their productivity. Given the prolonged recession, where both industrial doctors and personnel staff are coming under increasing pressure to return the afflicted to a healthy state, some doctors seem to be moving away from the former ideals of *clinical time* that prioritizes a 'natural' recovery, and are adapting to the demands of *industrial time* that constantly seeks, even for a therapeutic process, the principle of *efficiency*.

Under these new circumstances, the clinical principle of *Typus melancholicus*—which recognizes depressed patients' diligence and strong sense of responsibility, and 'rewards' them by prescribing ample rest in the form of long-term sick leave (at times even a few years)—no longer seems to have the therapeutic validity it once held. In fact, people identified as *Typus melancholicus* are quickly losing their legitimacy as model workers and legitimate victims, and are increasingly regarded as simply 'stubborn, inflexible, and outmoded'; they are now expected to overcome their depression and adopt, through cognitive enhancement and affective management, a more 'malleable, adaptable, and ambiguous' self, as one psychiatrist put it. Given the new corporate demands, psychiatrists themselves often appear uncertain, beyond prescribing antidepressants, how to help these people reclaim their place in the rapidly changing landscape of work while trying to recover a secure sense of self. This heightened uncertainly in the clinical sphere, and increasing confusion in the realm of industrial health, makes one wonder if the Japanese form of medicalization of depression may have simply served to make people's sense of anxiety and vulnerability even more imminent and ingrained, while the socializing discourse about depression has functioned as a tool utilized by the pharmaceutical industry for cultivating the market for antidepressants—even though it is now painfully apparent to both doctors and patients how partial their effects may be.

* * * *

The situation in Japan prompts us to ask how psychiatry can go beyond being a mere apparatus for managing negative affect and boosting productivity, at a time when depression is increasingly and globally regarded as a problem of labor. While recent developments in Japan suggest that there are no easy answers, what may give us hope is the fact that the kind of socializing discourse about depression in Japan is not an isolated occurrence but part of a global movement that is increasingly problematizing the psychopathology of work. In the 1990s, when I discussed overwork suicide with audiences in North America, it was often regarded as a kind of cultural exoticism, prompting listeners to ask me how Japanese could be so unreflective as to work themselves into depression, even suicide. In 2009, however, a spate of suicides among employees of France Telecom made news headlines internationally, as suicides attributed to work stress and the increasing pressure faced by the company's employees. Rising rates of suicide and depression in the workplace have also raised concerns elsewhere in Europe—including Germany, Italy, and Greece—prompting a debate about the rapid changes brought on by globalization and neoliberalism, the perils of privatization, the collapse of lifetime employment, and the crisis in social welfare, all of which some see as destroying their way of life. An idea that seemed strange to many in the 1990s—that one could be driven to depression by work stress—is now becoming recognized as a global reality, as people increasingly experience the effects of economic meltdown with a resulting "sense of vulnerability in being part of a world system" (Lupton 1999: 49). Given such concerns, we have to ask what kind of role the socializing language of depression can play in the expanding web of psychiatric practice, what place it has in the emerging movement for 'global mental health,' and what possibilities exist for local articulations about everyday distress to help generate a new theoretical framework in psychiatry to address the *social* nature of depression.

## Acknowledgments

This chapter is adopted from my book *Depression in Japan* (2012), which is based on ethnographic fieldwork I conducted in and around Tokyo in 2000–2003, when I did interviews with patients and doctors and participant-observation of everyday clinical practice at various psychiatric institutions. This was complemented by follow-up interviews in 2008–2009 as well as a decade of close observation of the changing scenes of depression in Japan. This study was supported by JSPS Grant-in-Aid for Scientific Research (No. 24300293).

## Note

1. '*Kokoro no kaze*' (a cold of the soul) is a term that was used to popularize the idea of depression in Japan in the early 2000s. '*Kokoro*' in Japanese denotes soul, mind, and heart.

## References

Hacking, I. 1999 *The Social Construction of What?* Cambridge, MA: Harvard University Press.

Kitanaka, J. 2012 *Depression in Japan: Psychiatric Cures for a Society in Distress.* Princeton, NJ: Princeton University Press.

Lupton, D. 1999 *Risk.* London and New York: Routledge.

Petryna, A. 2002 *Life Exposed: Biological Citizens after Chernobyl.* Princeton, NJ: Princeton University Press.

*Yomiuri Shimbun.* 2010 'Utsu Hyakumannin' Kage ni Shinyaku? Hanbaidaka to Kanjasū Hirei (New Medication behind 'One Million People in Depression'? Ratio of Sales to Number of Patients). January 6.

Young, A. 1995 *The Harmony of Illusions: Inventing Post-Traumatic Stress Disorder.* Princeton, NJ: Princeton University Press.

---

Both the expression and management of distress need to be understood in relation to the specificities of healing and care in local contexts. While the diagnosis of depression and its treatment with antidepressants may help to change working conditions in Japan, scholars point out that in the United States, the individualized management of 'neurochemical selves' (Rose 2003) serves to reinforce existing social structures (see also Dumit 2010; Jenkins 2010). The United States has the highest use of psychopharmaceuticals worldwide. People use pharmaceuticals to feel happier or less anxious, to have more concentration and stamina, and to unwind during a night out. Emily Martin shows that even sleep has become "a complex management project" (2010: 205; see also Wolf-Meyer 2012).

Sue Estroff's seminal ethnography, *Making it Crazy* (1981), illustrates how the side effects of psychopharmaceuticals have contributed to local understandings of stigmatized mental illnesses. This ethnography describes the late 1970s in the United States, when many people with mental health problems were de-institutionalized and heavily medicated. Estroff's unconventional field method of taking the drug Prolixin, so that she too would display the extrapyramidal symptoms of foot stomping, tongue thrusting, and uncontrolled limb movement, allowed her to emically experience people's negative reactions and therefore, to more fully understand the experience of being mentally ill and out on the streets. Estroff was able to show that part of what the general public considered to be the outward appearance of a 'crazy' person was, in fact, the side effects of the medications they were taking to control their conditions. The stigma of mental illness, as it was understood at that time, developed simultaneously with the increased prescription of Prolixin.

## Talking Therapy

The second case study in this chapter brings us to a very different setting. In Argentina, wellbeing is promoted through psychotherapy rather than through the use of pharmaceuticals. P. Sean Brotherton describes how in Buenos Aires his informant Gaston, who moved from a small rural town where he had been prescribed antidepressants, was advised by his psychiatrist to go to a psychoanalyst. While in many settings counseling, psychotherapy, and psychoanalysis are unfamiliar and may not be considered appropriate interventions, Brotherton explains that in Buenos Aires, there is a very high concentration of practicing psychoanalysts, and so seeing an analyst is a normative practice in the city: it is seen as a tool for general wellbeing. As Gaston says, "It's about learning to live your life."

---

### 5.2 Psychoanalysis in Buenos Aires

*P. Sean Brotherton*

"There was just a dark cloud that seemed to perpetually hover over me," said Gastón Fernández, 38, who moved to Buenos Aires in the mid-1990s from a small town in Santa Fe Province. "I just didn't care about anything or anybody," he lamented. After diagnosing Gastón with manic depression, a psychiatrist from his hometown put him on several medications. Notwithstanding the fluctuating dosages and changing medications, his general feelings of apathy never abated.

Shortly after arriving to the city to pursue university studies, Gastón sought out a new psychiatrist at one of the local public hospitals. "Do you have a therapist?" the psychiatrist asked during their first consultation. The question caught Gastón off guard. "I had always heard and read about psychoanalysts. I had seen them on TV and in newspaper articles. In high school, we were required to take psychology courses, which covered Freud's work fairly extensively," Gastón recounted. "But being from a small town, there was not really much access to that kind of therapy there." The psychiatrist explained that while medications could prove efficacious for certain people, psychotherapy could perhaps address other root causes that contributed to his illness. Gastón thus began attending biweekly sessions of psychotherapy.

A self-described Freudian analyst, the psychotherapist used their sessions to explore Gaston's inner experiences and thoughts to gain insight into the unconscious determinants of his behavior and how his past might be influencing the present. This clinical approach echoed the teachings of Sigmund Freud, who began developing the field of psychoanalysis at the turn of the twentieth century. In *Introductory Lectures on Psycho-Analysis* (1966 [1915]), Freud outlined his theory and therapeutic method for addressing how human behavior, experience, and cognition are largely determined by irrational drives that reside in the unconscious mind. According to Freud's model of the human psyche, conflicts that arise between the conscious view of reality and unconscious (repressed) drives result in psychological phenomena such as narcissism, hysteria, dreams, and the development of sexuality. Treating such psychopathologies, Freud advocated, could be achieved through the skilled guidance of an analyst, whereby a patient could gain 'insight' into bringing irrational (unconscious) material into consciousness. Freud made clear that as a clinical methodology, "(n)othing takes place in a psychoanalytic treatment but an interchange of words between the patient and the analyst. The patient talks, tells of his past experiences and present impressions, complains, confesses to his wishes and his emotional impulses. The doctor listens, tries to direct the patient's process of thought, exhorts, forces his attention in certain directions, gives him explanations and observes the reactions of understanding or rejection which he in this way provokes in him" (1966: 20–21).

Although psychoanalytic approaches have declined steadily worldwide, Argentina boasts the second-largest community of practitioners affiliated with the International Psychoanalytic Association (IPA), and has the highest number of Freudian analysts, with an equally high number of analysts who follow the doctrine of Sigmund Freud's disciple, Jacques Lacan. Only 19 years after the translation of Freud's complete works into Spanish, the development, integration, and proliferation of psychoanalysis in Argentina can be traced to the founding of the Argentine Psychoanalytic Association (APA), on December 15, 1942, as a non-profit academic institution. The founding members of the APA, including Argentine nationals and immigrants who had fled persecution in war-torn Europe, were shaped by a philosophical approach "in which individuals interested in the social as well as the personal domain of human liberation [turned to] psychoanalytic theories, including Marxism, to provide an understanding of the institutional and subjective limits on human productivity and pleasure" (Hollander 1990: 894).

Given the prestige of psychotherapy in Europe, it quickly became an "accouterment of upward social mobility and status" in Buenos Aires in the twentieth century (Hollander 1990: 897). Significant portions of the urban middle and upper classes sought out private sessions. Many of the analysts of the APA, however, also worked in public hospitals, and thus were able to utilize their training and expertise with patients from working-class backgrounds. More than a century after Freud's writings, the field of psychoanalysis has significantly transformed to include a diverse assemblage of theories, methods, treatments, concepts, and applications. It incorporates different schools of thought and methods (such as psychological stages, theory of archetypes/collective unconscious, death drive/jouissance, gestalt therapy) developed by different psychoanalytic theorists and practitioners in different countries.

Universal diagnostic categories for mental health, such as those found in the International Classification of Diseases (ICD) or the Diagnostic and Statistical Manual of Mental Disorders (DSM), often emphasize behavioral therapy and medical, surgical, or pharmaceutical interventions. Given that scientific medicine, neuroscience, biopsychiatry, and genetics have become the dominant models and organizing logics for mental health interventions in much of the world, how do professionals and laypeople in Buenos Aires develop alternative notions of what constitutes health and wellbeing?

## *Psychoanalytic Culture in Buenos Aires*

"[G]hosts appear when the trouble they represent and symptomize is no longer contained or repressed or blocked from view" (Gordon 2008: xvi). Such 'ghosts,' to borrow sociologist Avery Gordon's (2008) conceptual trope, abound in countless Argentine narratives of suffering and trauma, personal and collective, and in various public and private testimonials of loss, betrayal, longing, and nostalgia. While some of these sentiments are consciously articulated, others circulate at the level of repressed thoughts, sublimated feelings, and artifacts and fragments of past experiences. Argentina's history is characterized by prolonged and cyclical periods of economic growth and bust, militaristic violence, state-engineered disappearances, political instability, and debilitating austerity measures. Such apparitions, sometimes fleeting and intangible, are made visible and find fertile ground in psychoanalytic sessions.

The practice of psychoanalysis is almost ubiquitous in contemporary Argentina (Plotkin 2001). The leading authority on the history of psychoanalysis in the country, Mariano Ben Plotkin, argues that "(t)he culture of psychoanalysis runs so deep in Argentina that Argentines seldom reflect about it" (2003: 2). For instance, in January 2014, Argentina's Minister of Economy, Axel Kicillof, addressed the media about the country's rampant inflation and speculation on an impending economic meltdown. He accused the opposing political parties of generating 'psychosis.' Referring to the capital flight from the country's banking system, he said: "Economic phenomena can have this magic. . . . They are self-fulfilling prophecies, results of a herd mentality which have no real cause" (Romero and Gilbert 2014). Kicillof's psychoanalysis-inspired diagnosis of Argentina's economic problem was intelligible to a population for which the "everyday use of psychoanalytically inspired neologisms and the explicit references to psychoanalysis made by politicians and even generals . . . has become a *weltanschauung* [philosophy or view of life]" (Plotkin 1998: 271).

Argentina's psychoanalytic *weltanschauung* is particularly salient when read alongside empirical figures by the World Health Organization showing that the country has the highest number of practicing psychologists per capita in the world. Buenos Aires, the country's capital and largest city, has the highest concentration of practicing psychoanalysts; a residential area known as 'Villa Freud' is so called for its high concentration of psychoanalysts and psychiatrists. According to a 2009 national survey conducted by TNS Argentina, 32 percent of respondents have had a psychological consultation at some point (Moffett 2009).

Since the inception of the APA in the early 1940s, psychoanalytic approaches have been integrated into the facets of contemporary everyday life in the country, from the mundane to the spectacular, in the media, popular culture, political speeches, medical discourses, discussions of politics and the economy, and individual narratives about the body, health, and the state. Seeking psychotherapy is a normative practice. Dominant references from psychoanalytic theory—the unconscious mind, hysteria, defense mechanism, theory of the personality structure, Oedipal complex, fetish, and so on—have been popularized and incorporated into everyday idioms in myriad social contexts. Prominent psychoanalysts are quoted in newspaper articles, comic strips depict characters receiving therapy, magazines have columns dedicated to psychotherapy, the daily news includes psychoanalytic commentary on political and economic affairs, and popular reality TV programs show celebrities being psychoanalyzed.

Beyond popular culture, the various mechanisms through which the discourses and practices of psychoanalysis are reinforced, evaluated, transformed, challenged, spread, and inserted into everyday social interactions are, in part, influenced by institutions for public and private education. Psychology as a unique discipline and profession was introduced in Argentina in the mid-1950s. The original curricula focused on educational, industrial, and clinical psychology to train specialized professionals to work on individual and social development. In the mid-1960s, psychoanalysis became more embedded in the curricula as more APA-trained teachers took up positions in higher education. From the 1960s onward, Freudian, Lacanian, and Marxist philosophy intersected, which was represented in significant changes in the psychoanalytic community in Buenos Aires; this included in the academy (e.g. Faculty of Arts and Letters, University of Buenos Aires), in professional organizations (e.g. Argentine Psychiatric Association), and in political organizations (e.g. revolutionary Peronism). At that time, the political neutrality of orthodox approaches to psychoanalytic treatment led to accusations of the leadership of the APA conniving with the rise of the authoritarianism (which took hold of the country in the last military dictatorship, from 1976–1983). This culminated in the splitting of the APA in 1971, and the founding of two new groups, based on ideological differences.

Juridical changes also granted more authority to psychologists to participate in the mental health field. Prior to 1985, psychotherapy was restricted to the practice of physicians. In 1985, a new law loosened this restriction to allow psychologists (with undergraduate degrees) to also practice, although there are still no regulations on the national level for the accreditation of psychotherapists or the regulation of their practice. Despite these juridical changes, there remains a clear epistemological distinction in the training and clinical practice of psychologists and psychiatrists. Not all psychologists or psychiatrists in the country are sympathetic to psychoanalytic approaches, nor are the relationships between the different professional groups always antagonistic (Lakoff 2005). Gaston's psychiatrist, for example, was very comfortable in recommending him to a psychotherapist.

Buenos Aires is thus home to a very heterogeneous field of therapeutic techniques yet to be fully documented. Psychoanalysis in Buenos Aires forms part of and has produced a distinct 'epistemic culture,' a term coined by sociologist Karin Knorr-Centina (1999) to describe a specific body of knowledge, values, expertise, and practices. Given the impossibility of presenting a definite or authoritative account of the Argentine (or more specifically, *porteño*, a person who lives in the city of

Buenos Aires) experience of psychotherapy, I draw on Gastón's singular experience because it reflects many of the themes in other interviews I conducted in Buenos Aires in 2011 and 2013. These interviews revealed how psychoanalysis provides an interpretative and therapeutic framework through which individuals can come to imagine, narrate, and understand their bodies, health, and wellbeing in highly specialized ways. Through an ethnographic examination of psychoanalysis, I engage with Freud's position that "you cannot be present as an audience at a psycho-analytic treatment. You can only be told about it, and, in the strictest sense of the word, it is only by hearsay that you will get to know psycho-analysis" (Freud 1966: 20).

## *Psychoanalysis as a Tool for Wellbeing*

"I have been with my therapist for more than 10 years now," Gastón remarked. "Of course, this is an out-of-pocket expense for me because my therapist works in the private sector. But she has a sliding scale for payments for people like me, with low incomes. We have worked through so many things together, more than just my depression. My therapist is in regular contact with my psychiatrist to talk about my breakthroughs and progress. These sessions are not about *fixing me.*"

As Gastón explained, his biweekly sessions, each 45–60 minutes, were open conversations. He did not attend each session with the intention of 'treating' something or with a specific goal in mind. Sometimes he recounted events or thoughts he had been contemplating since they had last met, and the dialogue proceeded very organically. His analyst often reminded him of things he had previously said, many of which he did not remember. They would dwell and reflect on these moments. Gastón felt the sessions provided him with an empathic space to express himself freely, uncensored; there he could be heard, learn, and, in his words, "move away from the literal facts of what I say to reflect on what structures those very words, and my own way of reacting to that structure. You learn that repeated behavior can be a product of this structure, in your unconscious."

"Psychoanalysis is one way to produce self-knowledge (*autoconocimiento*)," Gastón replied when I asked why he pursued this form of therapy. He had also tried Lacanian therapy and cognitive behavioral therapy. But he was adamant that, although he did not accept the entire 'Freudian package,' his Freudian analyst worked the best for him. His analyst made it clear that there was no true interpretation of Freud, and the model itself was malleable. Gastón told me that his sessions were central to his weekly schedule; even when he traveled for work he used Skype to keep his appointments.

> It is about learning to live your life, and how to interpret things that happen in your life. Going to therapy is part of my life now. It is part of the way I think. You reflexively learn about yourself through your own experiences. This is not knowledge produced by hearing or learning 'what you are like' from other people who may know you, and, perhaps, are too lenient with certain aspects of your behavior—or from those who don't know you and have preconceptions of what you are like or, worse, make superficial judgments about you. My psychoanalyst presents no solutions and offers no absolute answers, although she definitely proposes alternatives that can produce healthy outcomes. That is why I think it is a major part of my overall health.

What crystallizes in Gastón's narrative is the way in which psychotherapy is both a comprehensive *theory* about human nature, experience, and development, and a method of *treatment* for psychological problems.

★ ★ ★ ★

Shortly after World War II, the World Health Organization defined health as a state of complete physical, mental, and social wellbeing and not merely the absence of disease or infirmity. While much debate and controversy has ensued on the feasibility of such an idealist conceptualization of health, we need to account for the diverse experiences, like Gastón's, whereby health is not merely a desired outcome but a process that one actively works towards. This provokes us to ask: in the psychoanalytic tradition, what constitutes health, or the notion of a healthy subject, why, and to whom? This case study addresses this by exploring the complex entanglements of different therapeutic regimes of care in the field of mental health and physical wellbeing.

Since its inception, psychoanalysis has always had its detractors. As Freud cautioned, "Persons who are impressed by the visible and tangible . . . [n]ever miss an opportunity of voicing skepticism as to how one can 'do anything for the malady through mere talk' " (1966: 20). However, Gastón concluded, "Luckily, I think Argentine society is increasingly open to considering the individual as a whole, comprised of body, soul, and mind. Physical health and mental health are intertwined; you need to have a balance. For me, being in touch with my psyche is important to my physical health." Psychoanalysis and the associated therapeutic techniques, which identify diverse psychopathologies and offer circumscribed interventions to address them, have produced an idiom through which different social actors and institutions articulate concerns about physical and mental wellbeing, as well as address social, economic, and political change.

Diverse psychoanalytic traditions in Buenos Aires offer a critical space for collective lament, at times in a highly politicized fashion. The ghosts that haunt Argentine narratives often move beyond discussions of the individual psyche to implicate the state, governance, questions of social justice, and the collective unconscious. Narratives such as Gastón's offer important insight into how psychoanalysis as a practice and 'epistemic system,' broadly speaking, can be a conceptual tool for thinking cross-culturally about health and wellbeing.

### *References*

Freud, S. 1966 (originally published 1915) *Introductory Lectures on Psycho-Analysis.* New York and London: W.W. Norton and Company.

Gordon, A. 2008 *Ghostly Matters: Haunting and the Sociological Imagination.* Minneapolis, MN: University of Minnesota Press.

Hollander, N.C. 1990 Buenos Aires: Latin Mecca of psychoanalysis. *Social Research* 57(4):889–919.

Knorr-Centina, K. 1999 *Epistemic Cultures: How Science Makes Knowledge.* Cambridge, MA: Harvard University Press.

Lakoff, Andres 2005 *Pharmaceutical Reason: Knowledge and Value in Global Psychiatry.* Cambridge, UK: Cambridge University Press.

Moffett, M. 2009 Its GDP is depressed, but Argentina leads world in shrinks per capita. *Wall Street Journal.* October 19.

Plotkin, M.B. 1998 The diffusion of psychoanalysis in Argentina. *Latin America Research Review* 33(2):271–277.

——— 2001 *Freud in the Pampas. The Emergence and Development of a Psychoanalytic Culture in Argentina.* Stanford, CA: Stanford University Press.

——— ed. 2003 *Argentina on the Couch: Psychiatry, State, and Society, 1880-present.* Albuquerque, NM: University of New Mexico Press.

Romero, S., and J. Gilbert 2014 The influential minister behind Argentina's economic shift. *New York Times.* January 26.

---

## Escape

Beyond the use of pharmaceuticals and the use of psychotherapy, as described in the first two cases, social suffering and distress associated with living under difficult conditions, over which people have very limited control, have been associated with the widespread use of alcohol, tobacco, the irregular use of prescription drugs, and the use of illicit drugs, resulting in addiction and premature death (Raikhel and Garriott 2013; Singer 2008). Singer (2008) argues that in aggressively pushing a means for escape, the producers of both legal and illegal psychotropic drugs, alcohol manufacturers, and 'Big Tobacco'—the multinational corporations that dominate the production, supply, and marketing of cigarettes—contribute to the maintenance of the existing unequal structure of society. Singer's critical analysis echoes that provided by Mark Nichter and Mimi Nichter, who, in the final case study of this chapter, show how Indonesia's tobacco industry advertises specifically to men, encouraging them to smoke a cigarette together as a way to maintain emotional balance, free of "excessive worry and negative thinking." The authors emphasize that these advertisements are effective because losing control is not only a loss of dignity in Indonesian society, it is also dangerous: the loss of control makes one susceptible to malevolent spirits. Advertisements feed into these cultural notions about stress and emotional balance by promoting smoking to deal with emotionally charged issues, such as losing a job or having to manage a girlfriend's nagging. Mark and Mimi Nichter write: "The tag line is usually on the same theme: 'Don't give in, just enjoy a cigarette and it will pass.'"

## 5.3 Promoting Smoking in Indonesia

*Mark Nichter and Mimi Nichter*

Smoking is pervasive in Indonesia: over 60 percent of men and approximately 3 percent of women smoke. The tobacco industry in Indonesia utilizes aggressive marketing strategies. The industry conducts extensive market research to identify what themes are most effective for selling cigarettes to different market sectors. Many themes in tobacco advertisements are transnational—themes of sophistication and success, masculinity, friendship, independence, sexiness, and having fun. But as we illustrate below, more subtle themes link smoking in Indonesia to cultural values and collective anxieties, the control of negative emotions, protection from malevolent forces, stress management as a form of self-medication, and enhancement. In these many ways, advertisements have established smoking as a normative practice with social utility (Nichter et al. 2009).

One of the most basic of Indonesian concepts—that influences what clothes people wear, the shampoo they purchase, how they judge the efficacy of medication, and their choice of cigarettes—is *cocok*. This Bahasa Indonesia concept has been translated as compatibility, but it means far more. The term is used in many contexts and does not just designate what is compatible, but what is suitable. To say something is cocok is also a statement that something is 'befitting' and 'healthy' in the original sense of the term, something that contributes to one's own sense of wellbeing. In Dutch, the term *gezondheid* does not index what is healthy for all people because what is healthy for you may not be what is healthy for me. In Indonesia, a medicine that is cocok for me is not just efficacious, but a medicine that has few side effects when taken. Cocok designates a medicine as beneficial to one's person, not just useful for treating a particular pathogen. If the medicine is given to someone else for the same complaint and if does not prove effective, it may not be cocok for them. Following this reasoning, when a medicine does not work, one does not conclude it is necessarily the wrong medicine for a specific disease or that it does not work because of drug resistance. Rather, it may simply not be cocok for them. One finds a similar concept termed *hiyang* in the Philippines (Hardon 1991). In the case of shampoo, a product that is cocok enhances the luster of one's hair; in the case of clothes it brings out one's best features. By way of a compliment, someone else may say, "yes, that is cocok for you."

So what does this have to do with cigarettes? A lot. It is thought that there is a brand of cigarette that is cocok for each type of person. One person's brand may be a heavy, strong, clove cigarette (*kretek*) that matches their persona; another person, who has a mild personality, may find a light, filtered cigarette cocok. Cigarettes are cleverly marketed to index subtle sets of associations and not just appeal to social significations such as social class. One's brand may be very personal, leading that person to refuse offers of other brands of cigarettes from friends. While researching smoking cessation in Indonesia, one of the only culturally appropriate ways of turning down the offer of a cigarette was to say, "Sorry, it is not cocok for me, I smoke my own brand or hand-rolled cigarettes."

The concept of cocok is exploited in what is a very sophisticated marketing strategy that juxtaposes the quest for a cigarette that is cocok with the importance of being brand loyal as a social value, once one finds the right cigarette. One respondent jokingly described this to us as the difference between searching for a girlfriend and finding a woman who was cocok with him, marrying her, and remaining loyal—even if one was tempted to try new brands.

Smokers search for a suitable brand of cigarette with the idea that one exists for everyone; you just have to find it. If you smoke and cough, or feel some irritation when you are smoking, it's a sign that it is the wrong brand—that brand is not cocok for you. If you are not a heavy smoker and become ill, and you are told that the reason for your illness is your smoking, you may be smoking the wrong brand. If, on the other hand, you smoke a cigarette that has long been cocok for you and experience symptoms like fast breathing after smoking, this may be a sign of ill health. Thus smokers interpret doctor's advice not to smoke as "You are too sick to smoke now. Don't smoke until you get healthy enough to start up again." Indeed, when a smoker gives up smoking, others may consider this a sign of ill health. Resuming smoking signals that things are now okay; the person has recovered from his illness and is strong enough to smoke again.

As people age, their cigarette brand may change. Young men may smoke light cigarettes (e.g. the popular youth brand, Mild A) and graduate to stronger ones later. But, once a person finds his cigarette brand, the one cocok for him, the tobacco industry encourages him to be loyal as it defines him as much as he defines the brand. Loyalty to the brand is linked to loyalty as a cultural value.

As anthropologists working in the field of tobacco control in Indonesia for 12 years, two of our biggest challenges were the idea that 'smoking was cultural' and that a brand of cigarette that is cocok will not harm the smoker if he smokes in moderation (10–12 cigarettes a day). A third was the dedication that smokers often showed for their brand. Their brand became a part of their self-identity.

An article and editorial in the journal *Tobacco Control* focused attention on one particular cigarette advertisement in which someone extends a hand to a person trying to board a moving bus with the tag line, "It is better to die than to leave an old friend behind." An analysis of the advertisement ties it to the timing of its appearance (during Ramadan), and suggests that the advertisement reminded people not to forget about smoking during the month of ritual fasting (Sebayang et al. 2012). Whether this was the intent—given that most Muslim smokers continue to smoke each evening when they break their fast—the core message spoke to loyalty.

Smoking in Indonesia is also tied to the core value of togetherness—*ramai*. Cigarettes are marketed as a primary way of sharing a collective experience—enjoying a family gathering, a party, hanging out with friends, or breaking the fast during Ramadan. Refusing a cigarette (or any consumable product) offered out of friendship is deemed impolite and is socially awkward. Smoking is a way of passing time with others and sharing silence, and when smokers try to quit, they often speak of feeling isolated or left out. Smoking is advertised as a means of enhancing people's inherent sociability and providing them with self-confidence to interact and speak. Those attempting to quit smoking often say that they feel insecure and socially awkward; they no longer fit into their familiar flow of social interaction. A non-smoker may be encouraged to smoke by other family members to fit in. One of our key informants, for example, was challenged by his father for being a non-smoker. How could he not smoke when all of the men in his family for the past three generations were smokers? How could he feel part of his family when everyone else smoked? He was advised to at least smoke a mild cigarette.

This may seem vaguely familiar to the Western reader—cigarettes are readily recognized as having social utility (Nichter 2015). Even those who do not consider themselves smokers in the US may smoke at parties to interact with others while drinking, to look like they are doing something. There is, however, a deeper cultural dimension of smoking that speaks to an entirely different type of social utility.

In Indonesia, smoking is promoted as an aid to assist a person to be emotionally balanced. In Central Javanese culture, in particular, considerable attention is placed on men staying in emotional control, ideally free of excessive worry and negative thinking, and calm at all times. To lose control is not only a loss of dignity, but is dangerous. When a man loses control, malevolent spirits may impose their will on him. Control is admired as a virtue, and young men test their ability to control bodily functions like sleep and appetite. Daydreaming while bored is also dangerous, as one's thoughts wander and one's mind is easily swayed by forces beyond one's control (Ferzacca 2002). Smoking can assist a person to be calm and not daydream or dwell on life's problems.

Cigarette advertisements tap into this network of associations. Advertisements use the yin-yang symbol and offer a tag line promising balance in one's life and smooth relationships. They promote smoking to deal with emotionally charged issues like a girlfriend's nagging or continuous disagreements. The tag line is usually on the same theme: "Don't give in, just enjoy a cigarette and it will all pass"; smoking will help these relational worries go in one ear and out the other.

The same is true for stress. Smoking is promoted as a way of coping with a wide range of everyday stressors. Unemployment among the young is high in Indonesia. "Don't stress," cigarette ads tell young men, "there is nothing you can do, so just enjoy" with a cigarette. If you feel you are being manipulated and not appreciated in your job, ads suggest that a man should laugh it off, have a cigarette, and just wait, because one day you will be the puppet master. If you encounter obstructions in life, rise above them by smoking—a cigarette break will help you find a way through or around your problem.

Cigarettes are marketed as a 'drug food' that will help one think more clearly and creatively and be an antidote for over-thinking and feelings of anxiety when the solution to a problem is not apparent.[1] We were repeatedly told by informants that when they could not concentrate or experienced inertia, they smoked so creative new ideas would come. "Having a cigarette makes me feel refreshed," stated one respondent, "much like taking a bath." In ads, successful professionals are shown, with copy that asserts that smoking inspires their thinking.

Coffee shops with Wi-Fi are commonplace in urban Indonesia and are frequented by creative people who are seen at the forefront of change. Smoking is ubiquitous in coffee shops and one of the most likely places one encounters women smokers. As in the US decades ago, smoking is a symbol of women's empowerment in Indonesia. While cigarette ads do not yet feature women smoking, they do present fashionable, modern, upscale women in environments where men are smoking, where they are shown to be approving of the behavior.

Indonesians who are compelled to quit smoking, often because they are very ill, find it difficult. One reason people offer for not being able to quit is their reliance on smoking to cope with *stres*, a term which in Indonesia indexes an ambiguous state that includes worries, concerns, and anxieties about one's future. Smoking is commonly depicted as a means of controlling stress in movies and late-night TV advertisements for cigarettes. Smoking is associated with gaining control, not of becoming dependent, and the tobacco industry has gone to great lengths to distinguish cigarettes from drugs that represent uncontrolled dangerous states. Cigarettes are not spoken about as addictive, and terms referring to addiction, *ketagihan* and *kecanduan* in Bahasa Indonesia, were reserved for alcohol and illegal drug use. The association of smoking with control is so strong that mental institutions hand out cigarettes to patients to help keep them under control.

Cigarettes are not just thought of as a drug food good for the qualities of enhancing and controlling; they are also associated with the unique taste of Indonesia (cloves) and, by extension, cultural identity. Many brands of cigarettes are marketed as good enough for the global market yet having a distinctively local taste. They are as Indonesian as the blends of spices used in food. Smokers are encouraged to remain loyal to their brand, but are also enticed to "just try" what appears to be an endless flow of new brands. Travelers encounter new brands in the same way as they encounter new foods, as one informant remarked: "Well, you want to just try it and see what the taste is, like a dessert. You do not want to eat or smoke too much of it, just a taste for the experience. For the smoker, it is like a hobby."

But there is more to smoking than being a hobby. The same informant described his smoking as like eating rice, the staple food in Central Java. "I can eat many things and have a full belly," he said, "but I do not feel satisfied if I do not eat rice. Rice has to be there. It is the same with smoking. If I do not smoke, I feel there is something missing. I am not satisfied." His statement is revealing on many levels. Not only is smoking normative for him as a habit (and no doubt an addiction, given that he smokes 20 cigarettes a day), but he is deeply influenced by the physiology of smoking. It affects how and when he eats, how foods taste, his thirst and liquid consumption, and feelings of satisfaction. From other men, we heard more extreme comments: "I think I can live without food for some days, but never without cigarettes."

Later, the same informant called attention to a bodily state mentioned commonly by people trying to quit smoking. The informant spoke of *mulut kecut*, a sour taste in the mouth that occurs both when one feels hungry and after eating a rice meal. "I always smoke when I'm hungry and I get this taste in my mouth . . . and I smoke after every meal to avoid *mulut kecut* as well as to feel satisfied (*puas*)." Medical anthropologists pay close attention to cultural experiences of bodily sensation, as it reveals a lot about the embodied states that people seek and avoid. What is notable about *mulut kecut* is that rice, the very staple of life, is not enough to feel satisfied. Without a cigarette, even the experience of a good meal is soured. This feeling of lack that smokers experience is a bonanza for the tobacco industry.

* * * *

In the Maluku Islands, Indonesia's 'spice islands' with the highest cigarette consumption in the country (over 70 percent of men), while writing this discussion, we watched a fisherman light a cigarette after his lunch of rice, fish, and *sambal*. This is the same meal we have had twice a day now for weeks. At this moment, we can well appreciate the appeal of a "refreshing" taste after lunch. Yet it all seems rather paradoxical, the way things have turned out. From the fifteenth through the eighteenth centuries, all of maritime Europe was in competition to find faster routes for reaching these very islands. Whoever reached the Malukus first could make a fortune by bringing back spices to preserve and add flavor to Europe's bland diet, to disguise food of dubious quality, and to be used as medicines for ailments ranging from flatulence to the plague. Today, tobacco companies have reached these same remote islands and successfully sell a variety of brands of cigarettes as drug foods that add a little spice to life and the local diet. Kretek and menthol cigarettes are imported and nutmeg, cloves, and mace are exported—mace for use in the manufacture of Coca-Cola. And so, the flow of global trade continues as it has for centuries, and with it the exchange of one drug food for another.

### Note

1. We use the term 'drug food' to denote any addictive substance consumed as part of one's daily regime and taken with the intent of enabling one to work, think, or interact with others in a desirable fashion (Quintero and Nichter 2011).

### References

Ferzacca, S. 2002 A Javanese metropolis and mental life. *Ethos* 30:95–112.
Hardon, A.P. 1991 *Confronting Ill Health*. Quezon City, Philippines: Health Action Network.
Nichter, Mimi 2015 *Lighting Up: The Rise of Social Smoking on College Campuses*. New York: New York University Press.
Nichter, Mimi, S. Padmawati, M. Danardono, N. Ng, Y. Prabandari, and Mark Nichter 2009 Reading culture from tobacco advertisements in Indonesia. *Tobacco Control* 18:98–107.
Quintero, G., and Mark Nichter 2011 Generation RX: Anthropological research on pharmaceutical enhancement, lifestyle regulation, self-medication, and recreational drug use. In *A Companion to Medical Anthropology*. M. Singer and P. Erickson, eds. Pp. 339–355. Malden, MA: Wiley-Blackwell.
Sebayang, S., R. Rosemary, D. Widiatmoko, K. Mohamad, and L. Trisnantoro 2012 Better to die than leave a friend behind: Industry strategy to reach the young. *Tobacco Control* 21:370–372.

---

The focus in Indonesian marketing suggests the relevance of a gendered approach to managing stress. While it remains the case that men are more likely than women to binge drink and smoke, particularly in low-income settings, and that these behaviors are common especially among unskilled, unemployed, and underemployed men and service workers. In Australia in the 1950s and 1960s, pharmaceutical companies marketed A.P.C. powders—a combination of aspirin, phenacetin and caffeine—to women with the slogan "Stressful Day? What you need is a cup of tea, a Bex, and a good lie down." Over-the-counter sales of A.P.C. combination drugs with phenacetin were banned in Australia in 1979 and in the US in 1983 because of the association of phenacetin with kidney disease and various cancers. In the 1970s, Virginia Slim cigarettes were marketed to women as a symbol of liberation, with the slogan "You've come a long way, baby." Today, women in low- and middle-income countries are subject to similar strategies, and tobacco smoking and alcohol are increasingly being marketed to women (Kaufman and Nichter 2010).

* * * *

What can anthropologists contribute? We have described how linguistics can help us unpack the semantic networks involved in cultural notions of stress, and by focusing on the social use of the term, we can gain a better understanding of the socioeconomic and political dimensions of what's at stake in everyday life. Beginning with the concept of idioms of distress, we start by listening to people describe their feelings of stress in their own terms, while making sense of their often fragmentary, tentative, and sometimes even contradictory meanings (Nichter 2010), rather than trying to translate their symptoms into *DSM* categories. Applied anthropologists who simply draw on such biomedical categories risk contributing to a medicalization of the problem at hand. Instead, anthropologists might work with alternative community-based approaches, and come to a different understanding of people's emotional responses to relational, financial, and other social problems. At the same time, we should not shy away from the biological stress reactions. Some anthropologists are engaging in interdisciplinary studies which incorporate biomarkers for stress (elevated cortisol levels measured in saliva) in their study designs, generating novel insights on the complex biocultural mechanisms through which household tensions and societal conflicts affect body and mind (Kohrt et al. 2014).

Cultures have developed ways to deal with these stresses, and there may be positive benefits to the existing methods of handling stress: addressing bewitchment can bring about resolution to tensions in social relationships, and expressions of *nervios* or *susto* can draw attention to inequalities. Of course, societies find ways to distinguish between stress and madness, but if we better understand idioms of distress, we can see that even the most severe mental health problems might be approached differently. While not idealizing or glamorizing non-biomedical approaches, given that people are highly stigmatized and ostracized in some cases, anthropologists can bring a greater contextual understanding, one that takes seriously local ways of knowing and expressing stress and illness.

## References


Asch, Timothy, Linda Connor, and Patsy Asch 2015 [1980–1981] *A Balinese Trance Seance and Jero on Jero: A Balinese Trance Seance Observed.* Watertown, MA: Documentary Educational Resources.

Baer, Roberta, Susan C. Weller, Javier Garcia de Alba Garcia, Mark Glazer, Robert Trotter, Lee Pachter, and Robert E. Klein 2003 A cross-cultural approach to the study of the folk illness "nervios". *Culture, Medicine and Psychiatry* 27(3):315–337.

Beattie, John 1967 Consulting a diviner in Bunyoro: A text. *Ethnology* 5(2):202–217.

Brunner, Eric, and Michael Marmot 1999 Social organization, stress and health. In *The Social Determinants of Health.* E. Brunner and M. Marmot, eds. Pp. 17–23. Oxford, UK: Oxford University Press.

Cartwright, Elizabeth 2007 Bodily remembering: Memory, place, and understanding Latino folk illnesses among the Amuzgos Indians of Oaxaca, Mexico. *Culture, Medicine and Psychiatry* 31(4):527–545.

Connor, Linda, Patsy Asch, and Timothy Asch 1996 [1986] *Jero Tapakan: Balinese Healer. An Ethnographic Film Monograph.* Los Angeles, CA: University of Southern California.

Dumit, Joseph 2010 Pharmaceutical witnessing: Drugs for life in an era of direct-to-consumer advertising. In *Technologized Images, Technologized Bodies.* Jeanette Edwards, Penelope Harvey, and Peter Wade, eds. Pp. 37–64. New York: Berghahn.

Ecks, Stefan 2014 *Eating Drugs: Psychopharmaceutical Pluralism in India.* New York: New York University Press.

Estroff, Sue E. 1981 *Making It Crazy: An Ethnography of Psychiatric Clients in an American Community.* Berkeley, CA: University of California Press.

Favret-Saada, Jeanne 1980 *Deadly Words: Witchcraft in the Bocage.* Trans. Catherine Cullen. Cambridge, UK: Cambridge University Press. Originally published in French in 1977.

Geschiere, Peter 2010 Witchcraft and modernity: Perspectives from Africa and beyond. In *Sorcery in the Black Atlantic.* Luis Nicolau Parés and Roger Sansi, eds. Pp. 233–258. Chicago, IL: University of Chicago Press.

——— 2013 *Witchcraft, Intimacy and Trust: Africa in Comparison.* Chicago, IL: University of Chicago Press.

Good, Byron 1977 The heart of what's the matter: Semantics and illness in Iran. *Culture, Medicine and Psychiatry* 1:25–58.

——— 2010 The complexities of psychopharmaceutical hegemonies in Indonesia. In *The Pharmaceutical Self: The Global Shaping of Experience in the Age of Psychopharmacology.* Janis H. Jenkins, ed. Pp. 117–145. Santa Fe, NM: School for Advanced Research Press.

Guarnaccia, Peter J., Roberto Lewis-Fernández, and Melissa Rivera Marano 2003 Toward a Puerto Rican popular nosology: Nervios and ataque de nervios. *Culture, Medicine and Psychiatry* 27(3):339–366.

Jenkins, Janice H., ed. 2010 *The Pharmaceutical Self: The Global Shaping of Experience in the Age of Psychopharmacology.* Santa Fe, NM: School for Advanced Research Press.

Kaiser, Bonnie N., Kristen McLean, Brandon Kohrt, Ashley Hagaman, Bradley Wagenaar, Nayla Khoury, and Hunter Keys 2014 Reflechi twop—Thinking too much: Description of a cultural syndrome in Haiti's central plateau. *Culture, Medicine and Psychiatry* 38(3):448–472.

Kaufman, Nancy J., and Mimi Nichter 2010 The marketing of tobacco to women: Global perspectives. In *Gender, Women and the Tobacco Epidemic.* Jonathan M. Samet and Soon-Young Yoon, eds. Pp. 105–136. Geneva, Switzerland: World Health Organization.

Kirmayer, Laurence J., and Allan Young 1988 Culture and somatization: Clinical, epidemiological, and ethnographic perspectives. *Psychosomatic Medicine* 60(4):420–430.

Kohrt, Brandon A., Daniel J. Hruschka, Holbrook E. Kohrt, Victor G. Carrion, Irwin D. Waldman, and Carol M. Worthman 2014 Child abuse, disruptive behavior disorders, depression, and salivary cortisol levels among institutionalized and community-residing boys in Mongolia. *Asia-Pacific Psychiatry* 7:7–19. DOI: 10.1111/appy.12141.

Martin, Emily 2010 Sleepless in America. In *The Pharmaceutical Self: The Global Shaping of Experience in the Age of Psychopharmacology*. J.H. Jenkins, ed. Pp. 187–209. Santa Fe, NM: School for Advanced Research Press.

Mysyk, Avis, Margaret England, and Juan Arturo Avila Gallegos 2008 Nerves as embodied metaphor in the Canada/Mexico seasonal agricultural workers program. *Medical Anthropology* 27(4):383–404.

Nichter, Mark 1981 Idioms of distress—alternatives in the expression of psycho-social distress—a case study from South India. *Culture, Medicine and Psychiatry* 5(4):379–408.

——— 2010 Idioms of distress revisited. *Culture, Medicine and Psychiatry* 34(2):401–416.

Ong, Aihwa 1987 *Spirits of Resistance and Capitalist Discipline: Factory Women in Malaysia*. Albany, NY: State University of New York Press.

Prince, Martin, Vikram Patel, Shekhar Saxena, Mario Maj, Joanna Maselko, Michael R Phillips, and Atif Rahman 2007 No health without mental health. *The Lancet* 370:859–877.

Raikhel, Eugene, and William Garriott, eds. 2013 *Addiction Trajectories*. Durham, NC: Duke University Press.

Rose, Nikolas 2003 Neurochemical selves. *Society* 41(1):46–59.

Singer, Merrill 2008 *Drugging the Poor: Legal and Illegal Drugs and Social Inequality*. Long Grove, IL: Waveland Press.

Summerfield, Derek 2012 Afterword: Against "global mental health". *Transcultural Psychiatry* 49(3–4):519–530.

Watters, Ethan 2010 *Crazy Like Us: The Globalization of the American Psyche*. New York: Free Press.

Weller, Susan C., Roberta D. Baer, Javier Garcia Alba de Garcia, Mark Glazer, Robert Trotter, Lee Pachter, and Robert E. Klein 2002 Regional variation in Latino descriptions of susto. *Culture, Medicine and Psychiatry* 26(4):449–472.

Wolf-Meyer, Matthew J. 2012 *The Slumbering Masses: Sleep, Medicine, and Modern American Life*. Minneapolis, MN: University of Minnesota Press.

Smoke Free Homes Initiative, 2013. Kerala, India.

## About the photograph

*Tobacco-related mortality in India is among the highest in the world. In this photo, a young girl in Kerala State in South India reads a poster which explains the dangers of smoking and the harms of secondhand smoke to women and children. This social movement, spearheaded by anthropologists and public health specialists, has reached her village. In this poster and through various activities, community members are asked to join together and ban smoking inside their homes. As part of this intervention, school children are provided education about tobacco, and are asked to talk to their fathers about not smoking inside the home to protect their family members from the harm of secondhand smoke. Ethnographic research in Kerala, in which I was involved, revealed that daughters were able to "touch the hearts" of their fathers, and were able to convey a stronger message than their mothers. I took this photograph when I working as a Co-Principal Investigator on Quit Tobacco India.*

—Mimi Nichter

# 6

# Bodily Resistances

*Elizabeth Cartwright, Anita Hardon and Lenore Manderson*

The body is a canvas, both physical and abstract, that is used to demonstrate who we are, where we belong and how we are taking a stand in the world. In this chapter, we deal with issues surrounding bodily resistances. We begin with the tensions inherent in acts of rebellion and resistance, and the desire for particular kinds of social belonging. Tensions are created via bodily rebellions at both the individual and social level, simultaneously creating difference and distance, as well as acceptance and inclusion. Tattoos, scarification, and circumcision are among the bodily practices that individuals use to signify something about themselves; so too are the ways we wear our hair, cover our bodies (or not), perfume ourselves, and otherwise decorate and manipulate our corporeal selves. In some instances, where an individual refuses to undergo a particular practice (female circumcision, for instance), the social risks are high even though the practice may increase other bodily risks. With each of the particular ways that individuals make these flexible, artistic, sometimes desperate statements about who they are, different constellations of bodily strengths and weaknesses are created that have both social and health implications. Identities are not constructed in isolation. They emerge in social networks that shape their way of being in the world in relation to each other, and to the institutions that they encounter.

Resistance is often directed at groups of people and institutions and the rules and regulations that they have put in place that shape people's lives through coercion or consent. Health care programs in particular have implications for the way people live their embodied lives. We are encouraged to eat less, move more, and refrain from smoking, and to engage in elaborate accounting mechanisms to ensure that our health care costs are reimbursed. In resisting such prescriptions and proscriptions, people make statements about authority, power and agency, and identity. Smokers cut themselves off from their 'non-smoker' friends, for example, while gaining friendship and belonging with the other 'smokers.' People with eating disorders learn new rules and ways of being within different social worlds where denying nourishment, binging, and purging are valued, while leaving behind the deeply social acts of sharing mealtimes and the sensual pleasures of savoring food. People may refuse medication and reject other behavioral prescriptions in acts of resistance to the particular subjectivity of the body to a diagnosis, its treatment regimes, and the power of medical practitioners, biomedicine, and often family. They may seek the support of others in their acts of resistance, or be influenced by social movements that encourage them to adopt certain health practices, while resisting others. Others embrace medical and surgical procedures in ways that defy normative views: the decision of a man or a woman to pay for a surrogate in order to parent is a resistance to hetero-normative and biologically based ideas of human reproduction. What can we learn about the ways that people act in response to medical and public health advice, and how they interact with health care institutions, and how and in what ways they resist?

Anthropology has a particularly strong history of theorizing small-scale rebellions and everyday acts of resistance (Abu-Lughod 1990; Scott 1985), and describing the forms of resistance to biomedical ideas and practices has been an important trajectory of inquiry in medical anthropology over recent decades. It continues to be a fertile ground for research (Marshall 2012; Nahar and Geest 2014; Sivaramakrishnan 2005). In the end, one must ask, what does it matter? Does it do any good to resist? Who is harmed? And conversely, what is changed for the better? What new configurations of care can be imagined once the dominant 'health' paradigms have been challenged?

Anthropological studies of resistance provide us with a starting point to tease out the essential components of everyday forms of resistance that can then be applied to matters concerning health, healing, and bodies. In his now-classic volume, *Learning to Labor: Why Working Class Kids Get Working Class Jobs* (1977), Paul Willis described the ways in which young men in the United Kingdom rebelled against formal schooling, and by doing so destroyed their chances for obtaining a good education and so the possibility of getting a job that was not tied to physical labor. The brilliance of Willis's work was in describing the verbal and non-verbal ways that these young men enacted their small rebellions with educational institutions that, in many ways, neglected their needs. The stated goal of the educators with whom Willis worked was to encourage youth to study hard, succeed in school, and so proceed, via university or other sorts of advanced training, to white-collar jobs—this goal was only partially conveyed to the working class 'lads.' By skipping school, dressing and talking in rebellious ways, and taunting those students who were good at their studies, the 'lads' sealed their own fate of failing high school. The lads' primary goal—to establish strong friendship networks while in high school, so that they would have friends on the shop floor when they went to work in factories—made good sense when looked at from their perspective. Studying took valuable time away from engaging in the kinds of adventures and 'larks' that created social bonding among their peers, which would be essential for their social survival at school and beyond school, once they had to start working as the breadwinners for their families. These young men had insight, if limited, into how the educational system worked. They knew they would never get high-paying jobs without degrees, but they also knew how to establish the social bonds—and so the social capital—that would allow them to get what they needed to survive in the world of industrial workers in 1970s Britain. In addition, perhaps there was also a sharp understanding of the implications of class in Britain at that time: they probably would not have had an equal opportunity to gain employment in the areas of their training, even with a higher education, and so saw the futility of staying at school.

James Scott, in *Weapons of the Weak* (1985), continued exploring and elaborating Willis's ideas of how less powerful groups in society have or don't have insight into how they are being controlled. Scott maintained that systems of power are rarely fully hegemonic. If, drawing on Gramsci's work (1992), hegemony is defined as the material and ideological methods used to force, trick, or subtly pressure groups of people to comply and so ensure their consent or compliance, then resistances can be found and described at many levels. Hegemony is rarely, if ever, complete, as it is enacted through a mixture of control and consent. To truly create a revolutionary moment, people who are oppressed need to gain material and intellectual control over the ideas and the resources being used by those in power, and to take them for their own. These understandings of hegemony and everyday acts of resistance can be fruitfully applied to understanding bodily resistances as they are acted out via resistance to the dominant biomedical and social ideas of particular situations. Following Scott (1985), such resistance may include foot-dragging, feigned ignorance, passivity, slander, and pilfering, and other similar acts, providing often subtle ways of rebellion and subversion for people who lack formal avenues to oppose others' authority over their lives. As Sivaramakrishnan (2005) points out, the more complete analysis of protest includes not only everyday forms of resistance, but also an analysis of everyday enactments of power.

Biomedicine's hegemonic dominance varies by situation. While it is the primary form of curing in many resource-rich countries, and for those of the upper classes in poorer countries, many of the world's poorest people have limited access to biomedicine and further, depending on their health status and the information they have about various approaches to treatment, they may reject it as a curing option. Biomedical technologies generate their power through a nearly all-encompassing system of visualization, diagnosis, and bodily manipulation. The institution of biomedicine, in its fullest expression, is found in high-tech hospitals where patients can be monitored, controlled, and treated using this logic of healing. As the enactment of biomedicine moves further from resources of all kinds, its grip on the local logic of curing is loosened; its hegemony is diminished.[1] Moreover, there are as many forms of biomedicine as there are places where it is being used. Resistance in each of those situations comes in different forms that are telling of the particularities of biomedical hegemony, local knowledge, traditions, resources, and inclinations.

A particular poignancy arises when considering acts of resistance in the context of ill health; the stakes are high, emotions accentuated. Acts of rebellion, resistance, dissimulation, and non-compliance can be seen variously as ill founded, ignorant, enlightened, or brave. To go against the locally accepted norms of maintaining health, being cured of a malady, or managing bodily processes elicits strong feelings from onlookers and puts the individual or groups of individuals engaged in the resistance into a fraught position, both bodily and socially. For instance, women in the US and other heavily biomedicalized places who choose to give birth at home can be subjected to stories of risk to the health of their babies, ridiculed by biomedical practitioners with whom they come in contact, and censured for less than optimal birth outcomes. Legislation also controls what people can access or refuse, limiting the options available to them. Going against the norm, especially when it is a biomedical norm, puts individuals into the uncomfortable social position of asserting their right to control their bodies and to opt out of biomedicine's grip.

Aengst's work (2014) in Ladakh shows how women consciously resist local norms that prohibit out-of-wedlock sexual relationships. When women in this situation become pregnant, their private resistances are exposed bodily through their pregnancies. The women with whom Aengst worked tried to hide their pregnancies for as long as possible. The difficult choices the women faced—in the context of the highly negative stigma that their ex-nuptial pregnancies carried, included of shaming themselves and their families—led them to endure the risks of pregnancy without any medical attention and sometimes, tragically, taking on the responsibility of killing their child. Narratives of how women dealt with their competing needs of sexual pleasure and intimacy, in the absence of the protective elements of social acceptance of these acts and/or of effective birth control, are not unique to Ladakh (see Briggs 2007), but they clearly show how a particular bodily act can put into motion both physical and social reactions that place individuals in extremely difficult situations.

Hilary Graham's (1976; 1987) research on the social ramifications of smoking provides a rather different perspective of resistance. The idea that smoking was a danger to women's health, and to their pregnancies, was just coming into the public conscience at the time of her original research, and women for the most part saw smoking as less a risk to themselves and their fetus than doing heavy housework or having sex in pregnancy (Graham 1976). Women reasoned, too, that they would become irritable if they gave up smoking and this would cause problems with their spouses, families, and friends. Seen from this vantage point, the women's different assessment of health risks makes sense. In a later article, Graham wrote about a "backcloth of poverty" that shaped the choices women made, which included resistance to health messages to give up smoking (1987: 51). The women who smoked, in contrast to non-smokers, were less likely to have a friend in whom they could confide and had problematic relationships with their mothers; their partners were often unemployed, and they were worried about money. Tired, stressed, and lonely, women used smoking with a cup of tea or coffee to take time out from the work of caring. For these women, Graham argued, women's smoking made sense in the "hidden world of caring in poverty . . . a cigarette can be the only purchase that women make, and (smoking) the only activity that women do, just for themselves" (1987: 55).

In this chapter, the case studies illustrate situations where individuals rebel in ways that seem, on the surface, to be self-defeating; for instance, we see people rebelling against medical authority by continuing to smoke despite clear evidence of its impact on their health, and young women practicing self-starvation. These acts of resistance reveal what matters most to individuals and those with whom they identify and interact in particular circumstances. In the first case, Megan Wainwright describes a Uruguayan woman named Julia, who has chronic obstructive pulmonary disease (COPD), diabetes, cardiovascular disease, and an array of aches, pains, and sufferings. Her resistance to her doctor's command to stop smoking is couched in the fact that in her life she has given up many other things because of her medical conditions; she can no longer have sugar, salt, or bread. She draws the line at giving up her cigarettes too. Julia views cigarettes as her one 'guilty pleasure,' a pleasure that poses the least threat to her health. This is not a view shared by her physician or, over the years, by her husband.

---

## 6.1 Rebellion and Co-Morbidity

*Megan Wainwright*

In 2010, while conducting ethnographic research on chronic obstructive pulmonary disease (COPD) in Uruguay, I met 61-year-old Julia. Over time, I learned that COPD was just one of several conditions she lived with, including diabetes, gastric reflux, hypertension, cardiovascular angina, mild renal disease, and chronic back and leg pain. Many of these are considered smoking-related illnesses. Julia had smoked on average 15 cigarettes a day for 46 years, six more years than her relationship with Fernando. She also had long-standing relationships with a plethora of health professionals whom she saw so regularly that her adult children referred to the hospital as her 'other home.' This is where my story begins.

### *October 2010*

I was in a public outpatient respiratory clinic in Montevideo when a physician invited me to meet a patient who was developing respiratory illness and still smoked:

*Julia:* I quit smoking for seven months but started again. My husband has many health problems and his situation made me start up again. I don't smoke during the day when at work, but my husband smokes indoors so I smoke with him after meals, but less than before.

*Doctor:* Your complexion is looking really lovely since you've cut down on the cigarettes.

*Julia:* Yes, people have been telling me that.

*Doctor:* Have you noticed a change?

*Julia:* Not really . . . I don't have as much phlegm.

*Doctor (to me):* In 2009 they found that the small airways were affected, but it wasn't quite obstruction.

*Doctor (to Julia):* This is the time to do something, you are starting to develop a chronic disease (COPD). (She picks up Julia's chest X-ray and looks at it in the light.) You're as beautiful on the inside as you are on the outside! I don't want to scold you, I'm just saying, each time you take a cigarette think about it, do I really want this cigarette?

*Julia (to me):* If you come to the house you will meet not one smoker but two! My husband will not quit. With all his health problems he has a lot restrictions put on him, so he says, "they're not going to take this from me as well."

## *November 2010*

Upon entering Julia and Fernando's four-bedroom apartment within a cooperative housing development, I noticed the walls, once white, but now, from decades of smoking, yellowy brown. That afternoon, the conversation around health and smoking was particularly lighthearted.

*Fernando:* We know that smoking is killing us, but we're just racing each other now, to see who will be the first (to die). With so many years together, we do not want to go one without the other. What they are doing now with smoking is prohibitive (smoke-free laws), and that makes you want to do it more . . . We come from a generation that lived through many struggles, the dictatorship for one, and we always fought against it. Well this is the same. Doctors need to know that. Anything that is forced, annoys us. The policies may work better for those that came after us.

*Julia:* There are too many impositions. Well ok, it's harming me, but the thing is . . . smoking is a satisfaction for me, and I smoke a third of what I smoked before, now I smoke one every once in a while . . . I can't eat sugar, salt, pasta, bread, pastries, a sandwich. There comes a time when you ask yourself, why am I living if I can't do anything? I can't run, I can't walk, well a little but I get tired and I have to stop, that limits you as a person . . . Even though it is bad for me, (smoking's) the thing that does me the least harm. Because if I eat sugar I can go into diabetic coma.

## *January 2011*

I learned over the phone that soon after my November visit, Julia had had a heart attack. On this visit, they argued profusely and the mood, unlike our previous encounter, was tense. It was the only time I heard Fernando comment on Julia's smoking in a concerned and disapproving way:

*Julia:* I'm still smoking like before. I've not cut down at all, I smoke even more than I did. I'm stuck here in the house all day (she quit work) and all I do is smoke.

*Fernando:* She smokes, and I mean she smokes! She smokes more than me!

*Julia:* Yeah, let's say I was smoking 10, but now I'm smoking like 30 a day. When I first got back from that episode, I just didn't have the energy to get up and move. I slept and slept and slept and I mean really slept. Fernando felt like I was getting depressed . . . I felt like I'd been mistreated in the hospital (describes how doctors missed the diabetes and questioned her chest pain). I know I'm a difficult patient.

*Fernando:* She's a difficult patient because she doesn't listen to the doctor. She doesn't follow what they say . . . The patient also has to help themselves.

*Julia:* I take my medications.

*Fernando:* You do whatever you like.

*Julia:* What, because I smoke?

*Fernando:* Ah there you go.

*Julia:* Ah, well! I cannot handle any more (restrictions) . . . There's nothing else to take from me. No Coca-Cola, none, just water. You watch everyone eat and you can't eat anything. All those things, for what? I take 25 pills a day. Now I'm here, inside . . . I would like to keep on living yes, at least a year and a half more, for my granddaughter's 15th . . . The truth is I'm tired . . . I'm not able to do things, because I'm not able to breathe, to move, more and more . . . My nerves (stress, anxiety) are the worst. My nerves kill me, there's so many things that have happened in this period. We've never had economic problems, but you know, my diseases, his diseases.

*Fernando:* Me! I'm in the best shape out of anyone.

*Julia:* Oh shut your mouth.

*Fernando:* Can I speak? Right. You have to focus on the solutions, not the problems. (to me) I cannot accompany her in this depression.

*Julia:* He says he's a healthy person . . . he's had heart problems for a year and he has emphysema which he sees a respiratory physician for, and every two months he has to go for biopsies from his stomach! All that affects me, he thinks it doesn't, but yes it does affect me. The thought that he may have something.

*Fernando:* First of all I am not a sick person, I have a disease, but I am not sick. I have a disease and I live with it. If you ask me how I am, I'll say "better than you." I don't feel sick. It's all in here (in your head).

*Julia:* No, it's not all up here (in your head). He's talking about fighting and facing what you have, that is something different . . . I'm not the type to stop everything for any little thing. But now, I just feel unwell. Maybe my body has just gotten tired, maybe I've given it too much of a beating.

## *August 2011*

In the months I lived in Tacuarembó, Julia experienced no more acute episodes. During this visit they smoked and smoked and smoked, without any discussion of it or comment made to the other. The mood and views expressed echoed my first visit. Spontaneously Fernando started the following conversation:

*Fernando:* Her way of doing things, her way of living life, has helped us overcome the difficult things in life. I don't think the medications have helped as much as that.

*Megan:* The way of thinking?

*Fernando:* That rebelliousness that we have against things, helps you confront it, not complain about it even when you feel awful. And even though the smoking is making the problem worse and worse, she is in much better shape than others with this problem of COPD that might not even be smoking. She rebels against it, not only against COPD, but against a society that's always telling you, "you can live better" . . . What I'm saying is not a scientific view.

*Julia:* That rebellion that you have, it goes as far as stopping you from quitting smoking. Because I want to live, I don't want to extend my life. Because you start to analyze all the things the doctors prescribe to you and tell you and you realize it is all, no, no, no. You're surviving but not living . . .

*Fernando:* She enjoys it, she experiences smoking with intensity . . . We know that we contradict ourselves, we know that cigarettes are harming us yet we keep smoking . . . but we are rebels (smiling).

*Julia:* I'll tell you what, this thing is a reality. If I quit, will my COPD go away? Will my angina? No, I have it. Can I make it worse by smoking? Yes. But would quitting cure it? No, it's not going to leave my body. So why sacrifice? The sugar, yes, I sacrifice because I don't want to go into a diabetic coma, you can become blind, I don't want that. Hypertension too, I don't want to have a stroke.

*Megan:* Do you see COPD as serious?

*Julia:* In the long run yes, it's not letting me breathe, but that's also because of the angina. Recently I found out that 40 percent of my artery is blocked . . . That's because of COPD and because of having smoked all of my life. Yes you can reverse it a bit by doing exercise to help the blood flow better, but I can barely walk.

*Megan:* Because of shortness of breath?

*Julia:* Yes, shortness of breath, and recently I found out the arteries in my legs are all blocked, not because of the heart but because of the diabetes.

Later in a pile-sorting exercise, Julia elaborated on her experience of living with multiple illnesses.

> I am a plague machine! I'm all plagued (laughing). I kill myself laughing about it all. There are people who for all these problems would be like, "oh, no, I can't take it." I take it as natural. I

have to take care of myself, and I do. Shortness of breath I have when I walk, when I do things, that's permanent. That comes not just from COPD but from the other as well (the heart). It's very hard to say what side it is from.

## *February 2014*

Two and a half years later, Julia had 'made it' to see her granddaughter turn 15. She was even more homebound and outspoken about not having 'any quality of life.' Both she and Fernando were smoking their loosely hand-rolled, filterless cigarettes incessantly and Julia reiterated that she had no intention of quitting. She spoke for the first time about not prohibiting her kids from smoking because she was certain that would just have made them want to smoke more. However, she told me:

*Julia:* With my grandkids, I tell them not to start smoking, to look at me and see what cigarettes can do to your health.

*Fernando:* It's not what they can do, it's what they *do* do.

*Julia:* I am not telling them they can't smoke, but I am offering myself as an example. It's one thing to be an example and another to prohibit.

*Fernando:* The biggest effect cigarettes have is reducing pulmonary capacity . . . You're going to be 40 and you're going to feel old.

* * * *

The more I met people of Julia and Fernando's age, the more I saw them and their views as illustrative of smokers of their generation—the generation of leftists that fought against a brutal 11-year military dictatorship, a dictatorship which restricted, oppressed, killed, and disappeared, and which gave Uruguay the highest political incarceration rate in the world (Andrews 2010). Julia and Fernando's prediction suggests that, as times change, newer generations will have a lessened reaction of rebellion against restrictions such as tobacco laws.

Julia's and Fernando's narratives merge in and out of each other, as fluidly as the smoke emanating from their shared cigarettes. In Julia's doctor's appointment, she expresses that her husband will not quit because he has many other prohibitions, yet I only ever heard her express her decision not to quit in this way. Similarly, taking illness as something 'natural' was a strong narrative of Fernando's, which Julia espoused to different degrees too on various occasions. The stark contrast was of course the period of acute illness—the heart attack, and the ensuing chronic disability—shortness of breath, fatigue, pain in the legs. In this episode, the risk of death was not so funny anymore. And Julia's physical state, and probably an acute episode of depression, wore away at her ability to keep up her spirit as a *luchadora* (fighter), and her 'laugh in the face of illness' way of being in the world—a way which both Julia and Fernando explain to be a help rather than a hindrance to their health. Even Fernando, under the threat of acute illness, stops joking, at least for some time.

At the time of fieldwork *el tabaquismo*—tobacco use or addiction—was actively defined as a chronic disease with physical, social, and psychological causes (Muñoz et al. 2009; Steinberg et al. 2010). Although I was at first critical of this medicalization, I saw its enactment in people like Julia. Her smoking history is filled with moments of having quit, of having cut down, and of having increased dramatically. Like the respiratory infection of chronic lung disease, the heart attack of heart disease, and the infected toe of diabetes, the smoking 'relapse' is the acute state of a chronic illness—addiction. Her illness of addiction oscillates within a chronic and acute spectrum in relation to all her other chronic conditions. Her leg pain, breathlessness, and fatigue keep her homebound, and she experiences an acute increase in her addiction. Similarly, the acute heart attack sets off an acute state of depression with chronic repercussions. The line is blurred, constantly, when trying to attribute a symptom to a biomedically defined disease. Her inability to walk is the meeting point of breathlessness, pain, and physical deconditioning. Her shortness of breath is likewise the intersection between damaged lungs, a sick heart, and an anxious state.

Julia and Fernando did not fear their own deaths, but they were terrified of each other's. Similarly, Julia did not fear breathlessness and by-proxy smoking, but she was scared of eating too much sugar or salt and ending up in a coma, having a stroke, or having parts of her body amputated (like her brother whose legs were amputated because of diabetes). She weighed up the costs and benefits of each 'restriction,' and COPD commanded the least immediate action. In her mind, there were much more frightening things on the horizon. While she did not fear becoming dependent on an oxygen tank, she did fear depending on others. For one reason or another, COPD did not hold this prospect. Of course, many people did depend entirely on others to dress, wash, and go to the toilet because of COPD. But Julia's airflow obstruction was not yet severe, and she had never been hospitalized for acute respiratory infection—a common feature of COPD, making it a perfect example of a so-called non-communicable disease which in lived experience blurs the line between infectious and non-infectious (Manderson and Smith-Morris 2010). While these exacerbations of COPD were not part of Julia's present, they could very well be part of her future.

In global public health discourse on the prevention of chronic disease, smoking as a risk factor is framed as an individual behavior or lifestyle choice. Julia's story shows that to understand her choice, we need to attend to her generation, her political views, her relationship with her husband, and her co-morbidities, across time. Failure to do so would mask the complexity of her relationship with cigarettes and the very real chronicity of her addiction.

### *Acknowledgments*

Enormous thanks go to Julia and Fernando (pseudonyms), for their frankness, generosity, and friendship. Many people and institutions helped make this research possible (see: http://etheses.dur.ac.uk/7270). Time for writing this case study was supported by the NIH-funded SASH Programme at the University of Cape Town, South Africa.

### *References*

Andrews, G.R. 2010. *Blackness in the White Nation: A History of Afro-Uruguay*. Chapel Hill, NC: University of North Carolina Press.

Manderson, L., and C. Smith-Morris 2010 Introduction: Chronicity and the experience of illness. In *Chronic Conditions, Fluid States: Chronicity and the Anthropology of Illness*. L. Manderson and C. Smith-Morris, eds. Pp. 1–18. New Brunswick, NJ: Rutgers University Press.

Muñoz, M.J., M.F. Galeano, J.B. Garrido, and W. Abascal 2009 *Guia nacional para el abordaje del tabaquismo*. Montevideo, Uruguay: Ministerio de Salud Publica.

Steinberg, M.B., A.C. Schmelzer, P.N. Lin, and G. Garcia 2010 Smoking as a chronic disease. *Current Cardiovascular Risk Reports* 4:413–420.

## Pathophysiologies of Resistance

Various pathologies complicate the notion of bodily resistances. While political resistance often entails such negative things as arrests, beatings, financial loss, and loss of prestige, there are feedback loops of a different order when talking about internal manipulations of the body. Julia's heart conditions, diabetes, general ill health, and damaged lungs create a personally unique constellation of pathologies accompanied by particular sensations, pains, and emotions. The addictive properties of the cigarettes she smoked created a chemical dependency and physical desire that interacted with her current constellation of pathologies. As Nichter notes, the decision to smoke is also based in the cultural meanings of smoking and the social milieu of the smoker; parents, peers, and friends who either smoke or don't play into the decision, as do ideas about style, 'appropriate' times to smoke and media images of 'sexy' smokers having a fun time smoking (Nichter 2003; see also Chapter 5). Julia's decision to resist medical advice, and to manage her husband's eventual concern too, was couched in terms of her making a decision and of having agency over her actions. How will this decision affect her as she ages and as her breathing becomes more difficult, and her heart, kidneys, and other organs suffer more damage from the out-of-control glucose and insulin levels in her body? How will she understand her progressively deteriorating body and what meanings will she create out of her personal health trajectory? How can we understand her 'decision' to continue smoking, in the light of the severe addiction caused by cigarettes? It is not easy to stop smoking. Most people fail. In understanding resistance, we need to do examine the microdynamics of power. In this case, such analysis means acknowledging that nicotine has power.

Another important point to consider while thinking about these case studies is the idea of *time*. In the preceding case study, Julia felt that she had only 'a couple of years' left to live and she made her decision to continue smoking based on that notion. In reflecting on the decisions people make that counter medical advice, it is not clear that they prioritize their health and bodily capacity on a long-term basis, nor that they see the links between present activities and behaviors and their possible health consequences. Neither is it the case that people have capacity to engage in meaningful forms of sociality. There are short-term gains, too, that people consider, for instance being 'someone you can go have a smoke with,' or, as in the case that follows, being 'the thinnest girl in the room.' How are those daily victories situated within bodily resistances that are viewed as harmful and un-healthy by biomedical practitioners?

In the case study that follows, presented by Megan Warin, we consider community resistance, and how collectivities reinforce individual behavior. A foci of anthropology has been on the micro-politics of the interactions between protesters when resistance occurs in groups. Rosario (2014) draws attention to the intimate social details of the interactions of protestors in her research of a long-term political protest staged in Puerto Rico. As an anthropologist, she lived with the protestors and carefully described the micro-politics of decision making, group inclusion and exclusion, and the exigencies of everyday life. It is at the small group level where many of these resistances occur, especially when considering individuals who are institutionalized for their unaccepted behaviors. The interactions between resisters are deeply telling; the ways in which power is asserted both by those resisting and those seeking to control them, the plotting of acts of defiance, and the descriptions of how those are carried out in hospitals, jails, boarding schools, and nursing homes, is rich terrain for medical anthropology.

Megan Warin is concerned with the particular socialities of a group of women who have anorexia, a common eating disorder that often co-occurs with bulimia. Anorexia is a serious psychiatric illness, marked by the refusal to eat, or to eat such miniscule amounts of food so as to prevent the person from maintaining an adequate, healthy body weight. Bulimia nervosa is characterized by recurrent binge eating in combination with some form of unhealthy compensatory behavior (Arcelus 2011). It is estimated that less than half of patients diagnosed from anorexia fully recover; the remainder experience recurring bouts with the illness over many years. Anorexia was once seen to be a young women's disease in the 'North'; now women and men, in many different social contexts, are recognized as suffering from it. The refusal of food is a form of bodily resistance that is, because of organ failure, particularly lethal (Gooldin 2008; Hardon and Moyer 2014; Rich 2006).

In the case study, Warin describes how the relations between young women with anorexia in a hospital unit are part of the social process of learning how to be a 'good' anorexic patient. The competition and complicity among the women who are the central players of this case reflect the processes of group pressure, strategic friendships and learning by watching. The power structure of the institution shapes the forms of resistance taken by the patients, and the rebellious actions of the patients shape the institution's responses via the rules, regulations, and surveillances that are put in place to control patient insubordination.

## 6.2 Relatedness in Anorexia

*Megan Warin*

As I walked across the asphalt car park to Estelle's suburban ground flat, I could hear the sounds of heavy metal music beating out of her open front door. It wasn't often that I heard music in people's homes. I knocked loudly to register my arrival. A figure appeared, opened the door, and invited me in. I was immediately struck by Estelle's distinctive look. Her shoulder-length hair was matted into dreadlocks, and the ends were dyed bright pink, some entwined with silver jewelry. Her fringe was pulled back by clips that sparkled in the light, accentuated by the glitter eyeshadow dotted around her eyes. She was dressed in 'club' gear, wearing two overlapping Lycra black tops (the outer with the animated Japanese character of *Astro Boy* on the front), and a blue hooded unzipped top. She repeated the layering with two skirts, one long, tight-fitting black skirt with purple shoes peeping out, and a shorter red skirt over the first layer. Her shared flat (with her friend 'bulimic Ben') was equally distinctive. It reminded me of a typical teenage/student style place—posters on the walls, a chair acting as a stool, a couch covered with a blanket, ashtrays placed precariously on armchair rests, a TV, and a stereo.

I was visiting Estelle to hear about her experiences of anorexia. She was 14 years old when she was given this diagnosis, and for the next six years it consumed every facet of her life. Sometimes sitting on the lounge room floor, and other times in her bedroom, Estelle told me her story of recovery. Through much laughter and cigarette smoke, she recounted a story that encompassed the highs and lows of anorexia, and highlighted the complex relationships that people with anorexia form with this disorder and each other. This case study weaves her narrative with the stories of other women with whom I conducted fieldwork, describing their desires to become a 'better anorexic,' their intense competition with others, and the difficulty of leaving anorexia behind. For many, anorexia was not an illness but a friend (if not abusive), a marker of distinction, a social support, someone that was always there for them.

### *Anorexic Wannabes*

Estelle had never met anyone with an eating disorder before she was referred to the eating disorder unit at a major public hospital in an Australian city. She vividly remembers her first visit as an entrée into a world that she had only heard of, a world that required a particular entry card. Whilst she was waiting to be assessed in the patient lounge room, she met another young woman who had just completed a bed program—an intensive bed rest program designed to increase people's weight and give them 'time out' to address underlying issues. Killing time, and unable to resist, this young woman struck up a conversation with Estelle: "You're so beautiful—you look just like a little fairy—I wish I could look like you." Estelle was surprised by her comment, perplexed as to why this woman would want to look like her. "I was freaked out," Estelle exclaimed:

> I was heaps thin and this girl, although having finished the bed program, was still quite thin. I looked at her and said, "No, *I* look disgusting—you look really nice." I hadn't realized at the time that I had really offended her and she took me saying "you look nice" as saying "you look really fat"—all these little things I had to learn after being in hospital about the whole anorexia thing.

Estelle was fast learning about the desires that many people with anorexia have to become more accomplished with their eating disorder. Some wanted to become the best anorexic—walk into a room, sweep it with their eyes, and know immediately that they were the thinnest in that room. Others like Estelle were trying desperately to not belong—to recover and leave, and have limited or no contact with any person with an eating disorder ever again. Many though were caught in a space of trying to do both, to rid themselves of what they described as the 'hell of anorexia'—the shame, the guilt, and the depression—but were not willing to give up what it afforded them. Staff repeatedly joked about the number of times they had heard this inherent contradiction from patients: "They want to get rid of anorexia but they don't want to put on any weight." The people with anorexia put it another way—"I want to get rid of anorexia but I don't know who I'd be." They were fearful about leaving the web of social relatedness—the belonging, the power, the secrecy, and the safety that anorexia provided.

## *Being Anorexic*

After her initial assessment, Estelle was admitted to the psychiatric ward for an eight-week period. Now that she had the legitimating 'rubber stamp' of a diagnosis, she began to learn how other in-patients used it as a form of social connection:

> I found that there was a whole culture behind anorexia—once I got in there—there was this whole culture thing—they all seem to stick together and they have their laxative abuse and stuff like that but it's almost like a trade secret—like it was really bizarre for me. Because I'd never really known anyone with eating disorders while I'd had it but I knew about this whole thing because I'd been warned about getting sucked in to the whole anorexia culture and the girls that didn't want to get better.

Because Estelle wanted to leave anorexia behind, she was called an 'outside anorexic' by other in-patients with anorexia. This was a disparaging term that marked her as Other, as somehow distanced from the powerful collective and ties that the others shared. I asked Estelle to explain 'outside anorexia': How could you simultaneously be diagnosed and treated for anorexia, but be 'outside anorexia'? Estelle explained that she had complied with the treatment program: "I was there eating all my meals and stuff and like trying to be really good, and the nurses were coming in my room and searching for food because they wouldn't believe me—they thought there was something up." The nurses were suspicious because it is very common for people with eating disorders to exercise secretly behind closed doors, hide food, and vomit into bins. Many are admitted to hospital programs not because they want to get better, but because relatives have pleaded with them to get help, or they are so sick they require medical management. Estelle was different because of her compliance, and because of this, her authenticity as a 'real' or 'true' anorexic was challenged by other in-patients—"you're not really an anorexic," they taunted. "That," she said, "for some reason offended me because I was like 'well what the fuck am I then?'—if I'm not a real anorexic and not a real, normal person—what am I?—why should I be a real anorexic anyway?"

Estelle was caught between her own desire to get better, taking expert medical advice on board to recover, and being maligned by other patients for doing so: "I wanted to fit into that culture when I went to the hospital because I wanted to fit in with everyone and they all saw me as this freak that wants to get better and at first I felt really rejected by them . . . I was leaving the thing [anorexia] so they didn't like me that much." Even though Estelle's non-conformity puzzled the group, they tried to seduce her back into the 'collective ways of anorexia.' They were surprised to hear that she hadn't used laxatives or speed to lose weight, so they explained the practice to her, tempering their information with anonymity—"we haven't really told you about this, okay?" Estelle was 'being let in' on a secret.

## *In 'the Club'*

Anorexia is unlike most other illness categories or disorders because people with this diagnosis think of themselves as different. In psychiatric wards, participants didn't see themselves as 'crazy' like other in-patients with a range of mental disorders, but as distinctive and extraordinary. They didn't always come together to support each other in recovery, but to support each other in being a better anorexic. On-line networks exemplify this desire to maintain anorexia, and this is what is so puzzling and perplexing for family members, friends, and treatment teams. Why do people with anorexia form these powerful relationships with others with the same diagnosis? What propels them to continue their suffering?

Estelle's experience demonstrates that anorexia is not simply a psychiatric label. People wanted to embody the symbolic power of a medical diagnosis, but they then transformed anorexia into a powerful network of connection and belonging—it became 'a religion,' 'a competitive sporting team,' and 'a fascinating game.' There were those who were in, or, like Estelle, out of 'the club.' Sonya explained the almost suburban and mundane inclusion:

> I suppose I sort of see it as like an anorexic club—it's just like everyone is pretty accepting of you if you are sick, everyone's friends and stuff like that and there's a bit of competition going on—it's sort of like you put in your membership and it's really hard to get out.

> *How do you get into the club—what's your membership?*
>
> It's sort of like a coffee club—you meet, you discuss the illness . . . you have these friends—you're suddenly in this group and people understand you . . . for me it was like my life had just been nothing for such a long time—that was all I knew, it was all I did and I didn't have any outside life and so suddenly I'm meeting all these people who don't have any sort of life either so it was just perfect—what else do we have to talk about—nothing but anorexia.

The metaphoric representations of anorexia as a club signal the workings of desire, seduction, power, and danger that coalesce in this community. Moreover, they point to anorexia as a pivot of as well as a vehicle for relatedness (Warin 2010: 97). This club was not always articulated through language. Participants intuitively knew who belonged to the club or not: they could point to strangers in a crowded shopping mall or on a downtown bus, and distinguish different types of 'skinny,' 'naturally skinny or anorexia.' They 'just knew' who was 'in' by their bodily comportment—'the dead look' in their eyes, the sallowness of their upper arms, and the heaviness of their gait. Recognition of another 'anorexic' was an invitation to form a relationship. It initiated an exchange of comforting words across the plaza on the university campus, and at other times it allowed for comparisons and exchange of purging techniques.

Creating relatedness through sharing what Estelle called 'trade secrets' was most clearly displayed when people participated in treatment programs, for here the sense of belonging to the 'anorexic' practice was most important and yet most at risk (from the threat of anorexia being exposed and thus disempowered). In-patients drew 'battle lines' against those who wanted to help them. "People [with anorexia] build such a wall around themselves," said Rita, "such a safety net of anorexia, that anyone who comes along and tries to threaten that has got to be viewed as an enemy." The 'enemy' included the doctors and nurses who were seen as taking away the relationship people had with anorexia. Kitchen staff were likened to the old woman in Hansel and Gretel—"they were out to get us skinny kids—scheming to make us fat." Participants would ask the nurses, "what kitchen witch is on today that's spread this butter on so thick?"

Undercover practices of resistance were in place—either solitary resistances or collective, organized attempts. These strategies of 'trickery' and tactical priorities were essential to maintaining a group identity, enabling 'sick patients' to garner power against the constant surveillance of the clinical staff.

In one of my field sites (an in-patient eating disorder program in a general psychiatric ward), those with anorexia were given private rooms in the direct line of surveillance of the nurses' station. Estelle recalled how those on lengthy bed programs "became very, very devious" as a "self-preservation mechanism." There were many 'tricks' which in-patients laughed and joked about, and people took great pride in telling stories in hushed and excited tones about the most ingenious of places where they could 'stash' unwanted food and drinks (such as in the ceiling panels or scraped into duvet covers). Audrey was terrified of having to drink a liter of apple juice a day, and would try to disguise it as urine in the bedpan, to be unwittingly taken away by the nurses and washed down the sluice.

In this setting, there were strict rules about interaction with other in-patients with anorexia whilst on a bed program. Interaction was curtailed, but handwritten letters were secretly passed from one door to another, the writers swapping stories of how to exercise without being noticed. And those who were allowed the freedom of the ward sometimes established groups of cheating, banding together like a relay team. Rita described how she was included in a network of food avoidance without any warning—the fact that she had an eating disorder automatically meant she was an accomplice in a 'network of deception':

> We're a secretive lot . . . this young girl tried to set up a little network of accomplices—like a support system to help her cheat. She even shoved a bag of nuts into my hand that was her supper. We were just walking past each other in the corridor and she shoved a bag of nuts in my hand and kept walking and . . . so I was actually aiding and abetting her by doing that . . . she was so desperate to lose those nuts. She didn't want to put them in her body.

The secrecy that operated within the clinical setting propelled people with anorexia together and encouraged intimacy between them. Secrecy was a distinctive feature of anorexia and its performance enabled participants to hide anorexic practices and so insist that they 'were fine.' Some suggested that there was a "secret language of eating disorders," a language that was articulated through a range of bodily practices that were only known and shared amongst those with anorexia. Although some participants chose not to engage with the collective secrecy, all knew of its existence and power from their own day-to-day strategies of concealment.

### *Anorexia as Distinction*

Secrecy operated to mark differences between those who had anorexia and those who did not, and so created status and prestige for those who practiced it. Those who maintained anorexia described feeling a sense of superiority over everybody else they encountered. Anorexia was a productive and empowering state of social distinction (Bourdieu 1984). It was not experienced as a debilitating illness, but a "gut-wrenching emptiness" that was viscerally "unique," "heroic," "a power trip," "a thrill," and "a high" in which people felt "indestructible" and "superhuman."

While secrecy marked distinct boundaries between those who did and did not have anorexia, there were other clear markers within anorexia. There were internal divisions, where some competed to be the best anorexic within the group. Like Estelle, many spoke about the insidious levels of competition, particularly when people came together in treatment settings. Sonya loved the sense of achievement that anorexia gave her, but was devastated coming into an eating disorder service "because there are so many other people who weigh less and they're better than I am at it." Once in hospital, her goal was "to be the sickest—to be the best at it. But the best is dead—that's when you win—that's the best you can be being anorexic is when you die." Because of this competitiveness, Sonya thought the eating disorder service was a pretty "dangerous place—[it fuels] a real competitive spirit."

People referred to 'pure' or 'true' anorexia: weight loss by total control through almost total abstinence of food and drink. 'Pure anorexia' was not only an important part of distinguishing them from other people, but also from other eating disorders, which were seen as lower down the scale. Estelle described people with bulimia as "having a harder time" because of the shame and embarrassment associated with "the whole throwing up and bingeing process—whereas with anorexia I was proud of the fact that I didn't eat like everybody else."

Estelle's entry into this world demonstrates the relational matrix of anorexia. As Janet Carsten (2000) showed in her ethnography of Malay kinship, relatedness is more than blood ties of kin, and comprises multiple elements. In anorexia, relatedness concerns the everyday exchange and concealment of practices, secrecy, and competitiveness with other in-patients, friendships formed through sharing a common diagnosis, and the creation of disunity by marking those who want to leave anorexia as 'outside anorexia.' These complex ties of relatedness speak to boundaries and belonging, and are central to the powerful production of everyday networks, allegiances, and identities in which anorexic worlds circulate.

### *References*

Bourdieu, P. 1984 *Distinction*. London, UK: Routledge and Kegan Paul.

Carsten, J., ed. 2000 *Cultures of Relatedness: New Approaches to the Study of Kinship*. Cambridge, UK: Cambridge University Press.

Warin, M. 2010 *Abject Relations: Everyday Worlds of Anorexia*. New Brunswick, NJ: Rutgers University Press.

---

This case study highlights Estelle's self-awareness of her desire to be part of the 'anorexic club.' Being part of the anorexic club means participating in secret acts of resistance to the therapeutic norms promoted in the eating disorder clinics to encourage healthy eating and recovery. The members of this group trust each other in the sense that they will not betray each other.[2] Being anorexic gives Estelle a sense of belonging and identity, and yet she is ill at ease with her membership in this form of resistance. She weighs the benefits of her sense of belonging, from others who engage in the same forms of food restriction and purging to attain an ever-more-slender body, with a revulsion at the *techniques du corps* (Mauss 2006 [1934]) necessary for its attainment.

Resisting the size and shape of her present body becomes a way to engage with a social group with whom she can co-construct her identity. The performance of anorexia revolves around visually assessing how thin one's body is, and how it compares with others in one's immediate environment—whether that is physical or virtual.

Marcel Mauss (2006) published his famous extended essay "Les Techniques du corps" in 1934. In it, he described bodily movements, and how even the most 'natural' of movements like walking and sleeping varied significantly between cultures. Mauss noted that humans learn bodily movements by imitation and instruction, and that they give meaning to their everyday gestures, postures, and movements. The non-verbal clues that Warin describes as the young women hide their food, engage in forbidden forms of exercise, and showcase their bodies to either reveal or disguise their emaciated conditions, can be understood as 'techniques de corps.' In resistances like anorexia and smoking, individuals turn onto themselves, using their own bodies as sites of refusal, rejection, and opposition. Sometimes the counterpoint—the object of opposition—is in plain view; other times, the oppositional subject—the institution, subject, or ideology—is inherent or muted. But in these cases, the individual who exercises resistance by refusing food, refusing medicine, or refusing support, turns on his or her own body. The body is the one accessible site of resistance to which they have access.

## Resisting Institutions

An alternative form of resistance is that of lodging a complaint within a bureaucratic system, although very different circumstances—relationships of power and access to resources—come into play in such cases. Bureaucracies and institutions of all types have protocols for receiving negative feedback, and the health care system is no exception. As medical systems change, so too does the ability of those using the system to protest against such things as the prices of service, the quality of care, or the prevalence of medical errors. Here, resistance is not internalized or passive; rather people are able to access pathways of opposition in order to force their case. In the following case study set in Puerto Rico, Jessica Mulligan describes the labyrinthine complexities of trying to obtain what one needs through interacting with large bureaucratic institutions both public and private.

---

## 6.3 Governing by Complaint

*Jessica Mulligan*

I left Acme's gleaming glass office building and drove east. San Juan sprawled; its highways and shopping centers held the outlying suburbs in a concrete embrace. Twenty-five minutes later, the highway abruptly ended, dumping cars onto a two-lane road peppered with traffic lights, box stores, and new housing developments surrounded by mud. Crumbling brick chimneys stood in the empty fields between developments, a reminder of the sugar plantations that once dominated the landscape.

Acme is a pseudonym for the private health insurance company where I worked as a compliance manager and anthropological researcher; Acme contracted with the U.S. federal government to offer Medicare benefits to eligible beneficiaries in Puerto Rico. As a territory of the United States, Puerto Rico participates in the federal Medicare and Medicaid programs, both of which were substantially privatized in the late 1990s and early 2000s on the island.

My destination that summer afternoon in the mid-2000s was Río Grande, a small town on the northeast coast of Puerto Rico located close to *el Yunque*, the rainforest and tourist attraction. There, I met two brothers—Don Enrique and Don Ignacio—who told me about their experiences with the newly privatized health care system. In their neighborhood, the two-room concrete houses were tightly packed on small lots; some had been modified with a second floor, a new shade of paint, or religious ornaments. Other houses were not aging so well and showed cracks in the cement walls or appeared to be abandoned. The brothers lived behind a cemetery; the government relocated them to the site when their former home was expropriated for the construction of the highway.

We gathered on their porch and talked over the hum of Mexican ballads interrupted by reggaeton from passing cars. The brothers called out to neighbors, joked about lost loves, and showed me pictures of their children on the mainland and on the island.

Don Ignacio was far more jovial than his brother, Don Enrique, who sat in a wheelchair, shirtless in a pair of shorts with a bandage on his left foot. The white gauze covered his toes and ankle. He explained that he was to have his foot amputated on Monday.

A small cut had developed into gangrene. In the house, they showed me several blood glucose monitors that Acme had sent. They were frustrated with their private health insurance company and could not understand why it kept sending the wrong machine. They wanted the old one with the large-print display, but that machine had stopped working. The brothers both had Type 2 diabetes and neither was able to monitor his glucose levels regularly.

The brothers did not know it, but Acme had contracted with a new durable medical equipment provider as a cost-saving strategy and the transition to the new provider was fraught with complications: orders were botched or never delivered and customer complaints were flowing into the Acme offices. Each time the brothers called Acme to complain, a member of the Customer Service Department opened a grievance in the electronic complaints system and a new glucose monitor was sent to their home.

Don Enrique said he was thinking about the impending operation. His pensiveness contrasted markedly with Don Ignacio, who fried up seafood-stuffed empanadas and insisted that I sample his homemade hot sauce. He showed me big cans of processed bulk food purchased at the new Sam's Club, explaining that the small amount of money he received from Social Security goes much farther at Sam's.

Don Enrique said he had known something was not right with his foot, but he let it go. He put off seeing the doctor for too long. He winced occasionally and mentioned the pain. He blamed himself.

The brothers' case was no anomaly. At this time, the most common complaint at Acme was a delay in delivery of medical equipment. Acme had determined that local durable medical equipment (DME) companies were overbilling and inefficient, so they contracted with a Miami-based DME company to handle all of their orders. Unfortunately, the Miami company was unfamiliar with the topography of the island and lacked the local knowledge to navigate unnamed streets and unnumbered houses (especially in rural areas). They expected that physicians in Puerto Rico would be very similar to those in South Florida; after all, they spoke Spanish in both places. They did not realize, however, that significant differences existed in prescribing customs between the island and Florida. When the new DME company received an order that contained 'insufficient' information (for example, if it lacked a diagnosis code or justification for non-standard equipment), the order languished while overwhelmed customer service representatives tried to obtain the missing information. Physicians' offices were not used to having their orders questioned and did not have the staff to deal with the added administrative burdens imposed by managed care. Ultimately, Acme's attempt at economization and increased efficiency produced a barrage of complaints at the organization, which too often, left beneficiaries receiving suboptimal health care.

These unintentional consequences stem in part from the contradictions inherent in transforming the medical care of the poor, the disabled, and the elderly into a profit-generating enterprise. Clearly, there was a lack of fit between market goals (saving money through a more standardized approach to product delivery) and health objectives (getting the brothers a glucometer that they could read and use in order to manage their diabetes). This example also illustrates how market-based reforms to public services increasingly call on citizens to behave as consumers; the only redress available to the brothers was to complain to Acme through their corporate complaints process.

## *The Right to Complain*

The question of how privatization reconfigures the manner in which health plan members register discontent and pursue resolution is ultimately a question about how privatization transforms citizenship. In a market-based health care system, the most effective way to register one's complaint is to join the health plan offered by the competition. Consumers are supposed to express dissatisfaction through purchasing decisions like opting to join a higher-performing health plan or to visit a better-rated doctor. But members of private Medicare plans like Acme can and do complain with more than their feet. In order to protect beneficiaries' rights, the federal government has mandated that private Medicare plans create extensive complaints-processing procedures that are subject to regular audit and verification.

Part of my work at Acme was to ensure that the plan complied with the federal regulations for complaints processing. Being found in compliance hinged on establishing a process for the receipt and resolution of complaints—it did not necessarily involve making a fair or just determination in any particular case. In my time at Acme, individual reviewers were not rewarded for denying coverage, nor were staff instructed to ignore complaints. Instead, complaints were processed vigorously. Even so, the complaints system exhibited a paradoxical quality. The consumer complaints system *appears* to foster a democratic means for beneficiary voices to be heard and due process to be protected. But in practice, the complaints system actually hollowed out (Gledhill 2005), or severely limited, beneficiaries' rights and avenues to redress because it narrowly channeled (1) *how* beneficiaries were permitted to complain, (2) to *whom* they may complain, and finally (3) *what* kinds of issues were deemed legitimate and actionable. Frequently, the reasons that motivated a beneficiary's complaint—issues such as a desire to be treated with respect or to have their unique health-related needs be recognized (like needing more than the standard blood glucose monitor)—were not addressed or recognized within the corporate complaints process. The complaints system also served as a safety valve for containing member and provider dissent, which in turn allowed the health plan to continue rationing care through utilization review.

## *Novel Forms of Liability and Complaints*

Customer service representatives answered calls piped directly into their headsets and typed customers' information into their computers. Many of the calls were prompted by the receipt of written communication from the health plan. The plan was required by law to issue denial notices to members if a physician submitted a claim and the health maintenance organization (HMO) decided not to pay. The denial notice included a description of the member's rights and instructions on how to file an appeal. The denial notices were often confusing for beneficiaries and ended up generating complaints even when no payment was being sought from the member by either the health plan or the service provider. The process also inadvertently created an adversarial relationship between providers and patients. The following complaint illustrates some of the ironies of these processes that were put in place to safeguard consumers' rights.

Doña Hermina called Customer Service and complained about a notice she received in the mail. She was liable for $55 for a procedure that Acme had determined was cosmetic and not medically necessary (and therefore not covered by Medicare). Doña Hermina claimed this was a mistake—the podiatrist, Dr. García, never told her she would have to pay out-of-pocket. The customer service representative phoned Dr. García's office and asked for his version of events. The provider, annoyed, said he had already been called about the case. The customer service representative could not reach a quick resolution, so she transferred the case upstairs to the Appeals and Grievance Department.

In contrast to the customer service representatives who were packed into a large room with small cubicles arranged side by side with no dividers in between, the appeals and grievance coordinators worked behind much larger desks where they could spread out the case files in privacy. The appeals and grievance coordinator classified the complaint as an appeal and resolved it by sending a letter to the member. The letter stated that the doctor would not charge the member for the service but that Medicare does not cover cosmetic procedures and next time, she would have to pay.

The podiatrist, Dr. García, after receiving multiple calls from the health plan regarding this case, became quite angry and vowed to file a grievance against the member. He sent a letter to Doña Hermina with a copy to the HMO claiming that he would sue Doña Hermina for libel and defamation of character. The manager of the Provider Relations Department eventually intervened. I sat in with her when she called Dr. García; we spoke to the doctor for 25 minutes. His primary complaint was that his reputation had been insulted. When the patient came to his office and tried to pay the $55, he refused to see her and much less take her money. He said his daughter can spend that amount on makeup in a weekend. He includes a reminder of the amount owed on patients' appointment cards, but if they cannot pay, he writes it off at the end of the year on his taxes.

Ironically, had the HMO not been mandated by the government to protect consumers' rights by sending out a denial notice, the conflict would never have erupted. Protecting the member's rights caused the complaint as new billing practices and notions of liability supplanted the customary practices that many doctors in Puerto Rico employed in their offices. To resolve the issue, the provider relations manager complimented the physician, assuaged his ego, and explained that the HMO was only investigating because the organization was compelled to do so by federal law. After venting his frustrations, Dr. García agreed to call off his lawyer. Doña Hermina phoned that day, agitated and crying. She couldn't sleep for her nerves since receiving the doctor's letter. The appeals and grievance coordinator told her that the doctor would drop the issue and would she please pick another podiatrist from the provider list. She agreed and vowed to never visit Dr. García again. The appeals and grievance coordinator made a note of the outcome and closed the case in the electronic tracking system.

The complaints process in this example was both dysfunctional and overzealous. The corporation attempted to protect the rights of its members in compliance with federal regulations, but the complaint investigation itself aggravated the situation as multiple Acme employees called the doctor and the member, thereby transforming what was a minor billing misunderstanding into a much larger problem. The physician's reaction—though rare—draws attention to how the initial denial notice and the subsequent investigation ignored local norms for handling billing discrepancies and re-scripted the interaction between the provider and the patient into the idiom of complaints, which created a newly adversarial and litigious relationship. Was Doña Hermina meaningfully protected in this interaction? Unfortunately, she ended up more aggrieved at the conclusion of the process than she was at the beginning.

### *Unregistered Complaints*

Some of the complaints I encountered while conducting life history interviews never figured in the official registers of managed-care organizations. Consider how Don Luis from Caguas discussed his dissatisfaction. Don Luis did not want the interview to be recorded, so his story is excerpted from my notes. We spoke in his immaculately clean living room, but outside in his front yard overgrown blades of grass reached as high as the windowsill. He explained it was the only way to keep his neighbors' dogs out of the yard.

Don Luis was previously on government insurance through a private plan but had nothing good to say about it. The main thing that the private plan did, in his view, was humiliate poor people. He told a story about a blond psychiatrist from San Juan. She kept an office in Caguas that she visited once a week. Don Luis made his appointment and was told to be there at least by 8:00 AM. He arrived at 7:30 and waited for the doctor who showed up at 2:00 PM. When finally Don Luis was called to be seen, the doctor did not look him in the face. She did not examine him. She simply pulled out her prescription pad and started writing. So Don Luis asked her if she was planning to examine him. She said, oh, so you must be a doctor if you know so much about medicine. He said, I'm not a doctor, that's why I have to come to you. But I do deserve a proper examination. Throughout his story, he spoke to the doctor with the polite, formal address, *usted*. By the way, he asked, do you know how long these people have been waiting for you? The situation escalated and Don Luis ripped up the prescriptions and made a scene in the packed waiting room. He said that the doctor also yelled at those people and told them if they're not happy, they can leave. Don Luis defended the people gathered there: just because we are poor, he said, we do not deserve to be treated like this.

Don Luis explained how this experience and his other troubles obtaining mental health care caused him to drop his privatized government insurance plan. When he received notice of his annual coverage renewal appointment at the medical aid office, he never went. He said it was just as well. One psychiatrist expected patients to line up starting at three in the morning to see him. Then, they might have to wait half the day to actually be seen. He didn't like any of his psychiatrists—they gave him too many medications.

Here we can see another avenue in market-based health systems for patients who are not content with their health care. They have the choice to leave public programs and may elect to pay for private health services instead or forego treatment altogether. In her research on Medicaid managed care in New Mexico, anthropologist Leslie López calls this "de facto disentitlement" (2005); beneficiaries fall off the rolls of public programs due to the barriers created by annual recertification procedures or other bureaucratic hurdles. In some ways we can see what Don Luis did as a protest. However, he never filed a formal complaint, even though he insisted that his rights had been violated. Therefore, his grievances about being disrespected, needing more than medication management, or his inability to obtain timely access to care were never heard by the private plan or the government. The actions he took were severely circumscribed and individual in scope. Instead of seeking redress as a citizen, he acted like a customer and took his 'business' elsewhere. Although there is obviously a conflict between the physician and Don Luis, they have one important thing in common. If people are unhappy with their medical care, they are 'free' to choose another provider. Don Luis now just receives his mental health care from his primary care physician (PCP). He likes his PCP—the man knows how to listen. He currently takes Zanax for his mental health problems. Sometimes he still falls into depressions, but he makes it through okay. The doctor told him to leave the house when he feels something coming on, go for a walk or a drive, just change his environment. Don Luis said this usually works for him. He has a pickup truck and sometimes he'll start driving and end up an hour away in Ponce.

While Don Luis seems at peace with his decision since it allowed him to maintain his dignity and self-respect, this is a man who was once hospitalized against his will and diagnosed with schizophrenia. When he dropped his insurance plan, he no longer received coverage for his prescription drugs. And so his unregistered complaint has also resulted in unmet health needs.

## Acknowledgment

This case study is excerpted from Jessica M. Mulligan, *Unmanageable Care: An Ethnography of Health Care Privatization in Puerto Rico* (New York: New York University Press, 2014).

## References

Gledhill, J. 2005 Citizenship and the social geography of deep neo-liberalization. *Anthropologica* 47(1):81–100.

López, L. 2005 De facto disentitlement in an information economy: Enrollment issues in Medicaid managed care. *Medical Anthropology Quarterly* 19(1):26–46.

---

As Mulligan shows, corporate forms of health care delivery can severely limit how individuals might lodge their complaints, and the corporations define what counts as a legitimate complaint. Akhil Gupta (2012) describes this process in rural India, where the bureaucratic enactment of health care is inefficient, wasteful, and at times absurd. Like Mulligan, Gupta emphasizes the importance of affective relations between those in power within biomedical institutions and those receiving care: "If institutional reform is to be successful, attention has to be paid to generating new narratives that alter the affective relations between the state and its poorest citizens" (2012: 35). Those affective relations can be very important to maintain from a practitioner's point of view as well. In more litigious settings, patients are more willing to pursue malpractice lawsuits when they feel animosity towards a practitioner; in countries where medical errors are handled via systems of educational remediation not monetary castigation, or where they are overlooked because of their recurrence, practitioner-patient relations play out differently (Cartwright and Thomas 2001).

We are in an era that is characterized by an ever-increasing, but always incomplete and changing, reach of biopower, a term Foucault (1980) used to refer to both the disciplining of individual bodies and the regulation of populations. Biopower is constructed via technologies that monitor patients, visualize biological process, and gather bio-data. These data provide clinicians with information about individual patients, and add to databases that reify and refine our understandings of human physiology. Such things as PET scans, MRIs, ultrasound sequencing, and nuclear medicine efficiently penetrate the human body, mapping our interior terrains. Even mobile phones and tablets have the ability to geographically locate us and transmit readings of our vital signs and ultrasound images, and these technologies are rapidly becoming even more powerful and more sentient. As we will see in Chapter 15, biomedicine moves across geographical and cultural spaces in a myriad of ways, making the study of how it is accepted and how it is resisted reflective of particular times and places.

In this chapter, we have described various kinds of resistance to biomedicine. The case studies indicate possible new directions in medical anthropological inquiry that can profitably pay attention not only to what people are saying, but what they are doing and how they are doing it. Anthropologists are well placed to make sense of the everyday acts of resistance, and the tactics that people use to creatively resist hegemonic institutions (de Certeau 1984). New methodologies, together with increasing ease of obtaining photos and video footage, are changing the way we can ask these kinds of non-verbal research questions (Cartwright 2013).

We should ask ourselves how medical anthropology, as a discipline, takes a stand for or against the particular enactments of biomedicine that we study. We need to question the ethical considerations that come from our particular positions via the powerful institution of biomedicine, whether we align ourselves with the subaltern classes or become handmaidens of biomedicine, whether there is a middle ground, and how our research either reinforces or questions biomedicine and its authority in any given situation.

## Notes

1. As we see in Chapter 15, the ideas and technologies of biomedicine travel and at each stop they are understood incompletely and interpreted differently.
2. For an introduction and references to secrecy as embodied practice, see Hardon and Posel (2012) and Davis and Manderson (2014).

## References

Abu-Lughod, Lila 1990 The romance of resistance: Tracing transformations of power through Bedouin women. *American Ethnologist* 17(1):41–55.

Aengst, Jennifer 2014 Silences and moral narratives: Infanticide as reproductive disruption. *Medical Anthropology* 33(5):411–427.

Arcelus, Jon 2011 Mortality rates in patients with anorexia nervosa and other eating disorders. *Archives of General Psychiatry* 68(7):724–731.

Briggs, Charles L. 2007 Mediating infanticide: Theorizing relations between narrative and violence. *Cultural Anthropology* 22(3):315–356.

Cartwright, Elizabeth 2013 Learning to use a new medical technology: Excerpts from a video-based study. In *Visual Research: A Concise Introduction to Thinking Visually*. Jonathan S. Marion and Jerome W. Crowder, eds. Pp. 69. London: Bloomsbury Publishers.

Cartwright, Elizabeth, and Jan Thomas 2001 Risk, technology and malpractice in maternity care in the United States, Sweden, Canada and the Netherlands. In *Birth by Design: Pregnancy, Maternity Care and Midwifery in North America and Europe*. Raymond De Vries, Sirpa Wrede, Edwin R. van Teijlingen, and Cecilia Benoit, eds. Pp. 218–228. New York: Routledge.

Davis, Mark, and Lenore Manderson, eds. 2014 *Disclosure in Health and Illness*. London: Routledge.

de Certeau, Michel 1984 *The Practice of Everyday Life*. Steven Rendall, transl. Berkeley, CA: University of California Press.

Foucault, Michel 1980 *Power/Knowledge: Selected Interviews and Other Writings, 1972–1977*. New York: Pantheon.

Gooldin, Sigal 2008 Being anorexic: Hunger, subjectivity, and embodied morality. *Medical Anthropology Quarterly* 22(3):274–296.

Graham, Hilary 1976 Smoking in pregnancy—attitudes of expectant mothers. *Social Science & Medicine* 10(7–8):399–405.

——— 1987 Women's smoking and family health. *Social Science & Medicine* 25(1):47–56.

Gramsci, Antonio 1992 *Prison Notebooks*. Joseph A. Buttigieg, transl. New York: Columbia University Press.

Gupta, Akhil 2012 *Red Tape: Bureaucracy, Structural Violence, and Poverty in India.* Durham, NC: Duke University Press.
Hardon, Anita, and Eileen Moyer 2014 Medical technologies: Flows, frictions and new socialities. *Anthropology & Medicine* 21(2):107–112.
Hardon, Anita, and Deborah Posel 2012 Secrecy as embodied practice: Beyond the confessional imperative. *Culture, Health & Sexuality* 14(Supp 1):S1–S13.
Marshall, Wende Elizabeth 2012 Tasting Earth: Healing, resistance knowledge, and the challenge to dominion. *Anthropology and Humanism* 37(1):84–99.
Mauss, Marcel 2006 [1934] *Techniques, Technology and Civilization.* New York: Berghahn Books.
Nahar, Papreen, and Sjaak van der Geest 2014 How women in Bangladesh confront the stigma of childlessness: Agency, resilience, and resistance. *Medical Anthropology Quarterly* 28(3):381–398.
Nichter, Mark 2003 Smoking: What does culture have to do with it? *Addiction* 98(Supp 1):139–145.
Rich, Emma 2006 Anorexic dis(connection): Managing anorexia as an illness and an identity. *Sociology of Health & Illness* 28(3):284–305.
Rosario, Melissa 2014 Intimate publics: Autoethnographic meditations on the micropolitics of resistance. *Anthropology and Humanism* 39(1):36–54.
Scott, James C. 1985 *Weapons of the Weak: Everyday Forms of Peasant Resistance.* New Haven, CT: Yale University Press.
Sivaramakrishnan, K. 2005 Some intellectual genealogies for the concept of everyday resistance. *American Anthropologist* 107(3):346–355.
Willis, Paul 1977 *Learning to Labor: How Working Class Kids Get Working Class Jobs.* New York: Columbia University Press.

Luis the Son, 2007. Tiahuanacu, Bolivia.



Note: Originally published in Jerome Crowder, 2013. Becoming Luis: A photo essay on growing up in Bolivia. *Visual Anthropology Review* 29(2):107–122.

## About the photograph

*After helping his family in the fields all morning, Luis returns to the house and waits while his mother, Basilia, prepares a midmorning meal of quinoa with fresh milk. He sits on the bench, where his uncles usually sit when the family eats together in this room, and talks about life with his mother on the farm. He shrugs, saying, "I see my mom nearly each week. Either she visits me or I see her here, sometimes she comes with grandpa to church, but usually I come back out after school lets out. Now that I'm 11, she says it's time for her to return to El Alto and find a job. Life here is hard, but there it's just expensive."*

*By 2011 Luis and Basilia were living in their own home in Urbanizaton Bautista Saavedra (El Alto) near his aunts and uncles and across the street from his former school. At 15, Luis decided he wanted to make money rather than attend school, so he became an apprentice to his uncle, a carpenter, who lives next door. Together they make furniture, doors and windows, which are sold in various markets across El Alto, such as the largest one in Villa 16 de Julio.*

—Jerome Crowder

# 7

# Chronicities of Illness

*Lenore Manderson, Elizabeth Cartwright and Anita Hardon*

Chronic conditions illustrate the complex relationships that contribute to the risks of ill health and disease, and shape their outcomes. As we have already illustrated, social organization, economy and politics, as well as health systems factors, interact with pathogens and biology to determine the epidemiology and chronicity of disease. These factors influence episodes of comparative illness and stable health, the progress of disease, decline, and functional impairment. Many people with chronic conditions also have other unrelated ongoing health conditions, frequently associated with the same personal and socioeconomic preconditions. At the same time, many co-occurring conditions interact biologically and in relation to structural factors. As we noted in Chapter 1, Merrill Singer (2009) developed the concept of syndemics to capture these multiple interactions, where two or more diseases interact at the level of biology, but also, concurrently, the diseases and structural factors interact, with the synergy resulting in an especially high prevalence and/or different mutations and expressions of the diseases (see also Ostrach and Singer 2012; Singer 2014; Singer and Bulled 2014). Such syndemics arise especially in extreme settings, as occur in periods of war, in fragile states, or in societies with marked inequalities. But outside of these circumstances too, people may experience a number of chronic conditions, often co-occurring and interrelated, and they struggle to manage these and head off complications. The result, for many, is a downward economic spiral and growing social exclusion, as accrued poor health inhibits workforce participation; as care, treatments and medications eat up available financial resources; and as poverty and illness combined erode family resilience.

In a single chapter, it is difficult to capture the chronicity and co-occurrence of long-term health problems, given the diversity of chronic diseases, including both those that are infectious and those that are non-communicable. In the following, we concentrate on three major themes where an understanding of social and cultural context helps to make sense of the persistence of particular diseases. We first consider, as a single case study, two chronic, infectious parasitic diseases that are highly prevalent in impoverished environments; the case study related to this illustrates how the politicization and economics of global health policy and medical interventions continue to impede their control. We then turn to chronic non-communicable diseases—the social perceptions of these diseases is that they are unavoidable or inevitable because interventions have failed as a result of their inability to address the structural factors that contribute to the development and influence the progress of diseases. In the third case study, we illustrate the impact on identity when a chronic disease is out of control: in this example, when uncontrolled diabetes results in amputation. In the final case study, we examine how long-term degenerative conditions impact social life, and explore questions of agency and the role of social support in enabling people to work around the impositions placed on them in ways that help them head off further decline.

## The Chronicity of Infectious Disease

Many infections—most upper respiratory tract infections, for instance—are commonplace and of nuisance value only, with little or no long-lasting effects. However, a significant number of infectious diseases are treated today, as in the past, with fear and stigma (Briggs and Nichter 2009; Brown and Kelly 2014), with severe morbidity and the high mortality rates characteristic of spectacular epidemics, their equally spectacular measures of control, and their devastating socioeconomic impact: consider, as examples, cholera, yellow fever, new emerging diseases such as SARS (severe acute respiratory syndrome), H1N1 (swine flu), and MERS (Middle East Respiratory Syndrome), Marburg, Ebola and other hemorrhagic viruses (see Chapter 15). While these infectious develop rapidly, various infectious diseases are chronic, and while fear of infection and its associated stigma pertain, people live with these infections, often at a subclinical level, for decades. Tuberculosis and HIV infection are common all over the world, and the challenges associated with adherence to medication and monitoring are precisely because of the combination of their chronicity and infectivity.

Chronic infectious diseases may be diagnosed long after the time of infection, hence the focus on their prevention, or they may result in generalized lethargy rather than frank illness. Their signs and symptoms may be subtle and confusing, and so these infections are frequently untreated, leading to continued transmission. Large numbers of people worldwide suffer from under-diagnosed parasitic infectious diseases, generally transmitted and most prevalent among people living in extreme poverty. They are increasingly referred to as 'neglected diseases of poverty,' with the populations afflicted, as well as the specific diseases, largely ignored by global medical science and industry (TDR 2012; Manderson et al. 2009). In the following case study, we discuss two of these parasitic infections, schistosomiasis and filariasis, for both of which drugs are readily available and safe, and have been available and used for decades. The diseases, in theory, are able to be controlled if not eliminated. But the chronic conditions of local living circumstances, the chronic underfunding and limited resources of health services, and chronic problems related to state governance, all inhibit the effectiveness of control programs.

Some 250 million people are infected with schistosomiasis (bilharzia) worldwide, the majority living in poor rural communities in sub-Saharan Africa. Parasite eggs from the feces or urine of an infected person are transmitted to snails in slow-running water used for everyday purposes—personal hygiene, children's play, washing clothes and household utensils, watering animals, irrigation and fishing. Swimming, wading or standing in this water, people are constantly at risk of infection and re-infection from parasites shed by the snails. The infection is often unnoticed, marked only by a brief period of fever, rash or itching—everyday changes in bodily wellbeing that are rarely considered worth a visit to a clinic. But the parasites multiply, and cause serious problems over time: anemia, malnutrition, pain and lethargy, impaired growth and development in children, impaired school performance and work, infertility, and in some cases, cancers and liver and kidney failure. Health education campaigns and community participation programs exhort people to avoid contact with infested water and use toilets, but this is unrealistic where people lack alternative water resources, taps and toilet facilities, and where the availability of the requisite infrastructure depends on state investment for its provision and maintenance. Further, infection in humans is often maintained because animals are reservoirs, and because economic development programs create new ecologies of infection rather than reducing risks, by disrupting water flows, as occurred for instance with the construction of the Aswan High Dam in Egypt (1967) and the Three Gorges Dam in China (2006).

The second disease considered in this case study is filariasis, which like schistosomiasis is a chronic parasitic disease, affecting some 120 million people in parts of Africa but also in South and Southeast Asia and elsewhere. Mosquito-borne parasitic infections and arboviruses, such as malaria, dengue, yellow fever, lymphatic filariasis and onchocerciasis (river blindness), often co-occur with schistosomiasis, highlighting the interplay of social and biological environments. Filariasis is transmitted in urban and densely settled rural areas where mosquitos breed in stagnant storm water drains, pit latrines and swamps. Health education campaigns for filariasis exhort people to avoid being bitten, but this is difficult. Depending on the species, the mosquito may rest indoors and feed at night, or bite both day and night, so limiting the effectiveness of preventive methods like bed nets, and people are at risk of being bitten and infected as they go about routine activities. As with schistosomiasis, people are often asymptomatic at time of infection. The first signs of filarial infection—again rashes and itching—are easily ignored by people who live in unsanitary environments with limited ways of ensuring hygiene. The disease manifests years after infection, when it becomes profoundly debilitating and stigmatizing: lymphedema, elephantiasis, scrotal swelling in men, and breast and vulva swelling in women. These conditions limit people's capacity to participate in social and economic life, and cause extreme social exclusion (Person et al. 2009; Weiss 2008).

Drug treatment for both schistosomiasis and filariasis is effective. But drug treatment deals with infections in individuals, and where infections remain in a community, people are at risk of re-infection. As a result, countries have implemented mass drug administration programs, when entire populations are treated concurrently to interrupt transmission and so break the cycle of infection. Often mass drug administration programs for schistosomiasis and filariasis include additional drugs to treat endemic soil-transmitted helminth infections (worms), with the drugs donated by pharmaceutical companies and able to be administered by school teachers, local pharmacists or community health workers. But the logistics are complex because of the need to ensure the concurrent treatment of all people in target communities, and so the programs are rolled out as if they were military operations. The process is complicated too where populations resist participation because of their lack of involvement in governance, limited understanding and suspicion of the medications, and confusion about the rationale for particular treatment regimes. Further, as Tim Allen and Melissa Parker illustrate in the case study below, weak health systems at national and local levels, difficulties in procuring and distributing drugs to ensure adequate and continuous supplies, and the leakage of freely available drugs from public health posts into the private sector, interfere with the scale up, success and sustainability of these programs. Medical anthropologists working closely with communities may not be surprised by these institutional and community problems, but others involved in the mass administration of drugs have reacted strongly to the criticism associated with the difficulties in delivering such programs, and their putative success, fearful that the resources directed to these 'neglected' diseases will be withdrawn if operational challenges are exposed.

## 7.1 Mass Drug Administration for Neglected Tropical Diseases

*Tim Allen and Melissa Parker*

Since the turn of the century, there has been a dramatic change in global and national responses to a range of chronic infections, increasingly referred to as the Neglected Tropical Diseases (NTDs). The term 'NTDs' was first coined in 2005 by scholars and activists who were lobbying to move them to the center of discourse on heath policy in affected regions, particularly in parts of sub-Saharan Africa. A case was made that the 'other diseases' mentioned in the Sixth UN Millennium Development Goal should focus on the NTDs, because an effective strategy was at hand and the results would be dramatic.

The number of afflictions grouped together under the label of 'NTDs' has expanded, partly because of successful attempts in attracting funding and assistance from international donors and drug companies. The WHO currently highlights 17 NTDs caused by a variety of pathogens: viruses (dengue and rabies), bacteria (buruli ulcer, leprosy, trachoma and yaws), protozoa (Chagas disease, human african trypanosomiasis and leishmaniasis) and helminths (cysticercosis, dracunculiasis, echinococcosis, foodborne trematodiases, lymphatic filariasis, onchocerciasis, schistosomiasis and soil-transmitted helminthiases) (http://www.who.int/neglected_diseases/A66_20_Eng.pdf).

A sub-group of these afflictions has been a particular focus of attention, with some scholars and activists arguing that they can be easily treated, controlled and, in some cases, eliminated with a combination of available tablets. In a key paper, Molyneux, Hotez and Fenwick (2005) suggested that lymphatic filariasis, schistosomiasis, intestinal helminthiases, onchocerciasis, and trachoma could be targeted by a program of mass drug administration (MDA) with praziquantel, albendazole, ivermectin and zithromax. The medications could be sourced from major companies, notably GlaxoSmithKline, Merck and Pfizer.

A remarkable momentum developed, with drug companies offering free tablets for MDA to people living in endemic areas. The drugs were deemed to be safe and efficacious, and the aim was to treat everyone, especially children, at annual intervals; for this purpose, considerable funds were allocated by the Gates Foundation, DFID, USAID and other donors. Essentially, large-scale, vertical distribution systems were introduced, with medications delivered through ministries of health to community drug distributors. Integrated treatment for multiple infections was simultaneously introduced in several countries, notably Uganda (2006/7) and Tanzania (2007). It was anticipated that these countries would provide models for others to follow.

Spectacular claims soon started to be made about what had been achieved, and fearing that it might be sidelined, the World Health Organization decided to seize the initiative. In 2010, the newly established NTD Division released its first global report on NTDs, presenting an overwhelmingly positive assessment. The following year, the WHO published a 'Roadmap' which outlined control, elimination and eradication targets to be reached between 2012 and 2020. This was followed by the London Declaration of 2012. The World Bank, the Gates Foundation, 13 of the world's leading pharmaceutical companies, and government representatives from the US, UK, United Arab Emirates, Bangladesh, Brazil, Mozambique and Tanzania participated in a joint meeting, held under the auspices of the WHO, at the Royal College of Physicians in London. They committed themselves to achieving the NTD Roadmap Goals, with donors offering hundreds of millions of dollars, and pharmaceutical companies offering indefinite supplies of free medicines.

In this context, anthropological research has proved to be contentious. Evidence from places where people have been receiving the treatments has raised doubts about some of the grand claims being made, especially about the possibilities of disease elimination. Fieldwork carried out at over a hundred sites in Uganda and Tanzania shows that the effect of integrated MDA programs are very mixed. In some places, the uptake of drugs for NTDs is high, especially where populations are contained, such as displacement camps in Uganda (where tablets have been provided together with food relief). There are also locations where understandings of MDA are such that there is willingness and even eagerness to be treated. An example is among fishing populations living on islands in the River Nile, to the north of Lake Albert in Uganda. Elsewhere, however, sometimes close by, there may be resistance or outright rejection of the tablets. Rumors circulate about the real purpose of distributing them. Fear of sterilization campaigns is common, as are local conceptions about the signs and symptoms of afflictions that counter biomedical perspectives. Overall, findings suggest that the drug take-up rates are below those necessary to systematically control infection (Parker and Allen 2013). Our anthropological findings on NTDs, drawn from fieldwork in Tanzania and Uganda, may be summarized as follows.

First, predominantly politically and economically marginal populations are the most likely to be infected with NTDs and the least likely to have access to information about the infections or therapeutic care. It is neglected people who have neglected diseases, and this social problem cannot easily be addressed by vertical biomedical interventions. Indeed, top-down delivery mechanisms are as likely to reinforce power relations as to undermine them. In Panyimur sub-county, north-western

Uganda, we found that MDA was failing in large part because the treatment program exacerbated antipathies among the fishing population towards district officials and towards the government. A crucial factor was the lack of respect shown by officials towards those who were supposed to benefit from treatment. There was disinterest in discussing the programs in public meetings or seeking support from local leaders, such as the traditional chief, leaders of religious groups, and local healers. Resisting disease control here was a form of what Scott (1985) has called 'weapons of the weak.'

This lack of respect for target populations reflects a broader concern about the orientation of MDA programming. Despite rhetoric emphasizing the need for community engagement and a list of sensible WHO guidelines, in practice MDA can end up being worryingly top-down and divorced from local realities.

Unpaid and poorly trained community drug distributors are often used. They are supposed to understand the purpose of persuading their neighbors to take medicines for diseases that they often do not know they have, and about which they may have very different local understandings. For those infected with *Schistosoma mansoni*, for example, it is possible to have no signs or symptoms of infection and to feel quite well during the early years of infection. In such circumstances, it is hard for community drug distributors (whose biomedical understanding of the signs, symptoms and etiology of the NTDs is typically quite limited) to persuade their neighbors to consume medicines on a regular basis. A similar issue arises with lymphatic filariasis. The drugs need to be given to people well before they develop symptoms. Also, in places like coastal Tanzania, the symptoms themselves are not normally connected with the parasitic infection, but with a range of other causes.

There are important differences between the kinds of tablets being offered and between the diseases being treated. Some tablets are welcomed. Albendazole is a good example. It tastes like a sweet and it has an observable effect. If it is given to people with particular kinds of intestinal helminthiases, the worms pass out in the stool. In contrast, praziquantel tablets are large, several have to be swallowed at once, and they smell horrible. In addition, they have side effects if taken by individuals heavily infected with *Schistosoma mansoni*. People may feel sick for days, and in Uganda, stories were circulating about people dying from the drugs. As a consequence, integrating treatment into a homogeneous and vertical delivery system is far from straightforward, and may be counter-productive. If one drug is disliked, none will be taken.

A lack of engagement with targeted populations, including poor explanations for the rationale for mass drug distribution and ineffective health communication about the signs, symptoms and etiology of NTDs, lends credence to conspiracy theories. Without such engagement, rumors about population control can escalate into dangerous situations. There were, for example, violent riots in Tanzania in 2008, initially triggered by attempts to implement MDA for schistosomiasis in schools.

Given the poor communication, ethical issues arise about the treatments, but tend to be brushed aside. Ethical issues apply particularly to the treatment of children, who are effectively compelled to take medications without their parents being properly informed or their teachers being properly briefed. Children have occasionally been observed being compelled to take the tablets after trying to run away, and angry parents berating the teachers is not uncommon. We have also observed the distribution in schools of the wrong tablets, the wrong combination of tablets, and the wrong health education leaflets for the medication. In certain places, such as Ukerewe Island in Lake Victoria, these problems have been overcome and observed take-up in schools is high. However, elsewhere, school distributions are partial. On occasion, unopened containers of tablets were found in school store rooms. It became apparent that teachers lacked the confidence or commitment to distribute them to the children as they had been instructed to do.

Rigorous assessment of MDA often does not occur. Even the most basic information tends to be guesswork, and treatment rates are commonly inflated. This became apparent from combining participant observation, careful collation of data from community drug distributors, and checking of reported drug uptake data as it passes from points of dispersal to national reports. For MDA to have clinical benefits, those being targeted have to actually consume the drugs being offered in the correct doses, and while this happens in some cases, in other cases it does not. Without knowing what is happening at very local levels, it is impossible to address this issue. In practice, it is mostly ignored.

Finally, ethnographic evidence from villages and anthropological assessments of MDA programs at district, national and international level has revealed that some of the literature emphasizing positive results, published even in well-respected medical journals, is linked to securing and sustaining grants. There is a pronounced tendency to de-politicize public health, with the roll-out of MDA clearly resonating with Ferguson's notion (1990) of the 'anti-politics machine.' Assertions emphasizing poverty alleviation are conflated with dubious data, and many passionate supporters of MDA end up believing their own rhetoric. This has been reflected in some defensive reactions to evidence from anthropological fieldwork showing that results are mixed. That has been especially so with respect to our findings on lymphatic filariasis in Tanzania (Parker and Allen 2014).

★ ★ ★ ★

To have any chance of interrupting the transmission of lymphatic filariasis so that elimination would be feasible, annual rates of drug take-up have been variously estimated at 65–80 per cent over four to six years (Mohammed et al. 2006) and 70–90 per cent over six or more years (Michael et al. 2004). In areas where the prevalence of microfilariae is greater than 10 per cent, such as coastal Tanzania, Michael and colleagues (2004) have suggested that coverage levels of 90 per cent or more may be necessary. Our research showed that there was no prospect of that target being reached. The highest rates of take-up we recorded were around 70 per cent near Muheza town in 2007, where there is a large hospital. Elsewhere in both Muheza and neighboring Pangani districts rates were often less than 50 per cent. For example, at one coastal site in Muheza district, a survey of 146 homes in 2007 found a self-reported drug take-up rate of 22 per cent, a drop from 47 per cent in 2004. A further survey in 2011 of 144 homes found a slightly higher self-reported uptake of 34 per cent for 2010. There is no way that lymphatic filariasis can be sustainably controlled, let alone eliminated, with these low levels of drug uptake. Moreover, actual consumption of drugs was undoubtedly even lower, because some recipients had taken the drugs from the distributors, but either did not swallow them, or only swallowed a partial dose. No serious effort had been made to communicate the reasons for MDA, and the vast majority of people had no idea that mosquitos were the vector of the disease, even after multiple rounds of treatment. The fact that taking the tablets had no discernible effect on those with hydroceles or swollen limbs confirmed to many that the drugs were at best defective, and at worse had a sinister purpose. There was no prospect of the situation altering without a radical change in approach.

We began publishing our research on lymphatic filariasis in 2011. Initially, we did so to illustrate the overblown claims being made about NTD control more generally. In response, two leading advocates of MDA, among those who claim that elimination of lymphatic filariasis is both possible and likely by 2020, wrote a ferocious attack in a medical journal (Molyneux and Malecela 2011). They did not explain why things might be better elsewhere or how targets for elimination might still be achieved. Instead, they argued, treating NTDs was a way of ending poverty, and questioning the effectiveness of the approach was ethically unacceptable; they characterized our work as "disrespectful to endemic countries" and "unethical and grossly negligent". Further, the authors referred to research at a "sentinel research site" in Pangani District, which they suggested contradicted our results—no such sentinel site existed. Subsequently these and other authors who were leading supporters of vertical MDA wrote a short piece in the *Lancet*, maintaining that anthropological insights into MDA should be dismissed as "cynical" (Molyneux et al. 2012).

Anthropological research on NTDs has, however, had impact, and becomes difficult to set aside when supported by clinical data. In December 2012 we were invited to a meeting at the UK's Parliament. At that event, the Director of Research and Chief Scientific Adviser at the UK's Department for International Development highlighted the dangers of exaggerating the results of treatment programs for NTDs. He referred to data on the control of leprosy, suggesting much greater reductions in prevalence (and therefore incidence) than was actually the case. The result was that donors reduced their funding for leprosy control in the mistaken belief that the disease was no longer a serious problem. He also noted that the targets being emphasized to secure support for control programs for other diseases could be similarly counter-productive. In the discussion that followed his talk, we agreed that it was unhelpful to use the term 'elimination' with respect to lymphatic filariasis—except in particular geographical areas where there is good evidence that this is technically feasible and realistic. In March 2013, our findings on lymphatic filariasis were foregrounded in a parliamentary debate, and our research and that of others has ensured that MDA programs are now paying greater attention to the specifics of local circumstances.

A commitment to combating NTDs is a hugely positive development, and that is all the more reason to avoid treating fund-raising rhetoric as facts, and to learn lessons from what has happened with such large-scale vertical schemes in the past. That means placing medical anthropology at the heart of programming. Anthropological studies confirm that MDA for NTDs can potentially assist poor and marginal people, while emphasizing the acute need for sober and rigorous assessment of locally specific effects and reported results. Far from being cynical, anthropological evidence confirms that this approach is sensible. There is an opportunity to help millions of people. It would be disgraceful to waste it.

### *References*

Ferguson, James 1990 *The Anti-Politics Machine: "Development," Depoliticization and Bureaucratic Power in Lesotho.* Cambridge, UK: Cambridge University Press.
Michael, E., M. Malecela-Lazaro, P.E. Simonsen, E.M. Pedersen, G. Barker, A. Kumar, and J.W. Kazura 2004 Mathematical modelling and the control of lymphatic filariasis. *Lancet Infectious Diseases* 4:223–234.
Mohammed, K.A., D.H. Molyneux, M. Albonico, and F. Rio 2006 Progress towards eliminating lymphatic filariasis in Zanzibar: A model programme. *Trends in Parasitology* 22(7):340–344.
Molyneux, D., M. Malecela, L. Savioli, A. Fenwick, and P. Hotez 2012 Will increased funding for neglected tropical diseases really make poverty history—authors reply. *Lancet* 379:1098–1099.
Molyneux, D.H., P.J. Hotez, and A. Fenwick 2005 "Rapid-impact interventions": How a policy of integrated control for Africa's neglected tropical diseases could benefit the poor. *PLoS Med* 2(11):e336.
Molyneux, D.H., and M. Malecela 2011 Neglected tropical diseases and the Millennium Development Goals—why the "other diseases" matter: Reality versus rhetoric. *Parasites and Vectors* 4:234.
Parker, M., and T. Allen 2013 Will mass drug administration eliminate lymphatic filariasis? Evidence from northern, coastal Tanzania. *Journal of Biosocial Science* 45:517–545.
——— 2014 De-politicizing parasites: Reflections on attempts to control the control of neglected tropical diseases. *Medical Anthropology* 33(3):223–239.
Scott, James C. 1985 *Weapons of the Weak: Everyday Forms of Peasant Resistance.* New Haven, CT: Yale University Press.

## Living Uncertainly

Changes in human settlement have influenced the prevalence and distribution of disease, and will continue to do so. In around 30 years, some 50 percent of people in Africa will reside in urban environments; already even greater proportions than this live in urban settings in middle- and low-income countries on other continents. Driven by this shift, by colonization, changes in patterns of work and industrialization, and by globalization, food produced locally has been replaced everywhere by commercially grown, manufactured and distributed products. Increasingly people subsist on a narrow range of foodstuffs, inexpensive but of poor nutritional value, and often aggressively marketed.

In association with urbanization and industrialization, cardiometabolic diseases—the clustering of obesity, diabetes mellitus and cardiovascular disease—are emerging as neglected diseases of poverty that are steadily outstripping infectious diseases in prevalence worldwide. These diseases were commonly referred to as 'lifestyle' diseases as they gained importance in the global North, but this terminology implies that people have a choice in how they live their lives. Regardless of setting, in high-, middle- and low-income countries, the decisions that people make about diet, physical activity and health are subject to environmental influences, social inequality, power, and politics. Unemployment, poverty and social disadvantage largely predict the chronicity of non-communicable diseases (Weaver and Mendenhall 2014), and the development of other co-occurring diseases. These social factors are all implicated in the increased incidence of obesity, diabetes and heart disease, and associated anxiety and depression. Similarly, the personal and social resources available to people—human capital and social capital—influence the likelihood of acquiring both non-communicable and communicable chronic diseases, and interfere with prevention and control, timely diagnosis, effective treatment, and ongoing care.

Treatment advances have altered the trajectory of many chronic conditions, both infectious and non-communicable, and in this context, early diagnosis can be life saving. There is considerable interest in early interventions to ensure that symptoms of disease are addressed when they are relatively minor, in order to prevent complications and early death. This is possible through effective screening and testing, despite that these technologies may either produce or resolve anxiety about non-communicable disease. People may undergo tests because they are concerned they have signs of disease, for example, or in response to a public health campaign; others undergo tests because of a family history of a condition. Of these, a proportion of people seek out information about risk to decide whether or not to act preventively, to gain information in order to identify potential signs and symptoms, or to make decisions about reproduction (Flaherty et al. 2014; for further discussion, see Chapter 14). Yet anxieties about the potentiality of disease do not necessarily result in behavioral change to prevent its development. People may argue low risk on the basis of local biologies (Lock 1993), or their own experience.

In many places across the globe, screening programs are simply unavailable, their cost-effectiveness in terms of public health gains is questionable, and access for the majority of the population is limited. Delays in diagnosis and treatment may also occur because early signs such as tiredness may be overlooked or dismissed, as we have suggested already, because of the lack of technology to diagnose with accuracy, because of poor access or resistance to attend available health services, or because of fear of the diagnosis. Health service factors, including inadequate and broken equipment, lack of laboratory services for pathology tests and results, and lack of follow-up with patients, all also affect pathways to care. Most cancers in resource-poor settings, for instance, are only identified when the symptoms are distinctive and the disease is advanced.

Everywhere people live with ongoing health and medical problems associated with degenerative changes and their complications, and with the continued fear of the return of a disease, as occurs for people who have had cancer, or a repetition of a life-threatening health event for those who have survived cardiac arrest or a stroke. Cancer's chronicity in particular is framed by fear of recurrence, regardless of age at onset, and people live precariously in a time defined by the years of their survivorship. Even the language around illness and recovery—of remission, not cure—captures the uncertainty of outcome and the limits to our understanding of such diseases. Cancer, too, draws attention to the ways in which biopathology is situated. Family history, genetics, various demographic factors including age and gender, food and diet, environment and exposure to known carcinogens, are all risks for disease (see also Chapter 14). Further, as Julie Livingston (2012) has so clearly illustrated for Botswana, personal- and community-level social and economic factors, area of residence, and country capacity, all also shape outcome, including timely and accurate diagnosis, and access to the technologies and finances for treatment, care and palliation. Paralleled only by HIV, cancer is also a political disease, with people advocating for its prioritization in research effort and funding, the delivery of particular medical services, and the provision of drugs. Breast cancer particularly, like HIV, is highly visible in public space, through screening programs, support organizations and fund-raising activities, and advocacy (Manderson 2011; 2015).

Because chronic conditions are by definition persistent, even if variable in their severity and impact, everyday living is interrupted when poor health intrudes, as normal bodily functions are undermined and as bodies are transformed physically. Individuals must seek to live with the fickleness of their health, restructuring social and economic activities and relationships, and modifying familiar routines and social roles. Illnesses, loss of capacity, repeated hospitalizations and institutionalization, and growing dependence on others place older ill and disabled people particularly in a frequently depressing and liminal space (Turner 2008; van Gennep 1960). This liminality, and embodied frailty, intersect with structural fragilities such as immigrant status (local or transnational), unemployment, geographic isolation and poverty, so further limiting people's social connectedness and the material and emotional advantages this yields.

While adjusting to bodily changes, chronic illnesses come to dominate the time of those affected. People must manage care regimens and medications, and monitor physical signs and symptoms that might augur complications and the need for further medical attention, diagnosis and treatment. People with health problems such as diabetes, COPD (chronic obstructive pulmonary disease) and heart disease may find their lives structured around the routines of self-care including diet, physical activity and medication; for those with diabetes, including insulin injections and blood sugar monitoring; for those with COPD, oxygen. In this way, people's homes are modified to incorporate biomedical equipment and accommodate impairment, with hand rails, ramps, lifts and blood pressure cuffs all converting domestic space into a hospital at home (Pavey et al. 2015; Sakellariou 2015). But there is a significant difference across countries and the affordability of such appurtenances. People in the global North can, in general, access regular supplies of insulin pens or oxygen, for instance, and may be able to join support groups to deal with the challenges of living with poor health. In resource-poor settings, people's lives may be cut short in the absence of access to medicine or its technologies, and they necessarily improvise, using plastic bags in lieu of purpose-made stoma bags; trolleys or wheelbarrows for mobility; or they simply crawl from one place to another.

As we have already indicated, diagnosis enables an individual to make sense of disturbing signs and symptoms and opens the door to treatment, whether it is potentially curative (as is the case for some cancers, or for leprosy, for instance), delays severe symptoms (with HIV and other cancers, for example) or prevents secondary disease, as with the early diagnosis of diabetes. In the process, it establishes an authoritative relationship between the patient and the health care provider—the doctor, usually, often with the backing of pathology results and imaging—and provides a way for the patient and provider to plot a future, or a choice of possible futures (Smith-Morris 2015). People may fear the diagnosis of certain long-term diseases, but may find relief because diagnosis allows them, their families and friends to make sense of particular signs and behaviors, and to begin to plan around the eventuality of changed health status.

In other cases, the development and diagnosis of chronic conditions may simply be regarded as part of an anticipated life-course, even when diagnosis releases the biomedical resources that are available, as Carolyn Smith-Morris illustrates in the case study that follows. Robin, aged 29, already has poorly controlled diabetes with complications, including kidney failure, and she is faced now with the likelihood of lifelong dialysis or, less likely, a kidney transplant and lifelong medication to prevent organ rejection. But Robin's poor health is the consequence not of one disease alone but an interplay of social and biological circumstances; her ill health illustrates what Singer (2009) describes as a syndemic. Robin is a Pima American, and her poor health is very much the product of a long history of colonial oppression, racism and continued structural disadvantages. Her particular health problems are the consequence of interactions in the prenatal environment, genetics, and contemporary social and economic limitations that influence her diet, activity, employment, mobility and care. Her diabetes, that is, is the result of various broad social determinants of (ill) health and the interactions of different physical health conditions.

## 7.2 Diagnosis and the Punctuated Life-Course

*Carolyn Smith-Morris*

On the Tuesday morning I last interviewed her, Robin awoke on the couch, having given her bed over to the three young cousins visiting her from Komatke, just a few miles down the road. Her teenage brother was draped over the reclining chair next to her, and there were two more pre-teens on the floor. The TV was still on, playing "The Big Country," and the flickering lights made the silent house glow in a strange half-sleep. Stepping gingerly through the crowd of blanketed forms, I greeted Robin and accepted her coffee and a seat at the table. I had only moments ago left a similar scene, in the house where I lived during this field season, so the circumstances of this interview did not require any substantive emotional or psychological effort on my part (in contrast to the first years on the reservation, when these intimate encounters were new and unnerving). Robin had already slept through the noise of her two youngest sisters' departure for school. A dark fleece blanket, the kind you can buy roadside or at the fair, this one with an eagle dancer on it, blocked the 10:00 AM light, so we sat in the TV glow talking over her federal assistance checks, her goal to get a high school certificate, and her health. This was our third such conversation as part of my ethnography of stress and diabetes in her community. Robin's grandmother would come home from work around noon and take her to an appointment at the clinic.

The house crowded with extended family, the sense of protection its walls gave from the harsh light and heat outside, even the sleeping arrangements, are all part of a fairly typical, although not universal, embodiment of 'life on the res.' Indian time is just one form of embodied resistance, and its naming a discursive strategy to stabilize a cultural system that is different from the colonial system. Life on the res is slower and driven by internal human rhythms. It involves a passivity about clock-dictated regimens, and an overt rejection of several more cultural characteristics of the Western European, industrial (now technological) complex. Robin's diurnal patterns, her willingness and ability to sleep wherever she lays her head (without sheet or pillow), and her unquestioning attitude toward the irregular company of extended family, are all features of reservation embodiment that strikes me as tethered to a remote, pre-colonial past. They are vestiges of communal Indian life somehow surviving to inhabit this partially modern landscape, a house in the middle of the desert.

The home we are in is one of the Bureau of Indian Affairs (BIA) housing units constructed with money from the Snyder Act of 1921, allocated "to replace and repair housing for the neediest of the needy: American Indians and Alaska Natives" (Bureau of Indian Affairs April 1, 2001). Then, many Pima lived in *olas'ki* (oh-loss 'key) or other hand-made housing. The new housing would include running water, electricity, and toilets with individual bedrooms and doors for privacy, security from theft, and clarity of ownership, and so also individualism. With the houses came a variety of ideologies into which each new generation would grow.

If you have ever driven through a reservation, you might recognize the rows of cinder-block houses, set apart in even increments oriented to a straight line rather than the natural contours of the former landscape. The federal system that oversaw housing for these Indian 'wards' of the US government, along with the disparate other mechanisms for Indian health care, education, and other priorities, had grown complex and tangled over time. Then, in 1975, came a momentous new piece of federal legislation, what is known as the Self-Determination Act. That law addressed health, education and governance, leading many tribes to write contracts for their federal funds, so that they could budget and manage those funds themselves rather than allowing control of federal employees over tribal funds, planning, and management. The new system has promoted tribal capacity for self-governance, many would say better than any other federal legislation since the colonial period. For Robin, this means she speaks with community members in local tribal offices when she needs housing repairs or assistance. A distant cousin helped her mother file an application for their current house (14 years ago), into which Robin moved as a 15-year old. Thirteen people now live in this three-bedroom, one-bathroom, 850-square-foot house; this density is characteristic. The home I left earlier that morning had eleven.

On Tuesday afternoon, Robin had an appointment with a nephrologist to determine her need for dialysis. She had been diagnosed with diabetes when she was eight or nine years old. As an overweight little girl, she raced, played soccer, and explored in the desert with her friends, but like most other American children, she also watched a lot of TV and ate junk food and soda. School programs taught her the facts of nutrition and portion size, but vending machines and vendors of fry-bread (a white flour and lard dough, shaped into 10-inch circles and fried) were more tantalizing. Her story is ubiquitous in America; the country is in the midst of an obesity epidemic because of industrial food's allure. But Robin's genetic code, and that of all Pima and many other Native Americans, makes her vulnerable to more than obesity. For those with the 'thrifty gene,' engineered through the millennia to survive the prehistoric feast and famine cycles of the Sonoran Desert, the constant feast of contemporary industrial Western life leads quickly to diabetes. Robbed in the 1800s of Gila River water so crucial to their agricultural lifestyle, the Pima sunk into decades of extreme poverty and dependence on the federal government, including commodity foods that were high in salt, fat, and packaged goods. Today, geographic isolation, federal food assistance, and widespread unemployment leave many without healthy food options (Smith-Morris 2004).

Robin is 29, more than two decades younger than most US Whites when they are diagnosed, which on average is at 55.4 years of age (2011 data). For the past several years, Robin has worked hard to change her eating habits and to embrace the concept of exercise, to protect the kidneys she can neither see nor feel. A person's kidneys serve to filter wastes from the blood that are produced by normal functions in the body. If the kidneys cannot perform this function, or cannot perform it adequately, the wastes build up and cause damage. But the kidneys can themselves be damaged, and in the case of diabetes, excess glucose in the blood stream damages the tiny blood vessels in the kidney. Damage from diabetes leaves them unable to function properly and can potentially lead to diabetic kidney disease. This is what Robin is facing, and the treatment options will be grim, especially for a 29-year-old woman: an indefinite and possibly lifelong regimen of thrice-weekly, three-hour dialysis, or to enter the multi-year waiting list to receive a kidney transplant.

Much of my ethnographic work has been about the Pima community of Gila River in Southern Arizona, where half of adults over age 35 have been diagnosed with diabetes. Because this community epidemic is rooted in both historic events and prehistoric genetic conditions, the Pima have had a long time to identify with this disease. And their reactions to the abundant diabetes prevention messages on the reservation, and to the tests they undergo, is intimately tied to this long history. For them, the term 'chronic' takes on new dimensions altogether.

To study diabetes in detailed and historical context requires us to view diagnosis as a punctuating moment within a long life-course. Focusing on that full life-course helps social and behavioral scientists to re-link the pre-, mid-, and post-disease states through which a person will live. We can examine how the diagnostic moment is all about encompassment or exclusion in health care, and how this moment may give the solace of understanding, but may also add a burden of label, stigma, limits, or exclusion.

Diagnosis is the nucleus of medicalization and the production of medical knowledge about diabetes (and every disease). Although there is consistent debate and revision to many diagnostic criteria, the process of diagnosing is inseparable from medical care (Smith-Morris 2005; Smith-Morris 2015). The American Diabetes Association (ADA) has set the thresholds for diabetes diagnosis at > 126 mg/dl fasting plasma glucose, or >200 mg/dl in either a casual plasma glucose or a fasting plasma glucose at the 2-hour mark in an oral glucose tolerance test, which involves consuming a glucose drink and watching the blood to see how the glucose is absorbed. Using these numbers, a doctor can say with reasonable certainty that the patient will develop diabetes-related complications within a few years. But since the Pima are known to have diabetes at higher rates than average, the ADA recommends they be tested more frequently, and at a younger age than others. If diagnosis can be made early, it is argued, its worst effects can be avoided. But what use is this process of diagnosis and prevention, if the treatment is unbearable or impossible, and if medications are not available and effective?

The importance of the anthropological project is to understand when and why patients embrace these diagnostic meanings, and when they don't. Understanding chronic diseases like diabetes requires a life-course perspective that situates disease within a greater world of competing priorities. There is a great colonizing potential in biomedicine to erase both individual and local cultural legacies, including those personal and creative adaptations that maintained lifelong balance long before biomedicine's arrival. The lifelong identity of illness carried by Pima Indians is imprinted in their bodies and behavior long before they might ever receive a physician's diagnosis of diabetes. Their identity is built on prehistoric foundations but is equally submerged in the contemporary diasporic world and biotechnological marketplace. Pima Indians must harness this identity and those personal and creative adaptations to survive the epidemic.

Beyond Gila River, the cost of an uncritical use of biomedicine's temporality is rising, and the conceptual battle over these concepts impacts real dollars and real lives (Manderson and Smith-Morris 2010). The temporal distinctions between 'chronic' and 'acute' set the direction for global health funding. "Chronicities of modernity" (Wiedman 2010) explain why so many developing communities undergo a transition to high rates of obesity, diabetes, and other metabolic problems as they begin to incorporate into capitalist, high-technology markets where the best paying work is sedentary, and the cheapest foods are saturated in sugar, salt, and fat. Temporal dichotomies of chronic–acute fail miserably to address any of the proximal causes of disease and, instead, shift both focus and burden onto individuals for understanding and treating what are community-based problems. And so rather than making smarter and healthier models for economic development, we repeat in community after community an evolution toward the endemicity of metabolic conditions. This evolution of endemicity happens when disease and related costs rise, and solutions are sought in existing capitalist and biomedicocentric structures, some of which may have helped produce the epidemic in the first place. Failure to come up with new systems and interventions means that old errors are then rolled out in greater force and breadth, entrenching the original (but flawed) infrastructure, leading to further rise in disease rates. "In these ways, overreliance on a biomedical paradigm actually contributes to longer and greater morbidity around the world by encouraging what could be called a passive discourse of chronicity in developed nations" (Smith-Morris 2010: 23).

In *Chronic Conditions, Fluid States* (Manderson and Smith-Morris 2010), the contributing authors frame the disease experience in new temporalities, pointing to the ongoing and sometimes life-long relevance of a disease condition to one's experience, and to the punctuations—e.g. diagnosis, inflamed symptoms, changes in treatment technology—that disrupt that chronic period. Life is a series of disruptions, small and large, all of which are related to one another through the memory and identity of each suffering individual. The Pima case study has always struck me as poignant on this aspect of chronicity, because the longevity of disease reaches back through time to parents (through the prenatal environment), through history (in changing federal policy that so defines the political-economic realities in reservation communities), and deep into generations (through genetic code). In this lifelong and culturally and historically embedded view, diabetes is just one punctuation in a life of chronic stress, strong or weak constitutions, greater and lesser wealth, and formative or traumatic life events.

Robin's kidneys are indeed failing, and she will have to start dialysis. The nephrologist is most worried about the poor circulation in her feet and the real possibility that she will face amputation of one or more toes if her blood glucose does not improve quickly. As a diagnostic event, this news might hit Robin with the overwhelming force of catastrophe. Diabetes is a disease process that can, at different temporal locations and by different speakers, be called a 'risk,' 'pre-disease,' 'gestational,' 'outright,' or fatal, but never 'in remission' or 'cured.' Yet when a diagnosis of diabetes is delivered, little of that complexity or controversy is talked about. For a person with high blood glucose, those controversies probably don't matter; what matters is health, how one feels, what one's future holds. For Robin, that future is likely painful, physically limited, and shortened.

But Robin did not see it quite that way. For her, diagnosis was an acute, short-lived disruption in an otherwise roughly flowing stream of life. The 'disrupted' portions of her life appear to me through ethnographic research. And so, it is we, the anthropologists, who draw attention to these 'liminal' elements of illness. Robin later relates to me that her experience of this diagnostic conversation was "nothing new." She had heard about these complications and possibilities before, and she did not experience them as 'new' when the doctor delivered an official diagnosis of renal failure and a treatment plan that would include dialysis.

The case of Robin's diabetes suggests that a de-privileging of acute diagnostic moments, and a greater sensitivity to the chronicities of daily adjustment, better reflects the lifeworld of Pimas. Very likely, this is true for all humans. "Continuity is not an illusion," as Becker exhorts, "the illusion is in our labeling of lifetime illness episodes as distinct from an otherwise disease-free life" (Smith-Morris 2010: 35). Many of my informants experience a continuous life: "I've been living wrong for so long that I think that it's not wrong. It's only me. Other people may see it, but not me" (Kyle, 29). Anthropological discourse that defragments will privilege continuity of illness and health together in lived experience, especially as these inform the identity of persons with chronic diagnosable conditions.

## Acknowledgments

This case is drawn from research conducted in the Gila River Indian Community between 2007 and 2009. I gratefully acknowledge the GRIC Tribal Council and my informants for allowing me to conduct this work, and the SMU University Research Council for financial support of the project.

## References

Bureau of Indian Affairs 2001 Code of Federal Regulations Title 25 — Indians. Chapter 1 — Bureau of Indian Affairs, Department of the Interior. Subchapter K—Housing. Part 256 — Housing Improvement Program, April 1.

Manderson, L., and C. Smith-Morris 2010 *Chronic Conditions, Fluid States: Chronicity and the Anthropology of Illness.* New Brunswick, NJ: Rutgers University Press.

Smith-Morris, C. 2004 Reducing diabetes in Indian country: Lessons from the three domains influencing Pima diabetes. *Human Organization* 63(1):34–46.

——— 2005 Diagnostic controversy: Gestational diabetes and the meaning of risk for Pima Indian women. *Medical Anthropology* 24(2):145–177.
——— 2010 The chronicity of life, the acuteness of diagnosis. In *Chronic Conditions, Fluid States: Chronicity and the Anthropology of Illness*. L. Manderson and C. Smith-Morris, eds. Pp. 21–37. New Brunswick, NJ: Rutgers University Press.
——— 2015 *Diagnostic Controversy: Cultural Perspectives on Competing Knowledge in Healthcare*. London: Routledge.
Wiedman, D. 2010 Globalizing the chronicities of modernity: Diabetes and the metabolic syndrome. In *Chronic Conditions, Fluid States: Chronicity and the Anthropology of Illness*. L. Manderson and C. Smith-Morris, eds. Pp. 38–53. New Brunswick, NJ: Rutgers University Press.

In the preceding chapter, we considered various ways in which people used their own bodies to resist public health and doctors' admonitions and social expectations for varied reasons. For some people, the need to adhere for life to medication is reason enough to resist. In some cases, as occurs with HIV medication, treatment breaks—'drug holidays'—may be suggested for clinical reasons, including to reduce long-term side effects or to ensure continued efficacy; others interrupt treatment because of cost and the tedium of the regimes. In an article on adherence to diet to control diabetes in Thailand, for example, people routinely 'cheated' between visits to doctors, and reverted to the advised dietary regimen immediately before a visit to the clinic to avoid being detected. For them, it was important that they, not their disease, were in control (Naemiratch and Manderson 2006). But failure to adhere to medical advice can have devastating consequences.

Below, Narelle Warren describes the difficulties experienced by Australian men when they end up with an amputation as a result of result of uncontrolled diabetes. Their consequent increased dependence on others, and the disability of living with one leg, seriously impact on their 'taken-for-granted' masculine identities. Reinforced in promotional material from the rehabilitation center, they aim to be mentally and physically strong; when they receive their prostheses, they train hard to regain their strength and so maintain their masculinity. This is at a cost: men do not discuss the emotional impact of their illness and amputation.

## 7.3 Amputated Identity

*Narelle Warren*

Masculinity is not a fixed state. Rather, it is determined and reconstructed by the physical, social and bodily context(s) in which men live. Cultural ideas about sex and gender, as well as age, ethnicity, social class, and geographic location, all shape how masculinity is understood on an everyday level. These factors in combination shape how men as well as women understand their health, with these in turn influencing their health outcomes. Serious long-term illness and its continuing effects on the body, including limits to activity, increasing dependence on medical surveillance and medication, and impairment, influence how men understand their masculine identity—their 'manhood.'

From 2004 to 2007, I worked with inpatients at four rehabilitation centers in Victoria, Australia, exploring people's adaptation to lower limb amputation (Warren and Manderson 2008; 2013); I still (in 2014) collaborate with an amputee support group. In the cases I present below (all names are pseudonyms), all involving below-knee amputation, I illustrate the ways in which amputation threatened many of the cornerstones of masculinity, as understood by study participants, not only in terms of their level of activity but also in terms of emotional expression.

I have never seen so many men cry. Australian men—most men—value emotional control, and link the performance of masculinity to the containment of their emotions. But of the 44 men with whom I spoke, almost all cried at some point in the interview. The emotional aspects of amputation was challenging for all participants—women and men—but especially for men. Although most underwent amputation for non-traumatic reasons, secondary to vascular disease or diabetes, amputation caused a rupture to men in terms of sense of self; it was emotionally traumatic in its *effect*. Yet, although the men recognized their heightened emotional vulnerability because of the recency of limb loss, they sought to deny or hide their feelings outside of the interview context. Most did not discuss their affective responses with their spouses, but focused instead on the practicalities of their living arrangements post-discharge. Fewer still talked with friends or others about the emotional impact of limb loss.

## *Roy: The Private Life of Emotions*

Roy Smith (80 years old, urban, lived alone) explained how he had become more 'sensitive' both in his amputated limb (the 'stump') and emotionally after his amputation, which occurred following a number of unsuccessful surgeries for varicose veins:

> If you have a leg off, you *do* care to lose it, but [you need] a certain amount of courage . . . (it) upset(s) you for a while, but (you) don't let your life stop, it must go on . . . I didn't take it too seriously 'til I start to think, or I start to see what's going to happen to me when my leg was gone. But um, but it's a hard thing to take into consideration that your, that your leg was there and now it's gone. You can't walk properly, you can't stand properly, so you feel very sensitive. Yeah. (pause) But I suppose in a way you give in, you know, to however it happened to you. And you take it as part of life. That's what I think.

Roy found it difficult to express how he felt after the amputation; this was complicated by the fact that he had never learned how to talk about his feelings: "I haven't done that sort of thing [talked about how he felt]. I don't do that sort of thing. I just take it as it goes really. But that's the way I was brought up." He drew on other life experiences in an attempt to make sense of his affective responses to limb loss: "There's a funny way you miss something like an old, an old mate (friend) . . . (*Is that how you're feeling about your leg?)* Oh, I would say so. Oh ah, you went through a war and go through all your sports and your life and you're laughing. And then it's gone." In drawing analogies with long-term friendships, and the loss of a number of these friends in the preceding decade, Roy sought to reframe his feelings of loss from a personal sense of despondency (which highlights personal vulnerabilities and countered his own understanding of masculinity) to become a different type of grief, one framed in a context of male friendship and loss.

Although he cried for most of our interview, Roy was unwilling to broach his feelings with his friends or his long-term partner out of concern that he would place on them an unwanted and unnecessary burden:

> There's no one who, no one sort of ah, who would want to, even would want to [talk about it], those things don't concern, people like that have worries of their own, troubles and that. They don't want to be burdened, it is a burden . . . That's the way I look at it. You just have to take it as it comes and that's that.

Roy was adamant that men were responsible to keep their emotions in check, and to keep their emotional life contained and private. He expected others to similarly manage their emotions, even though he recognized that there were times that were especially difficult:

> I don't show emotions, not like some people do. And I don't know how they feel when they're on their own, so . . . They might do [become more emotional]. I don't know. Not in front of me. (pause) Which is good, I think. (pause) You don't want everyone bawling around you, do you? (pause) But ah, the queerest thing is when you see those legs, you know, and they're not there. That's when it hits you.

### *Darren: Anger, Frustration and Bodily Rupture*

Unlike Roy, Darren Eaton (63 years old) was much more expressive in his response to his amputation—but only in terms of anger and frustration. He believed that his amputation was due to the failure of a podiatrist to exercise his duty of care to him, which left him feeling powerless:

> [The podiatrist] was well acquainted with the fact that I am a diabetic and I couldn't feel my toes and I couldn't see my toes adequately, so I really feel that he has a contributory part in this entire debacle . . . I do feel that he was guilty of, at worse, some sort of contributory negligence, and at the least, he is guilty of not taking his responsibilities to a middle-aged diabetic patient with the seriousness he should have . . . I mean I have lost a considerable portion of my mobility and my, and my leg, as a result of what I see to be his failing to do his job with a reasonable duty of care.

Darren lived alone in a small rural town (population 1,000); he had moved there from the city after retiring from teaching two years previously. He hated the town and was scathing about its long-term residents: "It's a rat hole, it's full of retirees and people who I don't know. I live there because the rent (for his house) is only ninety dollars a week and I'm absolutely independent . . . [and] I can play my bass trombone in the garden or in the house or whatever I want, but for a town to live in it's a really terrible place." He was profoundly isolated socially. He was unpartnered and only had two (female) friends, one living in a community about 60 kilometres away; the other, a new friend, in his town. But contact with both was largely by telephone.

Darren placed particular emphasis on being independent. His amputation therefore threatened his autonomy and potentially rendered him reliant on others in the community, challenged his understandings of who he was, and brought into question his capacity to take part in the activities that he enjoyed: playing music, cycling, and fishing. In thinking through the impact of amputation, he became quite despondent:

> Well it's a terrible blow to a person who felt he had another few years of trouble-free cycling, although I did look ahead and realize that my balance is becoming impaired because my vision is quite poor, so I bought a recumbent . . . and I love to watch their movies, and I love dancing, but I'm not going to do a lot of that, and there are a lot of things that I won't be able to do, like, you know, maintain my balance in the middle of a trout stream, try and catch fish, a lot of things. You see, you need good vision and coordination to maintain your balance, so I might, I might be in for a hell of shock. I'm not quite sure how it will work out. I might need the prosthesis plus a cane, if not two, I don't know. I don't know, but the town that I live in is not maintained . . . and so the footpaths don't get freshly concreted and it can be rather hazardous especially if you have a walking impediment. I may have a difficult time just walking down to the newsagent or walking to the supermarket . . . now I'm sort of dependent on my friend Marian.

Amputation both undermined Darrell's own independence, and he believed it would place pressure on his relationships with others: "I've made one friend in [his town], but she seems to be having a hard time because I've turned from a, a friend, a contributing friend, into a, almost a liability if you like, I mean short of half a leg now . . . " This concern echoed Roy's desire not to be a burden to others, and has its roots in masculine values of control, autonomy and independence.

Darren saw the amputation as more than the loss of his limb; it was a threat to his whole way of life: "I felt terrible about it, because it meant I was losing the fight. It meant that I wasn't going, going to win, going to save my previous life, eventually." At the same time, Darren's amputation threatened his bodily mastery, inhibiting his ability to manage his body: "I'm a slightly more than middle-aged incompetent male person who insists on living by himself, doing his own cooking and gardening and so on, and of course, I never really looked after my health to the nth degree." His pride in his body and sense of bodily wholeness too were threatened by the loss of his leg. Several days after our interview, Darren was given his new prosthetic leg in the amputee gym. Unlike the majority of inpatients, who gingerly held and then tried on their new leg, he threw his across the room, screaming abuse at the prosthetist and other inpatients. Later, he explained that the prosthesis offended every part of his bodily aesthetic; he could not contemplate it as part of his body. Darren felt that his personal ideals of independence, autonomy, bodily integrity, fitness, and strength—all perceived as masculine characteristics or attributes—were no longer present and would not be available to him in the future.

### *Dale: Masculinity Compromised*

Dale (44 years, rural, married) had cut his foot walking on the beach. His foot progressively became infected and ulcerated, and he was informed that the wound would probably not heal due to Type 2 diabetes. Dale underwent nine surgeries in an effort to cut away the necrotic flesh, until his surgeon decided that the only real option was to amputate his leg just below the knee.

Dale had not known he had diabetes at the time of the injury; he was diagnosed as his doctors tried to determine why he wasn't healing. At the time of the wound, Dale was the family breadwinner, with a well-paid managerial position in a large company and the promise of a new, higher-level position; his wife looked after their home and supported their ill (with cancer) adult daughter. He played football each week, and regularly socialized with friends over drinks or barbeques. He had, he thought, no reason to worry about his health. Amputation changed all of this. He was discharged from rehabilitation not to a well-appointed house in his city, but to government-supported housing in a lower-income suburb, close to the rehabilitation hospital. He was bitter:

> It's financially tough. We have gone from living on a thousand dollars a week cut down to whatever it is on this pension (about $300 per week, plus rent subsidies) 'til I start working again. And you don't budget for an accident like this . . . when you are at hospital for a long time, the money runs out, and when you haven't bargained for something like this to happen. We've lived week to week, and we have lived a good life too, but I suppose you don't put that money away thinking one day you will have your leg cut off, plus we were going to a much higher paid job and that's when we were going to start to look at doing something again like buying a house and that sort of stuff.

Dale concentrated on the practical, especially financial, impacts of his amputation. Most men in the study related their masculinity to their role as the main income earner within the family, consistent with broader Australian norms around work and family organization. Dale was continually in and out of hospital for about three months prior to amputation, and so could not earn an income and support his family as he had in the past. He registered for welfare assistance. This was extremely difficult because of the drop in income, and because he could not maintain his prior position in the family. He continually talked about returning to work: "I will be back on my feet again, doing what I am doing." Yet eight weeks after his amputation, he had developed diabetes-related peripheral neuropathy (loss of sensation in his extremities), retinopathy (vision loss), nephropathy (renal failure), and cardiovascular disease (hypertension, which later led to a non-fatal heart attack).

Dale found it difficult to talk about the amputation itself or about how he was feeling afterwards to anyone other than his wife. "You would probably be the first person that I have talked to about it," he explained. "I have dealt with my problems myself anyway. So [there's] no one I have really opened up to." This was related partly to how he was positioned within his wider family: "I have always been a strong sort of bloke, for want of a better word, they [family] have sort of seen me as a manager here and that sort of thing. It has come as a big shock to them."

The affective impacts of amputation were felt in other ways too. Dale described feeling intense pain immediately after his amputation; he had imagined that he would feel little pain and wondered if he was in such pain because of some sort of emotional weakness. He did not want others to think of him as being weak in any way and so, to maintain his self-presentation as mentally strong, he sought to hide his pain:

> They did the surgery and I woke up and I was in absolute agony, it was the most painful thing that has ever happened to me. I came out of surgery, I remember it, it was one o' clock, and I was in recovery till seven. At seven o' clock, they took me back to the ward and that was the start of the worst four hours I have ever had in my life. I had my uncle staying there, I was squeezing his hand, it was just absolute pain, I couldn't describe it to you. Evidently what can happen is it [the actual amputation] just throws the nerves out so much that the pain threshold is just unbearable and that is what happened . . . It was the worst of the worst. On a scale, rating in from 1 to 10, this was 20, double, off the scale . . . There were times when I think I blacked out . . . I don't think my wife ever knew.

Central to Dale's identity is a self-, and social, image as physically strong, emotionally resilient, and capable. These ideas resonate with hegemonic understandings of masculinity in Australian society (Connell and Messerschmidt 2005)—amputation challenged each of these conceptualizations in multiple ways.

* * * *

As Roy, Darren, and Dale demonstrated, with amputation, men experienced challenges to their identity in several ways. First, amputation threatened the affective dimensions of hegemonic masculinity: they experienced the amputation as traumatic and traumatizing, regardless of etiology. As they had perceived themselves as being mentally and physically strong, they expected that they should adapt to their changed bodily state with little distress and when they found they did not, sought to maintain their masculinity by repressing or avoiding any discussion of their emotional reactions to the amputation. Where they did express their feelings, they either drew on more 'macho' emotional responses, such as anger and frustration, or tried hard to avoid public displays of emotion, such as weeping. Second, amputation challenged their understandings of bodily wholeness and integrity; in this way, masculine values of competence and bodily ability were brought into question. These all unfolded before men were able to test their ability to 'do' gender once they had received a prosthesis; what emerged in follow-up interviews (not reported here) was that prosthetics allowed men to mediate their masculinity through maintaining a facade of wholeness and activeness. Finally, amputation was evidence of their declining health, and the lived vulnerabilities associated with this, which, by their very nature, contradicted local hegemonic ideals of manhood.

### *References*

Connell, R.W., and J.W. Messerschmidt 2005 Hegemonic masculinity rethinking the concept. *Gender & Society* 19(6):829–859.

Warren, N., and L. Manderson 2008 Constructing hope: Dis/Continuity and the narrative construction of recovery in the rehabilitation unit. *Journal of Contemporary Ethnography* 37(2):180–201.

——— 2013 Reframing disability and quality of life: Contextual nuances. In *Reframing Disability and Quality of Life: A Global Perspective*. N. Warren and L. Manderson, eds. Pp. 1–16. Dordrecht: Springer.

## Chronic Disease Prevention

In high-income settings, and in poor countries for those who can afford it, many chronic conditions can be treated. In the global North, the direct costs of treating chronic conditions, including medicines, are often limited for patients as a result of state subvention, but elsewhere the costs are born by individuals. Even so, everywhere chronic conditions impact on government budgets and health services, because of the need for lifelong monitoring and care, and the procurement and regulation of drugs; increasingly, these costs account for the greatest health expenditure in all countries. As we have noted, in resource-poor settings, health systems tend to be weak and governments have limited financial and human resources to manage complex conditions. In addition to the lack of trained health professionals, local health centers often struggle to maintain supplies of appropriate and affordable medications for current patients, and to maintain working equipment for appropriate monitoring; even scales to trace the weight of patients may be broken or simply stored away, unused, in cupboards.

Cardiometabolic disease (that is, diabetes, heart disease and obesity) and chronic respiratory disease are considered able to be controlled by 'healthy living'—a better diet, increased physical activity, and the cessation of smoking. 'Lifestyle' factors are routinely identified for health promotion and the prevention of chronic conditions, with an emphasis on individual culpability, reflecting a neoliberal philosophy that emphasizes personal action and obviates structural factors (Rose 2006). Yet as we have also suggested, in many societies health is compromised by structural inequality, poverty and disadvantage. Health promotional messages and medical advice to patients regarding changes to food intake and activity are often difficult for economic reasons, available time and patterns of sociality—the factors captured in the case study from Carolyn Smith-Morris, above. Gender norms and relations influence health outcomes across different societies too, both within personal relationships and because of the role of gender in determining the kinds of work people do (and whether they work), women's and men's relative access to and control of money, and other behaviors that are implicated in people acquiring and managing non-communicable diseases. There are differences in the freedom of men and women to participate in physical exercise, for instance, and in whether or not, and to what extent, they drink alcohol or smoke tobacco. Gendered norms influence the ability of people to seek treatment and care, take up health advice, manage sexual and reproductive health problems related to such diseases, and provide or receive support from others in bodily care. And as reflected in the previous case study, gender norms and relations influence people's capacity to adapt to the disease and so define a 'new normal' (see also Manderson 2011).

The ability of people to adapt to chronic, debilitating and degenerative conditions can be facilitated by their engagement with others who share their complaints, with whom they forge a sense of identity that allows them to work with and gain support in managing their problematic health. Here the idea of biosociality, dating from Rabinow's work (1996) on genetic commonalities, is especially useful in describing the outcome of a shared social identity that derives from biology, not only in relation to genetics and familial risk, but also to explain social groupings around diagnosed disease and disability (Renne 2013; Rose and Novas 2004; cf. Chapter 4 in which we discuss the limitations to biosociality). This concept has tended to be applied to large communities and their bio-identification, as occurs for breast cancer especially, but also diabetes, Alzheimer's disease and other conditions, through advocacy, fund raising and public awareness events and dedicated days, and online presence. Biosociality also provides a way to promote and understand the effectiveness of interventions in smaller groupings. In the case study below, Marjolein Gysels and Irene Higginson describe how women with COPD found support not through a large-scale consumer or patient group, but through the efforts of one committed physiotherapist. The women who experienced breathlessness—the visible and embodied expression of COPD—had little support and care. Against the background of their disability and suffering caused by breathlessness, the three women describe how they discovered the benefits of pulmonary rehabilitation, as a way to self-manage their chronic condition and so maintain acceptable quality of life. The case is about the social dimensions of care and management, and the role of social support and friendship, not the biotechnical ways in which people's health problems might be ameliorated.

## 7.4 Facing Up to Breathlessness

*Marjolein Gysels and Irene J. Higginson*

In an aging world, people live longer with multiple chronic illnesses and these are increasingly important causes of death. Progressive and incurable illness brings high rates of symptoms, especially towards the end of life. Breathlessness is one such condition, causing considerable suffering. It is associated with fear and anxiety, causes disability, loss of independence and social contact, and forms a threat to people's daily quality of life (Gysels et al. 2007). It is responsible for high levels of hospital admission, hospital death, and it is related to increased loss of will to live.

Breathlessness is defined by the American Thoracic Society as "a subjective experience of breathing discomfort that consists of qualitatively distinct sensations that vary in intensity." This definition appreciates the complexity of this symptom coming from interactions among multiple physiological, psychological, social and environmental factors, impacting on both the quality and the intensity of the person's perception of breathlessness, and possibly causing secondary physiological and behavioral responses.

Yet people can experience very different levels of breathlessness. Patients with minimal respiratory changes may suffer considerably, while others with severe disease may experience very little breathlessness. These varied experiences may relate to personal characteristics such as muscle strength and weight, psychological disposition and coping strategies, but also life experiences and cultural background.

## *Three Friends*

We present three cases of women to illustrate the complexity of the life worlds of people who experience breathlessness due to Chronic Obstructive Pulmonary Disease (COPD) in the advanced stages of illness. The accounts are in the first person; the 'I' in this context is Marjolein. The names of the women are pseudonyms. The three women lived near each other in South London. They were old school friends but they had lost contact when they left school. They found each other again in their sixties, all three suffering from breathlessness with advanced COPD. They faced similar challenges but they were exceptional because they were able to have an acceptable quality of life despite the distressing symptom of breathlessness and associated progressive disabilities.

### Liz

Liz came from a consultation with the respiratory physician at the hospital when I was recruiting people to participate in a study on the experience of breathlessness at the Cicely Saunders Institute. I still remember, when we had talked that first time and arranged a date for an interview, how she walked away towards the sliding doors out of the hospital. It struck me how energetic she looked with a bottle of oxygen in her one hand, slightly swinging it with the pace of her steps. Not quite the person I expected to suffer from breathlessness in the advanced stages of COPD. Two weeks later, Liz received me at her home for an interview. She was very cheerful, chatty and open, a pleasure to interview.

Before her breathing started worrying her she explained that smoking had been a routine part of her life—she described herself as "marching off to work with a cigarette in my mouth." She had a job in accounting, finding the work rather boring and retreating to cigarettes and coffees to get through the day. The first signs of breathing difficulties that she experienced were when she couldn't get up the stairs when the lift broke down in the office. Occasionally she had to get air from the window but she did not suspect that this was a progressive problem. She, like her father, had always had some asthma, and she thought she was becoming more like him.

Then one day she collapsed before going into work. Her colleagues at work were worried when she did not turn up and they called her daughter who found her in the doorway. She was diagnosed with emphysema. She realized that it was serious when she told her diagnosis to a woman at work who promptly started crying as her mother was dying of this same condition. Liz stopped smoking which she found terribly hard to do, "the addict of addicts" she called herself, still craving for a cigarette after a year. She started attending pulmonary rehabilitation classes but because these were difficult to get to from the part of London where she was living, she stopped going.

She deteriorated fast, at the same time experiencing problems in her relationship with her partner. He arranged a wheelchair for her so that she was mobile around the house but she became increasingly depressed. The doctor brought her back into contact with the physiotherapist with whom she had been training before. She was surprised by her poor condition and convinced her to work with her.

Liz explained that she was lucky and that the physiotherapist had rescued her. She had hoped for a lung reduction operation, but developed reservations against it as it is a high-risk operation and then learned that the operation was not possible in her case. Her struggles with routine activities echo the stories that others also told me: the good days and the bad days, the unpredictability of what a new day will bring, the sudden chest infections which paralyse all other activities.

For daily functioning Liz relied on a well-organized system of technologies. What she told about being able to function revolved around pacing herself, moving forward slowly, resting and persisting in one's activity. For example, going upstairs was possible as she had a large oxygen bottle for downstairs and one upstairs. But there was the problem of climbing the stairs, which she did in several times, sitting on the steps to get her breath back. She initially took a small bottle with her but had to save these for going out as they only lasted four hours and she was only able to gain three a week through social services. "Flipping things," she said, she depended on them but wished she did not.

Liz talked about the lack of information on where to turn to for advice, in case of doubts about medication, or when emergencies present. Applying for a mobility allowance, or when she needed the smallest of things from social services, was frustrating. She had been turned down for the allowance but reapplied, then waited endlessly, frustrated by inspections and argumentations.

Meanwhile breathlessness interfered with Liz's social life. As she could no longer join her friends in smoky pubs, they eventually stopped asking her to go out with them.

## Frances

Frances's story is one of hard work and persistence. She told about the triumph with the first signs of progress, and the emotions, when after months she could manage without her wheelchair.

Frances had worked her whole life in education, first as a volunteer, then in an employed capacity in a school. She spoke about her love for children, her own family of four, and how as an Irish woman, she saw herself first and foremost a mother, and carer for her husband. She was diagnosed 17 years before we met with COPD, and although she became somewhat slower in her daily activities, she could still lead a normal life. Four years ago the illness took a turn for the worse. One day she described how she was on her way to church, greeting the priest on the other side of the street, and while crossing to meet him, suddenly she was hit by breathlessness. She stopped in the middle of the traffic and she could hardly reach the pavement. The next day, the same happened; the day after it happened again. From then on, every time she stood up, every movement she made, was a major struggle.

Frances spoke of her regret of her habit of smoking: "I have nothing else to blame." She explained that she tried to stop smoking when she started to notice that breathing had become more difficult, but she lacked support; her friends teased her and pressured her to take a cigarette. She developed strategies by carrying cigarettes in her handbag herself, which she could then take out but never light. She managed to stop smoking for a few years, but then her husband had a stroke; when she saw him lying on the monitors in intensive care, she sought comfort in smoking a cigarette. Asking her what made her give up in the end she replied: "they gave me up." She stopped when her breathing had already put restrictions on her daily activities.

In interview, Frances reflected on what the doctor had told her when she was diagnosed—"It will take 10 years off your life," and of the meaninglessness of his words, thinking: ". . . well 10 years, I'd be too old to enjoy my life by then. But if I'd been told, that you're not going to be able to walk around your own home, and not play with your grandchildren . . . that would have made me give up smoking immediately."

After years of progressive breathlessness and disability Frances was referred to hospital where the doctor did not think she was strong enough for lung reduction surgery. He advised her to try pulmonary rehabilitation. Sitting in her wheelchair she was puzzled: "I can't walk, how am I going to exercise?"

Frances was the most disabled of the three women. She described how she has to get up at 5.30 AM if she has to leave the house at 10. This is because she has to stop and start constantly. She gets up, takes her medication, goes down to have breakfast, goes back up to her room to go on the nebuliser so that she gets her breath back to be able to have a shower. Then she has to recover, doing everything in slow motion as she calls it. Through time, she learned strategies to keep functioning despite the breathlessness. She found out how she reacts to steroids, when to take antibiotics, she has her inhaler and oxygen at hand, and last year when her family asked what she wanted for her birthday, she asked them to all club together so that she could afford a nebuliser. She spoke too of her wish to have a stair lift: sometimes she is on her way to the toilet upstairs but when she is too breathless to move, she can not make it in time. But social services have so far turned down her request, despite the support letter from the hospital.

### Susan

Susan lived with her grown-up son. When I rang her doorbell, it took awhile before she opened the door. Slowly she showed me into her living room and sat down, explaining that the prospect of being interviewed made her nervous, triggering breathlessness. She characterized herself as a nervous, jumpy person, someone whose thoughts were always faster than the actual events. She explained that this is exactly what breathlessness is caused by, plays with and thrives in.

When we met, Susan had learned to take control over her breathlessness and she showed me the posture and the pursed lip breathing the physiotherapist had shown her. Before she learned this technique, she was taken along a vicious circle, the anxiousness and breathlessness exacerbating each other until she couldn't breathe at all and finally, in a last panic, reaching for the phone to call an ambulance.

Susan had become aware that breathing was becoming more difficult but she ignored this for years, concerned instead with a seriously ill husband for whom she cared until his death, from chronic kidney disease, the previous year. After she began to experience continuous chest infections resulting in frequent hospitalization, she was diagnosed with emphysema. "COPD they call it nowadays, don't they?" She added, "it is self-inflicted with the cigarettes."

Susan explained that her breathlessness was 50 per cent due to her lungs and 50 per cent to her mind, and explained that the rehabilitation classes were "God sent." There she had learned the breathing and relaxation techniques that had become crucial to suppress the threat of breathlessness. It was there too that she met with Frances and Liz, and developed a close friendship; before, she was too frightened to leave the house.

Together the three women started up the Breathe Easy group, a patient self-help group to inform and educate others about the benefits of pulmonary rehabilitation and to keep up to date with the latest scientific developments. Twice a week they go to Tai Chi classes, which they discovered has a therapeutic effect on them. In addition, there they are with healthy people, taking part in normal activities, and in doing this they have learned not to be bothered about being different. They do the exercises sitting on a chair at the back of the class at their own pace. They enjoy the calmness the teacher creates with his voice, the movements and the concentration it requires.

## *Situating the Women's Stories*

Liz's, Frances's and Susan's stories reflect what others who suffered from breathlessness caused by COPD had also told us and which we could best capture by the metaphor of invisibility (Gysels and Higginson 2008). Breathlessness is invisible. It typically manifests itself gradually and surreptitiously, often noticeable only by the person herself and therefore easily wished away. The strange ways in which it presents itself to those it affects, blending in with other symptoms and emotions, discourages people from seeking help. Breathlessness is elusive since clinical assessment is not attuned to picking up subjective sensations that resist the conventional measurable criteria. People do not tend to present their problems to doctors as they may be perceived as futile and improbable, and therefore not legitimate. Guilt also plays a role here, as people who experience breathlessness want to avoid disapproval regarding their past of smoking, both from practitioners and from the wider society. Patients also find out that besides the diagnosis of their condition, labeled with a variety of terms such as emphysema, COPD, or chronic bronchitis, there is little information available about how to cope with breathing problems, when and where to go for help, and what to expect for the future.

The women whose stories we presented here are among the few people in our study who had succeeded to transcend the barriers to cope with this symptom as they learned to self-manage (Gysels and Higginson 2009). They were lucky to have discovered the benefits of pulmonary rehabilitation, which they had adopted as a way of life. Contrary to their initial expectations they found that they could train their way out of disability. This required the additional knowledge that the wellbeing they had achieved was not an end point but rather a precarious balance that needed to be maintained carefully with routine practice, and insightful daily management of their condition. This knowledge was based on previous experience; when this balance was tipped, there was the threat of being dragged along a downward spiral of deconditioning, bringing them back to their wheelchairs.

These insights based on people's experience of breathlessness in COPD, as well as in other conditions, helped to inform the development of the Breathlessness Support Service. The evaluation of this new integrated service for patients with advanced disease and refractory breathlessness in South London provided evidence for improved patient mastery and early introduction of palliative care (Higginson et al. 2014).


### *Acknowledgments*

The study was conducted at the Department of Palliative Care, Policy and Rehabilitation of King's College London. We thank all participants who contributed to this study. We also thank Cicely Saunders International whose funding made this research possible.


### *References*


Gysels, M., C. Bausewein, and I.J. Higginson 2007 Experiences of breathlessness: A systematic review of the qualitative literature. *Palliative Supportive Care* 5:281–302.

Gysels, M., and I.J. Higginson 2008 Access to services for patients with chronic obstructive pulmonary disease: The invisibility of breathlessness. *Journal of Pain and Symptom Management* 36:451–460.

——— 2009 Self-management for breathlessness in COPD: The role of pulmonary rehabilitation. *Chronic Respiratory Disease* 6:133–140.

Higginson, I.J., C. Bausewein, C. Reilly, Wei Gao, M. Gysels, M. Dzingina, P. Mccrone, S. Booth, C. Jolley, and J. Moxham 2014 An integrated palliative and respiratory care service for patients with advanced disease and refractory breathlessness: A randomised controlled trial. *Lancet Respiratory Medicine* 2(12):979–987.


---

The prevention and control of disease is marked by the need for continued new vaccines, as the initial success in the control of infectious diseases is undermined by drug resistance and virus and parasite adaptation and evolution, failures in drug development for treatment or cure, and the side effects of novel drugs that prevent their wide marketing. Even so, as we have discussed, many infectious as well as non-communicable diseases are now managed as chronic conditions, and life expectancy from all chronic conditions is increasingly extended. This management is largely achieved through drugs, and growing proportions of populations worldwide take medication for increasingly lengthy periods of time. 'Drugs for life' (Dumit 2012) now include multiple drugs for multiple conditions for people living very long lives. While this approach is problematic, it is a reality for people who, with diabetes and heart disease, depression and arthritis, for example, are willing and able to pay for beta-blockers for hypertension, statins, anti-coagulants, insulin medication, anti-depressants, anti-inflammatories for effective pain relief, present for regular medical monitoring and follow medical advice, and take advantage of ancillary health care like hydrotherapy pools, massage and counseling. These are not realistic options if a person is poor. For many people with chronic conditions, therefore, as illustrated for people with HIV, simply having sufficient food, of any kind, is difficult enough (Gwatirisa and Manderson 2009; Kalafanos 2010).

The drugs required for people with chronic conditions come at a high cost to national budgets, with growing expenditure on hospitalization and medication for chronic conditions over a growing period of time. Over 17 percent of the US Gross Domestic Product is spent on health, compared with around 10–11 percent in much of Europe, and half that or less in many poorer countries. Over time, unless underlying problems of inequality, poverty and social exclusion are addressed, these proportions may well increase further with the risk that governments will respond by increasing consumer and patient responsibility to carry the costs of their own health. We return to these themes of social structure, inequality and exclusion, as they affect people's health, in later chapters.

## References

Briggs, Charles L., and Mark Nichter 2009 Biocommunicability and the biopolitics of pandemic threats. *Medical Anthropology* 28(3):189–198.

Brown, Hannah, and Ann H. Kelly 2014 Material proximities and hotspots: Toward an anthropology of viral hemorrhagic fevers. *Medical Anthropology Quarterly* 28(2):280–303.

Dumit, Joseph 2012 *Drugs for Life: How Pharmaceutical Companies Define Our Health*. Durham, NC: Duke University Press.

Flaherty, Devin, H. Mabel Preloran, and Carole H. Browner 2014 Is it 'disclosure'? Rethinking tellings of genetic diagnosis. In *Disclosure in Health and Illness*. Mark Davis and Lenore Manderson, eds. Pp. 89–103. London: Routledge.

Gwatirisa, Pauline, and Lenore Manderson 2009 Food security and HIV/AIDS in low-income households: The case of urban Zimbabwe. *Human Organization* 68(1):103–112.

Kalafanos, Ippolytos 2010 All I eat is ARVs: The paradox of AIDS treatment interventions in Central Mozambique. *Medical Anthropology Quarterly* 24(3):363–380.

Livingston, Julie 2012 *Improvising Medicine: An African Oncology Ward in an Emerging Cancer Epidemic*. Durham, NC: Duke University Press.

Lock, Margaret 1993 *Encounters with Aging: Mythologies of Menopause in Japan and North America*. Berkeley, CA: University of California Press.

Manderson, Lenore 2011 *Surface Tensions: Surgery, Bodily Boundaries, and the Social Self*. Walnut Creek, CA: Left Coast Press.

——— 2015 Cancer enigmas and agendas. In *Anthropologies of Cancer in Transnational Worlds*. Holly F. Mathews, Nancy J. Burke, and Eirini Kampriani, eds. Pp. 241–254. London: Routledge.

Manderson, Lenore, Jens Aagaard-Hansen, Pascale Allotey, Margaret Gyapong, and Johannes Sommerfeld 2009 Social research on neglected diseases of poverty: Continuing and emerging fields. *PLoS Neglected Tropical Diseases* 3(2):e332.

Naemiratch, Bhensri, and Lenore Manderson 2006 Control and adherence: Living with diabetes in Bangkok, Thailand. *Social Science & Medicine* 63(5):1147–1157.

Ostrach, Bayla, and Merrill Singer 2012 Syndemics of war: Malnutrition-infectious disease interactions and the unintended health consequences of intentional war policies. *Annals of Anthropological Practice* 36(2):257–273.

Pavey, Amanda, Narelle Warren, and Jacquelyn Allen-Collinson 2015 'It gives me my freedom': Technology and responding to bodily limtations in Motor Neuron Disease. *Medical Anthropology* 34(5):442–455.

Person, Bobbie, Kay Bartholomew, Margaret Gyapong, and Bart Van den Borne 2009 Health-related stigma among women with lymphatic filariasis from the Dominican Republic and Ghana. *Social Science & Medicine* 68(1):30–38.

Rabinow, Paul 1996 Artificiality and enlightenment: From sociobiology to biosociality. *Essays on the Anthropology of Reason*. Pp. 91–111. Princeton, NJ: Princeton University Press.

Renne, Elisha P. 2013 Disability and wellbeing in Northern Nigeria. In *Reframing Disability and Quality of Life: A Global Perspective*. Narelle Warren and Lenore Manderson, eds. Pp. 39–59. Dordrecht, The Netherlands and New York: Springer Publishing Company.

Rose, Nikolas 2006 *The Politics of Life Itself: Biomedicine, Power, and Subjectivity in the Twenty-First Century*. Princeton, NJ: Princeton University Press.

Rose, Nikolas, and Carlos Novas 2004 Biological citizenship. In *Global Assemblages: Technology, Politics, and Ethics as Anthropological Problems*. Aihwa Ong and Stephen Collier, eds. Pp. 439–463. London: Blackwell.

Sakellariou, Dikalos 2015 Home modifications and ways of living well. *Medical Anthropology* 34(5):456–469.

Singer, Merrill 2009 *Introduction to Syndemics: A Critical Systems Approach to Public and Community Health*. San Francisco, CA: Jossey-Bass.

——— 2014 Syndemics. In *The Wiley Blackwell Encyclopedia of Health, Illness, Behavior, and Society*. William Cockerham, Robert Dingwall, and Stella R. Quah, eds. Pp. 2419–2423. Hoboken, NJ: Wiley-Blackwell.

Singer, Merrill, and Nicola Bulled 2014 Ectoparasitic syndemics: Polymicrobial tick-borne disease interactions in a changing anthropogenic landscape. *Medical Anthropology Quarterly*. DOI: 10.1111/maq.12163.

Smith-Morris, Carolyn, ed. 2015 *Diagnostic Controversy: Cultural Perspectives on Competing Knowledge in Healthcare*. London: Routledge.

TDR (Special Programme for Research and Training in Tropical Diseases) 2012 *Global Report for Research on Infectious Diseases of Poverty*. Geneva, Switzerland: World Health Organization.

Turner, Victor 2008 *The Ritual Process: Structure and Anti-Structure.* New Brunswick, NJ: Aldine Transaction Press.
van Gennep, Arnold 1960 *The Rites of Passage.* Chicago, IL: The University of Chicago Press.
Weaver, Lesley Jo, and Emily Mendenhall 2014 Applying syndemics and chronicity: Interpretations from studies of poverty, depression, and diabetes. *Medical Anthropology* 33(2):92–108.
Weiss, Mitchell G. 2008 Stigma and the social burden of neglected tropical diseases. *PLOS Neglected Tropical Diseases* 2(5):e237.

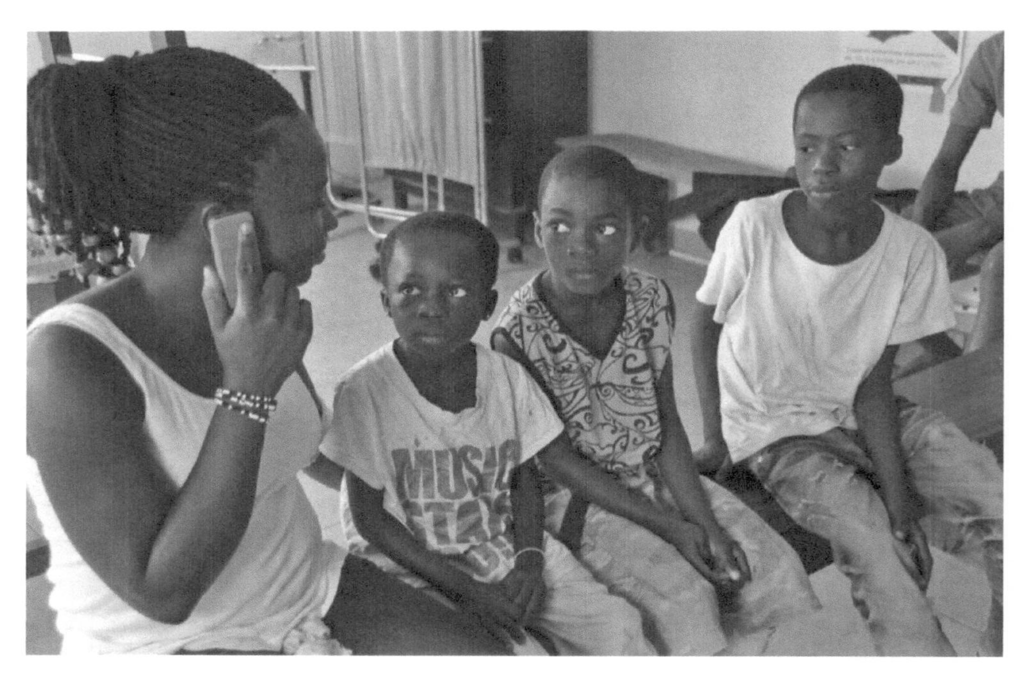

Cell Phones in Africa, 2012. Akwapim South District, Ghana.

## *About the photograph*

*Cell phones in Africa have great potential for disease monitoring and to facilitate decision making about health care, where women are required to consult their husbands before taking action. A woman in Ghana, visiting a clinic with her children so that she can be treated for a chronic ulcer, checks in with her husband via cell phone to inform him about the costs of treatment and how often she will need to visit the clinic. Nichter was the social science adviser to Stop Buruli Network West Africa, and was working with a local social scientist and a medical doctor to understand how cell phones might improve treatment outcomes.*

—Mark Nichter

# 8 Ways of Caring

*Elizabeth Cartwright, Anita Hardon and Lenore Manderson*

In this chapter we tack back and forth between critically thinking about the activities involved in and the technologies of caring, and reflecting on what it means to care for others—an act that is as ubiquitous as it is primal. Some kinds of care, undertaken at home, are short term in response to an acute and relatively minor infection. Longer-term care involves far broader tasks, as in caring for older people and for people of all ages with chronic health problems or communication and cognitive difficulties. We juxtapose the mechanical realm of increasingly high-tech medicine with human interactions that we carry out to physically and emotionally support one another. Our intent is to position 'technologies of care' as tools, implements, and prosthetic extensions of our desire to intervene with fate, to show love and compassion, to alleviate suffering, and to make our lived experiences of life and death better. We move from a discussion of the act of caring for others in the most basic, physically present kinds of settings, to a consideration of new ways of delivering care via systems of communication and monitoring of patients from a distance. We then discuss some of the complexities of implementing new systems of telemedicine, for instance, in very remote, culturally distinct regions of the globe. Finally, we turn to the very well-informed patients who interact with one another via the Internet and who are creating new forms of socialities based on their conditions and innovative understandings of their illnesses, possible treatments, and ways of getting the support that they need.

## Caring

One dimension of care includes the biomedical approaches to care in hospital and those that continue on as people achieve the status of outpatients. It involves logical continuums of finishing up medicines and the post-acute phases of the healing of bodily wounds (Mattingly et al. 2011), but, as we have noted in Chapter 7, it also includes more complex everyday routines in the face of chronic conditions, where care is needed for long periods, sometimes indefinitely, and where routines can increase in significance as more attention needs to be focused on them. To accommodate this, as we have already noted, dialysis machines, oxygen pumps, ramps, rails, lifts, and monitors are imported into home settings, converting the home to hospital and the home-carer—a family member—into an allied health professional (Sakellariou 2015).

The tasks of caretaking can seem to stretch out into the future in an interminable manner. Anthropologists have described the process of caring for an aging relative from a very personal vantage point, especially with respect to the long-term care needs of spouses and parents who were afflicted with Alzheimer's disease (Kleinman 2009; Taylor 2008). Kleinman's poignant description of how he cared for his wife of many years when she became ill with Alzheimer's, how he helped her move from room to room, dressed her, and fed her in an intricate dance that became increasingly complex as her condition worsened, clearly shows the intimate, moment-by-moment details that must be attended to just to make it through the day. Taylor's auto-ethnography of caring for her mother, as dementia took hold, highlights the evolving nature of caregiving when a patient is slowly deteriorating, as different resources need to be shepherded by the family, and as the emotional and physical demands of all aspects of care become increasingly difficult. Cultural differences abound in responsibility and configurations of caregiving, but the basic components of care, safety, carrying out of daily activities, and human presence surface recurrently.

Different physical and cognitive patterns of difficulties play out in different ways and warrant significant anthropological attention as our global populations live with chronic illnesses, and live for longer periods in states of serious ill health (see Chapter 7). The extremes of caregiving are seen in the frail, unhealthy elderly, who in various circumstances and settings are institutionalized and may have diminished cognition. The bodily exegesis of able-bodied caregivers putting the profoundly disabled elderly through their activities of daily living creates a situation of literally living the text of a life for another: moving limbs through space into wheelchairs, moving food onto spoons, into mouths, reminding to chew, reminding to swallow, toileting, bathing, and dressing. The work of such caring is often done by strangers, but also by children, spouses, other relatives and helpers, paid and unpaid. It is always taxing in many ways. These acts take their toll on caregivers, and result in frustration for those receiving care, for whom the loss of independence, often tempered with confusion, can be deeply distressing.

With medical technologies entering the home, and with Internet and smartphones enabling communication, more and more parts of care can be performed at a distance—that is, virtually. What is it that constitutes 'care' and how is that approximated, changed, and electronically reproduced across distances via technologies of communication? How is love transmitted through space and made real to those for whom we care? While there is not much new about long-distance relationships of one kind or another, the importance of virtual caring is growing exponentially as technologies of communication and perception increasingly act as prostheses between children and aging parents, homebound patients and medical personnel, and between patients experiencing similar problems. There is certainly a co-construction in the act of caring—there is the bodily participation, the touch, smell, and warmth that provides grounding and comfort. How do we convey these aspects through the screens and connectivity of our ICTs (information and communication technologies)? Greenhalgh and Swinglehurst (2011) have argued that ethnography is an effective methodology to use when looking at how ICTs are introduced into complex social systems, such as that of health care. They emphasize that the success or failure of the introduction of a new technology into a health care system is best understood via the small, everyday social interactions surrounding the use of the new ICT.

'Smarthome' technology is rapidly becoming available to those who have the resources; in the case of the elderly, it is particularly important as a tool for facilitating their independence and ability to stay in their homes for as long as possible. The QuietCareSM system, for instance, is a smarthome technology marketed by Intel-GE Care Innovations. It is based on the work of medical anthropologist Anthony Glascock and his colleague David Kutzik, a sociologist (2006). The QuietCareSM system is built on their cross-cultural, gerontological research, which led to the invention of a system of small motion-detectors that can be located in the homes of people who are elderly and frail, or disabled, to monitor their activities and to send out alarms when needed. The system uses sensors and WiFi to monitor Activities of Daily Living (ADLs)—bathing, dressing, using the toilet, eating, getting in and out of bed, and walking—along with the necessary

Instrumental Activities of Daily Living (IADLs), such as meal preparation, managing money, grocery shopping, making phone calls, doing housework, taking medications, and doing laundry (Glascock and Kutzik 2006: 60). The system 'learns' the regular movements of individuals as they carry out their normal daily routines and will alert the caregivers, who are able to remotely monitor the information from the sensors 24 hours a day, if something is out of the ordinary. These two researchers noted that for a technology such as this to be acceptable, one of its most important attributes must be that it is unobtrusive and that it monitors objects, not people. This system, and others like it, are being refined and made more effective, and are increasingly being used around the world.

In ethnographic research that explores the uptake of virtual caring technology, López and his colleagues (2010) describe the experience of being one of the remote 'caregivers' at the Catalan Red Cross Home Telecare Service in Barcelona, Spain. They point out some of the realities that go into the decision-making process at the central monitoring station when an alarm, such as that described for the QuietCareSM system, goes off in one of the homes in their Spanish system. Working at the central monitoring station, individuals who are oftentimes health professionals or para-professionals communicate with registered patients in the event of an alarm being triggered, muster neighbors or kin to physically check on the individual if no phone contact can be established, call ambulances, and perform other triage-at-a-distance. Similarly, technologies such as those described in the following case, by Thygesen and Pols, show how the Internet and various mobile devices are used to create networks of information, intimacy, and support for e-care networks of elder caretakers in Norway.

## 8.1 Care, Self-Management and the Webcam

*Hilde Thygesen and Jeannette Pols*

We all know these texts: Europe is aging, facing a tsunami of older people and people with dementia for whom a shrinking younger generation needs to care. Something should be done about this situation, these texts continue. The industry needs to produce technologies to efficiently support people who grow old and deal with disease at home, rather than in health care institutions. Older people should start caring for themselves with the help of these devices. They should become experts in 'self-management,' a term that has become the buzzword in European health care policy. As is usually the case with buzzwords, self-management means different things in different contexts. It is usually interpreted as the individual activity of caring for oneself or 'managing one's disease' with the help of professional information and devices. Implicit is the idea that older people will suffer from chronic diseases rather than acute conditions, strengthening the idea that lifestyle changes and prevention from deterioration are important activities.

Ethnographic studies are excellent ways to critically analyze buzzwords. Rather than blindly accepting that self-management is the way to understand how people care for themselves, we can study how people actually do this. Drawing on ethnographic data from a Norwegian webcam care project, we show what self-management comes to mean when we study what people *do* when they care. Our questions in this case are: who is the patient that needs to manage him or herself? Who cares, or manages, and who does not? What defines 'care' here?

### *SafetyNet and Its Patients*

SafetyNet (TrygghetsNett) is a public care service that uses a protected webcam system to aid persons living with a sick spouse.[1] The service started as an initiative of the municipality of Nøtterøy, in south-eastern Norway, directed at spouses of persons with stroke and dementia and is coordinated by professional nurses. Most people suffering from stroke and dementia are older than 65; their spouses too are older and suffer from different age-related diseases. Caring for a person with stroke or dementia can be a very demanding 24-hour-a-day job, even with home-based nursing services. Hence, the municipality targeted family caregivers as being particularly vulnerable. After a couple of years, the service was expanded to all 12 municipalities in the region and extended to include the parents of multi-disabled children and families of persons with alcohol or other forms of substance abuse. So the 'patients' involved in intervention were not the people with the 'medical problems' but their family caregivers. It was they who needed support for their role in the caring and management of the 'actual patient.' Care, in this context, is a burden that can turn caregivers into patients. Caring for others, rather than the self, is of course unavoidable when one's partner has a stroke or dementia, but it is overlooked in the discourse of self-management. Here we discuss the kinds of problems that the Norwegian caregivers experienced using the system.

*Professional caregiver (coordinator):* I have really experienced what it is like being a spouse/family caregiver [through my role as coordinator of SafetyNet]. We get [a lot of] insight into the lives of people. I used to think of family caregivers as pests and a bother. They never say 'thank you,' but always demand more. I thought, for instance, that men would never stay home to nurse their wives. But they do. And when the partner gets chronically ill, the world gets narrow for the spouse, too. We found that a lot of them were very isolated. We quickly realized that these people were very alone with their problems. At our meetings I often try to engage them with humor. "Are they [the partners with dementia or stroke] nice to you?" I asked at the meeting last week. Then someone answered that this was really difficult. "If anyone has had it with the sick person [gestures to the throat], speak up now," I encouraged them. And then a huge discussion started. They talked about feeling cornered, cut off, and tied up and that their efforts were not appreciated. Many felt that they couldn't do enough. Some were mourning. There were many bad feelings floating around the room.

The role of the caregiver is difficult, even if they do not suffer from a disease themselves. Their emotional investment and the lack of recognition of their care made them very different from a professional caregiver; it turned them into patients, too. The idea behind SafetyNet was to allow family caregivers to link up with and support each other. Through the combination of peer support and public service coordination, the idea of SafetyNet was to strengthen family caregivers in order to keep caring for their spouses longer.

*Hans (participant):* My wife did not want to move to an institution. So when we started with the senior center from 9 to 2pm she did not want to go. I had to trick her into going. It was like that all the time and it was very difficult. It took a long time before it got better. But it did get better after a while. She also got a permanent room in the nursing home, but she did not want to be there. It was a terrible struggle. And at that stage I was in constant contact with the girls at the SafetyNet base who said that "You just have to keep going, Hans!"

SafetyNet was developed so that the sick person would be able to live at home for a longer period of time. This was considered positive, as for many people living at home provided a higher quality of life than living in an institution. Additionally, it is cheaper for the municipality if the patient lives at home. Family caregivers are key to achieving this.

How did the family caregivers/patients support one another in the SafetyNet project? One of the first things that they exchanged was knowledge and expertise.

*Anna (participant):* Just an example. There was a man [in the group] who had a hole in his bladder. And he had a problem with the use of a catheter, about finding ways of connecting bags. He was left sitting at home because they [the home nurses] thought it was not possible. And he couldn't get out of the house. "Yes!" I told him, "It *is*

> possible!" So I could help him by telling how I had solved this with my husband. In this way we help each other.

Typically, the practical problems the group discussed were accompanied by complex moral demands and feelings. How to be a good caregiver was a common topic. Anna, again:

> We help each other, not the least psychologically. We all have periods of crying . . . My husband is going to the rehabilitation center on Monday [for three weeks of 'respite care' aimed to unburden the informal caregiver], and when it gets close I get like this [tearful]. He does not understand why he has to go, even though he accepts he has to. But he would rather be at home, he keeps saying. So I try to explain to him that in rehab he will get the chance to exercise every day. And I had to be tough and tell him that I need these three weeks. Now, after four years he understands. But it is not easy. He keeps telling me that he prefers to be at home.

Three weeks of respite care in the rehabilitation center helped to unburden the family caregiver, but turned out to be difficult to implement. Carer and patient can now *both* be identified as the patients—and their care needs can be at odds. Support from the SafetyNet group was of great value here. The recognition and the mutual support between members was an important success factor in the way care took shape. Sarah compares the relations within SafetyNet to those in a family: "We who are part of SafetyNet have actually become like a family. We get very close to each other and we know each other well." Yet the success of SafetyNet exchanges is characterized by participants *not* being family.

*Professional caregiver (coordinator):* Many of them have children. But children are children. They are not to be involved in trouble and care. This is an attitude that many old people have. They [the kids] have problems of their own, and should not be bothered with those of their parents. At first, I did not understand why they did not want to involve their daughters and sons in SafetyNet. But SafetyNet was something that was 'theirs.' Their relations to the children were possibly the only thing they had that was normal. It was also an assumption that the children would not come and visit if they complained too much.

SafetyNet participants believe that their relationships with their children should be free from concerns and burdens of care. They wanted to keep things light with their friends too, and SafetyNet provided people with a forum to share their problems.

*Sarah:* When I say that we in the forum [SafetyNet] are family I me . . . well, there is a difference compared to our ordinary friends. Firstly, they [ordinary friends] are fewer and when they are visiting we don't sit and talk about things that we are concerned about in everyday life that we find problematic. With our friends we are supposed to have a good time and live a normal life like before. When I am in contact with someone in the SafetyNet group, things are different because we can share our problems. Everyone has got something! Having lived here all my life means that I know a lot of people. But it is different in this forum. We have something in common. Sometimes when I feel like talking to someone I click on the 'want someone to call me' (*pratesyk*) icon on the display, and if someone gets in touch then it is always very good. Then we are both happy.

In SafetyNet relations, people could be caregivers, receivers, and friends:

*June (participant):* Hans arranges a summer party every year, and it is very nice. Everyone in the forum has a handicap, so the tolerance is high. Everyone helps each other as much as possible. Some do not like social occasions because they are concerned about spilling food when they are eating and those kinds of things. But it gives us safety to know that we are together and not alone.

Care had become an everyday activity for all. It included dealing with spilled food, but was also focused on not letting individual's disabilities get in the way of enjoying life. Patients and caregivers could be tourists, rather than patients and caregivers.

*June:* One other thing is that we have found another couple in the [SafetyNet] forum that we did not have contact with before. And we have traveled together every winter to Spain, to a place that is accessible, and that has Norwegian physiotherapists and nurses. For three years we have traveled with this couple. In the restaurants there they are used to people who sit in a wheelchair and that might need a bib on when they eat. This means a lot. We have had an agreement with a taxi that takes us on trips.

At the time of the fieldwork, there were 13 participants, 6 persons with a spouse suffering from dementia or stroke, and 7 persons who were 'solo.' The solos bothered the organizers of SafetyNet, because they stayed with their newly made friends and did not want to quit being a part of SafetyNet when their spouse died. The organizers worried that the solos were taking the place of other caregivers who might benefit from being a member of the group. The solos did not want to leave the SafetyNet community and friends that they had made during the difficult times of caring for their spouse. They had established new and valuable contacts with the members of SafetyNet and did not want to give up these friendships, but also, their role changed from occasionally asking and occasionally giving help, to having time on their hands—and motivation to spend it with their SafetyNet friends.

*Petra (participant):* I am in touch with someone in the forum every day, but not as often with the base [the professionals/coordinators of the network]. So it is the other members of the group that I am in contact with. I don't have just one or two—it varies who I am in touch with. Some are easier to talk with than others. And those in the solo group have more time than us, who are family caregivers. So it is easier to contact them. I think it is a good thing that the solo-people are taken care of. You cannot close them off despite the fact that they are no longer caregivers.

The idea that the informal caregiver needed support because their 'disease' was the disease of their partner implied that they were 'cured' once their spouse had died. But this was not how it was for the solos. Their friendships from the SafetyNet experience lasted and changed to meet their new needs once their partners had died.

*Jan (solo participant):* In particular I have one close friend. I met him through the [SafetyNet] forum. He started at the same time as me. We have been in Croatia and on the Donau together since we were widowed. We are in touch every day. Today he is at the eye specialist, but I expect that he is at home by now. I can see that he is logged on [the forum].

Jan had found new people to care for after his spouse passed away. Care shifted into something that made the members of the solo group feel good. It also assured the other participants of lasting and accessible care.

*Jan:* We in the solo group are much freer. We can decide everything ourselves. This is something that it not possible when you are two. We meet as often as possible. Last week two women [in the group] suggested a trip to the café. And then we ended up being seven or nine. Doing these kinds of things is very nice, it's cozy. And it strengthens the ties between us. When we are together we talk about all sorts of things.

*Petra:* We have all the time been conscious about keeping a light tone. This was not to be a site of mourning! It was to be, we could have a bit of fun actually. We were away for a bit [from the seriousness/grimness of the situation at home] and had something else to think about.

Caring and being cared for became a meaningful and 'fun' activity. It did not stop with the end of disease. Jan kept himself healthy and happy by continuing to care within friendly relations with people he was close to, because they had been through the same problems. Jan succeeded in giving his life a new purpose and in passing on that enjoyment to others as well.

### Conclusion

Care for others changed from being a burden to being something to live for. Care could incapacitate the family caregiver when it went one way, changing the former relation with the spouse completely. The family caregivers were emotionally burdened, on the one hand by the wish to care for their spouse, and on the other by their own limits and their lack of recognition and support. Their caring work turned them into patients. In SafetyNet, the caring relationship was different. People could switch between getting and giving care, and being part of a community with recognizable difficulties to live with. This made the new relations strong and valuable to the participants. Care for others emerged here as the way out of being a patient. Interestingly, the mutual care made the SafetyNet relations different from relations with family and friends. Here caring was not about medical aids and technology, nor about self-managing, but about supportive relations.

### Note

1. The SafetyNet website is a secure, online forum organized and coordinated by municipal services. The actual forum has two parts. The first is the written forum, where the members can write notes to be read by everyone in the forum. The other—oral/immediate part—is where you can make calls (like Skype) where you see each other on the webcam. This is one-to-one communication. When online a small panel shows the accessibility of the other group members. When logged in the names of the participants are listed, and on a small panel beside it the availability-status shows. If you are away on holidays or busy, you can click the 'unavailable' button and the other group members will see this when they log in. Or you can choose just to be 'logged in.' If you are in need of contact, you can click a button named *pratesyk* (in need of a good talk).

### References

Pols, J. 2012 *Care at a Distance: On the Closeness of Technology*. Amsterdam: Amsterdam University Press.

——— 2014 Towards an empirical ethics in care: Relations with technologies in health care. *Medicine, Health Care and Philosophy* 7: DOI: 10.1007/s11019-014-9582-9.

Schermer, M. 2009 Telecare and self-management: Opportunity to change the paradigm? *Journal of Medical Ethics* 35:688–691.

Thygesen, H. 2009. Technology and good dementia care. A study of technology and ethics in everyday care practice. PhD thesis. Senter for teknologi, innovasjon og kultur (TIK), Universitetet i Oslo.

Thygesen, H., and I. Moser 2010 Technology and good dementia care: An argument for an ethics-in-practice approach. In *New Technologies and Emerging Spaces of Care*. M. Schillmeier and M. Domènech, eds. Pp. 129–147. Farnham, UK: Ashgate.

---

Thygesen and Pols's case illustrates how new connections spring up among communities of caregivers when they have access to computerized communication; real friendships are established and new possibilities are opened up for the elderly individuals involved in the Norwegian caregiver support program. In her case study below, Tanja Ahlin describes the more 'naturally occurring' long-distance webs of medical decision making that exist between elderly parents and their far-flung children. Ahlin's study participants use telephones and computers to stay in touch, share information, and make decisions. Information about a family member's conditions are exchanged across continents. Requests and demands are made on both those at home and those who are residing around the globe, as care is strategized and enacted between physical locations where access to diagnostics, medicines, and trained practitioners all vary. In these instances we are not just talking about biomedically oriented care, for traditional and alternative treatments and practitioners are also an important part of this mix.

## 8.2 E-Care in Kerala

*Tanja Ahlin*

> "We knew she was gone before they reached Kochi," said Jane, a specialist nurse from Kerala, India, working in the United States. The death of her mother-in-law, Vimala, who was in her seventies, was sudden and completely unexpected: when she died, the family was seeking to obtain a visa for her to join Jane, her husband and their two young daughters in New York. At that time, the mother-in-law was living with Jane's husband's brother. Jane and her husband learned that she was sick when they called on the day that her visa paperwork was supposed to come through, and she was not able to speak on the phone. Jane told her relatives to take her mother-in-law to the hospital immediately. But the doctor who was supposed to receive her was on a leave, and so the relatives took her to Kochi, about two hours drive away. On the way, Vimala vomited blood. Jane was on the phone throughout this period, giving directions to keep Vimala's head straight and in position where she could drool, so prevent her from choking. Her advice was especially appreciated because she was a health care professional; her relatives had no experience with such emergencies. Today, Jane is still guessing about the possible cause of Vimala's death, but perhaps it was an abdominal aortic aneurysm, a common fatal event whereby the ruptured aorta in the stomach causes severe internal bleeding. "While they were driving, we kept asking, how is she doing, how is she, but we knew," Jane finishes describing her memories of the last time she took care of her mother-in-law, at a distance by phone.

Now in her forties and from a middle-class family, Jane is one of the many nurses from Kerala, mostly Christian, who have left their home country to try their luck abroad. One reason for migration of nurses, as of other professionals from Kerala, is related to the gap between high levels of educated people and poor availability of appropriate employment for them (cf. Chua 2014). Although there are no exact numbers of how many women, and more recently men, have done so over the past decades (see Percot 2006 for some estimates), families will often send a son to become a priest and a daughter to become a nun or a nurse, an offering of children to live in the service of God and society although historically, unmarried women who migrated abroad alone were viewed with suspicion and ambiguity. More recently, however, growing numbers of single and married women have left Kerala to work abroad, and parents now talk about them with pride. In some social classes it has become the norm for young girls, especially, to become nurses and migrate abroad. In Kerala, where there are little options for health insurance or pension available to non-public employees, lower- and middle-class elderly are particularly vulnerable. Educating at least one child as a nurse and sending him or her abroad has become a kind of informal insurance for parents who then need not worry about the high costs of weddings, home renovation and potential health care costs. As one widow explained, "because my children are abroad, I live happily here. They work and make money. They give me more than enough money and isn't it because of that, I am living lavishly? If they sit with me, will I enjoy my life? No." The money that her children provide is more than enough to buy her food and other necessities; her income in old age is secure. If she needs any help, in case of health emergency or anything else, she can still rely on her extended family members living nearby. While her children are always in her mind and she prays for them daily, she understands that they have their own children to take care of and is satisfied that they come to visit her during their annual leave. She was genuinely happy, very much in contrast to the popular belief in the devastating abandonment and loneliness of elderly people living alone after their children had migrated (Lamb 2009). I encountered very few cases where the migrating nurses had little or no contact with their parents. In a women's shelter, run by one of the local churches, I met a divorced woman whose daughter was a nurse in the Gulf. The two had no contact, but that was the case already before the daughter left Kerala to work abroad. Also, I found no parents of migrating nurses in the geriatric hospital I visited, through there were more than eighty patients there.

### *What Can Children Do from Afar?*

Let us return back to Jane's family. Jane, the second of three sisters, had been caring not only for her mother-in-law, but was also taking care for her own parents. An older sister lived in Kerala, only a couple of streets away from their parents' home, while the youngest one lived and worked as a nurse in Oman. All of them were married and had children of their own. Ideally, in Kerala, the youngest son and his wife would live with the parents, take care of them and as compensation inherit the house. By contrast, Jane's parents lived alone with a maid, and because Jane was the one who agreed to help them financially to renovate their home, she would eventually inherit it. When there is no son in the family, the responsibilities of care depend on agreements and personal engagement of daughters and their husbands. Calling her parents from the US by phone twice a day every day, Jane was highly involved with their everyday life, though this was not required by any social obligations. Jane and her parents relied on the phone as their communication tool of choice, because it was easy for the parents to use, while they had no experience of using a computer and so Internet-based webcam contact was not possible. Jane called me from New York after having returned from her night shift and calling her parents, and in our phone interview, she told me:

> I think it's all about your attitude, wherever you are. I think I am the one always there for them, whenever they need someone to stay with them, I am the one to find out what is the problem with them, and I am the one who is always calling the others around. I call them every morning and every night.

Jane was particularly concerned about her father, now in his late seventies, who had been suffering from Parkinson's disease for about eight years. At the onset of the illness, he had taken some allopathic drugs, "but only for one day," his wife told me. Jane's mother had not appreciated the side effects her husband experienced—he had been so sedated he would only sleep—and although she had spent a small fortune on the prescription and the medicines, she had thrown them away the same day. She had then talked to Jane about her idea of using āyurvedic drugs. Remembering those days, Jane told me that she knew her parents well and she understood what they needed; not knowing the person on the other side of the phone so intimately would, in her view, make such care at a distance much more difficult or even impossible. Her mother was well familiar with plants and had been preparing herbal remedies for years. So when they had proposed to try āyurvedic drugs, Jane had supported the suggestion. Almost a decade later, Jane's father's condition had improved significantly, such that he barely trembled, his memory was sharp, and he could see clearly and even write again.

Besides monitoring her father's chronic illness at a distance, Jane had also been there for her parents during an alarming health crisis about four years ago when her mother, now sixty-nine, had had a stroke. As Jane recalled in our interview, she was at work one day when she had a nagging feeling to call home although it was half past two in the morning in Kerala. She did not want to wake her parents, so she called her brother-in-law living nearby. Her call found him standing by the car in front of her parent's house, ready to take her mother to the hospital. At the nearest hospital, the staff said the doctor would only come in the morning. Jane's mother then received an injection and an X-ray was done, but then the family decided to take her in their family car to a hospital in Thiruvananthapuram, the capital of Kerala, about four hours drive away. All the while, Jane was on the phone, on and off, checking on her mother's condition and giving advice on what to do.

As a nurse, her knowledge was invaluable during this event, but there were clearly limits to what she could do for her mother by phone. Within several days, just at the time when her mother was discharged from the hospital, Jane was able to take a family emergency leave from her work and flew to Kerala. Her younger sister, too, flew home from Oman within a week, although she had recently given birth. Jane then stayed with her parents for over a month and learned from a professional physiotherapist how to massage her mother properly. By the end of her stay, Jane's mother was able to slowly walk again and today she has no problems at all, other than occasional slurred speech.

Jane's family was not the only one in which the children would fly home when a life-threatening health problem occurred. One of the nurses, working in the Maldives, told me of her colleague, also a nurse from Kerala, who returned home after her mother experienced a heart attack. But first, the nurse established a phone contact between the cardiologist in India who was treating her mother and a cardiologist in the Maldives who was working in the same hospital as she. The nurse reportedly did so because the two specialists "spoke the same language." In this way, she was able to confirm what exactly was happening and how serious her mother's condition was, and, on the basis of this acquired information, she decided to take a ten-day leave.

In contrast, when some older people I talked to experienced accidents or health problems such as diabetes and high blood pressure that were not life threatening, they would not always share that with their children over phone. This was irrelevant with respect to the geographic distance between the parents and their children, as the children could be living only one-hour flight or a two-day journey away. The parents decided to do so, they told me, so as not to worry their children unnecessarily, as there was nothing they could do to help from a distance. In this way, the parents were 'caring back' for their children, protecting them from unnecessary worry and perhaps feelings of guilt about not being able to help right then and there.

The fact that not all care can be provided through the phone also played a crucial role in Jane's decision to find a live-in maid for her parents. After her mother's stroke, Jane searched for and then employed a 'girl,' actually a middle-aged, childless Malayali woman, available to move in with her parents in order to cook, clean and do other chores as needed. Throughout my fieldwork, many people complained about the increasing difficulty of finding servants and maids in Kerala. Due to remittances sent home by Keralites working abroad, a significant proportion of whom are nurses, the standard of living has increased for all. This, according to my informants, made it practically impossible to find reliable servants, as the women who had carried out domestic work in the past would now receive their income from their children abroad. In line with the literature on global care chains (Yeates 2012), I expected that this lack of workforce in the domestic care sector would be replaced by migrants from some poorer North Indian states or other countries. However, although I could observe a significant number of men coming from Indian states such as Assam, Bihar or West Bengal as construction workers, I have not seen or heard of Indian women migrating to Kerala to work as housemaids. Several of my informants confirmed this observation, adding that there would be the problem of language (non-Keralites would not speak Malayalam) and, even more importantly, who could ever trust them? Jane's family was thus more of an exception than a rule among the families I met and interviewed, as most of them would not have a Malayali maid or a maid from a different state.

## *The Dying of Distance*

While Jane and her family relied on the use of the phone to stay in touch, quite a few families I encountered used an Internet-based calling service, namely Skype, which had become a household word. Those parents, like Jane's, who did not use this kind of communication told me that was because they did not know how to use a computer and therefore often did not even own one. However, in one family, I observed how Skype could be crucial in keeping in touch and providing care between family members. With two married daughters working as nurses in the United Kingdom, the parents in this family relied on the technical knowledge of their youngest daughter, also a nurse but still living at home while studying English to pass her IELTS exam. They used the free web calling service (Viber) to talk to both daughters abroad. Additionally, every day in the late afternoon or evening, they would set their laptop on the kitchen table to simply spend time with their middle daughter, her husband and their two young sons. Sitting with the youngest daughter and the mother of the family by the table, I was surprised by this: the conversation was far from the focused, static kind of communication between people sitting still by their computers and staring at the picture on the screen. Instead, the computers were the only things that were still; the conversation partners on both ends moved about the room, looking for something in the kitchen, playing with their small children, talking on the phone to somebody else, and so on. On that particular afternoon, the daughter in her house in the London suburbs was brushing her hair and preparing to go out, but she called to ask her mother for advice as she had forgotten to take her dose of some post-partum āyurvedic medicines. The mother, leaning across the table and looking straight into the screen, first scolded her daughter for being so careless about her health, then gave her precise instructions on how to proceed.

The computer became a window in space through which both the parents and the children felt as if they are "there, with them, in the same place," as one of the nurses I talked to said. Or as another mother told me, "since I can see (my daughter working in Saudi Arabia) and talk to her (every day, sometimes more than once a day), I never feel she is far away from me." For some moments at least, the distance collapses as parents and children not only communicate daily, but are also able to simply 'spend time together' and take care of and provide support to each other as if they were within reach.

## *Acknowledgments*

The material for this case study is based on my fieldwork in Kerala, India, which took place from January to March 2014. I carried out semi-structured in-depth interviews and participant observation of parent–children communication among seventeen families. Two interviews with the daughters working abroad were carried out over the phone and one over Skype. I also conducted interviews and observation at an English-language school, an old-age women's shelter, a geriatric hospital and a non-governmental organization tending specifically to the elderly. The interviews were carried out in English and/or Malayalam; the Malayalam interviews were conducted with the help of two interpreters or bilingual family members. I am most grateful to Dr. Leyanna Susan George and Dr. Harikumar Bhaskar for their invaluable help and support.

### *References*

Chua, J.L. 2014. In *Pursuit of the Good Life: Aspiration and Suicide in Globalizing South India.* Berkeley, CA: University of California Press.

Lamb, S. 2009 *Aging and the Indian Diaspora: Cosmopolitan Families in India and Abroad.* Bloomington, IN: Indiana University Press.

Percot, M. 2006 Indian nurses in the Gulf: Two generations of female migration. *South Asia Research* 26(1):41–62.

Yeates, N. 2012 Global care chains: A state-of-the-art review and future directions in care transnationalization research. *Global Networks* 12(2):135–154.

## Practicing Virtual Medicine

Medical anthropologists have become increasingly interested in how humans interact with technologies of diagnosis and healing. Clinical spaces in high-income settings, and those that have the necessary wealth in low- and middle-income countries, are filled with a rapidly changing amalgamation of ways to peer deeper into bodies, transmit images and texts, and facilitate communication between practitioners, patients, families, and communities. At the same time, technologies created as small, portable, durable, and sustainable tools for clinical use are increasingly cheaper to manufacture and easier to use. It is difficult to imagine a clinical tool that is not being re-imagined via new technologies of connectivity, screen availability, and minimization of size. Mobile technologies supplement the services available at fixed points (e.g. Telecare), as we discuss further below, and accelerate the transfer of clinical data and pathology reports and images, and support consultations, via email, What's App, Facetime, and Skype, between people and places. They extend biomedical knowledge, help retain clinical records, and increase quality care even as they amplify the panopticon of the clinical gaze (Foucault 1973).

Telemedicine is now a major player in the delivery of care worldwide for those with the resources to afford it. Starting in the early 1990s, the industry of telemedicine now connects practitioners, patients, and support groups to information bases, technical advice, and medical interventions (Sinha 2000). Industry blogs posit that telehealth will grow worldwide from approximately 350,000 users in 2013 to 7 million by 2018, with a concomitant rise in electronic medical visits from their 2015 level of 100 million. A blog posting on the website for Intel puts it this way:

> To control spiraling healthcare costs related to managing patients with chronic conditions as well as to navigate new policy regulations, 70 percent of healthcare organizations worldwide will invest in consumer-facing mobile applications, wearables, remote health monitoring and virtual care by 2018. This will create more demand for big data and analytics capability to support population health management initiatives.[1]

The complexities of healing, as described in the anthropological literature, inform us about the multiple levels of interactions between healers and their patients. Brigg's work (1994) with the Warao shaman of Venezuela, for example, makes explicit how curers gained agency over a distressing symptom such as pain from a wound, and then, through using traditional curing songs infused with knowledge of the social context within which the accident occurred, through words, music, and touch, created an emotional atmosphere that facilitated healing. This suggests that words have real power. Briggs argued: "I suggest that shamans gain power by collapsing discourses about pain and suffering undertaken by patients, their relatives, and the community at large into curative songs and practices" (1994: 164). The significance of clinical assessment, and the use in this context of sensory perceptions, sound, and speech, are no less relevant in a contemporary medical setting than they are for shamans and herbalists. Here, there are interesting questions for medical anthropologists: What parts of these embodied healing performances are still present in telemedicine and e-health? What parts are lost, and how is this corrected or countered as virtual medical care is extended from emergency obstetric to psychiatric care to surgery? How are these technologies appropriated into new configurations of embodied healing performances?

Annemarie Mol and her colleagues (2010) provide us with a new and interesting way of thinking about caring when they highlight the act of "tinkering." Resisting the obvious comparison between 'cold' high tech and 'warm' human caring, Mol and others show how people use technologies in unexpected ways, while caring for the unexpected. Tinkering occurs as caregivers make adjustments and compromises when the rigid protocols of biomedicine need to be enacted in emergencies or in cases where the necessary accoutrements of medicine are not available. As one moves further and further out from large, well-resourced hospitals and clinics, to settings where even the simplest of technologies are often not available, such tinkering becomes even more important.

Emergency personnel in rural areas encounter situations that often demand adjustments and creativity. Biomedical emergency medical systems are built on communication, transportation (ambulances, both ground and air), and a chain of practitioners trained to certain standards. While many situations are 'cookbook' or not unexpected, others are difficult in ways that cannot be anticipated, even for the experienced medic. When equipment fails or has been left behind, one needs to improvise with what is at hand: a oral-gastric tube used in place of the missing Foley catheter to empty an over-full bladder, a folding aluminium camera stand used instead of a Kendrick's Traction Device for a broken femur, a well-placed finger in lieu of sterile pads to stop a hemorrhage—one does the best that one can, with what is at hand.[2] Tinkering involves making small adjustments to equipment, pharmaceuticals, spatial arrangements, and ways of monitoring and documenting. Those caring for patients inevitably tinker, whether they are family members or medical professionals. Acknowledging this aspect of caring allows medical anthropologists to see beyond the written protocols to the rich environmental contexts in which humans interact with other humans, with the goal of caring for ailments both short- and long-term in nature.

## Changing Access to Technology, Changing Configurations of Care

As the above discussion suggests, electronic communication within and between clinical sites has become part of the clinical daily regimen, not only in areas with many resources but also on the periphery. Innovations such as telemedicine used to deliver care in isolated hospitals in the Peruvian Amazon and other remote regions demonstrate the value of the introduction of such technologies into various medical systems, but also how, when different ethno-physiological understandings coexist, unexpected changes occur in what clinicians and their patients experience in the process of digitization (Miscione 2007). In the Amazonian case, a telemedicine system was implemented but was only partially successful, as patients continued to delay coming in to the telemedicine clinic in favor of using their traditional healers and as practitioners struggled with intermittent Internet service and a record-keeping system that wasn't flexible enough to be used with the new technology. Implementing a new way for practitioners to access information changes attributes of the medical personnel from the point of view of the patients; seeing the new technology in use may confer more prestige, fear, or confusion on the part of the patients. Another aspect of implementing this kind of system is that it can provide practitioners with new ways of establishing agency over biological processes, through their increased ability to name conditions that were not diagnosable or known in the past, and through more effective and timely evacuation to better-equipped facilities.

Patients use the Internet to gain agency over their conditions. Increasingly patients undertake research into their own conditions, and arrive at the doctor's offices with the latest journal articles in hand. They expect to be involved in thinking through their diagnostic procedures, and they want to understand the options available to them, in order to decide on the treatment regimen. These informed or 'expert' patients may sometimes take the next step of working through social media to create disease nomenclature and arrange for new innovative treatments, as described by Gesine Hearn below. This movement of biomedicine can be thought of as a form of colonization, and like other forms of colonization it is incomplete, contested, and differently interpreted in each new place. Using the Internet and social media to exchange ideas about one's illnesses, to find fellow-sufferers and to seek out a range of possible treatments is something that did not occur before the widespread use of computers became common in the 1990s. The following ethnographic case study takes us to spaces that are newly colonized with things medical. Hearn highlights how Internet-based social groups come together and create common understandings and ways of enacting difficult-to-diagnose medical syndromes such as Chronic Fatigue Syndrome (CFS). Before the Internet acted as a communicative infrastructure for individuals suffering from these sorts of conditions, there was less awareness and certainly less group action with respect to identifying key symptoms and possible treatments. This creates a different kind of platform for patients to use to demand attention and treatments for contested illness syndromes like CFS.

## 8.3 Identities and the Internet

*Gesine Kuspert Hearn*

For ten years, I have followed the webpages of online self-help organizations for Chronic Fatigue Syndrome (CFS), Fibromyalgia (FM), and Irritable Bowel Syndrome (IBS). I was initially interested in how these virtual organizations viewed functional syndromes, contested diseases without a known etiology, and managed to establish research funding, diagnostic criteria, and public recognition for these diseases. I then compared US and German self-help organizations for CFS, FM, and IBS. Below, I focus on the formation of disease identities, virtual disease communities, and the presentation of the disease.

The uncertainty, invisibility, and chronic nature of these functional syndromes pose specific challenges for sufferers. Patients hope for a diagnosis and treatment, but due to the lack of any structural organic damage or pathology, biomedical diagnosis is difficult. Pain is a prominent symptom, but other physical symptoms include headache and fatigue. Sufferers often also complain about other psychological and neurological problems, including depression, anxiety, sleep disturbances, or trouble concentrating. As a result, doctors often consider these afflictions as being 'all in the head.'

In both Germany and the US, science-based allopathic medicine dominates the health care market. The American view of disease is predominantly mechanistic and based on the belief that external forces cause disease which, with American pragmatism and a 'can-do' attitude, results in an aggressive approach to diagnosis and treatment, including of these conditions. American medical textbooks emphasize the plethora of non-objective physical and psychological symptoms and the absence of an apparent pathology. The medical profession disagrees about the possible causes of functional syndromes, with arguments ranging from strictly biophysical pathologies, psychological and psychiatric problems, the interplay of biological, psychological, and social factors, to medicalization (Hearn 2009). The preferred treatment option combines anti-depressants and cognitive-behavioral therapy.

In contrast, in Germany, a more holistic view of disease prevails. Emotions are believed to play a pivotal role in health and disease, leading to a more open attitude towards alternative approaches and an emphasis on psychosocial factors of disease (Payer 1989). This more holistic view is evident in the German medical literature. 'Typical' psychological and social characteristics of patients with functional syndromes are often mentioned. Additionally, the German medical literature lists *psychosomatische Grundversorgung* (psychosomatic basic care), including psychotherapy and differential diagnoses, as an important treatment component for functional syndromes. Most medical articles suggest a biopsychosocial treatment approach.

Chronic conditions have a tremendous impact on the daily life and the identity of those who experience them. Chronic pain is particularly destructive and invasive; it tends to 'unmake' the normal world of sufferers (Scarry 1985). Chronic illness and chronic pain often result in biographical disruptions and 'assaults' upon self-identity (Charmaz 1991). However, patients can also re-organize their biographies and lives by assigning meaning to their suffering. Narratives help to create such meanings, and to create identities and communities.

## *Self-Help*

Self-help is about sharing first-hand experience of personal crises and coping with those crises. In the US, self-help groups were first formed in the 1940s, and they exploded in the 1970s with growing health care consumerism and rising health care costs. In Germany, self-help found increased acceptance in the 1980s; groups in Germany are partially funded by the German sickness funds and their work is supported by state-funded clearinghouses. Self-help organizations are either dedicated to a group of similar ailments or one particular disease; they are an explicit example of biosociality. Members are typically sufferers, but relatives, friends, or professionals working with these people also tend to join.

The Internet provides a new forum for self-help, enabling intensive and rapid exchange of information on a national and international level and providing a very useful venue for advocacy. Online self-help is particularly helpful for people who are geographically isolated or trapped at home by their afflictions or responsibilities (see also Thygesen and Pols, this volume). Today, many self-help organizations operate in virtual spaces. Self-help groups and organizations have had a positive impact, in particular on containing health care costs, but they have also been criticized for disseminating misleading information, undermining professional authority, negatively impacting patient–provider relationships, and fostering narcissistic and self-indulgent behaviors.

Increases in longevity and the resulting higher prevalence of chronic diseases have prompted the need for better informed and more active patients. Patient participation and cooperation have been encouraged, and this has led to changes in patient–physician relationships. Patients increasingly see themselves as lay experts and consumers of health care. The changing boundaries between experts and lay people are noticeable. Patients have successfully influenced diagnostic criteria, disease categorizations, research funding, and legislation.

In the US and Germany, patients with functional syndromes have established self-help organizations to support sufferers and lobby for the recognition of their ailments as serious organic diseases. These organizations do not differ in the presentation of the disease and the services they offer to sufferers, despite different health care contexts, different funding sources, and the different lobbying mechanisms in the two countries. Organizations in both countries validate the patients' experiences by generating narratives and thus meaning for their suffering. They also construct the face of the disease. Below, I discuss four leading US and three German online self-help organizations for CFS, FM, and IBS. I observed the websites of these organizations from January through March 2014.

The American self-help organizations started as local self-help groups and developed into national patient organizations, of varying size, revenue, and organizational structure. The main sources of revenue are membership fees and donations. Revenues pay for the websites, telephone hotlines, publication and dissemination of information materials, the organization of research symposia, awareness campaigns, and lobbying efforts. Some of the organizations fund research projects. The three German organizations, all established by sufferers, are *eingetragene Vereine* (registered organizations) and *gemeinnützige Vereine* (charitable organizations), and were the only national online self-help organizations. None provide information on the amount of revenues and expenses. The German organizations maintain websites similar in size and design to the American organizations. Information and advice is addressed to sufferers, doctors, and the public, and is provided through newsletters, journals, brochures, online publications, and telephone hotlines.

The explicit goal of these organizations is to assist and empower sufferers, and in doing so, carefully choose disease labels and use language that signals organic pathologies and the seriousness of the diseases. For example, the US self-help organizations label Chronic Fatigue Syndrome as 'CFIDS/ ME' (Chronic Fatigue and Immune Dysfunction Syndrome/Myalgic Encephalomyelitis). They also engage in advocacy, fighting for the scientific recognition of their diseases, more patient-oriented research, and increased research funding. They advise sufferers about medical, social, psychological, and legal aspects of their disease, and post official diagnostic criteria, personal illness accounts, new research findings, and current research grants and trials. Free articles can typically be found on the websites. Paid membership usually includes access to newsletters and addresses of 'good' doctors and local self-help groups.

## *Searching for an Identity*

Why do patients seek out these online organizations? The organizations tell us that people suffering from functional syndromes are frustrated, misunderstood, isolated, and alone. They feel victimized. No one understands the dramatic impact of the disease on their life, the pain, and the suffering they constantly endure. They are misunderstood because their diseases are invisible and often not validated with a medical diagnosis. They have seen many doctors and at best have received some psychological diagnosis. The sufferers insist that their problems are physical. They need answers for a life and an identity disrupted by an affliction no one can see or name. They need advice on how to navigate daily life; they need to know if they are eligible for social services; they need to know where they can find help for the mounting problems they have encountered since they have fallen ill. They want to know if there are others like them.

Sufferers first need to figure out if they have IBS, FM, or ME/CFS, and further, 'who they are.' The online organizations provide detailed information on the syndromes, detailed lists of symptoms, and online diagnostic tests. They ask, "Does this sound like you?" The organizations emphasize that these are "screening quizzes . . . not a real diagnosis." Symptom checklists help the sufferer assess how well his or her symptoms fit with the diagnostic criteria of particular syndromes. The hope is that the self-diagnosis will lead to a formal diagnosis. Fittingly, the Massachusetts CFIDS Association calls its guide to self-identification "A CFIDS Initiation" and lets readers know that "initiation rites into the CFIDS/ME club are pretty much the same" for all sufferers.

All organizations emphasize that official diagnoses can only be made by physicians. Here the organizations also provide help to people in the process of becoming a patient: sufferers can obtain educational materials for their physicians, direct physicians to the webpages of the organization, and most organizations provide lists of 'good' doctors—physicians who can and will diagnose CFS, FM, and IBS.

## *Acquiring an Identity*

Once officially diagnosed, the person afflicted with CFS, FM, or IBS becomes a patient with a specific disease. The feelings of fear, frustration, and confusion are now replaced with efforts to launch a new lifestyle. The misunderstood victim is whole again, a 'survivor' with a new identity: "you won't feel so fragmented!" But there is much adapting and learning to do: what sufferers characterize as developing a CFS radar, learning the medical lingo, catching up, managing counter-fear, getting informed, finding out what works, playing an active role in research, and standing up to doctors.

The message is clear. Patients with functional syndromes, although they now have a medical diagnosis, cannot simply let medicine take care of them. There is no cure for CFS, FM, or IBS. The diseases are chronic and available treatments only alleviate some symptoms and only for a while. The online organizations encourage patients to get and stay informed, and get involved; this will, they promise, empower new patients to manage their lives with these "complex, debilitating, and often disabling diseases." They urge patients to be active, to become experts of their diseases, and to make lifestyle changes, as the following quotes from webpages illustrate:

> Try to take an active role in your own health care. Obtain educational materials from your doctor and an organization such as IFFGD to learn more about IBS and how to best manage your symptoms.
>
> *(IBS Self-Help and Support Group 2014)*

> As patients, we are educating the medical establishment, one-on-one from the bottom up.
>
> *(Massachusetts CFIDS/ME & FM Association 2014)*

> Try not to let the initial stage of confusion and fear pull you down. Instead, as you add knowledge and connections with other patients, imagine yourself standing in the center of a circle. All the information you need is on the outside edges. As you pull the pieces from the edge toward you, you are in charge and getting continuously more knowledgeable.
>
> *(Massachusetts CFIDS/ME & FM Association 2014)*

The self-help organizations provide extensive information on living with the disease, including relationships, holidays, pregnancy, workplace, diet, and stress management. Personal stories of struggles and success by fellow patients help guide and reassure new patients. The website of the American organization for IBS posted 45 pages with personal stories on its website. The proposed adaptations are all encompassing, and space, place, identity, and daily life will all change if patients follow the well-meaning advice on the web. Patients are also advised to discuss such issues with their physicians. Websites include links to medical research centers and medical articles are posted on the webpages of all organizations. And the new patient should make him/herself available for furthering knowledge about the disease: "donate blood samples," "participate in research trials," "fill out surveys." Thus a person transits from puzzled lay person to well-informed citizen, victimized sufferer to empowered patient, knowledge seeker to object of knowledge formation.

Standing up to doctors and becoming a research object does not only benefit the patient—it benefits the whole community of sufferers. Being a patient with a specific disease validated by a medical diagnosis also means that you are now like 'us'—those who have the same diagnosis. Patients join an existing community and they acquire a new status as member of a disease community. They are now "people with FM," "Fibromyalgiker," "a patient with CFIDS/ME," an "IBS patient." There is "us" and "them," those who are ill and those who are not. Those who do not have the disease are not able to "relate to them as well as someone who has walked the same path"; they have the "view from the outside." New community members have to learn "the medical lingo" and "catch up" with the "old-timers." Speaking the new lingo will further separate them from others. To consolidate the community, activities for patients are offered: events, conferences, lectures, and advocacy. To affirm the community, celebrations take place: anniversaries, accomplishments.

## *National Differences*

US and German self-help organizations do not differ in the presentation of diseases and the services they offer. German organizations offer more free information, but the language is more impersonal. The formal *Sie* (You) is used on all webpages, and readers are rarely directly addressed. The organizations gather and synthesize medical information as a service to the patients. They refer to themselves as "your . . . self-help team," a team that is more "experienced." Such language implies a hierarchy of expertise. The patient seeking information is the layperson; the employee or volunteer of the organization is what Alfred Schütz called the "well-informed citizen" (1946). At the apex is the medical professional, 'the expert,' authoring much of the information for the German organizations. The advice from these professionals often includes alternative and complementary treatments, not surprising given a more holistic German medicine. Politically, the German self-help organizations engage in little more than awareness campaigns, due to different political structures and opportunities for political participation in Germany and the fact that they are partially funded by German sickness funds. The German organizations are also more internationally oriented. Information from and about foreign countries and translations of publications from American self-help organizations are posted on their websites, as are links to foreign research and medical centers, medical experts, and other self-help organizations. The German organizations explain the high number of American research articles on their websites as reflecting better funding of research in the US.

Most of the American organizations started as local self-help groups and then developed into leading national patient organizations. The main sources of revenue of the US organizations are membership fees and donations. Some of the organizations fund research projects. While a lot of the information on the webpages is free, more detailed information (such as addresses of 'good doctors') is only accessible to paying members. The American organizations are less internationally oriented and much more active in advocacy.

## Discussion

People afflicted with CFS, FM, and IBS, diseases with little biomedical validation, feel misunderstood, isolated, and embarrassed as their lives unravel due to the debilitating effects of these pain syndromes. They find support in online self-help organizations. The members of the organizations are like them. They seem to understand. And they have answers to all the pressing problems and issues sufferers face. Sufferers begin to rebuild their lives and identities: they are now people who belong with others like them; once officially diagnosed, they can legitimately claim a new identity as a patient with a particular disease; they are now members of a community of patients with the same disease. This is the story told by self-help organizations. The story generates the public face of the diseases, disease identities, and disease communities.

Initiation of the 'novice' takes place in the virtual space. The medical diagnosis in the real world affirms the new identity. The new patient then returns to the web to join the virtual disease community. However, successful treatment and getting healthy requires that their newly acquired identity has again to be re-constituted; new communities have to be found. This is particularly disruptive for people who were alone and isolated before, and it is not clear how connected patients actually feel. How involved are they with these virtual disease communities? How much interaction takes place? Does the story told by self-help organizations reflect the experiences of the patients?

Human problems are increasingly represented in biomedical terms. People expect answers from medicine for their afflictions and suffering. The self-help organizations firmly believe that CFS, FM, and IBS can be understood in biomedical terms, and thus cured. The organizations portray the image of a serious debilitating disease. However, those involved in the organizations, and present in the virtual community of sufferers, might just represent a more severely affected segment of the disease population. They might be people who lack social capital and coping abilities and thus need more help; or they might be sufferers who have more social and economic resources, such as Internet access and Internet literacy. If this is so, then whole disease categories and definitions might be based on the most serious cases of a 'patient elite' instrumental in establishing these. Here, the social construction of disease, disease identities, and disease communities takes place at the micro-level of virtual individual interaction, the meso-level of virtual communities, and the macro-level of interests organized online. The Internet is a powerful tool in the construction identities, communities, and diseases.

## References

Charmaz, C. 1991 *Good Days, Bad Days: The Self in Chronic Illness and Time.* New Brunswick, NJ: Rutgers University Press.

Hearn, G. 2009 No clue—what shall we do? Physicians and functional syndromes. *International Review of Modern Sociology* 35:95–113.

IBS Self-Help and Support Group 2014 Personal Stories. March 26. Accessed from http://www.aboutibs.org/site/living-with-ibs/personal-stories/

Massachusetts CFIDS/ME & FM Association 2014 The CFIDS initiation–A primer for new patients. March 26. Accessed from http://www.masscfids.org/resource-library/13-basic-information/105-the-cfids-initiation-a-primer-for-new-patients

Payer, L. 1989 *Medicine and Culture: Varieties of Treatments in the United States, England, West Germany, and France.* New York: Penguin Books.

Scarry, E. 1985 *The Body in Pain: The Making and Unmaking of the World.* New York: Oxford University Press.

Schütz, A. 1946 The well-informed citizen: An essay on the social distribution of knowledge. *Social Research* 13(4):463–478.

The new biosocialities that result from these kinds of self-help groups are a clear example of new ways that we are acting via our ICTs, as we are becoming more dependent upon these technologies for our everyday existence. What it means to be human is being modified, as we explore later (Chapter 15); the ramifications of this warrant a critical medical anthropological approach as they unfold. The virtual spaces described by Hearn are not only sites for exchange of understandings. Through interactions between people suffering from the same conditions, and through their reports of the kinds of treatments that they try out, new knowledge about diseases and treatments emerge.

## Looking to the Future

As new phases of 'cyborgification' occur, we integrate machines into our lives in ways very human (Manderson 2011). Rituals of diagnosis, intuition, and emotional reactions to the ICTs are just as important in the clinical realities of patients and practitioners, as are the digitalized and conceptualized patient data that is produced around the clock (Cartwright 1998; Murray 2012; Smith-Morris 2015). Anthropology's long history of understanding the embodied experience of illness as it is embedded in the subjectivities of everyday life gives the discipline a rich source of theoretical and ethnographic material upon which to draw, as we interrogate what it means to be a human in the midst of so many machines (Biehl and Moran-Thomas 2009).

Innovations in portable interpersonal and interactive communication equipment also address social and cultural barriers, and are improving the process of diagnosis. The expansion of the use of medical technologies allows doctors and other health providers to keep pace with rapid changes in clinical diagnosis and changes in ideas about care; clinicians in effect are being asked to change how they deliver care. Mobile devices such as smartphones, for example, are used for decision support, alerts and reminders, and information on demand, but also, with the use of peripherals, for data collection, such that smartphones and tablets are now the basic technology for monitoring and screening in pregnancy in many resource-limited settings (again, see Chapter 15 when we return to this theme). They review ophthalmic patients and test for visual acuity, deliver patient and physician education, and transmit advice during remote emergency care, so providing health professionals, wherever they are, with assistance in diagnosis and treatment. Along with the ability to gather vast amounts of bio-data and to transmit it from site to site (Ventres et al. 2006) comes the possibility of creating new databases of physiological information to augment our understanding of humans as they function under various environmental conditions—for instance, in extremes of cold, altitude, heat, malnourishment, and the presence of multiple diseases and co-morbidities (Singer and Scott 2003). We can now gather data on intertwined pathological states and record it in ways not imagined only a decade ago. The data generated by the constellations of new devices that can aggregate data across multiple platforms can put to test our present understandings of what 'normal' physiology is. New research questions and new ways of answering those questions are being facilitated through the innovations in caring discussed in this chapter; finding imaginative ways to build upon medical anthropology insights from the past to refine our knowledge of how we care for one another is our challenge for the future.

As caring for ourselves and for others becomes more and more virtual, our research methods have had to adapt. Virtual ethnographies, in which medical anthropologists participate in online forums to gain understanding of new care modalities and understandings of health and disease, can help us gain insights into new care modalities. New software allows the 'scraping' of content of such Internet sites and discussion forums, and content analysis of large data sets. In exploring such new technologies, anthropologists need to rethink ethics. How do we participate in virtual spaces as researchers? How do we present ourselves? And by what means can we communicate our findings back to the virtual communities?

## Notes

1. https://communities.intel.com/community/itpeernetwork/healthcare/blog/2015/02/04/part-ii-5-significant-health-it-trends-for-2015
2. Thanks to Mark R. Romero, nurse paramedic with the Lifeflight Network helicopter rescue team, for these insights.

## References


Biehl, João, and Amy Moran-Thomas 2009 Symptom: Subjectivities, social ills, technologies. *Annual Review of Anthropology* 38:267–288.

Briggs, Charles L. 1994 The sting of the ray: Bodies, agency, and grammar in Warao curing. *Journal of American Folklore* 107(423):139–166.

Cartwright, Elizabeth 1998 The logic of heartbeats: Electronic fetal monitoring and biomedically constructed birth. In *Cyborg Babies: From Techno-Sex to Techno-Tots*. Robbie Davis-Floyd and Joseph Dumit, eds. Pp. 240–254. New York: Routledge.

Foucault, Michel 1973 *The Birth of the Clinic: An Archaeology of Medical Perception*. Alan M. Sheridan, transl. London: Tavistock Publications Limited.

Glascock, Anthony P., and David M. Kutzik 2006 The impact of behavioral monitoring technology on the provision of health care in the home. *Journal of Universal Computer Science* 12(1):59–79.

Greenhalgh, Trisha, and Deborah Swinglehurst 2011 Studying technology use as social practice: The untapped potential of ethnography. *BMC Medicine* 9:45.

Kleinman, Arthur 2009 Caregiving: The odyssey of becoming more human. *The Lancet* 373(9660):292–293.

López, Daniel, Blanca Callén, Francisco Tirado, and Miquel Domènech 2010 How to become a guardian angel: Providing safety in a home telecare service. In *Care in Practice: On Tinkering in Clinics, Homes and Farms*. Annemarie Mol, Ingunn Moser, and Jeannette Pols, eds. Pp. 71–90. Bielefeld, Germany: Transcript Verlag.

Manderson, Lenore 2011 *Surface Tensions: Surgery, Bodily Boundaries, and the Social Self*. Walnut Creek, CA: Left Coast Press.

Mattingly, Cheryl, Lone Grøn, and Lotte Meinert 2011 Chronic homework in emerging borderlands of healthcare. *Culture, Medicine, and Psychiatry* 35(3):347–375.

Miscione, Gianluca 2007 Telemedicine in the Upper Amazon: Interplay with local health care practices. *MIS Quarterly* 31(2):403–425.

Mol, Annemarie, Ingunn Moser, and Jeannette Pols, eds. 2010 *Care in Practice: On Tinkering in Clinics, Homes and Farms*. Piscataway, NJ: Transaction Publishers.

Murray, Marjorie 2012 Childbirth in Santiago de Chile: Stratification, intervention, and child centeredness. *Medical Anthropology Quarterly* 26(3):319–337.

Sakellariou, Dikalos 2015 Home modifications and ways of living well. *Medical Anthropology* 34(5):456–469.

Singer, Merrill, and Clair Scott 2003 Syndemics and public health: Reconceptualizing disease in bio-soical context. *Medical Anthropology Quarterly* 17(4):423–441.

Sinha, A. 2000 An overview of telemedicine: The virtual gaze of health care in the next century. *Medical Anthropology Quarterly* 14(3):291–309.

Smith-Morris, Carolyn, ed. 2015 *Diagnostic Controversy: Cultural Perspectives on Competing Knowledge in Healthcare*. London: Routledge.

Taylor, Janelle S. 2008 On recognition, caring, and dementia. *Medical Anthropology Quarterly* 22(4):313–335.

Ventres, William, Sarah Kooienga, Nancy Vuckovic, Ryan Marlin, Peggy Nygren, and Valerie Stewart 2006 Physicians, patients, and the electronic health record: An ethnographic analysis. *Annals of Family Medicine* 4(2):124–131.

Funeral in El Alto, 1996. El Alto, Bolivia.



Note: Originally published in Jerome Crowder, 2007. Aymara migrants in El Alto, Bolivia, in *Ethnic Landscapes in an Urban World*, R. Hutchinson and J. Krase, eds. Pp. 181–195. Bingley, UK: Emerald Group Publishing Limited.

## *About the photograph*

*One morning I went to visit my friend Eusebio, a long-time resident and businessman in Villa 16 de julio, where I found his front door shrouded in black cloth. A traditional funeral dirge blared out across the neighborhood from a large speaker placed in the second-floor window above. Inside, the home was filled with smoke and packed wall-to-wall with family and friends. Surrounding a casket on the other side of the room were candles, large sprays of flowers, people quietly saying prayers and openly lamenting. The men sat on one side of the room, smoking and chewing coca leaves; the women congregated on the other, talking with each other and weeping. Eusebio had died the day before and I had come upon the wake in progress. Eusebio's family and friends stayed with the body for the next two days, before taking it to a cemetery on the periphery of El Alto. In the shadow of Nevado Illimani, the family recedes from the gravesite to receive condolences from friends and visitors who knew Eusebio.*

—Jerome Crowder

# 9 Endings

*Lenore Manderson, Elizabeth Cartwright and Anita Hardon*

Birth and death are wrapped in uncertainty and mystery, their timing and circumstances largely unpredictable. The two experiences are richly embellished; and where death occurs under propitious circumstances—and perhaps in an idealized setting—people honor their relationships to each other as they welcome the newborn and promise to nurture the child, or bid farewell to the departed. Many early ethnographers focused on death and its management as an example of an everyday life ritual that provides a means to comprehend local cosmologies and ontologies, life purpose and the divine. Anthropologies of death, and the responsibility of the living to ancestors, have provided insight into the relationships of humans with the physical and metaphysical; serious illnesses, dying and death sharpen the need for people to protect other individuals and the social body, and to manage the liminality of the person for whom death is imminent.

A number of early ethnographers mapped out cultural models of aging that have endured to the present. In these, characterizations of the elderly in small-scale societies often stand in sharp contrast with those of the aged in the highly industrialized global North, where social worth is more likely tied to productivity than, say, wisdom, spiritual engagement, or longevity and experience (Cohen 1998). In small-scale societies, it was suggested, older people were respected; with age, men gained power and authority with their accrued cultural and political wisdom; women enjoyed new freedoms once they ceased reproduction and were able to cast off controls over their sexuality; they too were able take on leading cultural roles. Lawrence Cohen (1998) reminds us that these are often nostalgic and rather romanticized accounts, and not all older people fared well in such settings. Colin Turnbull (1972) makes this point in his chilling account of the Ik of Uganda, for whom forced settlement had harsh impact on frail older people who were left to die by those who were younger, fitter and more self-sufficient. Moreover, economic development, state control and capitalism have all affected the roles of the elderly, the respect conferred to them, and the care with which they are both entitled and provided.

More recent works on aging and the elderly, published in various edited collections and monographs from the last decades of the twentieth century (Counts and Counts 1985; Lynch and Danely 2013; Sokolovsky 1997; Strange and Teitelbaum 1987) and in the journal *Anthropology and Aging*, demonstrate the complexity of aging as a process, and the variable experiences of older people, under different social locations and conditions, as the consequences of class, ethnicity, social organization, gender and geography. In this chapter, we explore this diversity, first considering how older age is experienced, and then turning to the end of life and the management of death. We begin by exploring how being old and frail merge and diverge as states of being. We highlight tensions around ideas of 'normal aging' and its medicalization in association with the emergence of gerontology as a unique field of medicine, and as psychiatry documented the deterioration of aging brains and minds. As Lawrence Cohen (1994; 1998) suggests, how aging and old age are understood, and how older people are cared for, are shaped by particular histories and cultural politics. Older people are sometimes, but not inevitably, infantilized as their independence, functional capacities, and capabilities are reconfigured. In the second half of this chapter, we consider different views about dying and being dead. We do not deal with infant or other early deaths here, but rather, primarily address death among adults. We look at the care of corpses, too, and so reflect on the fear of death and the stigma around dead bodies.

## Aging and Its Impositions

We all age, but the meanings of aging and being old are cultural as much as biological facts. 'Old' is defined by a number of factors, but these are not necessarily correlated. Physical aging, cognitive ability, memory and social roles are calibrated in ways that mark who is old and who is not, and whether and how a person participates in social and economic life. Being old tends everywhere to be tied to participation in production, and the capacity to contribute economically to a household, family, community or nation. Depending on the society, lineage, inheritance laws and ownership (of land, cattle or other resources), all shape the power or powerlessness that accrue with age. Further, in all nation states, the formality of retirement from paid work provides a line by which people are measured and valued, but there is considerable variability given that in some countries there is no longer a statutory aging of retirement, while in others it can be as early as 55 years. Further, Glascock and Feinman (1981) noted that in most of the 60 societies they reviewed, people are regarded as 'old' well before they begin to show signs of difficulties in everyday function and independent living. The older a person is chronologically, however, the less ambiguity exists: a person who is 90 is considered to be old irrespective of their capabilities. Value judgments about how people are categorized and treated tend to have less weight as people embody the characteristics of old age.

With age, given the wear and tear on the physical body of a life long lived, people face multiple assaults on their health and strength, and they must manage conditions that cause discomfort and develop degenerative conditions and complications. They may lose mobility and sensory acuity (of hearing, vision and taste), and tend to be afflicted by and diagnosed with growing numbers of health problems, some life threatening, others regarded as an expected dimension of aging (joint pain, for instance), although no less debilitating as a result. Many conditions prevalent in older age are painful—muscle and skeletal pain from arthritis, lower back pain and sciatica, for instance; other conditions may not be painful but result in distressing dysfunctions and disorders—incontinence, hearing and vision loss, sleep disruptions, physical instability, declining sexual function, and memory lapses, for example. In much of the world, these are 'normal' dimensions to old age; they are not illnesses nor are they signs of illness.

At the same time, aged and frail people may also be caring for their spouse, a child or a grandchild, or adjusting to life after the death of their partner or children. Sjaak van der Geest (2008) reflects on the ways that ideas about reciprocity within kinship relations ensured that older people were cared for in Ghana, but while old people expected their children to take on this care, this is not always the case. In his work in Uganda undertaken prior to the introduction of antiretroviral therapy, Alun Williams (2003) similarly highlighted the isolation and everyday difficulties of older people living in poor rural environments, who because of infirmity and weakness had reduced ability and lacked the energy to plant, harvest and prepare their own food, or to fetch water and fuel. Their children's migration from villages to urban centers, and the responsibility that they therefore had for their grandchildren and other small children, was once balanced by the sureness of remittances. However, the untimely deaths of many young adults from AIDS left these people struggling, without support. Ellen Block, in her case study in Chapter 4, similarly draws attention to the necessary renegotiation of roles of older people in Lesotho with changes in family structure due to HIV infection, the migration of young adults from rural to urban areas, and their migration overseas, all of which leave older people with far less kin-based support and extend the periods during which they are expected to provide care for younger family members (see also the case study by Tanja Ahlin, Chapter 8). Older people may be resilient in the absence of support from their children, and, with changes in life circumstances and social relationships, it is increasingly the case everywhere that older people may find themselves alone (van der Geest 1997; 2004b; 2008). Changes in function and communication, and physical and mental health problems, all cause the social worlds of many people to shrink, interfering with their capacity to maintain social and economic relations beyond close families, and so narrowing their access to networks that might provide them with support.

## Living Uncertainly

These are all somewhat negative visions of aging and being old, reflecting lives framed by loss, lack, limitations, decline, and debility. The alternatives, as we have indicated already, have been somewhat idealized portraits of aging when reduced expectations of economic contributions provide men and women with the opportunity to spend more time as cultural advisers and arbitrators, participating in social activities and pursuing spiritual paths. Aspirations of old age framed by such shifts in focus and activity pertain, but globalization, industrialization and labor migration, and increased longevity, have everywhere troubled the processes of aging and its opportunities. In these contexts, sometimes health and illness force the choices older people have; sometimes they do not.

France, 2013. Two events contrast ways of aging. In early January, in the heart of the northern winter, a 94-year-old woman was thrown out of a retirement home near Paris, and left outside the unattended home of one of her sons, because her bills incurred at the retirement home had not been paid. Two months later, a group of older women, aged 62–85, opened a self-managed women's only social housing project in Montreuil, Paris, with 25 apartments, of which 21 were for older women, four for students.[1] The dependency of one woman, and the violence against her, and the agency and vision of the other women, suggest not ways to avoid uncaring families or undignified life in an uncaring institution, but emphasize differences in capability.

As we wrote this book, our three mothers were very much older women, and the times we spent together writing were punctuated by personal enquiries and accounts of their health, welfare, care and wellbeing. As they showed us, capacity and health are plastic: age is not a simple line from strength to weakness, acuity to confusion, autonomy to dependency. And everywhere, old age is a time of negotiation amongst kin and with others. In the following case study from coastal Tanzania, Peter van Eeuwijk and Brigit Obrist highlight the fluidity of local understandings of 'old age frailty' in relation to the physical body and mental capacity. Local understandings of 'frailty' reflect judgments and assumptions about rights and responsibilities that are embedded in kinship and dependence, so affecting care arrangements and care activities.

## 9.1 Becoming Old and Frail in Coastal Tanzania

*Peter van Eeuwijk and Brigit Obrist*

Local understandings of 'old age frailty' vary. Among KiSwahili speakers in coastal Tanzania, East Africa, elderly persons (60 years and older) consider their aging as both advancement and blessing, in spite of the effects of HIV/AIDS, gradually unreliable social networks and few formal welfare structures. With aging, they may face a decline in their health, and so need increased support and changes in care arrangements. Yet such frailty, with gradual loss of bodily and mental strength, and increased episodes of illness, is again part of the normal process of becoming old.

Geriatric notions of frailty in older adults include weight loss, weakness, fatigue, pain, depression, and falls as recognized symptoms of physical and mental decline, which may lead to low levels of activity, poor endurance, and slowed performance (Boockvar and Meier 2006). This professional viewpoint varies considerably from local perceptions of aging and frailty. Aged and fit Americans, for instance, adhere to understandings of healthy old bodies and active aging, characterized by good health, physical and mental fitness, energy, and the absence of illness (Steadman et al. 2013). In contrast, older Indonesians allude to getting limp, exhausted and weak, unable to resist changes in appearance and impairments; they associate the decline in their mental strength and physical robustness with the fear of losing autonomy, mobility, and activity—but not with any classification of disease (Eeuwijk 2003). Everywhere, however, frailty represents an embodied change that is constitutive and defines old age (Laz 2003).

There is no distinct linguistic term for 'frailty' in KiSwahili language. Old persons express becoming frail by referring to *nguvu* (strength) in reference to both physical and mental qualities. Aging persons who do not rate themselves as frail speak of *nina nguvu* (I have strength). In contrast, elderly people with a frail body and mind allude to *sina nguvu* (I don't have strength) and refer to having 'less fluidity' in their bodies (e.g. in their joints). Temporary health problems of older persons associated with frailty are called *baadhi ya wakati* (in Zanzibar: some time [without strength]) or *wakati mwingine sina* (in Dar es Salaam: other time without [strength])—'sometimes with or without strength.' Here *nina nguvu* (having strength) dwindles for a while, but they recuperate. Older persons do not understand their 'frailty' as a fixed state or static illness category, nor do they see it as a sign of inevitable degeneration. Rather, an aging body and mind may fluctuate. This dynamic 'continuum of strength' ranges from 'full strength' to 'without strength'; the boundaries of frailty, health, and illness are porous and the health of individuals is fluid. Short illness episodes such as a leg broken following a fall caused by weakness may be associated with or follow a longer period of general physical weakness or mental exhaustion, but the elderly person may recover and regain strength.

These changes in degree and outcome of bodily and mental frailty are closely monitored by the older individual and by his or her family and other household members. Their assessment of 'performance in frailty' may differ from that of the aged person, and may trigger debates about providing and receiving care. Local Swahili understandings of 'frailty' refer to judgments and assumptions about rights and responsibilities formulated in terms of kinship ties and inter- and intragenerational dependence, and thus affect care arrangements and care activities. Kinship norms and values assign eldercare obligations to children, particularly younger women. This expectation of intergenerational care for elderly parents shaped most care arrangements in the Swahili study households, due to successful negotiations between parents and their children, strong emotional ties, and feelings of reciprocity during their life course. Even so, in almost one third of the households with which we worked, an older person (60 years and more) assumed the role of major caregiver for a frail, sick, or bedridden elderly person. These aged but healthy carers contribute to daily domestic work, activities related to illness and ailment, and offer what we might, in other contexts, term psychosocial counseling. Becoming old and frailer, as a result, has an immediate impact on care arrangements: new patterns of care have to be established. Other kin may join the household of a frail elderly person, or he may leave his own household to join that of his major caregiver(s), for instance, his oldest married son. In a very few cases, too, we found major non-kin carers, such as tenants and close neighbors. Old age frailty in this sense results ultimately in the formation of (new) care regimes. Moreover, aged persons becoming frail gradually move from being caregivers to care receivers, when they cease to be able to self-care and or contribute to the care of other family members.

This is complicated by the fact that the multiple morbidities of older people complicate reliable diagnosis and its adequate therapy. Geriatric skills are not yet widely developed in Tanzania, and biomedical health professionals, such as those working in most urban hospitals and rural health centers, do not provide appropriate therapeutic assistance for old frail persons (e.g. palliative care, analgesia, nutriments, and restoratives). And even if this range of therapies and technologies were available, few could afford them.

In the following, we present examples of old people with strength and without strength. The first case is of *Bibi* Asha (60 years, Muslim, married), a woman 'with strength' *(nina nguvu).* Bibi Asha lives with her blind husband in an unplanned settlement area in Dar es Salaam. She and her family moved in 2008 to the city from a village in rural Rufiji, into a house owned by her brother-in-law, where 16 other people, including her brother's family, reside. She is the main caregiver of her husband, who lost his eyesight in an accident some years ago in the village. In addition, she cares for her sickly and pregnant daughter, her only living child (from another husband), and for a young boy, a child of her husband's first wife who had left him and lives elsewhere. Bibi Asha commutes every day between her daughter's house (in the same quarter) and her own household, with her husband and daughter fully relying on her for care. Her family relies on her brother's financial resources, and he is the main decision maker for her family—a dependency that demands that she obeys her brother and his wife. She works hard every day and follows the Swahili norm that a wife has to provide good care of her husband to secure the marriage: she cooks for him, she accompanies him to the toilet, she prepares his bath, she washes his clothes, she sweeps the floor, she guides him where he wants to go, and she shops in the market, and prays for her husband at the mosque. This daily burden of care renders her tired and sometimes exhausted, but she is physically and mentally robust, ensuring she can fulfil these many daily duties. She considers herself as 'still energized,' a blessed strength which she attributes Allah's infinite power.

*Mzee* Hassan (male, 89 years, Muslim, widowed), in contrast, is 'without strength' *(sina nguvu).* He lives in the small rural town of Ikwiriri close to River Rufiji. He owns a dilapidated house and lives there alone. His very weak body, his blind eye, permanent pains in his back, and numbness in his swollen legs significantly limit him physically. Thus, he stays at home, he sleeps a lot. In the morning he sits in the warm sun in front of his house, and after a while, he goes for a sleep again in his house, in the shade of the veranda, or under nearby mango tree. Sometimes an old neighbor comes by for a chat about life's difficulties. Mzee Hassan relies mainly on his grandchildren, who come by regularly to visit, and his daughter, who helps him every day. He ate initially in the house of his son-in-law and then his nephew. The daughter and the grandchildren now cook and bring him food (e.g. fruits, vegetables), but sometimes when his daughter is busy, he has nothing to eat and may also have no water in the house. Nevertheless, when he had to be admitted to the hospital, the daughter accompanied him, brought food, guarded his empty house, and raised money for his treatment. His frailty and mainly his partial blindness keep him from participating in social, economic, and religious activities—he feels ashamed and embarrassed when other people see him. His daughter too is sometimes unable to fulfil all his needs because she is also struggling and cannot provide permanent and adequate care. For instance, he has prescriptions of medicine for back pain, but never had the money to buy the medication; his daughter also cannot afford it. Mzee Hassan considers himself now an old and very frail man, because he can no longer fetch water nor ride his bicycle to go to his farm. For three years he has not cultivated corn and bananas on his farmland and no-one watches over this land. Mzee Hassan considers this a very distinct sign of his advanced state of frailty.

Other people fluctuate, 'sometimes with and sometimes without strength.' Bibi Fatma (female, 74 years, Muslim, married) lives with her husband and his second wife (and her two children) under the same roof in a remote village in Rufiji District. She married only six years ago, when her aged husband divorced one of his three wives. She does the daily housework; the two children fetch water from far away. Her husband takes care of their farm, and she makes pots from local clay soil. When Bibi Fatma is ill and feels weak, her husband takes over some housework such as cooking, laundry, and cleaning the house. She had recently an accident and broke her leg; she still suffers from pain. Her husband called the local healer and paid for treatment; because Bibi Fatma prefers local medication, she did not go to the local hospital. Due to this accident, she had to stop making clay pots because she cannot carry the clay. When she feels seriously ill and weak (e.g. due to malaria), the children of the second wife bring firewood and support her husband on the farm. At times, when the children are not around, she asks the daughter of a neighbor to fetch water and pound grain (because this is a woman's work, as Bibi Fatma emphasizes). Despite temporary weakness and variable health, she feels she is still strong enough to make a steady and significant contribution to her household.

Two other people are 'losing and regaining energy' *(leo sina nguvu).* Bibi Zuhura (female, 62 years, Muslim, married) lives with her much older husband, one of her sons, the family of another son (including his four children), a nephew, her sister's grandson and a non-related young man (a football player), under one roof in an unplanned settlement in Dar es Salaam. She has six children (four boys and two daughters) and ten grandchildren. Bibi Zuhura worked as a government staff member at the Tanzanian Embassy in Kenya and Italy, which entitles her to a monthly pension. The family runs various businesses including a large commercial farm, which Bibi Zuhura ran after her retirement; she is the center of the family's economic undertakings and their control. Bibi Zuhura suffers from high blood pressure due to overwork, she says. One day, she collapsed during a business trip in northern Tanzania. This turned out to be a stroke. She lost her memory, speech, and bodily strength; one side of her body was paralyzed. She stayed for one month in a Christian hospital in Arusha. After she returned home, her youngest son, a football coach of a regional club, initiated a very strict physical exercise regime for her. He provided accommodation for a young football player of his club at his mother's house, and this young man supervised her daily exercise programs—using a bicycle machine, applying physiotherapy, and adhering to a strict diet. Three years later, she had recovered fully from the stroke—it was Allah who accepted her again and restored her full strength, she said. She does not feel frail any longer. She works now in the morning, and still follows the exercise program in the afternoon.

Mzee Emmanuel (male, 70 years, Christian, widowed) is a retired government employee and owns a nice apartment in a comfortable neighborhood in Dar es Salaam. Two sons, one daughter (who is a nurse and will marry soon), two grandsons (of a deceased son), and a housemaid live with him. He receives a monthly pension. His children bear his daily expenses (e.g. for food and transport) while Mzee Emmanuel supports financially his brother and one of the two grandsons. When he is at home (he often joins his church congregation), he works in his garden where he grows vegetables. Mzee Emmanuel has diabetes, but although he sometimes looks a bit frail and tired, he considers himself in good health and with strength. The health insurance, paid for by one of his children, covers his monthly diabetes treatment. The problem with his diabetes has its origins in the mode of treatment: he was prescribed oral medication, but his blood sugar level fluctuated, and he often felt very weak. At the diabetes unit in a hospital in Dar es Salaam, he then received insulin injections which he could apply at home. On the first day, however, he failed to estimate the appropriate dose to be injected due to his impaired eyesight, and he suffered from life-threatening hypoglycemia. He slowly recovered and moved back from injections to pills. After one year of increasing problems due to his diabetes, Mzee Emmanuel looks much better and stronger due to appropriate medication, traditional herbal medicine from a local healer (brought by his sister from his native village), rigid diet, and physical exercise. He walks every second morning, and this daily exercise helps control his blood sugar level. He also practices yoga every day in his bedroom by following a tutorial booklet. His voice is strong; he looks physically robust and is very confident with regard to his health.

★ ★ ★ ★

These cases from Swahili-speaking communities in coastal Tanzania show that frailty is not about being bedridden or disabled, or to physical attributes of old age, such as grey hair and wrinkled skin. It refers rather to the limited capacity to self-care due to debility, fatigue, impaired sight and hearing, and problems in standing up, bending down, walking, and carrying weights. Elderly persons without strength or with its temporary loss can neither care for others nor themselves; they become increasingly dependent on other care providers. This transformation leads to the establishment of new care arrangements or the reconfiguration of an existing one. This entails shifts in the allocation of social, economic, physical, and emotional resources at a household level, which depends in part on the old and frail person's social agency. Bibi Zuhura and Mzee Emmanuel, who lose and regain strength, illustrate the gradual appropriation of new global gerontological concepts such as 'active aging,' 'healthy aging,' and 'successful aging,' where frailty can be mitigated and even reversible through an application of biomedical therapies and health-related regimes. This change in understandings of aging is most common in the health practices and health status of predominantly better off, middle-class urban residents. They are able to develop a degree of resilience, in face of increasing frailty, beyond the capacity of poorer, rural Tanzanians.

## *Acknowledgments*

This contribution is based on two research projects conducted in Tanzania, funded by the Swiss National Science Foundation (SNSF). The research locations were Dar es Salaam and Rufiji District, Pwani Region (2008–2011, led by Peter van Eeuwijk) and Dar es Salaam and Zanzibar City

(2012–2015, led by Brigit Obrist). Jana Gerold, Vendelin T. Simon, Andrea Grolimund, and Sandra Staudacher conducted long-term fieldwork and collected empirical data with the project heads and Tanzanian field assistants. Both studies applied a combination of qualitative techniques (in-depth interviews, focus group discussions, direct observation, case studies), quantitative research methods (e.g. statistics, structured questionnaires), and documentary tools (film and photo documentation).

### *References*

Boockvar, K.S., and D.E. Meier 2006 Palliative care for frail older adults. "There are things I can't do anymore that I wish I could . . . ". *Journal of the American Medical Association* 296(18):2245–2253.

Eeuwijk, P. van 2003 Urban elderly with chronic illness: Local understandings and emerging discrepancies in North Sulawesi, Indonesia. *Anthropology & Medicine* 10(3):325–341.

Laz, C. 2003 Age embodied. *Journal of Aging Studies* 17(4):503–519.

Steadman, R., C.I.J. Nykiforuk, and H. Vallianatos 2013 Active aging: Hiking, health, and healing. *Anthropology & Aging Quarterly* 34(3):87–99.

---

People with long-term health problems such as diabetes, chronic obstructive pulmonary disease (COPD) and heart disease, as we have already illustrated (Chapter 7), must often structure their lives and tailor their activities around the practicalities of disease, treatments and the consequences of life-saving surgery—the need to manage a stoma and bag for elimination following a colostomy or urostomy, for instance; taking immunosuppressant medication following a transplant; constant care of a stump following an amputation (Manderson 2011). Although survival from potentially fatal conditions is increasingly common, accounting for increased life expectancy globally, the division between poor and rich people, in both poor and rich settings, pertains. Illness and impairments that impact communication, cognition and function strip people of the habitual ways in which they sustain social relationships and engagement regardless of age, but as we have suggested above, increasing frailty further limits the capacity of the very old and compounds these other embodied and structural limitations.

## Cognitive Impairment and Memory Loss

Increased longevity has increased the likelihood that people will develop conditions such as Alzheimer's disease and other dementias. Changes in cognitive ability, problems with speech and vocabulary, memory loss and mood disorders all occur with aging, causing frustration and distress for the person experiencing such problems themselves, in their interactions with others, and in the ability of others to communicate with them. Anthropological research has helped us reflect on our understanding of the cognitive, personality and behavioral changes that can be associated with aging. In the same way that autism spectrum and attention deficit disorders have become increasingly common diagnostic labels of various communication, behavioral and development disorders in children (see case study from Vu Song Ha, Chapter 2; Conrad and Bergey 2014; Kaufman 2010), so a putative diagnosis of Alzheimer's disease is often applied to older people, despite that the cognitive problem may be some other form of dementia, or may be a sign of an underlying physical problem (mental confusion in older people is sometimes the only sign of a urinary tract infection).

In *The Alzheimer Conundrum: Entanglements of Dementia and Aging*, Margaret Lock (2013) asks us to reflect on our own attitudes to aging; both she and Lawrence Cohen (1998) illustrate the ways in which attitudes to senility and frailty vary considerably, with shifts in function and capacity seen in some places as a sign that the person is moving to another world, in other cases as frightening signs of the growing loss of independence, autonomy and dignity, and impending death. How people deal with such shifts differs considerably. Ruud Hendriks (2012) trained as a clown in order to participate in and so describe one unique approach in an aged care facility in the Netherlands. Here, clowning was introduced as a way of enriching the lives of people with dementia, by alleviating mood, helping them connect to their social surroundings, and enhancing their quality of life: the approach built on the capacity of people to retain a sense of relationality and embodiment long after they were able to communicate verbally or to interact with others in predictable and socially acceptable ways.

Changing diagnostic categories influence which people are counted as experiencing memory loss, what this means to them and others, and how they are cared for. In the case study below, Annette Leibing draws on her research in Brazil to illustrate changes in the definition, etiology and risk factors of Alzheimer's disease, and in the range of preventive actions and treatment regimens delivered to people whom she refers to as "forgetful seniors."

---

## 9.2 Alzheimer's Disease in Urban Brazil

*Annette Leibing*

It is a rare opportunity for an anthropologist to observe and participate in the emergence of a new diagnostic category in a given society. In my case, I studied the beginnings of Alzheimer's disease (AD) in Rio de Janeiro as an anthropologist interested in geriatric psychiatry and as the founder and director of a research center and outpatient unit that became part of the local Institute of Psychiatry (IPUB). Although in the mid-1990s dementia was starting to be discussed more intensively within the Brazilian media, this occurred in a relatively abstract way, and few people seemed to know about it. Psychogeriatrics and related fields were only just becoming popularized among neurologists, psychiatrists, and psychologists, a new field and market within a rapidly aging country where youthfulness is strongly valued. Alzheimer's disease—the most common form of dementia—was a diagnostic category still in the making. There was only a fragile understanding of the disease, and its symptoms seemed to be especially dramatic in a landscape of aging wherein only recently a strong 'third age' movement had been implanted—a movement that generally propagated youthfulness, constant activities, and consumerism of the 'not old yet'—and in which decline and decay had an especially negative connotation. However, since symptoms of dementia were often linked to madness, a modern diagnosis of dementia also meant less stigma and new forms of attention by institutions and health professionals who had discovered 'the disease of the century.'

In the early years of Alzheimer's in Brazil, from around 1979, most articles in the press referred to research and discoveries on the molecular level, and, paired with great optimism, often described recent discoveries in the United States as being one step away from the final cure. Alzheimer's disease appeared to occur in an abstract, distant world of biomedical technologies far removed from the local realities of the families caring for the old and sick (although that was changing in urban Brazil). At the same time, the great optimism and trust in biomedical technologies went hand in hand with the notion of a 'mysterious' disease—about which almost nothing was known—affecting an alarming number of elderly.

A second phase started around 1993, when Alzheimer's disease first appeared as a 'national problem' in the Rio de Janeiro newspaper, *Jornal do Brasil.* The article emphasized the former ignorance of doctors and family members in diagnosing and dealing with the 'new' condition. From this point forward, AD was regularly characterized as a disease that could affect anybody, and numerous articles asserted that 1 million Brazilians were suffering from AD, without ever citing the source of these numbers. In 1993 too, probably not coincidentally, Tacrine arrived on the international market—a medication that was both a strong carrier of hope and promptly criticized due to its serious side effects. The pressures from patient and community groups on the government became stronger and, since politicians had recently discovered the elderly as a voter population, the contested medication was imported to Brazil.

Alzheimer's had the flair of a disease of the modern world. Increasing numbers of articles began to deal with risk factors that depended on individual behavior and risk management. In the late 1990s, the regular use of the brain was especially recommended, based on the idea that a 'brain reserve' would lead to a later onset of dementia. In contrast to the preventive measures and the strong link with cardiovascular health propagated today (see Leibing 2014), the etiology of Alzheimer's disease in the 1990s was unspecific—sometimes of genetic origins, but mostly due to bad luck and lack of education. Caregivers I interviewed often had a different idea. *Sclerose*, from the term 'arteriosclerosis,' was used to refer to two types of senility that manifest in old people's behavior. One was a gentle sliding away from the living to the dead, an in-between state where the older person was treated like a child; the other, especially when it included unpleasant, violent or anti-social behavior, was considered 'craziness' and was highly stigmatized. Very often sclerose was seen as the result of a person's life, stress and strain, or an especially heavy shock—something that is not specifically Brazilian.

In the first wave of research, conducted between 1997 and 2000, we asked family members whose relatives have been diagnosed with Alzheimer's disease two main questions: [1] how they understood the illness and [2] if they could tell us about the person. We found a surprising similarity when comparing these narratives to the way Harvard psychiatrist David Rothschild, and others, explained senility in the 1940s: a combination of hardship and certain personality traits that prevent an appropriate way of dealing with stress and strain.

Through interviews we had with the caregivers of older people with a dementia (who were members of the local Alzheimer's Association or attending the local specialized dementia center, the CDA), old and new models for interpreting signs and symptoms began to merge. Reminiscent of what Ian Hacking (1995) once called "semantic contagion," caregivers often referred to both the genetic and the life history of the family members they were caring for:

> Other relatives in my mother's family showed similar symptoms. But not only that. My mother became a widow at the age of 27. She lived for her children, grandchildren, then her mother got sick and she took care of her until she died. I've read a lot about it, I know about chromosome 21 and everything . . . But I think that dissatisfaction, sorrow, stress and the closing of an ever-smaller world may help towards self-destruction.
>
> *(Maria)*

> I believe it is a neurological genetic problem. That's what I've learned. Until now, these diseases are very little known by medicine . . . Maybe her isolation has started the disease. I am not sure. ( . . . ) She was independent, a hard worker. She bought and sold clothes and took care of the house, very well and alone, she was *caprichosa* (meticulous). She had a strong temperament, was dominating, revengeful like a good 'scorpion.' Her first son only left home when he got married with a woman as authoritative as his mother. She was extremely puritan, but did not separate from my father although she knew about his lovers. I think she liked him, but the bitterness destroyed her.
>
> *(Madalena)*

## *The CDA*

In the late 1990s, I was lucky to receive funding from the Brazilian Ministry of Education for the construction of a center combining multidisciplinary research, and a public outpatient unit for older people with mental health problems. This center, the CDA (Center for Alzheimer's Disease and other mental health problems for the elderly), rapidly became a reference point for the study and treatment of Alzheimer's disease. The CDA was part of the Federal University of Rio de Janeiro, and the director at the time insisted that the word 'Alzheimer's' became part of the center's name because "it was the disease of the moment" and, therefore, attracted funding and public attention. The opening of the center—the only specialized public institution in Rio de Janeiro at that time—became the subject of newspaper articles and TV talk shows, and was soon unable to accept all the seniors who needed psychiatric, neurological, and other kinds of psychosocial interventions. When the Brazilian *Reader's Digest* published an article about the CDA—and it seems that this was read in even the most remote parts of the country—a huge number of letters were sent to the center from people describing their older relative's symptoms, asking whether this was that *doença de Alzheimer* they had read about, and inquiring about the treatment at the CDA. I became the director of the center (not the clinical director)—a rather peculiar and contested decision made by the director of the Institute of Psychiatry of which the CDA was part.

The CDA offered multidisciplinary treatments, including musical therapy, psychology (including psychoanalysis and counseling), physiotherapy, occupational therapy, nursing, and family therapy, and had two anthropologists on its team. However, the focus was on diagnostics and drug prescriptions made by the psychiatrists. Alternative interventions were recommended by the doctors after their initial evaluation (and after the selection of patients for their own clinical trials)—partly because not much could be done, but also because families needed help with their loved ones, who were often in an advanced stage of the disease.

Although doctors in the 1990s claimed to treat patients with dementias, and articles in the media transmitted a general optimism, doctors were only able to medicate some of the peripheral symptoms of dementia. Psychiatric medications were prescribed for sleeping problems, psychotic symptoms, and dysphoria (e.g. depression and anxiety). Concomitant diseases, especially hypertension and heart diseases, were evaluated and medicated, and vitamin B and E were sometimes added to the treatment regimen as they were thought to likely be neuroprotective. But since the central symptom, memory loss, could not be treated, doctors were generally pessimistic about their patients. As one doctor explained: "We can only make it smoother for them, the last period of time, diminish suffering, it's the family who should make them feel good." In this sense, the central treatment was given by other health professionals to augment the patient's quality of life, and made them 'feel good,' while clinical authority remained with the doctors who could do little for those for whom they were caring.

Another psychiatrist who had just started to work at the CDA remarked that she had to rethink everything she had learned up until that point, and in essence, become a general practitioner again. "I am treating more the body than the brain," she said, referring to the fact that little could be done to slow down the neuro-degeneration in dementia, but that she was caring for the many concurrent diseases present in the elderly.

By 2005, however, the picture had changed. All four common Alzheimer drugs on the international market were available in Brazil—Tacrine was not prescribed anymore—although the inequities of access continued. One nootropic, Exelon, could be obtained for free from the State Secretary of Health; however, this was an extremely complicated process that took several months, and one that was almost impossible for anyone who was illiterate. Doctors, however, became much more confident in their ability to cure Alzheimer's disease. For instance, one neurologist interviewed made the following observations:

> With the technological advances, magnetic resonance, for example, one can make a more truthful diagnostic, one can now give names to different pathologies and patients who before were stigmatized as suffering from sclerose. Now [these patients] have a real disease and even potentially, depending on the moment, have a treatment. With the appearance of the medications, Alzheimer's became a treatable disease and up to a certain point, administrable.

This neurologist, who works at a public hospital and also has a private practice, considers Alzheimer's to be *treatable*, although she later confirms in the interview that the current medications only slow down the decline. This perspective differs significantly from those found in the mid-1990s, when the diagnosis of AD automatically indicated impending death. Now that dementia was diagnosed much earlier, the neurologist, like several of her colleagues, insisted on the importance of a multidisciplinary team, which was an emphasis that did not previously exist; it became a central issue since it was a necessary corollary to drug treatment. The people in this phase of new alive-ness based on hoped-for and, in several cases, observed increased alertness of individuals taking the medications ("she even is watching her favorite soap-opera again") needed occupational therapy or music therapy, and their families needed to know how to maximize their loved one's capacities, thereby enhancing their quality of life and general wellbeing.

Not coincidentally, those last two concepts—quality of life and general wellbeing—were also promoted by the pharmaceutical industry. A well-known psychiatrist tells me that:

> The way of understanding Alzheimer's has changed in the last years. This is because before, the principal issue for the diagnosis was cognitive decline and people only looked at cognitive decline quantitatively. But now, pharmacology studies started to pay attention more and more to criteria which before were secondary, which have now become more important in discussing Alzheimer's, such as activities of daily living, behavioral syndromes, costs of diseases, caregiver's

health—all these are now considered criteria of the disease and of its severity. And, of course, medications with approved efficacy emerged in the last ten years.

This quote reiterates the descriptions of Alzheimer medication produced with the pharmaceutical industry's spin. The new generation of Alzheimer medications still target memory decline, as intended by Tacrine. However, these new drugs are also designed to address broader implications like quality of life, and activities of daily living, previously considered to be peripheral symptoms, but now central to the diagnosis and conceptualization of dementia (Leibing 2009a; 2009b). One explanation for this change is that through the growing public and medical awareness of Alzheimer's, people are now diagnosed much earlier, and are therefore much more fit than patients were ten years ago. Another reason is that the efficacy of these drugs in treating cognitive decline cannot be proven, and increasingly studies are showing that the justifications used to recommend Alzheimer medications stand on shaky foundations (Whitehouse and George 2008: 13). Compared to cognitive decline, quality of life is a fuzzy subject, but now medications can be shown to work, even if we often do not know what exactly quality of life stands for. Finally, there was increasing attention towards using behavioral and psychological symptoms in defining and treating dementia.

In describing early dementia treatment in Brazil, I want to highlight how the emergence of a diagnostic category has an impact on people; it is "making up people" (Hacking 1995). However, names can have different meanings to different people, even within the same society. Research undertaken with my colleague Daniel Groisman in a shantytown (*favela*) has shown that dementia and its synonyms had no major importance for the older people or health professionals working in that environment. This was astonishing given that people living in a favela—and poorer Brazilians in general—are now living longer, and therefore, are especially vulnerable if the current models of dementia are true. Most seniors in the favela we worked in suffered from diabetes and hypertension—conditions considered to be important risk factors for dementia today—but also with formal education levels, which are lower than in other communities. The few accounts we received from our interviewees about 'strange old people' described them as being crazy. Here, within the symptom pool of possible signs of dementia, the psychiatric symptoms were in the foreground, while traditionally a diagnosis of dementia highlights cognitive impairment as the core symptom. These different ways of interpreting dementia result in divergent ways of living and relating to affected individuals and their families. Diagnostic criteria are not innocent—experts defining, health professionals applying, journalists writing about, and anthropologists analyzing criterion should be aware of 'what's in a name.'

## Acknowledgment

I am grateful to the many individuals and institutions who made this research possible: the Federal University of Rio de Janeiro and its Institute of Psychiatry, the funding agency CAPES, the older women and the health professionals in a well-known *favela*, the late Joao Ferreira da Silva Filho, the wonderful team of the CDA, Daniel Groisman, Rosimere Santana, and Lilian Scheinkman.

## References

Hacking, I. 1995 *Rewriting the Soul: Multiple Personality and the Sciences of Memory*. Princeton, NJ: Princeton University Press.

Leibing, A. 2005 The old lady from Ipanema: Changing notions of old age in Brazil. *Journal of Aging Studies* 19(1):15–31.

——— 2009a From the periphery to the center: Treating noncognitive, especially behavioural and psychological, symptoms of dementia. In *Treating Dementia: Do We Have a Pill for It?* J. Ballenger, P.J. Whitehouse, C.G. Lyketsos, P.V. Rabins, and J.H.T. Karlawish, eds. Pp. 74–97. Baltimore, MD: The Johns Hopkins University Press.

——— 2009b Tense prescriptions? Alzheimer medications and the anthropology of uncertainty. *Transcultural Psychiatry* 46(1):180–206.

——— 2014 The earlier the better—Alzheimer's prevention, early detection, and the quest for pharmacological interventions. *Culture, Medicine, and Psychiatry* 38(2):217–236.

Whitehouse, P., and D. George 2008 *The Myth of Alzheimer's: What You Aren't Being Told About Today's Most Dreaded Diagnosis*. New York, NY: St. Martins Press.

## Towards the End of Life

Worldwide, the care of the very old, and the care of others whose life is coming to an end as a result of disease and regardless of age, continues to be met informally largely through family-based arrangements, because of lack of alternative modes of care, because of the social value and economic advantages of enabling elderly people to remain in their own homes, and because of the wish by people and their families that they see out their days at home. Caring for people at the end of their lives is physically and emotionally challenging, and the decisions that family members must make at different points in time can be deeply distressing but also, it should be stressed, comforting (Kaufman 2005).

Everywhere spouses and adult children, women especially, are most likely to provide continuous care. But while personal relationships of carer to care-recipient—a wife, mother, sister or daughter, husband, father or son, for example—strongly influence who takes on the tasks of primary care, the capacities of families and particular individuals to give care is variable and fluid. The responsibility to do so is therefore influenced by ideas related to age and other responsibilities, gender, time, affective relationships, patterns of residence and household composition, family-based values, and ideas of obligation and responsibility, respect and duty. The socioeconomic context in which families and external support services are positioned further determines who is able to take care, and with what, if any, financial and practical support from others.

A prelude to such decisions of care giving and place of care is the communication to people of diagnosis, treatment pathways, prognosis and end of life. The conventions here vary considerably not only in the arrangements that people make to provide care for the frail elderly, but also in caring for people of any age whose lives may be foreshortened by disease (Davis and Manderson 2014). Communication about cancer illustrates powerfully the challenges that health professionals face with patients and their families, for ethical reasons (the right to know) and because of requirements, increasingly worldwide, that people provide informed consent to various procedures such as surgery, radiation, chemotherapy, and resuscitation. Inherent in this, there is a growing commitment that patients and their kin understand the implications of different interventions. In hospitals everywhere, medical and surgical interventions may carry considerable risks and limited gains, depending on the general health or frailty of the patient, and patients are expected themselves to participate in decision making and to assess different treatment paths against the alternative of palliation.

The concept of the right to know, including whether one might live or die, is not a universal. In other cultural settings, families may be empowered to make these decisions for their kin, and in the event of end-of-life decisions, while senior health professionals may initiate disclosure, who receives the disclosed information and how they understand the information imparted varies considerably (Bennett 1999; Kaufert 1999). Each decision about disclosure requires that an individual patient, clinical staff, and/or family members consider the costs and benefits of knowing the outcome of the disease or particular health event, and knowing what an intervention might entail. Joe Kaufert (1999) has described how, among Aboriginal Canadians, family members may resist discussing poor prognosis, palliative care, and impending death with a patient; they see their responsibility to be both to make decisions related to care, and at the same time, to protect the emotional health of the ill person as the most appropriate strategy to ensure their possible recovery or gentle passing.

The communication of a diagnosis of any condition, therefore, can be fraught with cultural, moral, and ethical challenges—who to tell, how much information, and with what spin. People are often fearful of the end of life, and put off talking about their own death. Further, prognosis can never be certain, and no-one can predict the transition from life to death, nor the moment of death; there is always uncertainty involved in discussions about outcome. And while there may be no known cure to a disease, no clinician or ritual healer can advise their patient of the rate of degeneration, the length of life still to be lived, the quality of that life, or its definite end. In the face of uncertainty, perhaps particularly with the increased secularization of contemporary societies in the global North, pragmatics override the optimism of cure. Increasingly, people are encouraged to complete advanced care directives on their own behalf or in relation to a family member no longer able to articulate their views, so pre-empting decisions about medical and surgical interventions, high-risk surgery, intravenous or tube feeding, and life support (Kaufman 2005).

## When Death Comes

Anthropologists have long engaged in questions of how dying is managed as a social process; in the timing of death and how a person might be regarded as and known to be dead; how dead bodies are managed and disposed, and souls put to rest. Medical anthropology extends to the social lives of dead bodies and those who are presumed dead. We write 'presumed dead' because this is not straightforward, as Margaret Lock (2002) has explored in her research in the United States and Japan. In the United States, death is defined as brain death, the complete and irreversible loss of brain function. In Japan, death is defined as the irreversible cessation of circulatory and respiratory functions, but this latter definition is also key to lay understandings of death, wherein time of death is linked to a person's 'last breath' rather than to the end of brain activity. In some settings, clinicians take control of the process of dying, determining whether to proceed with life support, for instance; in other settings, families take control over the timing if not the process of dying, and even in a hospital where the technical authority is with the clinicians, they must participate in decisions to hasten or prolong this process (Chapple 2002; Stonington 2012). This is not an easy decision, and people may spend months by the side of someone who is kept alive through intravenous feeding and other kinds of life-support, in a liminal state, neither quite alive but not yet dead.

Ideas of the proper or ideal place to die are strongly held everywhere, and home remains an ideal and common place of death, as illustrated for Ghana, the United States, and Japan (Becker 2002; Long 2004; van der Geest 2004a). Among the Pitjantjatjara of Central Australia, where people die, and the company in whose presence they do so, is framed by a deep commitment to home or 'country,' and in making decisions about dying, kinship relations take precedence over medical services (Willis 1999). Maureen Kirk and Lenore Manderson, in their collaborative work in Australia with regard to breast cancer, noted that Indigenous women have often chosen to reject or cease treatment, or to leave urban-based medical units to return to 'country' (their homes), not only because of their wish to be in country at the time of their death, but also because their commitment to kin and community overrode any personal need for pain relief, treatment, and care (Kirk et al. 2000; Manderson, Kirk, and Hoban 2001). Stonington (2012) similarly emphasizes the social values associated with dying at home in northern Thailand. Here, people may consider it unethical to withdraw life support for someone in hospital, but this does not apply at home—where life support would usually be technically impossible—because Buddhist beliefs regarding the relationship of the place of death to the quality of rebirth take precedence. Provided the family has maintained the ethical and ritual obligations associated with Buddhist and other local values, Stonington argues, home is the optimal place to die; hospitals, in contrast, are places of emergency, death, and danger.

Choosing *when* to die is an issue for people who prefer self-elected death, when their ability to function socially or physically is impaired to an extent that they feel they no longer have a good quality of life. While in some countries people may author their own death, their right to do so is highly controversial in many cultural settings and nations. In Switzerland and the Netherlands, doctors can grant a patient's euthanasia wish, if the patient is confronted with lifelong suffering, and Robert Pool (2000) has described how in one hospital in the Netherlands, physicians facilitated end of life of lung cancer patients by administering extra doses of morphine. This form of dying is classified as palliative care. The death-hastening effects of morphine sulphate are well known among hospice workers worldwide; where pain control and respiratory suppression are intermingled, patients may retain some agency over the timing of their departure.

Arlie Hochschild's (1983) insights into emotion work are particularly relevant here. Many people providing such care are recent immigrants, and poorly paid; care recipients and families expect that their labor includes kindness and care as they go about the mundane tasks of intimate care. Death and dying draw deeply on the emotional resources of those who are bereaved: family members and friends, and on the resources of the palliative care staff and others involved in their care, whether a person dies in an institution or at home. Death creates a chasm in the everyday lives and emotional security of those most closely affected; grief can be overwhelming and shocking. Funerals are designed around the inevitability of death and the need to support survivors. Consider the pervasive practices of sitting with, viewing, or handling the corpse, the disposition and displays of grief by renting garments, wearing clothes of particular colors (most often black or white), veiling, shaving hair or leaving it unkempt, social withdrawal or exclusion, the employment of professional mourners, and wakes and other ritual gatherings. The care of a corpse, the timing and appropriate disposal of the remains, funerary rituals and formal aspects of mourning, are never haphazard, therefore; the adherence to such protocol is often regarded as critical to social harmony and the aversion of disaster to the community at large, and even in highly industrialized settings where this may not be perceived, the rituals of death are tightly scripted and valued. All acknowledge death's impact on those who survive, provide ways to manage grief, and—depending on local ontology—might ensure the repose of the soul of the departed or an auspicious rebirth.

Several ethnographic examples illustrate this. In his classic monograph, *The Religion of Java* (1960), Clifford Geertz describes the rituals that are conducted as soon as possible after death; these involve stripping, washing, wrapping, and burying the body according to Islamic precept. Immediately following the burial, the first of a series of *selametan* (feasts) is held; these selematan continue at intervals until 1,000 days after the death, allowing those who are bereaved a structure within which to manage their grief. The aspired demeanor of the bereaved, supported through these rituals, is *iklas*, a state of 'willing affectlessness' and detachment, intended to reinforce an understanding that any death —regardless of circumstance or the age of the person who has died—is the will of Allah. Excessive displays of suffering are condemned; the ideal is restraint and emotional control. Such managed grief is not unusual, as Marilyn Nations and Linda Rebhun (1988) argue in relation to the logic of women in their comportment at the funerals of infants and small children in northeast Brazil; since "angels with wet wings won't fly," their restraint is a way to ensure their safe passage. Their analysis contrasts with that of Nancy Scheper-Hughes (1992), who in her classic book *Death without Weeping* argues that maternal love for very young and vulnerable infants is a luxury that extremely poor women cannot afford, or at least must control, given the compromised health and poor odds of their survival.

Funerals come at a cost. Funeral societies, which collect and distribute savings to pay for funerals, are common in many societies, although these have been commercialized in contemporary industrialized societies through life insurance and funeral cover plans. The care of the corpse at a mortuary or funeral home, the announcement of death, funerary services, burial or cremation arrangements and conduct, condolences, the care of graveyards, cemeteries, and shrines, are increasingly professionalized everywhere, and in any setting, certain people have ritual roles related to the care of the dead. In many cases, especially in industrialized settings, people's only experience of these rituals, their administration and the negotiations that shape them, may be when a close family member dies. Karen Nakamura's documentary short film *A Japanese Funeral* (2010) is an invaluable resource; she provides an intimate view of how a Japanese Christian family cares for the body of their deceased brother, accompanies his body to the crematorium and, finally, how they package and distribute his incinerated remains.

Medical anthropology therefore extends to the social lives of dead bodies, to the work of undertakers, the role of the dead in mediating commercial relationships, and how people create and recreate rituals to manage the materiality of the corpse and the bereavement of the living. In the following case study, Ruth Toulson describes how Chinese Singaporeans manage death. Deeply historical practices influence ideas of living with restless souls and evil spirits, and malevolent ghosts, capricious ancestors, and the 'undead' pervade people's homes and public spaces. In contemporary Singapore, however, living with death has led to professional transitions, and coffin carriers and gravediggers have been replaced by funeral parlors and personal shoppers for the dead.

## 9.3 Caring for Corpses in Singapore

*Ruth E. Toulson*

Singapore, June 2012

In the back room mortuary of a funeral parlor, Lavender and I worked together to prepare a corpse. The body, that of a middle-aged man who jumped to his death from the 16th floor of a public housing block, lay on the metal embalming table, naked apart from a small cloth that covered his genitalia. Tubing ran from the artery at his thigh to the pressurized embalming machine on the side table. A further tube drained his blood, which pooled in a bucket on the floor. He bore the marks of a hasty autopsy: his chest cut in a t-shape from shoulder to shoulder, then from clavicle to groin, the skin, with its inch of yellow fat, rolled back. The postmortem cuts and the fact that his neck was broken left his body opened out and flaccid on the table. Only his spine stood high: a fleshy ridge of red.

In the background, FM976 radio station played the top 40 hits and occasionally Lavender sang along. She folded fabric to pad the pulped flesh and shattered bones of the man's head, glancing occasionally at the man's driver's license, which rested on the side table, trying to match the crushed half face with the photograph. "Look, there's too much damage to wire," she showed me. "If I can just get the shape right, I can cover all this with wax, but it's going to take a while. Want to get sushi when we're done?"

In a bucket underneath the table the viscera soaked in Dodge Company cavity solution. Jade, the general manager, pushed open the door to the mortuary. "The family are back," she said. "How much longer?" Lavender was irritated. "It's always such a rush here. In New Zealand, when the funeral director rushes you, it's because he wants to get home, because his dinner's on the table. Here, it's the dead who are in a hurry." "Why's that?" I asked. "Well, supposedly, if the body isn't in the coffin before dusk, the dead will become angry. They'll come back and haunt you." Lavender smiled: "I don't really believe in ancestors or ghosts. What I do is for the comfort of the family."

Perhaps six people die each day in Singapore, a tiny city-state. Not many, but approximately the number demographers would expect for a wealthy country with an aging population. Singapore is a multi-ethnic country with an ethnic Chinese majority who constitute approximately 74 percent of the population, with the remaining Singaporeans either Malay or Indian. While their lives are integrated, death is largely segregated, dealt with on ethnic lines. And most Chinese Singaporeans will end up in Sin Ming Drive, a dead-end street in an out-of-the-way industrial estate, each road dedicated to a different 'dirty' business: to laundry, to welding, to car repair, to death. Avenue Thirty-seven is lined with funeral companies: Hock Heng Undertaker, Charity Casket, Ang Yew Seng Funeral Services, World Casket, An Lok Funeral Services, Fairprice Funeral, One-Stop Buddhist, and Asia Casket. Each business has a small front office and one or two parlors, each parlor large enough for a casket and a circle of mourners, the parlor's front open onto the street. In that street, I worked alongside Lavender and Jade to examine transformations in how death is ritually marked and the dead are mourned.

While the depiction of Lavender at work, above, is an everyday scene in the Singaporean business of death, it is also one that would have been unthinkable a mere decade ago. Read the older literature on death in Chinese societies and one is confronted by a world in which those who work with the dead are regarded as dangerously polluted by the task that they do. Watson (1988), in an account of a Cantonese village in the 1970s, describes "those who come with the coffin" as "the ultimate form of human degradation" (126). Villagers will not speak to them, touch them, or eat with them. Such is the power of their pollution that the cups and bowls they use are broken and later buried (125). They are recognizable by the odor of garlic on their breath, as villagers believe that they keep garlic cloves in their mouths to disguise the smell of death that clings to them (125). Watson suggests that most corpse handlers are opium addicts without kin or other means of support, who lived together in squalor in the back rooms of coffin shops.

Certainly, Lavender and Jade don't fit the stereotype of those "who come with the coffin," and not only because both are female. Jade is Mrs. Ang's eldest daughter, just turned 30, dark hair to her waist, undeniably beautiful. She is a mortician who loves to roller-blade and salsa dance, a relative newcomer to the world of death. After her father died, with her mother worn down by the burden of the business, Jade decided to leave her job in finance and work at Ang Yew Seng full time. At first there was resistance to a woman running the firm, particularly a young woman with new ideas. Some of the older men refused to work for her, telling her: "I've eaten more salt than you've eaten rice."

Lavender is just 23 years old, her hair tied up in a ponytail, black eyeliner perfectly applied. She sleeps in her make-up, she tells me, so as to not go out bare-faced if a call comes in the middle of that night. Today, she wears lilac scrubs, a macabre Hello Kitty appliqué stitched on the chest pocket, a bow-adorned skull and cross-bone kitten with heart-shaped hollows for eyes. She tells me she is Hokkien—Singapore's most common Chinese dialect group, who trace their descent from Fujian and Guang Dong Provinces—but she speaks little Hokkien dialect, only enough to make her friends laugh, and is more comfortable if we speak in English than in Mandarin. Her English is tinged with a New Zealand accent; she had just returned from embalming school there, where she gained a four-year degree in mortuary science. She had known at high school that she wanted to do this; she tells me, "It's a way to operate on people without being a doctor, and here, well, wanting to be a doctor is just too cliché."

Despite her professional qualification, Lavender's introduction to the world of death came from a very different time. Her first knowledge of embalming was when, as a child, she caught a glimpse of the preparations for a funeral, in the open public area underneath the high-rise apartment block where she lived. Seeing the dead body, lying under a sheet, while the embalmer worked with an embalming machine fashioned from a bicycle pump, she'd asked her mother why a doctor's clinic had set up on the deck. Then embalmers worked in the open air, preserving bodies while families watched. Lavender recalls, at her grandfather's funeral, chasing away cats that had gathered to lap his blood as it pooled into the storm-drain.

Now, Singaporean corpses are treated in sterile mortuaries, with Australian and American embalming technologies, and licensed professionals trained overseas. But that is not to say the world of death is entirely transformed nor that Singapore is an easy place to work in the funeral business. While highly trained young men and women have now largely replaced illiterate coffin-carriers, they have largely failed in their attempts to rebrand the business as a service industry like any other. Their necessary work is still treated with disdain and they continue to face discrimination. Jade's mother, for example, had begged her to reconsider her choice of occupation, telling her that "no good Chinese boy will ever love you if you work with the dead." No matter how many times she washed, Lavender's parents asked that she did not touch her infant sibling with her "filthy" hands. She often remarked that Singapore was no place for a highly trained professional with a four-year degree in mortuary science. She wanted to go back to New Zealand; she had a boyfriend there. They had fallen in love at mortuary school, exchanged glances across the corpses, their hands first touching as he passed her a trocar, the implement used to pierce the organs so that the powerful cavity solution can solidify the stomach contents, and guided her to pierce it through the abdomen to pump embalming fluid into the body's cavities. They dreamed of opening a funeral parlor together one day.

While Lavender was adamant that her work was for the benefit of the family, she had not entirely discarded notions of symbolic pollution. When we discussed her future plans, she shared her worries about becoming a mother. "Imagine," she said, "can you really picture a bump in the mortuary?" Pregnant women avoid the dead at all cost, in case proximity to the corpse causes miscarriage. The dead are a danger to the unborn as they share a state of vulnerable liminality. Following death, in the days before burial or cremation, the soul is pulled apart from the body. In pregnancy, over time, the soul becomes attached to the body of the fetus. The corpse may envy the fetus as it gains what the corpse loses. "What would you do?" I asked. "Tie a red ribbon around my middle, I suppose," said Lavender. The red thread is a common element of funeral ritual. It acts as form of protection, channeling pollution away from the body of the wearer.

If the body is symbolically polluted, one might wonder why embalm at all? The vast majority of Singaporean Chinese bodies are embalmed. This is in contrast to funeral trends in the US, for example, where embalming is falling out of fashion with the decline of open-casket funerals and wakes held in the home, increasing concern regarding the long-term impact of formaldehyde-based embalming solutions on the environment, and rising demand for funerals that are inexpensive and 'green.' Even more puzzling is the fact that the vast majority of Singaporean Chinese bodies are cremated, generally within a week of death, so embalming is not necessary for preservation. The choice to embalm is often motivated by the decision to have an open casket for much of the three- to five-day funeral. Funeral directors stress the importance of embalming in creating a 'memory picture,' of seeing the dead for one last time but now in peaceful repose. They suggest that, particularly in the case of an unexpected violent death, if family members do not see the body, their imaginings of it are often far worse than the reality. Embalming solution, because it plumps skin and lifts out bruises, can hide the signs of a long illness. Reconstructive embalming, where body parts are rebuilt, can erase injuries, hiding the traumatic nature of a death. This is particularly important in the case of death by suicide, which is considered shameful. A skillful embalmer may be able to hide the cause of the death and protect the family's reputation. Practically too, if bodies are not embalmed, by the third day, the body becomes less pleasant to view; the features begin to collapse inwards and the body may purge, the decaying stomach contents bubbling from the nose or mouth.

Family members often provide alternative narratives to explain their choice to embalm. An open casket is necessary so that the dead can take part in the funeral. The dead are not considered entirely dead, but exist in a liminal state, where the soul, often in a state of distress, hovers near the body until burial or cremation. Funeral rituals are not for the cathartic release of grief or to bring emotional closure, but for the good of the dead themselves. By ritual actions, the living transform a corpse into an ancestor. If they fail to perform rituals correctly, the dead risks becoming a wandering hungry ghost. Further, the purchase of the services of a skilled, overseas-trained embalmer is seen as a sign of the family's cosmopolitan tastes, a further element of conspicuous consumption. Through expenditure, families make visible the degree to which they loved the dead. Further, relatives often described elements of the encoffining process—washing the body, applying make-up, dressing in clean clothes—as final acts of love and care. The family members themselves once performed these acts. Now, funeral professionals take on this work but they are regarded as polluted by it.

## *Reference*

Watson, James L. 1988 Funeral specialists in Cantonese society: Pollution, performance, and social hierarchy. In *Death Ritual in Late Imperial and Early Modern China*. J.L. Watson and E. Rawski, eds. Pp. 109–134. Cambridge, UK: Cambridge University Press.

---

★ ★ ★ ★

We have written of the pain and grief associated with ailment and infirmity, frailty and loss of independence, dying and death. Elaine Scarry (1985), in *The Body in Pain*, wrote of the inexpressibility of physical pain: pain dominates the perceptivity of the sufferer, yet it is beyond others' apprehension. Her insistence of pain beyond language applies equally to emotional pain. We struggle to find proximate experiences by which to empathize and so minister care to others. We do not know, until we too are in that place, what it is like to lose control over bodily functions, or for confusion to reign—for memories, vocabulary, volition, and logic to recede and muddle. We capture poorly the anguish of the decisions that must be made about someone with a disease for which there are no more interventions, or of caring for someone who can no longer dress or feed themselves; we have few accounts of how these play out in families and communities where resources are few and the options especially limited. Without further accounts, we will continue to struggle to advise on and provide care and support for people at the end of life, and for those professionals and families who are part of this process.

## Note

1. "French retirement home throws out 94-year-old woman on winter weekend," http://www.english.rfi.fr/economy/20130107-french-retirement-home-throws-out-94-year-old-woman-winter-weekend; "The Babayagas' house, a feminist alternative to old people's homes, opens in Paris," http://www.english.rfi.fr/france/20130305-babayagas-house.

## References


Becker, Gay 2002 Dying away from home: Quandaries of migration for elders in two ethnic groups. Journals of Gerontology. Series B. *Psychological Sciences and Social Sciences* 57(2):S79–S95.

Bennett, Elizabeth 1999 Soft truth: Ethics and cancer in Northeast Thailand. *Anthropology and Medicine* 6(3):395–404.

Chapple, Helen S. 2002 Could she be dying? Dis-orders of reality around death in an American hospital. *Anthropology and Humanism* 27(2):165–184.

Cohen, Lawrence 1994 Old age: Cultural and critical perspectives. *Annual Review of Anthropology* 23:137–158.

——— 1998 *No Aging in India: Alzheimer's, The Bad Family, and Other Modern Things*. Berkeley, CA: University of California Press.

Conrad, Peter, and Meredith R. Bergey 2014 The impending globalization of ADHD: Notes on the expansion and growth of a medicalized disorder. *Social Science & Medicine* 122:31–43.

Counts, Dorothy Ayers, and David R. Counts, eds. 1985 *Aging and Its Transformations*. Lanham, MD: University Press of America.

Davis, Mark, and Lenore Manderson, eds. 2014 *Disclosure in Health and Illness*. London: Routledge.

Geertz, Clifford 1960 *The Religion of Java*. Chicago, IL: The University of Chicago Press.

Glascock, Anthony P., and Susan L. Feinman 1981 Social asset or social burden: Treatment of the aged in non-industrial societies. In *Dimensions: Aging, Culture and Health*. Christine L. Fry, ed. Pp. 13–31. New York: Praeger.

Hendriks, Ruud 2012 Tackling indifference: Clowning, dementia, and the articulation of a sensitive body. *Medical Anthropology* 31(6):459–476.

Hochschild, Arlie 1983 *The Managed Heart*. Berkeley and Los Angeles, CA: University of California Press.

Kaufert, Joseph M. 1999 Cultural mediation in cancer diagnosis and end of life decision making: The experience of Aboriginal patients in Canada. *Anthropology and Medicine* 6(6):405–421.

Kaufman, Sharon R. 2005 *And a Time to Die: How American Hospitals Shape the End of Life*. Chicago, IL: University of Chicago Press.

——— 2010 Regarding the rise in autism: Vaccine safety doubt, conditions of inquiry, and the shape of freedom. *Ethos* 38(1):8–32.

Kirk, Maureen, Celia McMichael, Helen Potts, Elizabeth Hoban, Deborah C. Hill, and Lenore Manderson 2000 *Breast Cancer: Screening, Diagnosis, Treatment and Care for Aboriginal and Torres Strait Islander Women in Queensland. Final Report*. Brisbane, QLD: Queensland Health.

Lock, Margaret 2002 *Twice Dead: Organ Transplants and the Reinvention of Death*. Berkeley and Los Angeles, CA: University of California Press.

——— 2013 *The Alzheimer Conundrum: Entanglements of Dementia and Aging*. Princeton, NJ: Princeton University Press.

Long, S.O. 2004 Cultural scripts for a good death in Japan and the United States: Similarities and differences. *Social Science & Medicine* 58(5):913–928.

Lynch, Caitrin, and Jason Danely, eds. 2013 *Transitions and Transformations: Cultural Perspectives on Aging and the Life Course*. New York: Berghahn.

Manderson, Lenore 2011 *Surface Tensions: Surgery, Bodily Boundaries, and the Social Self*. Walnut Creek, CA: Left Coast Press.

Manderson, Lenore, Maureen Kirk, and Elizabeth Hoban 2001 Walking the talk: Research partnerships in women's business. In *Geographies of Women's Health*. Isabel Dyck, Nancy D. Lewis, and Sara McLafferty, eds. Pp. 177–194. New York and London: Routledge.

Nakamura, Karen 2010 *A Japanese Funeral*. New Haven, CT: A Manic Films Production.

Nations, Marilyn K., and Linda A. Rebhun 1988 Angels with wet wings won't fly: Maternal sentiment in Brazil and the image of neglect. *Culture, Medicine and Psychiatry* 12(2):141–200.

Pool, Robert 2000 *Negotiating a Good Death: Euthanasia in the Netherlands*. New York: Haworth Press.

Scarry, Elaine 1985 *The Body in Pain: The Making and Unmaking of the World*. New York: Oxford University Press.

Scheper-Hughes, Nancy 1992 *Death without Weeping: The Violence of Everyday Life in Brazil*. Berkeley, CA: University of California Press.

Sokolovsky, Jay, ed. 1997 *The Cultural Context of Aging: Worldwide Perspectives*. Westport, CT: Bergin & Garvey.

Stonington, Scott D. 2012 On ethical locations: The good death in Thailand, where ethics sit in places. *Social Science & Medicine* 75(5):836–844.

Strange, Heather, and Michele Teitelbaum, eds. 1987 *Aging and Cultural Diversity*. South Hadley, MA: Bergin & Garvey Publishers.

Turnbull, Colin 1972 *The Mountain People*. London: Jonathan Cape.

van der Geest, Sjaak 1997 Money and respect: The changing value of old age in rural Ghana. *Africa* 67(4):534–559.

——— 2004a Dying peacefully: Considering good death and bad death in KwahuTafo, Ghana. *Social Science & Medicine* 58(5):899–911.

——— 2004b "They don't come to listen": The experience of loneliness among older people in Kwahu, Ghana. *Journal of Cross-cultural Gerontology* 19(2):77–96.

——— 2008 Resilience and the whims of reciprocity in old age: An example from Ghana. *Medische Antropologie* 20(2):297–311.

Williams, Alun 2003 *Ageing and Poverty in Africa: Ugandan Livelihoods in a Time of HIV/AIDS*. Farnham, UK: Ashgate.

Willis, Jon 1999 Dying in country: Implications of culture in the delivery of palliative care in Indigenous Australian communities. *Anthropology and Medicine* 6(3):423–435.

Buying Medicine in the *Mercado Negro*, 2007. La Paz, Bolivia.
© 2007, Jerome Crowder. Printed with permission. Original in color.

## *About the photograph*

*Over-the-counter medications (OTCs) are commonly sold in* tiendas *(corner stores/*bodegas*) across El Alto, the city that spreads out above La Paz, on Altiplano, Bolivia. In these small shops, residents purchase medicines by the pill, not the bottle. At times, these OTCs are mixed with* remedios caseros *(home remedies) to improve their potency. Shop owners explained that they purchase the meds in the* mercado negro *(black market) down in La Paz. When a neighbor told me he was going to restock his supply, I asked to accompany him to see for myself. This photo was made during that visit: stall after stall of medicines lining the street, at which one can purchase common items like paracetamol, Vicks Formula 44 and Alka Seltzer, made in Brazil, Perú or Chile.*

—Jerome Crowder

# 10
# Marketing Medicine

*Anita Hardon, Lenore Manderson and Elizabeth Cartwright*

Therapeutic remedies have long traveled across the world; merchants visiting the ports of Southeast Asia from at least the fifteenth century sold Chinese herbal medicines and herbal remedies from Egypt were used in medieval Europe. In this chapter, we focus on the marketing of modern medicines. We present concerns about medicalization that emerged in the 1970s, and offer critical analyses of the more recent global marketing of lifestyle drugs. Social science scholars who study these processes have, as a common point of departure, a strong theoretical sense of globalization. Through the global flow of modern pharmaceuticals, and the communication and advertising campaigns promoting their use, people worldwide are prescribed medicines to enhance their everyday lives and to treat routine variations in mood, such as feeling sad and irritable. But just how homogenous is this commodification of wellbeing? Do people around the world turn to the same pharmaceuticals, and do they do so in the same way? We outline how anthropologists who study local medicine markets reveal diverse forms of globalization. Medicines divert from their intended paths, and people use them creatively for their own perceived needs.

The modern pharmaceutical industry can be traced back to nineteenth-century chemists who experimented with compounds such as morphine and quinine (often on their own bodies), and discovered compounds that they could use to treat common disorders (Bensaude-Vincent and Stengers 1996). It became a major industrial sector after the Second World War when antibiotics were discovered, along with other highly effective compounds such as corticosteroids and antihistamines, resulting in high profits compared to other industrial sectors. Global pharmaceutical spending reached almost US$500 billion in 2003, with the United States of America and Canada accounting for almost half of it (Angell 2004).

Since the 1970s, social science analysts and activists have pointed to the medicalization of everyday life that has resulted from the global distribution and marketing of modern pharmaceuticals, and to the influx of modern pharmaceuticals into local markets all over the world (Healy 2012; Illich 1975). Four decades ago, Ivan Illich (1975) warned that normal processes like sleeping, aging, and dying were increasingly being defined as medical conditions in need of intervention. This process has made people more and more dependent on the pharmaceutical industry for managing their everyday lives, leaving them less able to deal with 'natural' processes. Feminists, such as those working in the Boston Women's Health Book Collective (1988) and the Women's Health Action Foundation of Amsterdam, were concerned that natural reproductive processes such as menstruation and pregnancy had also come to be seen as medical problems requiring medical intervention (Wolffers et al. 1989). Others were concerned that pharmaceutical fixes had become solutions for the ills of marginality (Ecks 2005; Nichter and Vuckovic 1994; Rozemberg and Manderson 1998).

## Questioning Need

Concerns about the uncontrolled distribution of large volumes of modern medicines in local markets and the aggressive marketing of these commodities (and their related high costs) led consumer activists and southern countries such as Sri Lanka and Mozambique, at the time lobbying for a new international economic order, to propose policies to limit the number of pharmaceuticals on the market to those for which there was a clear medical need and to regulate advertising (as decades later was done for tobacco). In 1975, the then Director General of the World Health Organization (WHO), Halfdan T. Mahler, responded to these concerns, announcing a shift in direction of the agency to ensure that essential drugs were available at a reasonable price worldwide (Greene 2014). In 1977 the WHO published the first essential drugs list, which included 250 drugs that had been proven to be safe, effective, and affordable, and for which there was a clear medical need (Laing et al. 2003). The medicines on the essential drugs list were generics, not brand name drugs.

In a formal letter to the WHO, Michael Peretz, the vice president of the International Federation of Pharmaceutical Manufacturers Associations (IFPMA), outlined reservations about the policy, arguing that by promoting generic drugs over brand name drugs, the WHO was exceeding its technical capacity to monitor drug standards. He argued that developing countries cannot assure drug quality, and that therefore medical professionals and consumers had to be able to rely on the reputation of companies that produce well-known brands (Greene 2014). Despite this opposition, the WHO went on to develop guidelines for countries to implement their own national drug policies based on the original essential drugs concept. Such policies would enable the public sector to provide much needed and affordable essential drugs, and would curtail drug advertising in the private sector. The underlying logic was that medicines were a health commodity, not a commercial good (Greene 2011; Kanji and Hardon 1992).

During the 1984 Executive Board meeting of the WHO, the IFPMA was confronted with a progress report of the WHO's Drug Action Programme (DAP), which advocated for the adoption of essential drugs in developing countries, and referred to the possibility of developing a pharmaceutical marketing code similar to the one developed for infant formula in the early 1980s. The US representative, Neil Boyer, asked the WHO not to embark on monitoring a private sector organization, pointing out that the IFPMA had its own marketing code (Hardon 1992).

## The Battle for Generics

In the late 1980s, the Philippines was one of the countries that intended to rationalize its drug market, containing over 12,000 pharmaceutical preparations, along the lines recommended by the WHO. The Ministry of Health adopted a list of 350 essential drugs, and that list was used as the basis for drug procurement by government hospitals and health centers. The parliament adopted a generics law, which stipulated that all drugs in the country be labeled with their generic name, with the brand name one font-size smaller than the generic name to help consumers determine the products' contents and compare prices (Hardon 1990).

The Philippines government was confronted with fierce opposition from doctors, multinational drug companies, and representatives of the United States government for trying to adopt the generics law and hence promoting generics over brand name drugs. Doctors defended their freedom of prescription, arguing that they knew best what drugs were most appropriate for their patients and that brands ensure the quality of products. Drug companies which belonged to the Drug Association of the Philippines attacked the plans of the Ministry of Health, claiming that they arose from carefully orchestrated moves by 'non-professional and self-appointed groups,' who had misinterpreted the WHO policy on essential drugs; they claimed too that the generics policy would lead to drug shortages. Two US senators, Richard Lugar and Alan Cranston, wrote a letter to President Aquino warning against "strict private controls," while emphasizing that "if decisions are made which jeopardize or penalize US firms presently doing business in the Philippines, the task of stimulating new US investments may become more difficult" (Hardon 1990). Despite these threats, the Philippines government implemented its generics policy. The government gave some leeway to companies by allowing them to print the brand name on the medicines' packages, as long as the brand was printed in a one digit smaller font than the generic name. Most companies pragmatically responded by printing the brand names in bold.

★ ★ ★ ★

Thirty years later, the medicine market has changed dramatically. Most major pharmaceutical companies produce generics along with their brand name drugs. In the United States lists of generic drugs have been adopted by insurance companies and managed care programs (Greene 2014). Even so, the promotion of brand name drugs in the US is among the most aggressive in the world, and it is the only country besides New Zealand that allows direct-to-consumer advertising of prescription drugs (Moynihan and Cassels 2005). Generic firms now dominate the world market, with India's pharmaceutical industry, including companies such as CIPLA and Ranbaxy, evolved into world leaders in the production of generic drugs. In response, northern pharmaceutical multinationals, which vehemently opposed generic drugs, are now also expanding their production of generics.

Joe Dumit (2012), in *Drugs for Life*, describes how American lives have been medicalized beyond anything that Ivan Illich could have imagined in the 1970s. Pharmaceuticals have become "integral to daily life in America: they help those on diets to have a Christmas dinner; they help schools fill up with attentive kids; they are part of our identities as well as our lives" (Dumit 2012: 181). He argues that 'lifestyle' medicines have become popular in the United States due to heavy investments by the pharmaceutical industry, not only in research and development, but also increasingly in communication campaigns that encourage people to see themselves as inherently sick and in need of medicines on a daily basis.

Dumit (2012) and Moynihan and Cassels (2005) outline how companies increase their profits by designing and defining disorders and dysfunctions for which their products can be used. This trend has been referred to as 'disease mongering,' though medical directors of pharmaceutical companies prefer to describe it as 'awareness raising.' Thus the new condition—'pre-menstrual dysphoric disorder'—is created as a label for women who feel irritable the day before their period, for which antidepressants are marketed as an effective treatment. Disease mongering also includes efforts to increase our attentiveness regarding very early stages of possible disease development. Indeed 'pre-treatment'—of conditions such as 'prediabetes,' peri-menopause, and precancerous lesions—is becoming more and more common. The question remains, of course, whether or not these conditions would progress to the diseases they might become, or if unnecessary, expensive, and possibly dangerous treatments are being prescribed.

Disease mongering is not only an American phenomenon, although direct-to-consumer advertising of prescription drugs facilitates such marketing. Stefan Ecks (2005; 2014) describes this trend in Mumbai, where an Indian pharmaceutical company promotes awareness of mental distress by distributing leaflets that contain a self-test. Ecks reports: "On the front, there is a big 'smiley' face encircled by the words 'Defeat Depression—Spread Happiness,' and the text underlines that the test is 'FREE!' . . . On the back, a list of ten items for self-diagnosis of depression is presented" (2014: 151). Those who experience five or more symptoms are advised to see a doctor. In presenting drugs as a solution for everyday sadness and feelings of worthlessness, companies project their drugs as safe and effective cures.

Science and technology scholars question the alleged safety of medicines, arguing that research designs are biased towards documenting the beneficial effects of the products. Science and technology studies show that the 'real' effects of chemicals as studied by biomedical scientists are not fixed but rather shaped by research methodologies and funding interests. These scholars have given us insights into how safety and efficacy claims are arrived at by pharmaceutical companies, showing how the intertwining of academic and commercial interests leads to an emphasis on the desirable effects of pharmaceuticals and to the downplaying of their risks (Applbaum 2009; Healy 2009; Sismondo 2008).

Controversies emerge when users are confronted with unexpected adverse effects. In *Medicines Out of Control*, Medawar and Hardon (2004) detail how the 2003 BBC Panorama documentary *Seroxat: Emails From the Edge* publicized some of the unadvertised effects of antidepressants on youths, including extreme anxiety, self-harm, and cases of suicide. Although the manufacturer initially downplayed this evidence, subsequent scrutiny of the original clinical trial data suggested that the drugs were indeed associated with the increased risk of self-harm. The UK regulators generally were not fast to act, as they were worried about starting legal fights with companies who can afford good lawyers (Herxheimer and Mintzes 2004; Medawar and Hardon 2004).

## Consumer Agency

Anthropological case studies of local medicine markets suggest that while multinational companies commodify health in similar ways around the globe, people appropriate pharmaceuticals for their own means and have their own ideas about safety. These studies seem to point to a form of 'weak' globalization (Foster 2008; Friedman 1995) which leaves space for consumer agency. The package insert, which provides information on how and why to use the drugs, the doctor's prescription, and the direct-to-consumer advertising do not necessarily determine what is done with them. As Susan Reynolds Whyte and colleagues (2002) pointed out, medicines are reattributed site-specific values in terms of their therapeutic potential or economic worth as they move from manufacturing sites to pharmacies, and from clinics to patients.

Following the flows of medicines to remote corners of the earth, anthropologists have described how modern pharmaceuticals are embedded in local cultures of health, where they acquire meanings that often diverge from the indications assigned by pharmaceutical companies to their products. Senah (1997) described how people in Ghana consider heat to be the main cause of measles, constipation, and children's stomachaches. To treat measles, people in Senah's study take Septrin (co-trimoxazole) syrup, a multivitamin syrup, calamine lotion, *akpeteshie* (local gin), and an herbal concoction given as an enema to 'flush out' the heat. Etkin (1992) described how Hausa considered bitter medicines to be dangerous for pregnant women, and bitter medicines such as chloroquine were used as abortifacients. More recently, Hardon and Ihsan (2014) have described how potent painkillers containing carisprodol, a habit-forming muscle relaxant, are used by sex workers in Makassar, Indonesia, to feel confident when they go out to seek customers. They say the medication makes them feel euphoric, helping them overcome the shame that they feel when doing sex work. Interestingly, the same off-label practice was observed in Hollywood 50 years ago when the painkillers were mixed with martinis as party drugs. Thus, in each local market, diverse groups of users attach meanings to products, at times converging with the purpose assigned to them by companies, and at other times creating completely new indications. These are then shared in local health cultures; all sex workers in Makassar, for example, know that Somadril makes you feel confident, and Hausa women know that chloroquine 'works' as an abortifacient, even though this drug is considered to be safe to take during pregnancy.

## Diversion from Intended Paths

Medicines are subject to strict regulations regarding their manufacture and sale. International treaties grant companies 20 years of patent protection for medicines, a period in which competitor companies are not allowed to make generic versions. After the patent expires, generic copies can be sold, but these have to be proven to have 'bio-equivalence'—that is, they must produce the same biological effects as the branded compounds. Moreover, medicines are generally categorized into several classes, which determine how they can be sold. Relatively safe medicines are assigned a status as over-the-counter drugs that can be sold in accredited outlets, while drugs that require clinical diagnosis are typically only sold to consumers who have a prescription from a doctor (Davis and Abraham 2013). However, ethnographic studies in settings where governments lack the implementation power to control local drug markets have shown how medicines such as Somadril and chloroquine are frequently diverted from their intended paths through pharmacies, informal stores, and markets where they are sold without a prescription (Hardon 1990).

In a study of the pharmacy sector in Mumbai, for example, Kamat and Nichter (1998) point to a proliferation of pharmacies. Out of the 75 pharmacies that they surveyed, 32 percent were less than two years old and 52 percent were less than five years old. High-density populations with a high incidence of common diseases guaranteed these pharmacies high sales figures. The authors further noted that in nearly all of these pharmacies, pharmacy attendants managed the front counter. The qualified pharmacists were either absent or busy managing the cash register. Pharmacy attendants reported high levels of confidence in advising patients, and stated that they rarely consulted the pharmacist. When the researchers asked them about the drugs' mechanism of action and side effects, they tended to be very vague. Kamat and Nichter (1998) noted that the sale of medicines in the pharmacies was influenced by aggressive marketing strategies. By offering attractive incentive schemes, companies encouraged the use of particular drugs. They noted too that most clients observed at the 75 pharmacies bought medicines without prescriptions. When they did present prescriptions, the researchers reported that they were often more than two months old; Indian pharmacists tended to return prescriptions to clients after selling the drugs, and patients were able to reuse them.

Hardon (1990; 1994) showed how medicines also flow to informal community groceries (called *sari-sari* stores in the Philippines), whose owners buy drugs at pharmacies and resell them on a *tingi* (small quantity) basis to the urban poor. Hardon conducted fieldwork on the ways in which mothers treat their children's common coughs, colds, fevers, and diarrhea in two urban poor communities in Marikina, a town on the outskirts of Manila in the Philippines. Hardon worked for a community-based health care program, which called on people to use local resources (rather than American drugs) such as ginger and guava for common coughs, colds, and diarrhea. Contrary to the advice of the program's community health workers, however, she found that when they were confronted with childhood illnesses, mothers would buy pills—painkillers, cough medications, and antidiarrheal drugs—from small neighborhood sari-sari stores, without consulting doctors, pharmacists, or their attendants. Such stores were not officially allowed to sell pharmaceuticals, but they do so, even today. Customers value sari-sari stores because they allow people who lack income to buy just a little bit at a time. The products they sell are over-the-counter medicines that are heavily advertised on local radio stations.

## Manipulating Pharmacies

In the first case study below, Emilia Sanabria describes the market dynamics in Brazil, where the retail pharmacy industry is very profitable. She points to a rise of US-style pharmacy chains, where pharmacists are mainly involved in selling standardized health commodities, while at the same time Brazilian-owned 'manipulation pharmacies' tailor medicines to their client's needs. Pharmacists working in manipulation pharmacies mix modern pharmaceuticals with herbal products to 'soften' their effects; they provide 'diet bombs,' for example, that contain amphetamines, tranquillizers, and a few herbs like chamomile and passion flower. The powdered pharmaceutical compounds that the pharmacists use are imported from India and China.

---

## 10.1 The Compounding Pharmacy in Brazil

*Emilia Sanabria*

Manipulation or compounding pharmacies are common in Brazil. Every shopping center in the northeastern city of Salvador has a franchised branch of a manipulation pharmacy chain, such as *A Formula* or *Flora*, and homeopathic, herbalist or dermatologically orientated manipulation pharmacies are widespread throughout the city.[1] Manipulation pharmacies grew out of the apothecary tradition, wherein chemists would make up formulas at the back of the shop. This practice re-emerged in Brazil in the early 1990s, and in 2009, the Director of the Bahian Council of Pharmacists (BCP), Dra Ademarisa, estimated that the sector held 10 percent of the pharmaceutical market in Brazil. It continues to expand.[2]

There are four categories of drugs in Brazil: imported brand-name drugs, generic drugs, similar drugs (non-generic copies of reference drugs estimated to make up 65 percent of pharmaceutical consumption) and the compounded pharmaceuticals produced in *farmácias de manipulação*. Roughly 90 percent of drugs are bought out-of-pocket in Brazil, often without prescription. Yet the retail pharmacy industry in Brazil sustains growth, averaging 14 percent a year (McKinsey & Company 2011), a fact widely explained by the expansion of the middle class and increase in disposable income.[3] With population aging and the growth of chronic disease, the universal health care system has struggled to meet the soaring demand for drugs. Hence the explosion of pharmacies in Brazilian urban centers, and the highly competitive and diversified forms they take. Large drugstore chains dominate the market because of their capacity to lower drug costs. As competition intensifies, customer rapport is developed to give a competitive edge, with loyalty-card systems, home-delivery services, pharmaceutical attention (to deliver personalized drug interaction evaluations), and the careful distribution across the city of specifically designed 'store formats' according to neighborhood.

*Farmácias de manipulação* capitalize on this trend toward personalizing pharmaceutical care. With the rise of national pharmacy chains, the pharmacy profession has been reduced to a merely mercantile role. Most pharmacies operate without a trained pharmacist (although this is, formally, illegal) and few request prescriptions to deliver even prescription-only drugs. The BCP strives to "reclaim the profession" and "return pharmacies to pharmacists," as Dra Ademarisa explained it. Their vision of pharmaceutical practice blends economic and public health objectives while creating a commercial differential through the personalization of service. The distinction between 'personalizing' and 'massified' is marked by strong class undertones in Brazil (Sanabria 2010).

### *Tailor-made Medicines*

From the outside, manipulation pharmacies present as clinical environments with glass shelves displaying a selected line of beauty products, herbal substances and dietary supplements. Behind the counter, a window offers a view into a laboratory where technicians go about their business. At the counter, patrons present their prescription, which is entered into the system, generating a time for pick-up. The aesthetic of these pharmacies tends to be the opposite of the drugstore pharmacy, privileging uncluttered surfaces and natural iconography (leaves, flowers, clouds, etc.). A door to the side of the counter leads to a corridor that runs through to the semi-solids lab (where gels or creams are manipulated), liquids lab (for syrups, solutions or suspensions) and solids lab (for capsules). According to the stringent regulations for manipulation pharmacies, these must be clearly differentiated and sealed-off environments. Each laboratory usually includes a vent, a workbench and a series of cupboards in which the substances to be manipulated are stocked. Here vats of emblematically allopathic products such as cortisone or anti-inflammatory drugs are stocked alongside herbal and mineral substances.

Manipulated remedies are distinguished into two categories: *formulas oficinais* and *formulas magistrais*. The former are single products (e.g. magnesium), or products derived from Brazilian pharmacopoeia (e.g. Calendula *officinalis* tincture); the latter are *receitas* (recipes), that is, they are composed of various elements. *Formulas magistrais* mix pharmaceutical compounds, minerals, herbs and vitamins and can be produced into solid, semi-solid or liquid preparations, depending on the purpose and mode of administration.

A key selling point is that of personalizing medicines. The idea is that the standardized packaging of drugs makes little sense given that each individual is unique and will metabolize substances differently, as one pharmacist explained to me:

> Her inflammation [pointing to the attendant] won't be the same as yours. It arises in different conditions, as a result of different factors, and thus cannot be treated with a standard dosage. Remedies have to be adapted to the individual, in function of their size, weight, or the type of agent that causes the infection. . . . Personalizing treatments means that you can give that patient the exact quantity, and thus there is less waste, which substantially lowers costs.

This tailoring to the patient's individuality is pitched against the standardized practices of the pharmaceutical industry, which produces drugs *en masse* without differentiation. As Elizete, a manipulation pharmacist employed in a national chain, explained:

> The manipulation pharmacy attends to the *individuality* of the patient. Industrialized medicines only exist in 10mg or 50mg. But what if a patient needs 23mg? This is not viable for the industry. . . . In addition, by manipulating a cream or a capsule you can combine various treatments in harmony, and in accordance with the specific biotype of the patient. And on top of that, there is the added value of pharmaceutical care; as a pharmacist you can check drug interactions and advise both patients and doctor on correct use of different active principles.

In this sense, individualized becomes what is not standard. In the manipulation pharmacies to which I was granted access (not the chains), personalization seldom involved the modulation of doses. What was more common was the dispensing of an exact course of treatment, sometimes known as fractioning, whereby patients acquired the exact number of pills in the right dosage rather than making up a course of treatment from standardized boxes which often require buying in excess of actual need. Otherwise, patrons sourced homeopathic formulas specifically made up for them, found dermatological products, or found a preferred mode of administration (such as a soft capsule instead of a caplet). All these practices were referred to as 'personalization.'

As a corporate body, manipulation pharmacists highlight the difference between the traditional figure of the pharmacist as a craftsman of medicines and drugstore vendors who, they argue, merely retail standardized pharmaceutical drugs. But how does this personalization function, in practice?

## *Prescribing 'Manipulated' Formulas*

Dra Janaina is one of the few gynecologists in Salvador who does not automatically prescribe hormonal contraceptives to treat premenstrual tension (PMT). While she favors hormone-replacement therapies for most of her menopausal patients—and often prescribes low-dose topically applied testosterone gels produced in manipulation pharmacies to boost sexual desire during menopause—she feels that PMT symptoms should be treated holistically. Like many doctors, she works mornings in a public hospital and afternoons in her private practice. I met her in one of the teaching hospitals where I was conducting research and was struck by the fact that she recommended Pilates classes and manipulated *receitas* (prescriptions) to the low-income patients she treated there. The recommendation seemed incongruous given the absence of Pilates classes in the peripheral neighborhoods where these women lived. In the hot overcrowded waiting area outside, I asked a group of women about

Dra Janaina's receitas. One explained that it was *uma maravilha* (marvellous), but far too expensive for her to continue. Another commented that it was *muito personalizado* ('very personalized') and a third cut in that she already cured herself with herbs and had not come *here* for *that.*

Dra Janaina explained that many patients she sees in her private practice consult for PMT. Each patient has "her own particular kind of PMT," she explains. While some feel anger, irritation and headaches, others experience depression, swollen breasts and hyper-sensitivity. Rather than treating PMT with anovulatory contraceptive drugs that suppress menstruation, she prefers to treat the different symptoms directly, using compounded formulas adapted to her patient's 'profile.' She combines these with recommendations for dietary changes, nutritional supplements, and physical exercise or relaxation techniques. For patients who suffer from bloating, nausea or fluid retention, she prescribes manipulated formulas containing potassium and *passiflora alata.* For those with breast tenderness, tiredness and acne she uses formulas containing vitamin B and essential fatty acids. For those suffering from menstrual cramps she prescribes formulas containing non-steroidal anti-inflammatories and magnesium. She explains that she works in close partnership with a manipulation pharmacy and that they adjust dosages and combinations together in response to the clinical results. This functions on a trial-and-error basis, as doses are adapted at follow-up visits in response to a patient's perceived clinical response. *Personalizado* is a leitmotif in Dra Janaina's description of her practice. This filters down to her attention to individual dietary habits, working patterns, stress and familial situations.

The application form to become a franchise of the compounding pharmacy chain *Farmô* reveals the kind of market that manipulation pharmacies aim for. The applicant is asked to evaluate other compounding pharmacies, shopping centers, gyms, dermatologists, plastic surgeons, endocrinologists and rheumatology clinics in the vicinity. Dermatologists, endocrinologists and nutritionists are the main prescribers of manipulated formulas according to several informants' evaluations. Gynecologists, as we have seen, also often make use of compounded pharmacies. So do homeopaths, and many holistic practitioners who use phytotherapeutic preparations, Bach flower remedies or essential oils.

Reinaldo, who works for a regional manipulation pharmacy chain, explains that his company produces a catalogue of suggested *recietas,* delivered to doctors across the city according to their specialty. Each catalogue contains a list of products organized by therapeutic indication and a range of dosage options or combinations that can be adjusted according to the patient's profile. Other chains, Reinaldo noted, operate with representatives who visit specialists known to prescribe compounded formulas, although it is formally illegal for these pharmacies to market their products to doctors. Analise, the owner of *Biophitus,* a manipulation pharmacy specializing in homeopathic formulas, estimated that 10 percent of the price of the formula goes to the prescribing doctor, encouraging them to recommend particular chains of manipulation pharmacies—although they tend to be more expensive, and do not necessarily offer the best quality, she noted with concern. Analise also pointed out that doctors were not well trained to adapt dosages: "Sometimes the prescription that arrives is erroneous, and you have to call the doctor up and say: 'Listen, if I may, this dosage is above the norm for this patient, what would you think if we changed it to another such dosage?' "

I asked several informants to explain how the personalization of the dosage is achieved in the absence of data on the basis of which to arbitrage. Answers to this question were vague, with some pharmacists gesturing to the future promises of pharmacogenetics, or emphasizing racial differences or differences in dietary patterns among the Brazilian population. Personalization as currently practiced remains fairly limited, but it has a kind of promisory aura to it, and a strong discursive function in a context where the 'standardized' is quite negatively marked. Analise joked that a pharmacist must be "like an octopus, because it is a profession that requires several arms: one to attend to patients, another to manage stock, one to deal with the bureaucracy of pharmacovigilance, yet another to supervise staff, one to keep up to date on new compounds, and one to engage in diplomacy with doctors, *e ai vai* (and so on)."

## *Stabilizing Objects*

The practices of the manipulation pharmacy make evident a broader concern with drugs' modes of administration and bioavailability, or action in the body (on bioavailability, see Sanabria 2014). Manipulation pharmacies capitalize on the possibility of delivering the precise quantity of an active ingredient for each patient, and a choice in mode of administration. Tailor-made capsules may combine several treatments, reducing the number of pills a patient needs to ingest daily and reducing the waste associated with multiple boxes of pills.

Manipulation pharmacies give pharmaceuticals *form*, connected to the ultimate desegregation of these objects (Sanabria 2009). This is evident in the case of hormonal implants in Brazil. Worldwide, a single-rod implant containing 68mg etonogestrel and marketed under the name Implanon is the most common implant. In Brazil, however, Implanon is fairly rare. Exploiting the possibility of manipulating drugs in small-scale laboratories, some clinics are producing their own hormonal implants. Although hormonal implants are imagined as very high-tech treatments (a vision reinforced by advertisements in women's magazines), their manufacture is entirely manual: they are handcrafted biotechnologies.[4] The manipulation pharmacy receives the powdered hormonal compounds from India and China through a wholesaler. The *hormônio* comes in large gray plastic storage containers and is transferred into a mortar and pestle, under a vent, where it is 'fluidified,' making its insertion into the silicon tubing easier. The tubes are cut to size against a yellow school-like ruler, from large wheels of silicon SIL-Tec. With tremendous dexterity, the tubes are dipped by bunches of five into the mortar, held and flicked repeatedly until the tube is filled. An instrument is used to pack the powder and the tubes are weighed before the other extremity is sealed with silicon glue. Although lab technicians work under a vented workbench, use gloves and a mask, they report having 'altered cycles' and irregular bleeding patterns which they directly associate to their exposure to *hormônio*. Because the *substancias*, as the technicians refer to them, have different weights, the SIL-Tec tubing is cut to different lengths, calculated such that each implant contains 50mg of active substance, releasing 4–5 mcg per day. "Estradiol is the lightest, *não é* (isn't it)?" the senior lab technician explains, and the tubes have to be filled more tightly to obtain the correct dosage.

Regulations related to this are intended to guarantee the quality of the ensuing drug. According to several manipulation pharmacists I interviewed, this extensive regulation is also the outcome of BigPharma's lobbying activities. Both Brazilian and foreign pharmaceutical corporations are said to actively campaign within national regulatory instances to impede the spread of manipulation pharmacies which, in populating the already densely saturated Brazilian pharmaceutical landscape, are taking away an important share of their market. But beyond profit per se, what is at stake here is the relationship large corporations create between active principles and clearly bounded, carefully marketed products whose names ultimately come to stand for their pharmacological components.

## *Drugs Made Natural*

Although she is careful not to name particular brands, Analise explains that manipulation pharmacy chains source their materials in bulk and reduce costs by importing products which are not always of good standard. The specificity of independent pharmacies, she argues, is that they source their products very carefully, with attention to quality. She sources synthetic active pharmaceutical materials used in her pharmacy from wholesalers in the south of Brazil—themselves often imported from Asia—but explains that she travels widely throughout Brazil to source quality herbal products, tinctures, teas, supplements, minerals, vitamins, essential oils and homeopathic formulas.

Many pharmacies mix emblematically allopathic pharmaceutical compounds with herbal extracts. At Analise's pharmacy, solids and liquids labs were stocked high with all manner of different substances: herbal, mineral and synthetic chemicals, including anti-inflammatories, steroids, and various hormonal compounds. The mixing of both 'natural' and 'artificial' substances was often relayed to me as explaining the success of this pharmacy genre. But more than offering a green-wash or natural touch to what are essentially allopathic treatments, this can be read as signaling that mixing natural and allopathic or artificial substances does not pose the same kind of problem as it might elsewhere. An extreme example of this 'mixing' was described to me by one of the gynecologists I interviewed, who was about to attend the funeral of a woman who had died of heart failure after taking a 'manipulated' weight-loss formula produced by one of the larger chains. An endocrinologist had prescribed her fenopropolex, a laxative, anxiolytics and "a few herbs like Chamomile or Passiflora to make the whole thing look natural and harmless." Such *bombas* (diet 'bombs')—used to accelerate metabolism, one woman told me—have been associated with several deaths in Brazil and are a cause of public concern. They commonly combine amphetamine-type anorectic drugs and psychoactive substances such as benzodiazepines. One study recently showed that over 85 percent of prescriptions for anxiolytics or anorectic drugs were delivered in compounding pharmacies, with 83 percent delivered to women (Nappo et al. 2012).

Following four child deaths in 2003 attributed to the use of manipulated formulas containing clonidine, and several deaths in 2004 associated with formulas containing levothyroxine, the Brazilian pharmacovigilance institute, ANVISA, introduced a series of regulations, including RE 1621/3/10/03 which forbids the manipulation of drugs with low therapeutic indexes. This regulation was later revoked, despite strong pressures from pharmacologists and public health specialists to maintain it. However, following further deaths associated with manipulated formulas in the state of São Paulo in 2004, a public inquiry was held; in one case, ANVISA had found significant disparities in the concentration of active ingredients analyzed from a single container, that traceability was a problem, and that contaminants were common. Further regulations were therefore introduced to strengthen quality control in these pharmacies and to ensure the standardization and homogeneity of the active ingredients of drugs distributed through manipulation pharmacies. Manipulation pharmacists I met argued that the regulations would do little to deter extreme practices such as the retail of 'diet bombs' and other formulas favored by private practice endocrinologists. However, these regulations were impeding the development of small-scale, community pharmacies, which often derived their business from dispensing homeopathic or herbal remedies. ANVISA's position is that given the difficulty in quality control of drugs dispensed in manipulation pharmacies, these should merely play a 'complementary role' such as when there is no industrialized drug alternative. This complementarity is, of course, open to interpretation, and the rhetoric of personalization is used to overcome this fairly loose regulatory constraint.

## *Notes*

1. Research on sex hormones, pharmaceuticals and sexual and reproductive health was conducted in Salvador da Bahia, Brazil, between 2005 and 2012. This involved ethnographic work in a range of pharmacies catering to low-, middle- and high-income neighborhoods (including several manipulation pharmacies). I met pharmacy sector regulators and members of pharmacist professional organizations. I also interviewed the national marketing directors of four major pharmaceutical corporations in São Paulo and followed the work of their regional pharmaceutical representatives.
2. An internal document, dated 2005, sent to me by ANVISA and entitled "Technological Viability of Safe and Efficient Manipulated Medicines Production," claims that the number of manipulation pharmacies more than doubled between 1998 and 2002 (from 2,100 to 5,200) and that the number of pharmacists trained in 'manipulation' rose from 8,710 to 14,560 over the same period.
3. This growth in disposable income is based on a highly unsustainable credit system. Even in pharmacies, patients can acquire products *parcelado* (with split payments) on their credit cards.
4. For a discussion of hormonal implant use and plastic surgery in Brazil, see Edmonds and Sanabria (2014).

## *References*

Edmonds, A., and E. Sanabria 2014 Medical borderlands: Plastic surgery, sex hormones and the remaking of the natural in Brazil. *Anthropology and Medicine* 21(2):202–216.

McKinsey & Company 2011 Perspectives on Healthcare in Latin America. Accessed December 1, 2014 from http://www.mckinsey.com/~/media/McKinsey/dotcom/client_service/Public%20Sector/PDFS/Perspectives_on_Healthcare_in_Latin_America.ashx.

Nappo, S., E.A.C. Aparecida, L.F.S. Moreira, and M.D. Araújo 2012 Prescription of anorectic and benzodiazepine drugs through notification B prescriptions in Natal, Rio Grande do Norte, Brazil. *Brazilian Journal of Pharmaceutical Studies* 46(2):297–303.

Sanabria, E. 2009 Le médicament, un objet évanescent: Essai sur la fabrication et la consommation des substances pharmaceutiques. *Techniques & Culture* 52–53:168–189.

——— 2010 From *sub-* to *super*-citizenship: Sex hormones and the body politic in Brazil. *Ethnos* 75(4):377–401.

——— 2014 'The same thing in a different box': Similarity and difference in pharmaceutical sex hormone consumption and marketing. *Medical Anthropology Quarterly*. DOI: 10.1111/maq.12123.

---

Sanabria's account suggests consumers' interest in individually tailored drugs that mix modern and herbal ingredients. While pharmaceuticals tend to be sold as standard formulations, in this example consumers are given options to fit the products to their own perceived needs. Although the retail pharmacists who engage in this practice must compete with the large pharmaceutical chains, their capacity to provide personalized medicine provides them with professional credibility and satisfaction. However, we also see that multinational companies are concerned about the impact of this on their sales, and so have begun lobbying for regulations to constrain compound pharmacies.

## Concerns about Counterfeit Drugs

Compare these market dynamics in Brazil with those in Nigeria, where multinational pharmaceutical companies lack interest in selling their brand name drugs because of the economic vulnerability in the country. Kristin Petersen (2014) describes how in her fieldsites in Nigeria local businesses fill the gaps by importing generic pharmaceutical compounds in bulk from China and India. However, many of these drugs are fake or substandard. The Nigerian Food and Drug Authority has tried, with limited success, to control the quality of imported drugs, accredit good quality drugs by printing hard-to-copy holograms on their packaging, and close markets where drugs are found. The government simply lacks the resources to police the pharmaceutical flows in this large country. Indeed, no government or international organization is currently capable of policing the counterfeit drug trade; this is a situation that will become increasingly severe in the coming decades if more concerted and effective means of enforcement are not established (Nordstrom 2007).

With an increasing number of producers of drugs in the global market, the circulation of counterfeit drugs has become an issue of global concern. In the following case study, Julia Hornberger shows how global programs against counterfeit medicines are implemented in a local police station in Johannesburg, and so how security, health and safety interplay. Hornberger describes the confiscation by police of cosmetic skin lighteners that they claimed were fake. The packaging of the products, with holograms, is designed to indicate authenticity. While these products are not medicines, they could well be so classified given their chemical compounds; if they were 'real' medicines, they would not be deemed safe. The sellers, three women, are held in a jail cell for a night and pushed to sign an admission of guilt, although the police could not file a case under the fake counterfeit medicines act because the products are legally considered cosmetics.

---

## 10.2 Policing "Counterfeit Medication"

*Julia Hornberger*

One afternoon in July 2012, when I was working with the Commercial Crime Unit of the South African Police (SAPS) in Johannesburg, Inspector Langa came beaming and sweating down the corridor of the offices of the Unit. He called out for some support from his colleagues: several boxes of seized goods had to be brought up from the back of the police pickup van in the garage. He was followed by three young women, whom he had arrested for selling fake creams and whom he now made help to carry up their own, seized goods. The women, soon identified by their asylum papers to be from Congo and Mozambique, were asked to sit down in the meeting room next to the boxes and plastic bags of all sorts of creams, shampoos, body lotions, soaps and ointments. They sat there quietly, looking distraught and worried. Soon, the head of the Unit Captain Mthethwa joined the scene. He whistled through his teeth, in awe at the amount of seized goods and in surprise at the unusual case at hand.

His group of police officers makes up the anti-counterfeit section of the Commercial Crime Unit. While in principle they are responsible for all types of counterfeit, their staple is pirated DVDs and counterfeit clothing. There were occasional cases of fake electronics or software, although these cases were investigated by the private investigators of the company that produces the original goods, with the police only involved to carry out arrests. In this new case, however, Inspector Langa had acted on his own initiative. He had received a tip-off from one of his informers that illegal creams were being sold and he had sniffed a chance to expand his unit's policing repertoire and up its performance record.

This case clearly involved the police venturing into new territory, the territory of security and health. The original impulse to go after the cosmetics was animated by a vague sense that there could be something wrong with the creams, lotions and other body products. Over time, the police officers constructed it as "counterfeit medication" case.

The fight against counterfeit medication is a relatively recent concern globally and nationally in South Africa. Although there is only anecdotal evidence about the extent of medications being willfully copied and traded, the concern that global health has come under growing threat from criminal intent on trading in fake medications has brought together an unusual set of international actors. In 2006 the WHO, Interpol, the pharmaceutical industry and a range of national drug regulators together launched the Declaration of Rome. The Declaration established the International Medical Anti-Counterfeiting Taskforce (IMPACT) to mobilize forces against "this new form of evil." The techniques employed were multiple but included tightening international and national intellectual property (IP) and counterfeit laws, increasing enforcement capacity in the field, and sharing intelligence and expertise.

This new concern brings together questions of security (police), health (global health actors such as the WHO and national drug regulators) and commerce (pharmaceutical industry) in new ways (Hornberger and Cossa 2012). It marks a shift in the paradigm of drug regulation. Traditionally, the production and distribution of medication has been subject to a drug safety paradigm, which assumes ignorance rather than criminal intent from defaulters, and implies that regulatory measures can discipline bad practices into best ones. This inclusive mode has been overtaken by a different, exclusionary mode. The assumption now is that counterfeit medication threatens the safe consumption of medication, and this cannot be defused through older disciplinary means. It entails an idea of a criminal venture which operates outside of regulation with profits derived from deceptive activities, which can be controlled not by education but by different, harsher means (Hornberger and Cossa 2012).

The convergence of security and health is an old topic (Foucault 2008), but finds an unexpected new rendition in the policing of counterfeit medication, especially as it interfaces with the commercial aspects of pharmacy. Through "counterfeit medication" as an emerging category, new forms of policing are enabled, but also policing modifies and makes applicable the idea of health to its work.

* * * *

Captain Mthethwa inspected the pile of allegedly fake cosmetics. "Okay, what do we have here. Yes, they are counterfeit. I know these creams; looks like skin lightening creams to me; they are not allowed to be sold here." And then to himself rather than anybody else, he said: "I need to call this guy from Pfizer. He'll help us. I better call him right away." With this he was off. Inspector Langa seemed a little perturbed by his senior's undertone of disorientation, and eagerly took to making a detailed inventory of the goods seized and to putting them into sealed bags. Later, trying to affirm some sense of direction, Captain Mthethwa explained to Inspector Langa, "I am still waiting for the Pfizer guy to come back to me. I think we make it a counterfeit medication case."

At first, it was not clear to me what Captain Mthethwa expected from someone from the pharmaceutical industry in a case of cosmetics. But it then dawned on me that he was hoping that the 'Pfizer guy' could potentially provide the police with an expert report confirming the fakeness of the cosmetics. The particular requirement to build a legally sound counterfeit and copyright case is that an 'expert report' is provided. Normally someone from the same industry provides the report, with descriptions and photographs of the original and the copy, highlighting the subtle or not-so-subtle differences. By reaching out to the 'Pfizer guy,' whom he had met at some training once, as the *expert* for this case, Captain Mthethwa was taking an extraordinary leap. How did he get from bottles of imported shampoo to counterfeit medication? It amounted to cobbling together a chain of disjunctive but telling associations, and the loose definitions of counterfeit medications developed to mobilize activity in many fields of interest. The main link was that cosmetics and medication were *somehow* both about the body. They are applied externally or internally, but all somehow are intended to make the body better, cleaner, prettier or (look) healthier. They can also have the adverse effect, that is, if the product is bad or dubious.

Some creams among the pile promised a skin lightening effect. Few were skin lightening creams proper, but rather, they featured a citrus fruit on their packaging, implying some kind of 'natural' bleaching effect through acidity. This was enough for Captain Mthethwa to throw into the mix what he knew about skin lightening creams. South Africa has a unique history with regard to skin lightening creams (Thomas 2012). In a public campaign in the 1980s, their skin-damaging effects had been exposed, but also, they had been coined, from a black consciousness point of view, as morally deficient and unacceptable. This campaign led to a complete ban of skin lightening creams even before the end of apartheid. This provided Captain Mthethwa here, as he kept mentioning them in the context of the seized cosmetics, with a firm cornerstone as to the cosmetics' illegality.

By making the link between the cosmetics, the damaging effect of skin lightening creams, and counterfeit medication, Captain Mthethwa was emphasizing not that the products were counterfeited, meaning wrongfully copied, but that their ingredients were toxic and harmful. He relied on an assumption that 'counterfeit medication' had a high propensity to be ineffective, or poisonous and harmful. And this assumption is global, underlying the formation of a coalition between the WHO, the police and the pharma industry to fight counterfeit medication, leading to the establishment of IMPACT.

Not long after IMPACT had commenced its activities, the coalition came under attack by global health activists (i.e. Médecins Sans Frontières) and country representations at the WHO level of countries with bourgeoning national generic production, including Thailand, India, Brazil, China and South Africa. Their suspicion was that under the guise of a concern for health and in the name of the WHO, the intellectual property rights of Big Pharma were being protected; new law-enforcement agendas and broad legal definitions of what counterfeiting entails were being used to limit the rights and legality of the generic industry. IMPACT responded by clarifying the definition of counterfeit medication and emphasizing that generics too could be victim to fakery. A good reason for the WHO to be worried about counterfeit medication, IMPACT argued, is that counterfeits were often of substandard quality and therefore harmful and toxic. The critical retort to these efforts to save the legitimacy of the coalition was that, if so, the WHO should only be concerned about substandard medication and not about legal or illegal copying, and certainly not in a coalition with police and pharmaceutical companies.

This standoff has stalled multilateral developments regarding the fight against counterfeit medication. Still the original collaboration of these otherwise disparate actors created sufficient momentum for various actors to continue to fight against counterfeit medication. For example, Interpol, which has its own medi-crime unit, has organized major exemplary operations across the world on policing the trade in fake medication. Coalitions that mirror the original composition of health, security and commerce are reforming at plurilateral and regional levels, where no global consensus is necessary to move ahead (for instance, the Trans Pacific Partnership Agreement). These coalitions are possible precisely because counterfeit medication has many potential meanings—it can be about drug quality and/or about IP rights. It can bring moral high ground to an IP agenda, and channel private resources towards a health agenda.

This blurring of what counterfeit medication is allows the police to expand its radius. Captain Mthethwa can use (potentially toxic) skin lightening creams to subsume all cosmetics under the new official category of counterfeit medication. Paradoxically, fake skin lightening creams, according to the logic that the fakes lack the proper dosage of active ingredients, would actually be *less* harmful (that is, less effective at bleaching). But what seems to be at play here, in Captain Mthethwa's efforts to make the case work, is improvisation rather than logic.

But a third aspect justified the seizing of these cosmetics as counterfeit medication. All the cosmetics derived from outside of South Africa. Most were identified on the packaging as produced in Tanzania or Congo. Moreover, they were being sold through the informal markets of the downtown streets of Johannesburg, an additional sign for the police officers that they had probably been smuggled into the country, without customs payment. Again, smuggling is considered a dimension of counterfeit products; in security parlance, smuggling is often used interchangeably with counterfeiting. For example, Interpol Southern Africa conducted a workshop to train police officers to recognize fake goods, and in so doing, drew a parallel between counterfeit and smuggled cigarettes. While smuggled cigarettes are mostly exactly the brand they say they are, and their quality is uncompromised (people can't easily be fooled about what they are smoking), they can be sold off cheaply if they have not been taxed. But similar motives and criminal techniques are assumed to underline the trade of both counterfeit and smuggled medication. Interpol links the business of counterfeit medication to transnational organized crime. The creams were coined as circulating in the realm of a transnational underworld.

We end up, then, with cosmetics as 'counterfeit medication' in a criminal category of suspicion and incrimination, heavily imbued and animated by the fact that these products are somehow about the body, are copies, of substandard quality, and smuggled. Only one or two of these criteria need to be met for a product to be a counterfeit medication, and to implicate all other criteria as well. And if a product 'qualifies' as counterfeit medication, it can be policeable and seizable, and the people who trade in them are likely to be harassed and arrested.

* * *

But I want to pause here to look more closely at the products that were seized. I have described how the enriched definition of counterfeit drugs works in a generative way. It allows the police to carry out and subsume other kinds of activities. But it would be one-sided to claim that Captain Mthethwa's effort to link the goods with counterfeit medication was just a freewheeling constructivist practice. Instead, something about these seized cosmetics actually invited such an understanding. Through their packaging, these products were already imbued with and played on the possibility of mistaken identity and posturing. While Inspector Langa was doing the inventory, I looked at some of the boxes, tubes, and cartons. I wrote a few names down and searched the products' origin later on the Internet.

Some of the products, especially the antiseptic ointments, carried holograms, little shiny squares that shimmered under the neon-light of the police office. They contrasted quite strongly with the old-style sun-bleached packaging of most of the products, including the very ointments to which they had been applied. Generally, holograms are applied to packages to protect them from being suspected to be fake. They authenticate the product. Like other inscription technologies such as serial numbers, 3D barcodes, radiochips and scratch numbers, holograms are a technological response to secure medication against counterfeiting. In certain countries they are already mandatory (e.g. China, Turkey, Saudi Arabia); many more countries will follow in this vein. Through additional track-and-trace technology, it should be possible to check the identity of the goods at every step of the supply chain. Here, security becomes the outstanding feature to differentiate the copy from the original. Through this, security is being turned into an outstanding economic value. With no other way to distinguish between a good and bad copy, the secure copy—the legal generic, for instance—is the more valuable and has an edge in a market in which there are all kinds of copies.

In this light, the fact that some of the cosmetics seized by the commercial crime unit carried holograms seemed remarkable. Although they were not medications, something seemed to warrant the need to communicate their authenticity. A hologram is not cheap and adds to the cost of packaging. The extra investment means that even in the market of small things, there is an edge to be had by providing security features. Yet, something did not fit. The high tech fanciness of the hologram, which means to elevate the product, jarred with the otherwise rather expendable and cheap packaging. This contrast overemphasized the security feature and instead of assurance, it became a suspicious feature, evoking the world of fake goods and implicating the goods. It raised suspicion too of the genuineness of the security feature itself. Thus while the security too had economic value, it needed to be communicated in a felicitous way, lest it have the reverse effect. This is what it did for the police officers: the exaggerated insistence that the products were not fake confirmed to the police officers that they were, enabling them to charge their traders for selling counterfeit medications.

* * * *

Even with the 'active' materialistic implication of the goods, none of this would have held in court. But this was not at stake. The 'guy from Pfizer' never showed up (the products were not from Pfizer, so he may have had little interest) and the case never made it onto a trajectory of formal prosecution. Instead, the case stayed in an extralegal realm where suspicion, the law as pretext, and incrimination by association ruled. While it might be important to have a law and legal category to frame the situation, it is not necessary for the law to take effect. So, just 48 hours after the goods were seized, by which time the women would have to have been properly charged, the police officers released them. But not without still getting something out of them. The police officers went to the holdings cells where the women had spent the night in custody. They offered them to go free if they paid a fine and made an admission of guilt. (This was a common practice they applied for traders of pirated DVDs, when they often also lacked evidence on which to build a sound case.) The women complied: they had no idea that the police had no case and would have had to release them anyway. But their situation was so precarious, and they took the opportunity to get out of prison despite the confiscation of their goods and the need to pay a fine. They signed the admission of guilt, and so gained a criminal record which would stand in their way were they ever to apply for citizenship in South Africa. In the informal interstices in which they tried to make a living, these women had no choice. But their admission of guilt allowed the police to bank the case as *successful*. The counterfeit case therefore translates into the formal realm, even though it was reduced to just a number in the arrest- and closed-cases statistics of the police performance sheet of the unit.

### *References*

Foucault, M. 2008 *The Birth of Biopolitics: Lectures at the Collège de France 1977–1978*. New York: Palgrave Macmillan.
Hornberger, J. 2010 Human rights and policing: Exigency or incongruence? *Annual Review of Law and Social Science* 6:259–283.
——— 2011 *Policing and Human Rights: The Meaning of Violence and Justice in the Everyday Policing of Johannesburg*. London: Routledge.
Hornberger, J., and E. Cossa 2012 *From Drug Safety to Drug Security: The Policing of Counterfeit Medications*. Johannesburg: African Centre for Migration and Society.
Thomas, L. 2012 Skin lighteners, black consumers and Jewish entrepreneurs in South Africa. *History Workshop Journal* 73(1):259–283.

## Creating New Markets for Traditional Medicines

Scholarly attention has focused on the production, prescription, and promotion of Western medicine. However, global medicine markets are filled with medicines produced in non-Western settings, emerging from non-biomedical systems of knowledge. China, Indonesia, and South Korea in particular subsidize the development and export of Asian traditional medicines, and food and drug authorities in the region have separate categories for traditional medicines, such as *jamu* (Indonesian herbal medicines) and Ayurvedic medicine. In tailoring the drugs to new markets, ancient recipes are reformulated (Afdhal and Welsch 1988; Pordié and Gaudillière 2014). In the third case in this chapter, Laurent Pordié describes how an Indian company reformulated an ancient Ayurvedic liver remedy, Liv52, into a new product called PartySmart. Pordié provides a fascinating account of what the company does when it realizes that there may be moral opposition against this hangover drug marketed to upwardly mobile young men, and how this opposition could affect the public image of the company.

## 10.3 How a Lifestyle Product Became a Pharmaceutical Specialty

*Laurent Pordié*

Vijay managed to convince three of his friends to join him for what he would later call "one of the most perfect weekends ever" in Pondicherry. This town in Southern India on the Indian Ocean is marketed to an ever-increasing number of domestic tourists. The striking contrast between the western and eastern parts of the city reveals an urban geography that seems to espouse the differences and hierarchies of the country at large. The noisy and crowded city of the west clashes with the quiet and wealthy colonial town in the east, near the sea. The irresistible charm of Pondicherry, demonstrated by the Indian tourists filling the hotels every weekend, is linked to its French heritage. But one does not only come to Pondicherry for its beautiful promenade and French legacy, nor for the famous Sri Aurobindo Ashram and Samadhi (grave) of the Mother—a French woman living there in the 1960s who for many years held ambitious discussions on the future of the human race with the Bengali philosopher Aurobindo. Paradoxes define the city: alcohol is cheaper than in most of the country and is frequently consumed by visitors. Excesses are not systematically hidden here. Activities such as visiting wealthy neighborhoods, showing acts of devotion to a female saint originally from the West and finishing the day in one of the many bars of the city, are not seen to be inconsistent with each other. In fact, they render Pondicherry wildly attractive.

Vijay and his friends knew that. They were interested in the party side of the city. A group of software engineers and highly educated NGO officers from the industrial and scientific hub of Bangalore in their late 20s, they chose to stay in a hotel in the colonial town, a location that reflected their social status. They had their first beer for breakfast upon their arrival in Pondicherry on a sunny Saturday morning of August 2010. Vijay said that he had forgotten the number of empty beer bottles they left behind them before they passed out in their hotel room late in the night, but insisted on recounting the following morning. "We were fine, fresh and ready to go again . . . we never forget to swallow this nifty little pill called PartySmart®," he said, and burst out laughing. According to the manufacturer's prescription, PartySmart® should be taken 30 minutes before the first drink as a preventative measure for the troubles experienced the morning after excessive consumption of alcohol. Indeed, Vijay and his friends very much enjoyed the idea of escaping a hangover in order to make the best of their weekend. The next evening, he confided, "was a rocking room party where we were joined by two girls and yet again PartySmart® did the trick!"

This drug was invented by a leading Ayurvedic[1] company in India and worldwide, The Himalaya Drug Company (est. 1930). It is a polyherbal compound that was, when first marketed in 2005, the latest addition to the firm's range of products targeting liver disorders. These products all involve the reformulation of classical Ayurvedic recipes, combining Ayurvedic and biomedical sources in their prospection mechanisms in order to create new, ideally global, Ayurvedic medicines for biomedically defined ailments. In a context of accelerated industrialization, the Ayurvedic industry is thus reinventing its remedies through a complex innovation regime involving multiple translations and the formation of heterodox epistemologies (Pordié and Gaudillière 2014).

This case study places PartySmart® at the center of an ethnographic enquiry that follows one main analytic thread pertaining to the ambiguous status of the drug in the firm's catalogue. What is the social meaning of PartySmart® consumption in a country where the use of alcohol is widely considered in negative terms? How can we avoid impairing the image of the manufacturing firm in this context? What kind of response should the firm make in order to continue benefiting from this market? This slippery terrain explains the pathway diversion of this drug, which moved out of the commodity state for a while, as well as its shift in status: the drug was turned from a lifestyle product into a pharmaceutical specialty. It is no longer promoted in the night clubs frequented by trendy and emancipated Indian youths, but through scientific marketing directed towards biomedical prescribers. The sketches below take as a basis shifting marketing strategies to demonstrate how an object moralizes lifestyle patterns, and gains agency and pharmaceutical credibility.

* * * *

The Himalaya Drug Company refers to PartySmart® as an innovation product, which was found "by chance" by the research team responsible for Liv.52®, the company's flagship product invented half a century earlier, in 1955. Indicated for a number of liver-related disorders, Liv.52® sells more than one billion tablets world-wide and over 13 million bottles of syrup annually. It is ranked seventh among the bestselling medicines in India—the only herbal preparation among the first 100 according to the IMS ranking of January 2013.[2] Liv.52® is one of the world's most enduring phytomedicines, supported by more than 260 peer-reviewed (but sometimes questionable) clinical studies; it is approved by the Swiss Medical Agency as a pharmaceutical specialty product and prescribed as an adjuvant in the treatment of tuberculosis in Russian hospitals, to cite just a few examples. Liv.52® sales have grown consistently for over half a century.

Making use of the legacy of Liv.52® and the firm's claimed mastery of liver disorders made obvious sense in the marketing of PartySmart®. Liv.52® "enabled Himalaya to achieve the status of the world leader in hepatic disorders. This expertise was harnessed by Himalaya in the development of PartySmart," reads the drug promotion website. After all, hangovers also involve the liver.

PartySmart® is backed up by a number of scientific publications and its pharmacology is not presented in Ayurvedic language, but exclusively in biomedical idiom (hepatic protection, enzymatic activities, accumulation of acetaldehydes, etc.). The company thus seeks 'scientific credibility.' The modes of legitimation used to promote PartySmart® are diversified, as shown by the geographical and temporal references chosen to promote the plants that form part of the formulation. All of the six plants constituting this Ayurvedic proprietary formula find their origins, therapeutic indications and effects either in ancient (Ayurvedic) texts or from healing practices outside of India (e.g. ancient Greek, Roman, Egyptian, Arabic).

The company omits an important fact from its promotional materials: only 0.2 percent of published scientific research concerns the treatment of hangovers, and the preliminary clinical trials showing positive results for treating or preventing hangovers have not resulted in further studies (Pittler et al. 2005), despite that 75 percent of people who consume alcohol suffer from hangovers on a regular basis (Piasecki et al. 2005). The ambiguous status of the drug therefore explains why The Himalaya Drug Company did not present itself as one of the world specialists in hangover prevention; the company sets great store in claiming its ethical approach and moral righteousness.

In a country where the use of alcohol is publicly considered in negative terms (Sharma et al. 2010), the marketing of PartySmart® raised some eyebrows. The main problem was that the promotion of this product was thought to encourage alcohol consumption and so damage the firm's image. Although a recent study proves the contrary—hangover relief does not encourage the subject to drink more (Piasecki et al. 2005)—the stigma of both alcohol consumption and the product itself in India was considered seriously by the firm's directors, as explained by the CEO:

> This [product] is not who we are. . . . It is not really our philosophy. Our philosophy is to heal people. We are against this idea, like you tell a guy, "listen you pop three of these and you go get wasted buddy, you know, we are with you." It is not our message; it is not a responsible message. Our message is "look after yourself, look after your body" . . . How I see PartySmart® in the scheme of things [at Himalaya] is a long story. . . . It was the outcome of some very good research out here, and they said "listen, we have this fantastic product that actually cures hangovers, why don't we sell it?" I said, yes let's do it. The Chairman is very clear, he said "look, I don't encourage drinking. If a guy wants to get smashed let him go to hell, we're not in this space, we should be treating people who are not well. . . . " I asked what about this product, and he says "fine, go ahead."

Even so, after its invention in 2005 and during the next few years, PartySmart® was aggressively promoted in bars and clubs frequented by middle-class, trendy youths, by a sister firm of The Himalaya Drug Company. It had its own dedicated website, which introduced the product, provided detailed research papers, a photo gallery of young people having fun and a list of alcoholic cocktails and their origins and detailed recipes. As with other products at the firm, PartySmart® is aimed at the middle and upper social classes of urban India. In many ways, this new configuration of Ayurveda resembles what classical Ayurveda was in the past. It was an elite medicine, initially serving royalty, practiced by higher castes, and, contrary to common assumptions, use of this medicine has always been relatively limited among the majority of the population, who prefer biomedicine or use 'folk healing' by default—a few exceptions, like skin diseases in some states, set apart. What is more, as far as alcohol consumption is concerned, one of the founding Ayurvedic texts written in the first centuries AC, the *Carakasaṃhitā*, underscores the fact that alcohol consumption is only acceptable among the higher strata of society. This is made clear in a chapter dealing with the treatment of alcoholism, which sets out the rules (*vidhinā*) concerning the consumption of alcohol and the category of persons for whom it is a poison (see Papin 2009: 443–444, 447).

PartySmart® is promoted directly to such high-end urban consumers in their contemporary manifestations, and not to the physically strained workers and poor villagers. The company initially classified the product as a lifestyle drug, an OTC (over the counter) product. However, due to the need to uphold the image of the product (and the company), the drug was withdrawn from the market in 2009 in order to relaunch it as a pharmaceutical specialty, available (in theory) through prescription only. The original colorful, flashy packaging was replaced at this point by a sober presentation in line with the graphic chart of the pharmaceutical range of products. In this way, PartySmart® looked like the other products belonging to this range when it was reintroduced in 2010. The slogan associated with the drug and present on the packaging was also revamped: in 2005 it read "Hangover free—naturally"; this was changed in 2010 to "Relieves the unpleasant after-effects of alcohol." This is a typical case of commodity pathway diversion (Kopytoff 1986) as the drug moved out of the commodity state before it returned, reintroduced in a new light. While the process had no effect on the content of the product, the drug is now a different object because its very status has changed:[3] PartySmart® became a pharmaceutical specialty.

PartySmart® is now marketed to biomedical prescribers and no longer directly to consumers—although the drug's latest website is the only one of the company to play dance music. These prescribers act both as intermediaries between the firm and the consumers and as agents conferring a new pharmaceutical credibility, a new order of legitimacy for a product once considered to be superfluous. In other words, they mediate social relationships and make them more acceptable. They help to draw a line between what is morally apposite and what is not. Of course, no one is taken in by this. The line is unstable and the frontier it embodies heavily porous. While this shift has consolidated the new pharmaceutical status of the drug, the aim was also to increase its consumers. As the CEO puts it, "We found that there are regular users and people who do not touch [PartySmart®]." These potential consumers could be reached through the help of doctors and their prescriptions. Since in India the law governing the sale of 'prescription drugs' is rarely, if at all, enforced, this move was also a clever strategy to extend this drug's market.

The Himalaya Drug Company insists on a professional register and a scientific relationship with the prescribers. The firm develops its market through aggressive scientific marketing, conducted in practice with the help of medical representatives, events sponsoring and other forms of advertisement. The marketing is largely directed towards biomedical practitioners, who, I was told, account for 60 percent of product sales in India and bring credibility to the brand. As a way to circumvent the taboo issue of hangovers, the focus has now turned to 'social drinking.' The discourses of the people in charge of PartySmart® at the company, that of medical representatives, as well as the leaflets they leave at doctors' offices, stress that for many people in today's India, social drinking is unavoidable (with colleagues or friends), so better help them to avoid hangovers in order to improve their quality of life and to be more efficient at the office the next day. The argument is built by citing biomedical papers that underscore the product's efficacy and its mode of action. Attempts have also been made to extend the product's indications by encouraging its use in the recovery days that follow extreme consumption of alcohol: "Sometimes guys who are very smashed end up in clinics so we tell the medical personnel to put them on PartySmart@ for a week because it helps to clean up the body. The mode of action remains the same," asserts a member of the company.

This new promotional strategy "did not impact OTC sales in any way," while "sales through prescribers are much better than before. . . . There was an initial spike and after that the sales stabilized." This was asserted by one of the leading personnel in the Business Department who also recognizes that PartySmart® ranks as one of the overall lowest product sales for the company. The problem, says the CEO, lies with drinking habits in India:

> Yes, and that is what the Chairman also says; we know PartySmart® works, we know it is a great product, we are not into giving a personal license to people who want to party. In countries where it is not taboo, then it is fine. Well . . . in India people drink to get smashed, not really to enjoy taste. That's the problem.

Despite strong scientific marketing efforts in 2010 to launch the reimagined product, the firm does not spend as much on marketing and promoting this marginal drug as it does for other pharmaceuticals. It is still a fringe product. PartySmart® continues to be seen as disconnected from the company products and ethos.

★ ★ ★ ★

The experience of Vijay and his friends corresponds to the original consumers' portrait, to people who use this drug as an adjuvant to night club partying. Today, this type of consumer in India is not the official target group for the drug, for they potentially undermine the image and ethical aspirations of the firm. As a solution to this problem, The Himalaya Drug Company chose to medicalize PartySmart® by ensuring its promotion through biomedical prescribers. However, the type of scientific marketing deployed by the firm—such as modification of the packaging, intervention of medical representatives, and social events—does not make PartySmart® a prescription-only drug. Anyone is free to buy it, not only in India but worldwide, in retail shops and over the Internet. Transforming a lifestyle product into a pharmaceutical specialty does not systematically change the profile of the users; at best it helps to reach a new clientele—namely, those in India who are supposed to be prone to social drinking, whether at business dinners or recreational meetings, but need a clear head the following morning in the office. This new clientele does indeed recall the social profile of Vijay and his friends. Although it uses new routes, The Himalaya Drug Company still targets the same social group, from whom—perhaps ironically—they expect different social practices.

### *Notes*

1. Ayurveda, the science (skt. *veda*) of longevity (skt. *āyus*), is a learned medicine stemming from the Brahmanic tradition, the doctrine of which is established in three founding collections composed in Sanskrit and attributed to Caraka, Sushruta (first centuries AD) and Vāghabta (6th–7th centuries). It is a humoral medicine found widely in the Indian subcontinent and today exported overseas.
2. IMS is the leading provider of information services for the health care industry, covering markets in more than 100 countries around the world. Stockist Secondary Audit (SSA) is an audit which tracks the stockist sales in India to the retailer (i.e. secondary sales).
3. On the (shifting) status of the pharmaceutical object see Pordié (2014).

### *References*

Kopytoff, I. 1986 The cultural biography of things: Commoditization as a process. In *The Social Life of Things: Commodities in Cultural Perspective*. A. Appadurai, ed. Pp. 64–91. Cambridge, UK: Cambridge University Press.
Papin, J. 2009 *Caraka Saṃhitā. Traité fondamental de la médecine ayurvédique. Tome 2—Les thérapeutiques*. Paris: Almora.
Piasecki, T.M., K.J. Sher, W.S. Slutske, and K.M. Jackson 2005 Hangover frequency and risk for alcohol use disorders: Evidence from a longitudinal high-risk study. *Journal of Abnormal Psychology* 114(2):223–234.
Pittler, M.H., J.C. Verster, and E. Edzard 2005 Interventions for preventing or treating alcohol hangover: Systematic review of randomised controlled trials. *British Medical Journal* 24(331):1515–1518.
Pordié, L. 2014 Pervious drugs. Making the pharmaceutical object in techno-ayurveda. *Asian Medicine* 9(1):49–76.
Pordié, L., and J.P. Gaudillière 2014 The reformulation regime in drug discovery. Revisiting polyherbals and property rights in the ayurvedic industry. *East Asian Science, Technology and Society* 8(1):57–79.
Sharma, H.K., B.M. Tripathi, and P.J. Pelto 2010 The evolution of alcohol use in India. *AIDS Behavior* 14:S8–S17.

## Be Your Products

In the last case study in this chapter, we consider the role of local distributors selling various herbal medication and nutrition supplements, which are relatively unregulated compared to pharmaceuticals. Multilevel marketing as a business model has grown rapidly all over the world to develop new consumer markets. Often utilized by direct-selling organizations, the model—sometimes refereed to as 'pyramid selling'—allows companies to rely on a network of distributors to grow a business by selling products or recruiting new members into the scheme. Typically, 'investors' pay a 'joining fee' to become an agent; this gives them the opportunity to recruit others to do the same. Agents are rewarded with 'incentives' or 'bonuses' depending on the number of other investors they recruit as their 'downlines' (Krige 2012). Network marketing companies tend to emphasize their anti-corporate style, claiming that they empower workers to be their own bosses and set their own hours.

Peter Cahn (2008), in a rare study of multilevel marketing, presents the intriguing case of Esperanza, one of the three million distributors of Omnilife nutritional supplements in Mexico. Between 2003 and 2006, Cahn conducted fieldwork with over 50 current and former members of the Omnilife sales force in the central-western state of Michoacan, Mexico. Cahn (2008: 430) reports that, "although the company likes to showcase the top earners in its promotions, the vast majority of distributors resemble Esperanza, who rarely receives any commission checks for her work. As I followed her through different phases of her career with Omnilife, I observed how completely Esperanza had internalized the tenets of neoliberalism: the free market rewards every entrepreneur without limit and individuals should help themselves rather than look to government handouts. Even as her efforts to cash in on capitalism's promise faltered, she refused to fault the system. Rather, she blamed herself and vowed to redouble her efforts to promote Omnilife and its products."

Alice Desclaux, in the final case study of this chapter, presents us with a peer of Esperanza's in Senegal. We meet Theophile, who in fact had wanted to be a health worker but failed to become a licensed doctor. Instead, he is an active member of a multinational, multilevel marketing company that sells herbal medicines, food supplements, and cosmetics. The entrepreneurial model of the company for which Theophile works reinforces health care practices whereby individuals are responsible for their own health. The products are designed for nutrition, beauty, and health, but they look like medicines with trade names, expiration dates, and detailed information on indication and dosage. These distributors sell their products by referring to their own experience in using the commodities: they are living evidence that the products work. In this case, the distinction between consumers and sellers of drugs disappears—their products are used both to prevent illness and to make a living.

## 10.4 Cosmopolitan Phytoremedies in Senegal

*Alice Desclaux*

Dakar is seen as a city of opportunities by the young men and women who come from rural Senegal and other West African countries. Economic opportunities are linked to the local development of the middle class, to substantial cultural and material exchanges with the Senegalese diaspora, particularly from France and the United States, and to the city's status as a regional capital for international agencies and companies. In a society shaped by Islam and by 'traditions' defined by their perpetuators as regional or ethnic, Dakar is also a bridge for globalized cultural influences. The development of alternative medicines under these influences shows that health management is a very dynamic field, beyond public health systems planned by international bodies and public policies for a country deemed a 'low-resource setting.' Theophile's experience, which I present below, illustrates how individual strategies for economic success are part of major social processes regarding the reinvention of medicinal tradition in a globalized perspective, the biomedicalization of everyday life and the emergence of a new category of lay health specialists. In this short piece, I describe how these three outcomes depend on the successful interpretation of globalized concepts and practices within the Senegalese culture and social system.

### *Becoming an Expert and Entrepreneur*

Theophile came to Dakar University to study psychology. His first wish was to become a doctor. Preparing his master's dissertation, he decided to join a health center and work with PLHIV (People Living with HIV). Independent, curious and determined, he managed to do an internship in the regional office of a UN agency during his studies. When I met him a year after he had completed his master's degree, he had abandoned psychology and joined a multinational network firm selling cosmopolitan phytoremedies. His main motivation, beyond earning a living, was his satisfaction in learning about care and prevention through the firm. He spoke enthusiastically about the products he sold, turning the conversation into a discussion between a seller and a client-to-be. His first words described the product's vegetal components: medicinal plants such as *Aloe vera*, spirulina, green tea, *Cordyceps sinensis* or ginseng, all included on the international 'best seller' list of phytoremedies, similar to 'heroic plants' named by ethnobotanists for their claimed efficacy. The products also contain micronutrients, including vitamins and minerals such as calcium and zinc. Just as a pharmacist would, Theophile started explaining the medical indications of each remedy that combined prevention and relief. However, he insisted that these 'products'—as he called them—were not pharmaceuticals and had no curative claims. In his presentation, he emphasized *nature*, i.e., natural products as opposed to drugs; *tradition*, i.e., the uses of these products by ancient civilizations and local tradition; and product *quality* relative to their production in the US and certification by an international committee. Theophile's narration and exchanges during our next encounters complemented the information I had gathered over the past decade in Burkina Faso and Senegal from other distributors from the same firm, who were also eager to explain their professional activity, patient narratives about these health resources, company documents, and in training sessions.

Multilevel network companies follow the direct sales model first developed in the US by Amway from 1959 and Herbalife in 1980. Companies in Africa sell products to registered distributors, who offer them at retail prices through face-to-face transactions. In Dakar, as in all of West Africa, two main companies with origins in the US and Taiwan sell imported phytotherapy products, nutritional supplements, cosmetic products and appliances for physical treatments (massages, vibratory or electric stimulation); other companies sell hygiene products and cosmetics as well as health products. Buyers may join a scheme that combines distribution, care and personal development: by becoming contractual resellers or 'distributors,' buyers become 'partners' who can purchase the exclusively distributed products for personal use at a lower price than an independent buyer. When joining a 'multilevel' purchase and sales system, they increase their benefits by mobilizing and involving new buyers/sellers and receive a percentage on the sales made by those distributors. Theophile's company now claims more than 20,000 distributors in Senegal, Mali and Guinea, and over 10 million distributors in more than 145 countries worldwide.

In Dakar, the products are stored and sold in the headquarters of the Senegalese branch, where distributors meet for introductory training and advanced sessions on the use of products and management. Newcomers, such as people who want to buy the products, are also invited to attend information and training sessions that are gateways to further contractual participation, and distributors actively invite them to participate when they are perceived as potentially efficient new buyers or resellers. During those sessions, managers, ranked according to seniority and selling performance, explain the meaning of the products, their composition and their use. Participants at information sessions include simple consumers, people in the process of contracting with the company, and first-level distributors. A sociality is developed about the sale of health products through group or competitive activities and regular meetings among distributors and managers. Nothing is mentioned about patient/user confidentiality of what is said; on the contrary, participants are encouraged to diffuse information as a way of advertising.

## *Reinventing Ethnomedicinal Tradition*

Aloe vera, the main medicinal plant advertised by Theophile's company, can be easily found nowadays in Dakar, and in the past ten years, its popularity has increased dramatically. Potted plants are a common item sold by nurseries and street vendors, who point out that, some time ago, only Cape Verdeans would grow it. Endemic in the Cape Verde Islands, the Aloe vera species seems to have been introduced by this population who immigrated to Dakar during the mid-twentieth century. They used the plant for magico-religious reasons to protect households and as a remedy for common ailments, through ingesting the pulp or application on wounds and burns. Catholic missionaries also included it in their pharmacopeia, and prepared medicinal drinkable solutions for Senegal's Catholics, who make up 10 percent of the population. But according to ethnobotanists, the main ethnic groups in Dakar, such as the Wolof or Lebu, did not use Aloe vera, and the recent demand for Aloe vera, which now encourages many vendors to state that this plant belongs to the Senegalese pharmacopeia, probably results from the firm's publicity. After being told about Aloe vera as a panacea or after purchasing the manufactured product, consumers turned to the least costly sources of the plant, encouraging local production and patrimonial claims.

Besides this symbolic reshaping of local ethnomedicinal tradition, the significance of locality in tradition has also been reinterpreted. Aloe vera is presented by Theophile's firm as a powerful plant used by various ancient civilizations such as Egypt, China and Mesopotamia. This legitimization by tradition seems to follow a cumulative logic. In the firm's discourse, the plant is considered powerful not because of its unique significance for a specific population, but because several foreign civilizations have used it. This kind of legitimacy puts forward the plant's biological dimension rather than its cultural aspect; the plant has not been empowered through a symbolic history particular to an ethnomedicinal tradition but owes its symbolic power to its biological efficiency, confirmed by its empirical uses throughout centuries and civilizations. In this perspective, like cosmopolitan New Age rhetoric, globalized tradition legitimates local knowledge and uses, which are more practical but less prestigious. This discourse on the value of ancient knowledge as a source for efficient phytoremedies, combined in other settings such as Europe with the valorization of exotic cultures and 'first peoples,' portrays a new hierarchical order in the recognition of cultures.

## *The Biomedicalization of Everyday Life*

The range of products presented by Theophile or by other network marketing companies in Dakar covers a broad spectrum, including cosmetic products, nutritional compounds and food products that contain an ingredient that has an impact on health and wellbeing (for example, Cordyceps cappuccino powder or hypolipidemic tea), under the general label of 'nutritional supplements.' The apparent hybrid nature of those products—designed both for nutrition and health—is particularly obvious when products look like medicines. Some are presented as capsules and pills in plastic containers, with trade name, phytochemical composition, indications, dosage and expiry date—precisely the information found on the packaging of pharmaceutics even though the word 'product' is used during interactions between trainers and users at the firm's headquarters. The trainers in their speeches recurrently assert that these products are not medicines, for legal reasons: only drugs with a marketing authorization can legally make a therapeutic claim, i.e., maintain that it prevents, treats or cures a disease. The pedagogic tools provided by the company for training sessions or for clients are very precise, therefore, but speeches for and by distributors are shaped by ambiguity; some distributors, consciously or not, strategically avoid infringing on the law while presenting very broad claims. The ambivalence turns to explicit double discourse, such as when a manager says during a training session in Burkina Faso: "Our product cures AIDS because it neutralizes the virus, but don't say that to a doctor."

During training sessions, when an open discussion follows a lecture presentation, participants share experiences on the product's use and sale. These discussions reveal how they recommend a product for the care or prevention of all kinds of physical ailments, from everyday wounds and stress, to ailments related to age and chronic diseases with costly pharmaceutical treatment, such as high blood pressure. Participants also present their products as the key to coping with the supposed difficulty in finding quality food in third world cities. In Dakar, for instance, they may advise the use of a compound made of extracts of North American fruits and vegetables to compensate for the scarcity of local vegetables during the rainy season. Distributors' discourse concerns diseases such as diabetes or rheumatisms, physiological variations, psychological dispositions, health transitions, and wellbeing, and does not spare any aspect of life, from 'the normal to the pathological.' It conveys medicalization—or pharmaceuticalization, depending on theoretical interpretations—not only through the wide range of topics and situations considered, but also because it plays with the symbolic attractiveness of pharmaceuticals by constantly denying it (Desclaux 2014).

## *Lay Health Specialists*

Distributors appear as a new kind of 'lay health specialists,' different from the 'community health workers,' active in the global South since the 1990s, who were chosen by communities and worked for its common wellbeing with very little compensation. Distributors may work for the company in addition to another job, sometimes in medical services, or work full-time, based on their ambitions and time availability. They are private entrepreneurs at the interface of nutrition or health specialists and 'expert patients,' since many of them first contacted the company for their own treatment. Educating people in health is an important part of their work since they earn money through developing the knowledge and skills of the distributors they supervise. Their professional skills first rely on knowledge about health care and phytoremedies, and social techniques to engage more consumers.

Distributors' interventions during training sessions attest to their lay knowledge, which mixes cosmopolitan notions of secondary-school anatomy and physiology with local nosology (illness concepts) and 'ubiquitous' concepts such as *anti-oxidants* or *stress*, used in biomedicine and CAMs (complementary and alternative medicines), popularized by multinational advertisements for cosmetics and food products. They focus on common local pathological entities such as *hemorrhoids* that, in West Africa, convey dense social representations of internal disturbances in the abdomen or the whole body and, according to local nosology, may refer to several etiologies related to social contexts and physical factors. Theophile explained that many people found relief for 'complicated hemorrhoids' through purifying treatments, and they later joined the company. Such vague ideas are particularly convenient for interpretation in biological, cosmopolitan alternative, and lay local knowledge systems. Some distributors invest time and energy to extend their biological and biomedical knowledge. Some evoke their own physical and moral transformation—as entrepreneurs and as healthy persons altruistically focused on people's wellbeing and health—and relate the two levels as two 'positive ways' of considering life.

Theophile has particular psychological skills that he uses to engage potential clients in a reciprocal exchange and turn it into a long-term relationship that will lead to the purchase of products and maybe to a 'partnership.' Partnership occurred with the 40 clients he convinced to be distributors, and who now allow him to earn better wages than a medical doctor in public health services. A manager in the same company showed his 'sociological skills' when he explained that his plans focused on groups who already had sufficient economic resources and an organized network to take advantage of previous social bonds. For instance, the extended families of the Bamileke in Cameroon easily support a distributor; women secretaries in administrations in Burkina Faso may sell products in their office; university professors may convince their students looking for performance enhancers during examination periods. Distributors also use relational techniques acquired through experience or exchanges with peers, which are to some extent similar to those of sales representatives of the pharmaceutical industry. They must also balance cooperation and competition among themselves in order to extend their networks. These lay health specialists combine new global and previous local patterns of health knowledge and social skills in very specific ways.

## *Barefoot Neoliberalism*

During the 1980s, developing countries trained barefoot doctors and set up community initiatives for managing prevention and care "where there is no doctor" (Werner and Thuman 1992). The biomedical private sector has expanded since the turn towards neoliberalism in the 1990s. The new sector of lay sellers of cosmopolitan alternative health products has developed more recently in West Africa, relying on the dynamism of persons with health-related skills and social savoir faire who do not find work opportunities in the formal system, despite being highly educated as in the case of Theophile. The development of an educated middle class and the influence of diaspora and transnational migrants offer an economically and culturally favorable setting for their activity. Their private aim, based on an entrepreneurial model, is free from any control from biomedical institutions, though central technologies and concepts of biomedicine are used to develop a field of activities and extend social networks based on phytoremedies. The individualistic model they support, focused on personal responsibility in health prevention and disease management, fits with the global consumer patient discourse, only partly endorsed by national health authorities and public policies in the global South, but diffused by international media and private firms. The hybridization between the local ethnomedicinal tradition conveyed by lay knowledge, alternative global representations, and knowledge about biological and biomedical sciences engender a powerful interpretation system about health, care and medicines. This assemblage, particularly efficient in a city like Dakar, is a vector of medicalization of everyday life and of change in the status of tradition. The professionalization of lay sellers/distributors who, like Theophile, manage to adapt transnational concepts to their social and cultural environment questions the importance of these 'barefoot alternative health specialists.' Although they have been studied much less than the associative 'expert patient' movement, they play a major role in a transnational cultural process that reframes local health landscapes, as part of neoliberal globalization.

### *References*

Desclaux A. 2014 Ambivalence in the cultural framing of cosmopolitan alternative 'medicines' in Senegal. *Curare* 37(2):53–60.

Werner, D., and C. Thuman 1992 *Where There Is No Doctor: A Village Health Care Handbook.* Palo Alto, CA: Hesperian Foundation.

---

* * * *

By not assuming that medicines follow prescribed pathways, anthropologists gain insights into situated market dynamics that shape both the accessibility and use of medicines. Actors involved in the treatment of common coughs and colds include mothers who monitor their children's health; husbands and neighbors who give advice; *sari-sari* store owners who buy a handful of drugs in pharmacies in town and sell these in smaller quantities at a small markup, thus earning them some meager income; companies which promote products by buying airtime on the local radio; pharmacies which tailor medicines to local needs and sell prescription drugs over the counter; and local businessmen who buy generics in bulk in India to meet local needs in places where Big Pharma has lost interest.

Collectively the cases point to a vibrant health and wellness market selling not only medicines but many other health-related products, such as nutritional supplements, diet bombs, anti-aging therapies, and hangover remedies. At a time when governments implore people to look after themselves, and structure health systems on this basis, as populations age and there are ever-increasing technological possibilities to extend life, marketing medicine presents constant new opportunities (Petersen and Lupton 1996).

While global policy efforts in the 1980s attempted to shape markets so they would provide a limited number of safe, effective, and affordable drugs that could meet medical needs, we are now confronted with much more complex market dynamics in the new millennium, with complicated global drug circuits, characterized by Kristin Petersen as "an assemblage of indirect and lateral paths that comprise transcontinental drug manufacturing and distribution" (2014: 20). With medicines and their raw materials flowing in so many directions, it is hard to control their quality. To make sense of complex medicine markets, anthropologists are starting to study international treaties, bilateral trade agreements, and national policies, and the accompanying local level sanctions and policing if rules are not followed. In these studies anthropologists observe that many health products found in local markets, such as Ayurvedic medicines, Indonesian *jamu*, 'nutrition supplements,' and cosmetics, escape international regulations and policies (Pordié and Hardon 2015).

Medicines and other health-related products are assigned value through marketing, which increasingly involves consumers who can give first-hand testaments of their positive experiences with the drugs. Companies, both multinational pharmaceutical companies and local producers of non-Western medicines, reinvent their products to meet the new needs of the growing middle class in some countries, while elsewhere they withdraw because of overwhelming poverty (Petersen 2014; Pordié and Gaudillière 2014; Pordié and Hardon 2015). At the same time, anthropologists show how, in these complex marketplaces, consumers appropriate products for their own needs, deciding for themselves if the products are effective and safe, and using their own notions of health and wellbeing. And they describe how local markets are populated by diverse categories of economic agents, who earn, or try to earn, a living through selling products, whether in pharmacies, small grocery shops, markets, and the homes of their friends and relatives. Through these channels medicines often are diverted from their intended paths and reformulated to meet local needs.

## References

Afdhal, Ahmad Fuad, and Robert L. Welsch 1988 The rise of the modern jamu industry in Indonesia: A preliminary overview. In *The Context of Medicines in Developing Countries: Studies in Pharmaceutical Anthropology.* Sjaak van der Geest and Susan Reynolds Whyte, eds. Pp. 149–172. Dordrecht, The Netherlands: Kluwer.

Angell, Marcia 2004 *The Truth about the Drug Companies.* New York: Random House.

Applbaum, Kalman 2009 Getting to yes: Corporate power and the creation of a psychopharmaceutical blockbuster. *Culture, Medicine and Psychiatry* 33(2):185–215.

Bensaude-Vincent, Bernadette, and Isabelle Stengers 1996 *A History of Chemistry.* Cambridge, MA: Harvard University Press.

Cahn, Peter 2008 Consuming class: Multilevel marketers in neoliberal Mexico. *Cultural Anthropology* 23(3):429–452.

Davis, Courtney, and John Abraham 2013 *Unhealthy Pharmaceutical Regulation: Innovation, Politics, and Promissory Science.* London: Palgrave Macmillan.

Dumit, Joseph 2012 *Drugs for Life: How Pharmaceutical Companies Define Our Health.* Durham, NC: Duke University Press.

Ecks, Stefan 2005 Pharmaceutical citizenship: Antidepressant marketing and the promise of demarginalization in India. *Anthropology & Medicine* 12(3):239–254.

——— 2014 *Eating Drugs: Psychopharmaceutical Pluralism in India.* New York: New York University Press.

Etkin, Nina L. 1992 Side effects: Cultural constructions and reinterpretations of western pharmaceuticals. *Medical Anthropology Quarterly* 6(2):99–113.

Foster, Robert J. 2008 *Coca-Globalization: Following Soft Drinks from New York to New Guinea.* New York: Palgrave Macmillan.

Friedman, Jonathan 1995 Global system, globalization and the parameters of modernity. In *Global Modernities.* Mike Featherstone, Scott Lash, and Roland Robertson, eds. Pp. 69–90. London: Sage.

Greene, Jeremy A. 2011 Making medicines essential: The emergent centrality of pharmaceuticals in global health. *Biosocieties* 6(1):10–33.

——— 2014 *Generic: The Unbranding of Modern Medicine.* Baltimore, MD: John Hopkins University Press.

Hardon, Anita 1990 *Confronting Ill Health: Medicines, Self Care and the Poor in Manila.* Quezon City, The Philippines: Hain.

——— 1992 Consumers versus producers: Power play behind the scenes. In *Drug Policy in Developing Countries.* Najmi Kanji and Anita Hardon, eds. Pp. 48–64. London: Zed Press.

——— 1994 People's understanding of efficacy for cough and cold medicines in Manila, the Philippines. In *Medicines: Meanings and Contexts.* Nita L. Etkin and Michael L. Tan, eds. Pp. 46–49. Quezon City, The Philippines: Hain.

Hardon, Anita, and Amelia Ihsan 2014 Somadril and edgework in South Sulawesi. *The International Journal of Drug Policy* 25(4):755–761.

Healy, David 2009 Trussed in evidence? Ambiguities at the interface between clinical evidence and clinical practice. *Transcultural Psychiatry* 46(1):16–37.

——— 2012 *Pharmageddon.* Berkeley, CA: University of California Press.

Herxheimer, Andrew, and Barbara Mintzes 2004 Antidepressants and adverse effects in young patients: Uncovering the evidence. *Canadian Medical Association Journal* 170(4):487–489.

Illich, Ivan 1975 The medicalization of life. *Journal of Medical Ethics* 1(2):73–77.

Kamat, Vinay, and Mark Nichter 1998 Pharmacies, self-medication and pharmaceutical marketing in Bombay, India. *Social Science & Medicine* 47(6):779–794.

Kanji, Najmi, and Anita Hardon 1992 *Drugs Policy in Developing Countries.* London: Zed Press.

Krige, Detlev 2012 Field of dreams, field of schemes: Ponzi finance and multi-level marketing in South Africa. *Africa,* 82(1):69–92.

Laing, Richard, Brenda Waning, Andy Gray, Nathan Ford and Ellen 't Hoen 2003 25 years of the WHO essential medicines lists: Progress and challenges. *The Lancet* 361:1723–1729.

Medawar, Charles, and Anita Hardon 2004 *Medicines Out of Control? Antidepressants and the Conspiracy of Goodwill.* Amsterdam, The Netherlands: Aksant.

Moynihan, Roy, and Alan Cassels 2005 *Selling Sickness: How the World's Biggest Pharmaceutical Companies Are Turning Us All into Patients.* Vancouver, BC: Greystone Books.

Nichter, Mark, and Nancy Vuckovic 1994 Agenda for a pharmaceutical anthropology. *Social Science & Medicine* 39(11):1509–1525.

Nordstrom, Carolyn 2007 *Global Outlaws: Crime, Money, and Power in the Contemporary World.* Berkeley, CA: University of California Press.

Petersen, Alan, and Deborah Lupton 1996 *The New Public Health: Discourses, Knowledges, Strategies.* Washington, DC: Sage.

Petersen, Kristin 2014 *Speculative Markets: Drug Circuits and Derivative Life in Nigeria.* Durham, NC: Duke University Press.

Pordié, Laurent, and Jean-Paul Gaudillière 2014 The reformulation regime in drug discovery: Revisiting polyherbals and property rights in the Ayurvedic industry. *East Asian Science, Technology and Society* 8: 57–79.

Pordié, Laurent, and Anita Hardon 2015 Drugs' stories and itineraries. On the making of Asian industrial medicines. *Anthropology and Medicine* 22(1):1–6.

Rozemberg, Brani, and Lenore Manderson 1998 'Nerves' and tranquilizer use in rural Brazil. *International Journal of Health Services* 28(1):165–181.

Senah, Kodjo Amedjorteh 1997 *The Popularity of Medicines in a Rural Ghanaian Community. Community Drug Use Studies.* Amsterdam, The Netherlands: Het Spinhuis.

Sismondo, Sergio 2008 How pharmaceutical industry funding affects trial outcomes: Causal structures and responses. *Social Science & Medicine* 66(9):1909–1914.

Whyte, Susan Reynolds, Sjaak van der Geest, and Anita Hardon 2002 *Social Lives of Medicines.* Cambridge, UK: Cambridge University Press.

Wolffers, Ivan, Anita Hardon, and Janita Janssen 1989 *Marketing Fertility: Women, Menstruation and the Pharmaceutical Industry.* Amsterdam, The Netherlands: WEMOS.

Haitian Mother Preparing Food, 2012. Port-Au-Prince, Haiti.

## About the photograph

*I took the picture in May 2012 in a displacement camp in Port-Au-Prince, Haiti, where thousands of people had moved after the January 2010 earthquake. More than two years later, in a makeshift tent more permanent than what displaced people may have wished, a smiling young woman stands out. She and her two young children were surrounded by their meager belongings. The image attempts to dignify her existence, despite conditions of extreme poverty, and reflects the resilience needed to survive in the midst of harsh circumstances. After all, they survived the earthquake. Yet, unintentionally, she contributes to the vicious cycle of deforestation, devastating floods caused by hurricanes, and more poverty. Charcoal, the culprit fuel, hidden under pots and pans and her only means to cook, is the result of scarce trees burned down for barter or cash in a land that becomes less fertile with time, but was once as green as its Dominican neighbor.*

—Arachu Castro

# 11
# The Anthropocene

*Elizabeth Cartwright, Lenore Manderson and Anita Hardon*

The environment is a major player in human health. It's not just the backdrop to what we are doing during our busy daily routines, but it is an important contributor to the way that our physical bodies function; the quality of the food we eat, the air we breathe and the water that we drink are all predicated on the quality of the environment in which we live. In this chapter, we switch our focus to thinking about the environment, how it has changed over the past few hundred years, and how those changes have real ramifications for the functioning of our human bodies. The term 'Anthropocene' was coined by an ecologist (Eugene Stoermer) and an atmospheric chemist (Paul Crutzen) as a way to describe the "influence of human behavior on the earth's atmosphere, lithosphere, and hydrosphere in recent centuries" (Haraway 2015a: 258). The ways in which people have altered the earth, down to its very core, will be read as a time of ecological collapse and destruction that will be written into the earth's geology, according to these scientists. Other scientists, in a number of disciplines including anthropology, have picked up the use of this term, each time nuancing its definition to fit the purview of their academic interests. Bruno Latour, in his Presidential Address to the American Anthropology Association, argued the term gives "another definition of time, it redescribes what it is to stand in space, and it reshuffles what it means to be entangled within animated agencies" (2014: 16). Donna Haraway has discussed the strengths of the concept of the Anthropocene as well as a more precise term that she puts forth, the 'Capitalocene' (Haraway 2015a; 2015b).

> It (The Anthropocene) would probably be better named the Capitalocene, if one wanted a single word. The mass extinction events are related to the resourcing of the earth for commodity production, the resourcing of everything on the earth, most certainly including people, and everything that lives and crawls and dies and everything that is in the rocks and under the rocks. We live in the third great age of carbon, in which we are witnessing the extraction of the last possible calorie of carbon out of the deep earth by the most destructive technologies imaginable, of which fracking is only the tip of the (melting) iceberg. Watch what's going on in the Arctic as the sea ice melts and the nations line up their war and mining ships for the extraction of the last calorie of carbon-based fuels from under the northern oceans. To call it the Anthropocene misses all of that; it treats it as if it's a species act. Well, it isn't a species act. So, if I had to have a single word I would call it the Capitalocene.
>
> *(Haraway 2015a: 259)*

The terms 'Anthropocene' and 'Capitalocene' help us to conceptualize the massive environmental changes that are occurring on our planet. In this chapter, we contextualize human health threats within this rapidly changing environment. As Merrill Singer puts it, "Given the demonstrated ability of life forms to thrive under extreme ecological conditions, what is at stake is not life per se or even the planet, but life as we know it, including the historic life of our species and that of many other faunal and floral inhabitants of Earth. In short, what is on the line is the very domain of our discipline and we along with it" (Singer 2009: 815–816). The importance of understanding the ramifications of the Anthropocene (or Capitalocene) to medical anthropology is vast; the health of humans and all living things on earth is affected at a systemic level as a result of our present, highly wasteful and environmentally threatening 'mode of production.' We are living on a planet that is currently experiencing the mass extinctions of some species and the overpopulation of others, as well as shortages of the elements that are critical for life—water, air, and food. As we progress through this chapter, we reflect on both our less polluted and more environmentally healthy past, and on the uncertain future possibilities that face us now.

Anthropology has amassed an amazing compendium of tribal and Indigenous people's wisdom with respect to the environment. Early ethnographies, such as Frank Hamilton Cushing's work in the nineteenth century, that we discuss below, preserved narratives of people living in harmony with nature, using natural cures, and depending on the movement of animals and the cycles of wild and semi-wild crops for their sustenance. Medical anthropology's deep commitment to understanding Indigenous people's ways of healing, traditional approaches to herbal medicines, natural ways of giving birth and environmentally respectful religious beliefs has added to that wealth of knowledge about human–environment interactions. Situating that understanding in the political, social and (more recently) climatic upheavals of the last couple hundred years and processes of industrialization that have polluted the air which we breathe, the food that we eat, and the water that we drink provides us with insights into how quickly environments can change and how thoroughly Indigenous knowledge can be lost. In this chapter, we move from the late nineteenth century, through the twentieth century and into the uncertain future as we explore the intertwined topics of health and environment in the Anthropocene.

## A Starting Point

In introducing the case studies, we start with a brief exploration of work done in cultural anthropology during the nineteenth century. Cushing (1981) spent years among the Indigenous groups that he studied, most notably the Zuni in the American Southwest, while working for the Bureau of Ethnology in Washington, DC. His description of Zuni surgery and their use of botanicals in curing provides very early documentation of the power of Indigenous knowledge of healing practices:

> The old surgeons took up one after another of the straight lancets, and with them dissected away the proud flesh and other diseased tissue, removing it cleanly without wavering vein or artery or tendon, until they had fairly exposed the bone itself. . . . Finally, the openings were filled up or rather stuffed, with the pinion-gum softened by warmth of the breath and in the hands that were the while kept constantly wet with the red fluid. More of this gum was spread on narrow strips of cloth, and with these the wound was neatly closed as with adhesive plaster. The entire foot was sprinkled or thickly dusted over with the yellow pollen and root-power, and then bandaged with long strips of the old rags as neatly as it would have been bandaged by a surgeon among ourselves.
>
> *(Cushing 1981: 222–223)*

Cushing's descriptions of how the Zuni understood disease and the workings of their bodies, and how they used the elements available to them in their immediate environment to heal themselves when necessary, is a good starting point for our consideration of the role of the environment in human health and wellbeing. In the extract above, Zuni medicine men cure a wounded warrior by excising his rotted leg flesh, irrigating the wound with herbal mixtures, avoiding arteries and tendons as they cut and cleaned. They prescribe a healthy (and meat-free) diet, and cure the man. The ways in which we see the environment as both the source of illnesses and the source of cures has been a thread of investigation that continues to provoke new ways of describing and theorizing in medical anthropology. We emphasize the interpretive process of understanding illness within the multidimensional patchworks of smells, shapes, sounds and tactilities, and the flora, fauna and landscapes that make up our environments (Desjarlais 2003; Feld 1990; Seremetakis 1993).

## Places Full of Meanings

Places are full of meanings, none more potent than those places seen as causing harm to human health and happiness. Narratives of places often reveal antagonistic relationships with our environment; we huddle around the campfire to tell tales about ghosts residing in haunted places. Wandering souls, bad spirits who continue to people the world just past the edges of what we can see, cause interference in our lives from the perspective of many people. The Amuzgo Indians of Oaxaca, Mexico, spoke about places where bad things happened to people as being full of emotions that could cause bodily harm and illness to anyone who came in contact with them (Cartwright 2001). The environment of the Oaxacan village was a constant source of morality stories made tangible by the very visible effects of the illnesses that were produced there. An illness narrative that Elizabeth Cartwright's Amuzgan study participants told her came up as they walked by the doorstep where someone had recently been murdered—the fright from that event caused an infected eye in her neighbor Leno; he had walked past the place of the murder late at night, and by the next morning, his eye was red and puffy and his vision was clouded. When she asked what had happened, he told her about the young boy who had been sitting on that particular doorstep, a few months prior, when two men assaulted and murdered him with their machetes. They killed him, Leno said, because he was dealing marijuana. The boy's fright from the attack stayed in that place, making it dangerous, and the fright (*susto*) had caused Leno's eye to swell. The moral evaluations of the actions that led up to the event were instantiated in the place by the narrative; they were made real in another way by Leno's swollen eye. The frights (sustos) and angers (*corajes*) that the Amuzgos experienced continued to linger in specific places in the village, causing illnesses long after the original negative events had occurred (Cartwright 2007). What does it mean to move through a village feeling the residual presence of the emotional upsets and pains of past occupants of the area? Like Keith Basso's (1996) descriptions of the ways that Western Apache used their environment as a source of emplaced narratives, the Amuzgos elaborated their environment in a way that made deep connections between their personal health, the social health of the village, and the health of their physical surroundings.

What happens when that environment changes in unpredictable ways—ways that change where food grows, how much food will be produced, and when it will be ready for harvest? Sharon Stephens (1995) described how the Sami reindeer herders of Norway reacted to learning that the lichens, grasses and mushrooms that their herds depended on for food had been irradiated with nuclear fallout from the catastrophic nuclear explosion in Chernobyl, Ukraine, in 1986. Although Norwegian health officials told the Sami not to eat the meat from their reindeer, to be a Sami was to eat reindeer, and so many of them continued to do so. Children and pregnant women mostly avoided the dangerous meat in the early years after the Chernobyl disaster, but older adults and the elderly were willing to take on the health risks, and continued consuming the reindeer meat. For the Sami, to be part of the environment and part of the natural cycles of consumption was an integral part of life itself.

This attention to the environment emerges in the place-contextualized descriptions of health threats from such things as climate change and pollution. In the following case studies consider the importance of how place is narrated: that is, how meaning created through stories is reflective of a local environmental ethos deeply entwined with understandings of health and illness. Indigenous people of the twenty-first century are the subject of the first case study written by Noor Johnson. Johnson describes how the Inuit people with whom she worked view 'nature' in present-day, ever warmer, Canada. We then move to Italy, where Roberta Raffaetà, writing of polluted and toxic environments, and the perceived effects of chemicals on human health, takes us into a discussion of microenvironments and emotions, those ephemeral, often-invisible components of the world that may cause serious allergic reactions. Finally, we move our gaze to the future, to the extremes of inhabitable spaces and places, and in this context, we consider two case studies of one recent disaster—Hurricane Katrina, one by Ben McMahan and one by Vincanne Adams—that illustrate what Singer calls the "pluralea of interactions" that are creating a perfect ecological "storm" of new diseases emerging from a profoundly polluted and out-of-kilter environment (Singer 2009: 797). We first turn to Johnson's case study.

## 11.1 Inuit Health in a Changing Arctic

*Noor Johnson*

Clyde River, or *Kangiqtugaapik*, which in Inuktitut means a 'nice little inlet,' is a small hamlet on the northern coast of Baffin Island in Nunavut Territory, Canada. At 70 degrees latitude, four degrees above the Arctic Circle, Clyde River enjoys summer days when the sun never sets and winter days when it never rises. To a first-time visitor accustomed to large towns and cities, this fly-in community appears a very contained place. The hamlet's physical infrastructure consists of dirt roads lined with prefab housing, a Northern Store that sells groceries, clothing, and equipment for more than twice the cost in Toronto or Montreal, an Anglican church, a community hall, and—since this is Canada, after all—a hockey rink. To the nearly 1,000 primarily Inuit residents who live there, however, the settlement of Clyde River is just a small part of a much larger landscape that includes tundra, fjords, mountains, and sea ice. In this vast terrain beyond the settlement, families create summer campsites where they fish and gather berries, grandfathers take grandsons hunting for their first seal, and relationships between people, animals, and the land are reaffirmed.

Since the 1950s, when the federal government encouraged residents of the eastern Arctic to give up their semi-nomadic lifestyle, Inuit in Clyde River have accommodated and adapted to significant change. Much of this came about through the increasing influence of social policies such as the introduction of a formal, Western education system, health care provision, and wildlife management protocols that regulate hunting of certain marine and terrestrial species such as polar bear, bowhead whales, and narwhal (toothed whale). More recent social and environmental changes stem from increased mining and oil and gas activity, increased shipping, the presence of contaminants in Arctic food webs, and the complex ecosystem impacts of climate change. In various ways, these policies and activities brought Inuit into a system of governance and economic practices that seem alien and often irrelevant to local lives and experiences.

Knowledge of how to subsist in a harsh environment sustained Inuit for millennia. Inuit knowledge is experiential; traditionally, they learned from older family members and by observing and interacting with the Arctic environment and animals. After a generation of children—now adults—was sent away to residential schools in the 1950s and 1960s, this system of knowledge transmission was disrupted. Demographic growth has further complicated traditional knowledge transmission: around half of Clyde River's population is under the age of 18, making one-on-one instruction in land-based activities difficult. Additionally, equipment needed to travel on the land is expensive; few households have the income to invest in snowmobiles, outboard motors, hunting rifles and ammunition, and camping equipment. Those who do hold a full or even part-time job often find it difficult to balance wage labor with time on the land.

In spite of these changes, however, the land and animals of the Arctic tundra and sea ice continue to play a significant role in Inuit subsistence and sociality. In Clyde River, ringed seal, caribou, Arctic char, narwhal, and polar bear remain important food sources in many households and are a healthier and less expensive alternative to the limited options at the Northern Store. Land-based skills and activities also remain an integral part of Inuit identity, as does sharing and eating "country foods." Hunting, fishing, gathering berries, and spending time on the land are therefore simultaneously material and cultural practices, central to the individual and collective health and wellbeing of Clyde River residents.

Because opportunities to spend time on the land are limited for many residents, community-based institutions play an increasingly important role in connecting people to land-based skills and knowledge. The Ilisaqsivik Society of Clyde River, whose name means 'to recognize oneself' in Inuktitut, is a health, wellness, and community development organization that offers a wide range of services for families and individuals of all ages, including programs that implement cultural and therapeutic activities out on the land. Every summer, for example, the organization runs a two-week healing and wellness camp, attended by several hundred people, that offers discussions on topics related to the experience of colonialism, such as residential schools, alongside Inuit knowledge workshops on topics such as how to clean and prepare seal and caribou skins to make clothing. In the winter, the Ataata/Irniq (Father/Son) program pairs experienced hunters with male youth for a two-week trip to fish and hunt seal and caribou. These programs are an important way for many Clyde River residents to continue to spend time on the land in the context of family and community. Although Ilisaqsivik's role in facilitating land-based experience is a relatively new development in Clyde River, these experiences offer a sense of continuity in a context of rapid social and environmental change.

## *Climate Change, Inuit Health, and Social Policy*

Over the past half-century, temperatures in the Arctic region have risen at twice the global average. Warming has been accompanied by a dramatic loss of multi-year sea ice and a shift in the period of sea ice cover, with freeze-up occurring later in the fall and break-up earlier in the spring. Inuit depend on sea ice to travel to hunting and fishing areas, and utilize traditional knowledge of ice and snow conditions to determine when it is safe to travel (Gearheard et al. 2013). Increasing instability of ice and changes in its behavior have made travel on the sea ice more dangerous, increasing the risks of hunting in spring and fall, and forcing many to adopt longer, more costly, and sometimes more dangerous land-based travel routes.

Increases in extreme and rapidly changing weather conditions present another climate-related hazard. In Clyde River, elders and hunters have observed that the weather in springtime (March—May) changes much more rapidly now than in the past, increasing the likelihood of getting stuck for long periods of time in hazardous conditions (Weatherhead et al. 2010). Elders report that they can no longer draw on their knowledge to predict the weather, reflecting a wider concern in Inuit society about the viability of traditional knowledge given the rapid pace of social and environmental change. Additionally, climate change is affecting the distribution and abundance of some of the animals that Canadian Inuit hunt, including caribou, polar bear, and fish species and bird species. Complex ecological dynamics related to melting sea ice and thawing permafrost also create synergistic impacts on the level of toxins and heavy metals in the Arctic food chain. Combined with the challenges of travel in an unstable sea ice environment, these impacts raise concerns about food security for Inuit households, many of which already face limited access to country food.

Thawing permafrost has created infrastructural challenges in many Inuit settlements, including slumping roads and cracks in housing foundations. Increased storm severity and coastal erosion due to sea ice retreat, particularly in Alaska, is literally causing some settlements to fall into the sea. While these communities have sought assistance, there is currently little in the way of social policies or legal protections in place to support the high cost of rebuilding and relocating (Shearer 2012). In Canadian Arctic communities, permafrost melt is forcing construction crews to drill deeper when putting in pilings for new homes, taxing the limited resources allocated to address a chronic housing shortage.

These issues directly and significantly affect health and wellbeing of individuals and families in many Inuit settlements.

Inuit health in a changing climate is shaped significantly by social policies and regulations such as hunting quotas that limit adaptation options. In the past, for example, Inuit responded to changes in the Arctic climate by shifting their hunting to focus on different species; under a quota system, these kinds of adaptations are likely to be restricted (Wenzel 2009). Inuit are concerned, as well, about the potential for national and global biodiversity and climate change policies to create new restrictions on hunting. Some conservation and animal rights organizations have lobbied to have the polar bear listed as 'threatened with extinction' through the Convention on International Trade in Endangered Species (CITES), a global treaty that regulates trade of certain animal species. Inuit Tapiriit Kanatami, a national Inuit organization in Canada, argues that this designation would not have any clear benefit for polar bear conservation, but would harm Inuit livelihoods and infringe on their hunting rights.

## *From Knowledge to Action*

In the Canadian Arctic, reclaiming local ownership of knowledge production has been an important part of decolonization. In 2005, Clyde River residents founded the Ittaq Heritage and Research Centre, an organization that supports community ownership of and involvement in research projects in the community. Through Ittaq, community members have collaborated with researchers from the National Snow and Ice Data Center in Boulder, Colorado, to install local sea ice monitoring stations and portable weather stations to better understand the changes that are occurring, drawing on traditional knowledge to site these instruments and analyze the data they collect (Gearheard et al. 2013; Weatherhead et al. 2010). Community residents use the information generated by the stations to make informed decisions about travel on the sea ice.

At a global level, Inuit knowledge of climate change contributed to the Arctic Climate Impact Assessment, which included a chapter on Indigenous knowledge (Huntington et al. 2005). Inuit knowledge, including interviews with Clyde River elders and hunters, played a central role in a landmark human rights petition, submitted in 2005 to the Inter-American Commission for Human Rights. The petition was led by the Inuit Circumpolar Council (ICC), an Indigenous peoples' organization that represents the 155,000 Inuit from Alaska, Canada, Greenland, and Chukotka, Russia, in global political arenas such as the United Nations and the Arctic Council. The petition made a case that the human rights of Inuit, including rights to subsistence and to bodily health and wellness, were being violated by the United States in its refusal to regulate its greenhouse gas emissions. Although the Commission rejected the petition, the widespread media coverage it gained helped build a case that climate change infringes on human rights.

Although documentation of Inuit knowledge and community-led monitoring initiatives have been important, when I was in Clyde River in 2009–2010, I encountered a sense of impatience with research and an interest in moving towards tangible, concrete action. "Is there anything that can be done about climate change?" one man, an active hunter in his 50s, asked me. A woman in her early 30s told me that elders were tired of talking about climate change because they had already shared their knowledge extensively and had seen little in the way of policy action or response. Global and scientific climate change discourses are also at odds with some of the ways that residents understand and relate to change. For example, Inuit observe natural cycles such as fluctuations in wildlife populations over time. Some residents explained to me that recent changes in the environment may be part of a cyclical process. Other community members made connections between environmental indicators, such as more flowering plants and shifts in the number and range of animals and fish, and evangelical religious discourses of repentance and healing (Johnson 2012). A Western scientific perspective, in contrast, correlates similar indicators with climate change.

Rather than research on climate change, then, Clyde River community members were interested in initiatives that would build and strengthen social ties, assist with traditional knowledge transmission and teaching land-based skills to youth, and support other health and wellness needs such as suicide prevention, access to healthy foods, drug and alcohol abuse prevention and treatment, and counseling and mental health services.

To give one example: in the fall of 2010, I accompanied a group of Inuit women from Clyde River and the nearby community of Pangnirtung on a four-day berry-picking trip. The trip was funded by a Canadian federal agency, Health Canada, through a program focusing on climate change and health adaptation. The stated purpose of the trip, described in the proposal that Ilisaqsivik submitted to Health Canada, was to document Inuit women's observations of climate change, since much of the research on Inuit and climate change has focused on men's hunting activities (Dowsley et al. 2010). For the participants, however, the trip was about building social ties by spending time together on the tundra and bringing home as many berries as possible to share this delicious and culturally important food with their families.

The women spent long days patiently filling buckets with crowberries and wild blueberries. While they picked, they shared stories and talked about challenges and hardships they had faced and overcome, associated both with past life on the land and with living in permanent settlements in the present. Climate change was never discussed overtly, but these activities were a way for the women to process change together and to strengthen their networks, reflecting a grassroots, holistic perspective that understands that social and environmental changes are inextricably connected.

## *Conclusion*

Inuit in the Canadian Arctic, along with other Arctic Indigenous peoples, have been among the first to observe and experience significant changes in Arctic ecosystems brought about by global climate change. Changes such as melting sea ice and permafrost and unpredictable and extreme weather have direct impacts on Inuit health and wellbeing. Far from experiencing these changes as uniquely environmental, however, Inuit view them in the context of a longer history of social change associated with life in settlements under governance systems that often fail to consider or understand local priorities and perspectives.

Inuit understand that to respond proactively to change requires working to heal and address past and present traumas and challenges, such as those incurred by social policies like the residential schools program or the presence of alcohol and drug abuse in their communities. It also requires maintaining strong social connections and working to ensure that land skills are passed on to younger generations, while being flexible and adaptive in the face of changing social and environmental conditions. Institutions like the Ilisaqsivik society play an important role in supporting these various wellness activities and initiatives. Flexible social and health policies that strengthen Indigenous and local institutions will help ensure that bureaucratic governance systems support rather than interfere with the ability of community members to prioritize and act on their own understandings and visions of change.

## *References*

Dowsley, M., S. Gearheard, N. Johnson, and J. Inksetter 2010 Should we turn the tent? Inuit women and climate change. *Etudes/Inuit/Studies* 34(1):151–165.

Gearheard, S.F., L.K. Holm, H. Huntington, J.M. Leavitt, and A.R. Mahoney 2013 *The Meaning of Ice: People and Sea Ice in Three Arctic Communities*. Montreal, Canada: International Polar Institute.

Huntington, H., S. Fox, F. Berkes, and I. Krupnik 2005 The changing Arctic: Indigenous perspectives. In *ACIA Scientific Report*, Arctic Climate Impact Assessment. Pp. 61–98. Cambridge, UK: Cambridge University Press. Accessed from http://www.acia.uaf.edu/PDFs/ACIA_Science_Chapters_Final/ACIA_Covers_Final.pdf.

Johnson, N. 2012 "Healing the Land" in the Canadian Arctic: Evangelism, knowledge, and environmental change. *Journal for the Study of Religion, Nature and Culture* 6(3):300–318.

Shearer, C. 2012 The political ecology of climate adaptation assistance: Alaska Natives, displacement, and relocation. *Journal of Political Ecology* 19:174–183.

Weatherhead, E., S. Gearheard, and R.B. Barry 2010 Changes in weather persistence: Insight from Inuit knowledge. *Global Environmental Change* 20:523–528.

Wenzel, G. 2009 Canadian Inuit subsistence and ecological instability: If the climate changes, must the Inuit? *Polar Research* 28(1):89–99.

---

In the context of a changing Artic climate, the Inuit are finding new ways to build their social community through re-learning traditional forms of hunting and food gathering. Healing the ills of alienation from the land, compounded by a sedentary lifestyle and unhealthy food choices at the local stores, is achieved through engaging in hunting, fishing, and berry picking—all activities that provide natural foods and opportunities for elders to narrate Inuit ways of living to younger members of their community, in ways that engender pride and wellbeing in the participants. The old ways of sensing and making sense of the environment in the Arctic are changing with global warming, thus introducing uncertainty into the lives of the Inuit. The environment of the Arctic is a bellwether for how our world's climate is changing, shifting, and becoming more fierce and unpredictable, as we will see later in this chapter.

## The Hostile Environment

Government agencies, anthropologists and other scholars, and lay publics have all played a role in monitoring the environment, introducing and enforcing legislation to reduce risk, and attempting the clean-up of areas which are known to be dangerously polluted. There has been growing attention, in these different domains, to the short- and much longer-term effects of environmental toxicity, and, at times, evidence of the suppression of evidence that might hold particularly companies or governments culpable. Anthropological and other social research on the Chernobyl nuclear accident mentioned above (Petryna 2003), the gas leaks in Bhopal in 1984 (Fortun 2001), and on the leaks at Fukushima nuclear plant following the Tohoku earthquake and tsunami in 2011 (Furukawa and Denison 2015) all illustrate how power is played out in the most shocking environmental disasters.

But human health is not only negatively affected by such dramatic examples of pollution. Air, ground and water pollution, food contamination, toxins used in manufacturing such as, historically, asbestos and lead, and radiation, all create risks to human and other animal health. Industrialization and development, and in this context, the use of toxic chemicals in industrial practices—in mining, agriculture, and manufacturing—all impact people's health. Health is affected, for instance, by outdoor air quality from vehicles, smokestack emissions, fossil fuel used for heating, and from direct exposure to and drift from farm chemical sprays. Additives to food to extend shelf life, a necessity where food is shipped across continents, can also compromise health unless tightly regulated. Indoor air quality, including in relation to the preparation of food, and noise and air pollution in working environments, all too have negative impact on human health, leading to public sensitivity of the environment as pathogenic. We constantly move through dangers of all kinds: from micro-dangers like allergens and toxic chemicals to macro-dangers like fast-moving vehicles, hurricanes, tsunamis, forest fires and earthquakes.

In the following case study, Roberta Raffaetà describes a region in Italy that has been severely polluted from manufacturing industry and commercial agriculture. It is also an environment that is increasingly urbanized. The chemicals in the environment, both outside and indoors, are perceived by local residents to play a role in causing their allergies. Allergy sufferers struggle in vain to get away from the chemicals by physically removing themselves from the polluted environment; their desperation is palpable. Italian medical doctors also pay significant attention to toxic interpersonal conflicts that may cause allergies, and they routinely prescribe psychological consultations and psychiatric medications for allergy sufferers. The environment writ large and small is layered in, around, and through us. When that environment is toxic, the sensitivities and reactions that result can be devastating. In her case study, Raffaetà takes us deep inside the disease, to investigate the microclimate of people's homes and psyches. The way that we react to what is going on around us in the most intimate of spaces is of primary concern for her participants suffering from allergies.

## 11.2 Environmental Pollution and Allergies

*Roberta Raffaetà*

Allergic symptoms are increasing globally. Epidemiologists estimate that by 2015, 50 percent of the global population will suffer from some kind of allergic reaction. Despite its widespread occurrence, there is still much debate about the causes of allergies (Raffaetà 2011a). In my own field research, conducted between 2004 and 2008 in Verona (a medium-sized city in northeast Italy), environmental pollution was identified as the main cause by people suffering from an allergy (Raffaetà 2012; 2013). Allergologists, on the other hand, usually refused to recognize environmental pollution as an important cause of allergies and explained allergic patients' 'obsession' with environmental pollution simply as a sign of psychological problems (Raffaetà 2011b), and after consulting an allergologist, people with symptoms of allergy often ended up with a referral to a psychologist.

In biomedicine, the 'hygiene hypothesis' suggests that allergies are provoked by living in a world where bacteria and parasites are significantly reduced by public health interventions such as the wide use of antibiotics and mass-scale vaccinations. Lifestyle changes such as improved hygienic conditions, few siblings, and much time spent indoors are also implicated. All these factors have reduced the opportunity for individuals to get infections and so be exposed to a wide variety of organic substances. According to the hygiene hypothesis, the lack of bacterial and parasitic infection and the scarcity of biodiversity put our immune systems into a state of inactivity. The immune system, in this context, triggers randomly against otherwise harmless substances such as pollens, latex or nuts, provoking allergic symptoms. This hypothesis is consistent with very recent attempts to think about health and illness in terms of a microbial balance. The human microbiome is defined as the ecological community of various microorganisms contained in the human body and on its surface. Recent studies highlight the crucial role of the human microbiome in the regulation of many vital functions, including those related to immune reactions.

The hygiene hypothesis deals with the alteration of the human microbiome caused by a modern lifestyle that has lost contact with parasites and bacteria, the so-called Old Friends in the biomedical literature. Unfortunately, it neglects what I term the 'New Enemies,' new entities represented by the wide array of chemicals and toxins. These, too, have an impact on the human microbiome.

The city of Verona, and its surroundings, are located in an area that experienced massive industrial development after the Second World War. While agriculture continues to play an important role, small farms have been sold to larger agricultural companies and agriculture has increasingly become industrialized and intensive, changing the organic composition of soil and air, the appearance of the landscape, and the ways in which people live. Laborers, traders, service workers and white-collar workers have replaced peasant farmers, and animal breeders. A strange smell—a mixture of dung, cattle feed, fertilizer and fumes—pervades the plains surrounding the city. The rivers, once part of the social and economic life of this region, have become polluted and are no longer used by local residents. People are now in frequent contact with various substances and technologies that have altered practices, habits and relationships to the human and non-human world; new chemical particles and toxins have appeared. New molecules have been introduced through technological processing (such as the introduction of GMO); the effects on human health of which are still under debate. Moreover, technological substances introduced in people's daily lives are usually standardized in their composition, so aggravating the problem of biodiversity depletion. These changes surely matter for people suffering from an allergy.

Environmental pollution strongly influences people's daily strategies to prevent, treat and live with their allergies. Below, I demonstrate that environmental pollution matters for people suffering from allergies. Allergic symptoms prompt us to think outside psycho-sociocultural *or* physiological categories. Allergic bodies bring to light a different order of reality, one that ignores comforting dichotomies and that can be defined as the order of a *body-in-relation-to.*

### *Bianca*

9 AM, 3 November 2004; the waiting room of the allergology department of a public hospital. Outside it was very foggy, lightly raining and still quite dark, a typical autumn early morning in Verona. About 15 people were waiting for their appointments. Some had already been tested

but had to wait a couple of hours more for further testing. Apart from reading some old magazines, people in the room were quite bored. While waiting, they were glad to chat with me. I approached Bianca, a 68-year-old widow who described herself as "an housewife who feels well when she stays at home. I am not one of those women who like to go around. . . . " However, she continued, recently her beloved house had transformed into her worst nightmare. During the summer, the wall behind the kitchen tiles had been painted. The day of the painting, she had cleaned the doors of the kitchen with bleach, but they were particularly dirty so she also used a strong cleaning powder. While she was busy cleaning the doors, her nose began to itch. A few minutes later, her eyes and her tongue began burning. Initially she thought that the cleaning powder might have caused the reaction. She went to the emergency room but, after some tests, the doctors told her that the reaction was not caused by the cleaning powder and that her symptoms would disappear in few hours. The next day, patches of eczema appeared on her back. While these responded to a topical cream, the burning in her nose, tongue and eyes continued to worsen: "The burning is so intense, it is so disturbing. When I go to bed I cannot pull up the sheets, I have to keep them below my waist because the burning is so painful." She also began to notice a strange smell on herself and in the house:

> I do not know what this smell is. I smell it on me, but nobody else can smell it. When I am at home [she sadly laughs], I continue to smell the walls, the furniture. I open the fridge, and from there also comes that smell. But I do not know what that smell is! I think it comes from the walls because I smell it when it is humid, I sense it is in the wall, in the tiling on the floor and in the furniture. I can smell it on me, too. I ask my daughter and other people to smell me [pause] also when I pick up the phone and I raise the receiver to my ear to speak I perceive the smell. What is it? Does it come from my mouth, from my body?

Bianca looked desperate. She said she could not stay inside her house anymore. When she gets up in the morning, she tries to go out as soon as possible to go to shopping:

> Even if I do not buy anything, it is an excuse to go out. I try to keep myself busy and to stay out until midday. At midday I come back home. If it is sunny I open the door. Some days, however, if I open the door I feel the smell even more intensely. Maybe this is because of the humidity, which comes from outside, and gets on the walls, contributing to the smell.

She cooks her lunch and waits until three o'clock, when her daughter—who works in a bar—comes back home. Soon after her daughter's arrival, she leaves again and goes to the public gardens. But when I meet Bianca, it was getting cooler in Verona and the days were getting shorter. Bianca was anxious because for the last week she had not gone to the gardens. "What's the point of getting cold? It is so wet, there is a lot of humidity. As soon as it is sunny it is ok, but after that. . . . I cannot continue such a life. I do not know what to think anymore; I do not know where to spend my days." Her nights were also difficult. With the winter approaching, in the evening she had to switch on the heating for a couple of hours, but then she switched it off during the night, otherwise her nose, tongue and eyes would begin to burn.

Bianca was convinced that chemicals in the paint caused her condition. One of her friends told her that other people had the same reaction after having painted the walls of their homes. But none of the medical doctors could find anything wrong with her and she was left without care: "I do not know anymore what to think, where to go. . . . I even went to the Ministry of Health. They told me that chemical allergies are difficult to trace and that I have to be patient. I've been waiting, but I am still here, feeling stupid, and the winter is approaching." She also went to the company that had painted the kitchen to express her concerns. "They told me that it might be true, they did not say that the paint is harmless! They said that I might be sensitive to some ingredient. They said that I am the first person to have these symptoms. And then they gave me the list of the components of the paint, but I don't think that the list was complete." The high number of chemical compounds, their ubiquitous nature in her house, and the various interests involved all complicated Bianca's quest for care. The uncertainty of the diagnosis, the need to reshape her routine and the poor recognition of her bodily symptoms were making Bianca's life very difficult: "I was never sick, it is the first time I have ever had a serious ailment. I might be laughing now while we're talking [pause] but if you could see what is inside me [pause], this condition is going to give me a nervous breakdown."

Bianca's daughter and neighbors accused her of suffering from a psychosomatic disorder. Bianca refuted this, affirming that what she felt was "true and real." How 'real' was environmental pollution in Bianca's life? She had visible bodily reactions—the eczema and the constant burning sensation in her nose, eyes and tongue, and she had made very tangible modifications to her daily routines as a result to avoid contact with what she believed was producing the negative effects in her body. These embodied alterations stand as 'real' against the politics of multiple regimes of scientific and biomedical 'truth.'

## *Erika*

It was 4 PM, and I was in a bar in the city center to meet Erika. I easily recognized her from her description, and we introduced ourselves; we ordered something to drink and started to talk. Erika told me how she had suffered from allergies since childhood. As an infant, she had severe dermatitis that covered her entire body, and was treated with large doses of cortisone. This provided temporary relief. Subsequently, she was diagnosed as suffering from mite allergy and an allergy specialist prescribed a cycle of immunotherapy. This, too, had limited effect. Finding no cure, after some years she turned to a homeopath who prescribed pills to detoxify her. The detoxification process forced her to bed for three days, with weeping sores and high fever. According to her homeopath, this reaction was caused by the elimination of toxins that had accumulated over the years from exposure to chemicals in her food and the environmental. Since then, she reported feeling much better, although her allergies had returned. At this point, with the help of a psychotherapist, she became aware of the fact that she 'polluted' her body by an excessive attachment to her mother; Erika's father died when she was 10 years old. After modifying the relationship with her mother, Erika said that her symptoms had diminished. She then encountered a naturopath who suggested that she join a breathing meditation group: "Initially my symptoms became very bad. I developed eczema everywhere, my breasts and genital areas included. After the initial rash, though, I healed. It was as if I had spat everything out, I was clean. There were no more toxins, there was nothing. Now I can even scratch my skin when I am stressed; my skin is resilient." Erika told me that she was starting to enjoy her allergy-free life. Even so, she confessed, one allergy—to cats— persisted. She told me that if one can understand the real cause of an allergy you should be able to eradicate it. A friend of hers had successfully resolved her son's allergy to dogs, which had a psychological basis. Erika tried to identify how cats could be linked to her personal history but could find no connections.

As we continued to chat while walking to our cars, Erika reflected on life's unpredictability and told me how her sister died when Erika was 14. "I think that certain things have to happen, that's all. In the case of my sister's death, for example, there was no logic. It was really unbelievable, if you think. She was driving her scooter and a cat was in the middle of the street. Probably she tried to avoid it . . . she crashed and died from her injuries. She wasn't going fast at all. Imagine, dying for a cat!" At these words I felt the blood freeze in my veins. I felt trapped in a game that I did not want to play; quite naively, I assumed that one of the rules of a good researcher was to try to not interfere too much with the research participants' lives. But the connection was too apparent. I looked at Erika and I could not stop myself from saying: "Erika, do you still ask yourself why are you allergic to cats?!" She shuddered, looked at me in wonder and brought her hand to the mouth. Bewildered, she stared at me and whispered, "How could I not think of that . . . " About two months later I received a phone message: "Thanks Roberta. I am not allergic anymore to cats! Hugs Erika."

Following the traditional model of the division of psyche and soma, Erika's narrative about the causes of her allergies could be divided in two main types of explanations: one is material and the other is psychological. In the first case, she refers to 'toxins,' 'pollution' and 'chemicals' with reference to the physical world; in the second case she speaks of 'toxins' produced from an unhealthy family relationship or unresolved traumas, toxins produced in the socio-affective environment that were the products of a relationship between her and other people. The two analytical categories are, however, not so clear-cut. For example, when Erika recounts the breathing technique to remove toxins, it is not clear if she considered the toxins as from the physical environment or from the complicated affective relationships with her mother. The two are so interlinked that for Erika, there was no need to distinguish them: she spoke both of "spitting out toxins" and to "stress." The two were blended in her account. This was not just a matter of language; Erika's skin spoke even louder than her words, in her mind, inseparable from her father's death, her mother's suffocating attachment, and the cat that caused her sister's death.

Her life history and the emotions attached to it were literally toxic, embodied in her flesh, receptors, bone marrow and the millions of fibers attached to it.

### *The Sustainability of Bodies-in-Relation*

By taking seriously the embodied experience of people with allergic reactions, who describe their illness as linked to environmental conditions, it is possible to depict the environment as a complex set of relations between various elements. The 'reality' of the impact of pollution on people's bodies emerges in the complex assemblages of sensory perceptions, discourses and practices. Suffering from allergic symptoms is to enact—both through the flesh and the self—a breakdown in the relationship with what is outside one's body and one's self.

Allergy helps us think about the human condition as an unfolding field of relations between the human and non-human world. These relations are not separated, but constitutive of what is going on throughout the organic world. Between emotions, organic tissues of bodies and toxins ,there is a system of "mutual constitution from which no particular element emerges as the origin, predetermining term"(Wilson 2004: 19). The relation between the elements, rather than the elements themselves, determines the final configuration. An allergy is not just a metaphor or a representation of an emotional reaction to external relationships; it is one of the various shapes taken at the level of the skin, the nose, the eyes, the tongue and the body (Raffaetà 2012). Sensitivity in touching, eating or smelling certain substances does not *represent* some emotional state; it is a direct enactment of toxic relationships.

Tim Ingold (2011: 70, 71) argues that "(t)hings *are* their relations . . . 'the environment' might, then, be better envisaged as a domain of entanglement." A crucial question, however, is rarely addressed: which relations are sustainable for us? What is the threshold of sustainability for human beings? Anthropology does not simply argue that all people are in relation with others—leaving things politically ambiguous; anthropology's role is to investigate the place of human beings within these relational webs. As Latour (2004: 5) has asserted, "political ecology has nothing to do with nature," in the sense that what is usually understood as 'nature' will survive no matter what transformations take place. The real problem is the survival of human beings in shifting landscapes. It is up to us to decide if we want a sustainable future or if we want to become allergic to whatever surrounds us.

### *References*

Ingold, T. 2011 *Being Alive. Essays on Movement, Knowledge and Description.* London and New York: Routledge.

Latour, B. 2004 *Politics of Nature. How to Bring the Sciences into Democracy.* Cambridge, MA and London: Harvard University Press.

Raffaetà, R. 2011a Understanding allergy: A review of relevant studies. Curare. *Zeitschrift für Ethnomedizin* 34(3):182–192.

——— 2011b The allergy epidemic, or when medicalisation is bottom-up. In *On Bodies and Symptoms. Anthropological Perspectives on their Social and Medical Treatment.* S. Fainzang and C. Haxaire, eds. Pp. 59–77. Tarragona, Spain: URV Publicacions.

——— 2012 Conflicting sensory relationships. Encounters with allergic people. *Anthropology & Medicine* 19(3):339–350.

——— 2013 Allergy narratives in Italy: 'Naturalness' in the social construction of medical pluralism. *Medical Anthropology* 32(2):126–144.

Wilson, E. 2004 *Psychosomatic. Feminism and the Neurological Body.* Durham, NC: Duke University Press.

In contrast to viewing the home as a dangerous place, MacPhee's work (2012) on household health practices in Southern Morocco focuses on how women's daily routines are replete with small protective acts within the environment of the home that range from dietary choices to prayers for the protection of health and wellbeing. Humans hedge against dangerous environmental encounters through a myriad of acts of protection; they are constantly alert to the sensory signals around them. The anthropology of the senses adds to our understanding of the environment by including the ways that sounds, tastes, and smells are made culturally meaningful over time and through processes of embodied discourses. Seremetakis (1993) talks about this process of meaning-making as *commensality* that she defines as "the exchange of sensory memories and emotions, and of substances and objects incarnating remembrance and feeling." Incarnate (from the Latin *incarnare*, 'made flesh'), past frights, pollutants, and contact with bad things are written into persistent eczemas, swollen eyes and inflamed spaces of the lungs. These are meaningful commentaries on how humans suffer within their lived and perceived environments.

## The Bigger Picture

Increasingly, scholars are grappling with how human health is affected by variations in the environment, most recently due to economic development, climate change, and human and animal mobility and habitat. A growing number of studies of health and the environment are now falling under the appellations of 'One Health,' 'Ecohealth' and 'Microbiome' studies (Wolf 2015). This work has built on and extended the biomedical and social research conducted from the 1970s on 'tropical diseases,' of which those that are vector-borne highlight the relationships between humans and their natural and built environments (Manderson et al. 2009). Deep understandings of the connectivity of humans and other living things is emerging from research on parasitic infections, pandemics, re-emerging and newly emerging diseases (Lowe 2010). Inter-species infections and environments modified through the twisted and polluted pathways that characterize so many parts of the globe are resulting in new disease panoramas that are best understood via multi-disciplinary teams that include social, biological and medical scientists. Anthropologically, we are challenged with the idea that there are patterns of culturally understood health risks present across these newly conceptualized environments.

Olson (2010), in her work on the ecobiopolitics of outer space travel, provides a window into thinking on this enlarged environmental scale. The NASA workers who Olson describes conceptualize themselves with respect to the harsh environments of outer space as they encounter it through the technologies of survival at their disposal. The spacesuits, spacecraft and all the attending space technologies are the outer milieus that enable human biological life to be extended into territories where it formerly would not have been possible. Along with that extension of life into new terrains comes a different form of normal that Olson calls "space normal" (2010: 172). The physiological stresses that the astronauts endure are captured via the data-gathering systems of the space shuttle and, importantly, they are understood and analyzed in conjunction with the functioning of the shuttle itself. The health of the astronauts is thus intertwined with the health of their transportation vehicle. Risk is calculated across time via the progressive changes in the functioning of both the humans and the machines that transport them into space. We return to the notion of risk below.

In the following case study, Ben McMahan takes us to coastal areas of the United States affected by hurricanes, oil and gas exploration, urbanization and pollution. The slow-moving changes of the environment are constantly in the process of creating new environmental dangers. The Gulf Coast that McMahan describes is an environment that is slowly shifting, vulnerable to the rapid onslaughts of regular and extreme atmospheric events like hurricanes and flooding. The technologies of environmental control used along the Gulf Coast, the levees, the building codes, the atmospheric tools of prediction, are used, under normal conditions, around the clock; their functioning becomes especially critical during emergencies when they are essential for human survival.

## 11.3 Reading the Environment

*Ben McMahan*

> Every storm is important, but some storms they just have water over your baseboards—and I'm not trying to minimize that—but I'd take that over this flood we just had . . . but I have not talked to anyone that's ever seen anything of this magnitude. I'm talking back 80 years . . . I've seen many many talks, especially from meteorologists . . . they had models and they had slides . . . of where category 1 storms would be, category 2, category 3, category 4, and category 5, and where it would flood. . . . What I understand is that since this hurricane they have now quit judging water levels . . . they just can't do it—because no matter what category storm it is, it only matters what the surge can be . . . and this [Ike in 2008] proved it . . . they can't really predict what kind of surge it's going to be.
>
> *(Political official, Southeast Texas)*

The US Gulf Coast stretches for approximately 1,630 miles, from its southernmost point on the US/Mexico border in Texas, to the southern tip of Florida. There is a wide range of social and environmental systems along this arcing coastal landscape, including highly developed vacation and tourist resort communities, sprawling urban/suburban metropolitan zones, and undeveloped 'natural' spaces, many of which are within state and federally protected wildlife areas. There is also considerable sociocultural diversity, tied to complex and overlapping histories of numerous ethnic and social groups. People with African American and Afro-Caribbean, Cajun, Chinese, European, French, Indian, Mexican, Native American, and Vietnamese heritage have settled along this coastline, as waves of new arrivals have joined established groups.

The region has a long history of resource extraction industries—of timber, agriculture, seafood, and most recently, oil and gas. In Louisiana and parts of Texas, these industries expanded to drive economic growth and demographic change: the increased demand for a stable workforce in emergent and developing industries stimulated population growth and community development,[1] with a pronounced effect on the local physical environment. In Southeastern Texas and Southern Louisiana especially, oil-and-gas-related activities have become the economic mainstay of the region, with both good and bad impact socially, economically, and environmentally.

Hurricanes are prominent in both social and meteorological histories. Hurricane season comes every year from June to November, when the entire coast is subject to the possibility of a direct hit or a regional disaster. The experience of hurricanes is an important part of life on the Gulf Coast, and can act as a galvanizing force through the collective trauma of a disaster or crisis, or the anticipation of a storm season, even if no serious storms actually affect the community. These impacts are also tied to the local history and context of the community, and the experience of hurricanes is contingent on interactions within these nested social, economic, environmental, and historical contexts. In the unlucky event of a severe storm layering onto pre-existing social inequality, environmental degradation, or failures of government, the consequences can be catastrophic.

Hurricanes are best understood as complex assemblages of variable experiences, shaped by historical and environmental context, political and economic motivations, senses of place and regional identity, all of which play out in an increasingly chaotic sociotechnical system that lies at the intersection of social–environmental systems and natural meteorological hazards. Like many other 'natural' disasters (cf. unnatural disasters, Jackson 2005), the potential threat and the strategies to mitigate risks are tied to patterns of human intervention into 'natural' systems, in an effort to harness or deflect the power of the natural world. These interventions place ever increasing populations under threat of natural, social, environmental, or technologically mediated risks, especially when the protection systems fail, or other modifications alter the landscape in ways that further amplify risk (cf. Bürgi et al. 2004).

The threat of Gulf Coast hurricanes and their aftermath came into sharp relief during the 2005 hurricane season, when Katrina and Rita hit the Gulf Coast in quick succession. This pair of storms highlighted general gaps in preparedness capacity, and the levee protection system, as well as specific problems associated with ongoing social inequality and environmental degradation in the region. Hurricanes may be a persistent reality for the Gulf Coast, but not every season leads to an imminent threat. The possibility of crisis requires communities and government institutions to remain vigilant and prepared, lest they be caught off guard at a time of disaster. Hurricane Katrina demonstrated the consequences associated with the intersection of social conditions, political realities, and an environment of risk, and is a stark reminder of the consequences of the intersection of a 'natural disaster' with widespread and persistent social inequality and governmental inefficiency.

Widespread environmental modifications associated with the oil and gas industry (channelization, canal building, dredging, wetland degradation, etc.) made an indelible mark on the landscape over decades of intensive modification and development. Other interventions are visible in municipal and residential contexts, such as infrastructure projects using levees and seawalls or elevated homes, designed to protect households and communities from the threats associated with seasonal hurricanes and coastal flooding. This is a complicated dynamic, as oil industry activity degraded the local environment and amplified the potential damage of hurricanes and coastal flooding, while simultaneously enriching many of the residents living in these same communities. At a larger scale, the oil and gas industry has driven economic growth at community or regional scales, but Gulf Coast communities have suffered the consequences of this growth through coastal land loss, wetland fragmentation, pollution, and environmental degradation.

This highlights the complex way that local and regional risk-scapes are produced, and the clustering and accumulation of diverse risks, including economic volatility, social disruptions, hurricane threats, and coastal flooding, are important parts of the story of hurricanes and disaster in this region. To understand the everyday life of risk on the Gulf Coast is to see the system as a twisted network of interactions, of layered effects, and accumulating experiences. 'Social' systems and institutions are embedded within a 'natural' landscape, but this distinction is problematic, especially in a region where economy, environment, and social structures are so entangled.

The convergence of prediction technology, population management techniques, disaster preparedness and response planning forms the basis of the modern emergency preparedness project. Modern emergency preparedness strategies and logic are focused on the future in the present—the way in which descriptive statistics and predictive modeling can help officials better understand the potential of a possible storm, and therefore enact plans to better protect a population from harmful outcomes. Predictive technologies, meteorological models, and historical storm track data act as an organizing force for an otherwise unwieldy scientific phenomenon (hurricanes), and allow for order to be made of the chaos of storm seasons.

Storm models and predictions are designed as 'objective' measures of observable phenomena that lead to better understanding of storm patterns and better predictions of the possible tracks, storm potential, and outcomes of hurricanes. The manner of how models and data are used, the power embedded within their deployment, and the compulsory power and governance over populations they can engender through their production are also key components of a more careful look at these data and models. As predictive technologies become more accurate, increasingly complex, and better developed, it is important to understand how they are operationalized and deployed, and to understand the productive power or compulsory control they purport to hold over populations.

Modern emergency management strategies emphasize scientific predictions and bureaucratic recommendations (to evacuate: when and where), instead of relying on accumulated knowledge and experience. Discourses about preparedness quickly morph into discourses of personal responsibility in terms of following government recommendations, with the associated judgments of character and intelligence when these recommendations are not followed. Residents of coastal communities can choose to embrace this transition, but this discussion goes beyond simply passing judgment on those who do not respond to emergency management dictums, or otherwise ignore 'rational' decision-making processes in the face of a looming disaster. Expertise is no longer the sole province of credentialed experts, as citizens can empower themselves through this process of consuming (existing) and producing (novel) scientific data and models, as well as conveying this knowledge and expertise to friends or family.

### *Seasonal Hurricanes and Layered Effects*

National media, federal emergency management, and the general public typically treat individual hurricanes as unique and isolated events. But hurricanes are cumulative crises owing to the layered effects of spatial and temporal overlap between acute events and long-term changes to the environment, social disruptions, and institutional breakdowns. Emphasis on the acute over the chronic captures the particulars of a given storm, but this ignores the long-term effects that are embedded within the social and physical landscapes, and the cumulative effects linked to multiple and overlapping events.

Technological intervention in a built environment helps mitigate the acute threats that hurricanes pose, although these interventions do less to address long-term chronic changes to the landscape. These interventions serve a similar role to the maintenance strategies required for dealing with chronic health problems in individuals, and the parallels to chronic health are an instructive and intuitive link (cf. Thorne 1993). Parallels between managing for chronic landscape change and chronic health problems include the maintenance strategies required to deal with chronic effects, the threat that acute episodes pose to general (environmental) health, and the destabilizing effect that an acute event can have when layered onto a chronic problem. Whether as a metaphor or as a guide in designing policy, visualizing the landscape in terms of chronic environmental health helps focus on the holistic issues that shape clusters of experience, rather than acute interventions that target specific issues.

The chronic and (relatively) slow-moving nature of coastal environmental change also highlights one of the fundamental concerns about these systemic shifts within an environment of layered and overlapping risks. Solutions are typically oriented towards immediate tangible solutions to acute and fast-moving problems. The water is rising and we need you to elevate your home. The channelized river is at an ever-greater risk of flooding, and we need to build a seawall and bulkhead the river to mitigate this risk. There is a storm threatening the region, and we need you to evacuate. In isolation, each of these events has a solution that is unique to the context in which it is experienced. But by looking at the context of the US Gulf Coast and the environment of risk of hurricanes, from the perspective of the layered effects of hurricanes, it becomes clear that solutions to these layered crises will require similarly complex responses. If the wetlands are crumbling and storm surge flooding poses an ever-greater risk, simply elevating homes or evacuating for each storm is a set of stopgap measures. If the flooding continues to worsen over time, you will have to elevate your home even more or evacuate even more often. Building more and more homes and businesses in a hurricane-risk zone increases recovery and insurance costs, and at some point, people within and outside the region will question whether it is sustainable or cost-effective to repeatedly rebuild in an area under persistent threat.

A number of barriers exist to developing holistic responses to these challenges. One barrier is the sociopolitical context of environmental change on the US Gulf Coast in oil and gas country. Some environmental changes are directly attributable to the impacts that decades of oil and gas extraction activities have had on the landscape, such as extraction and subsidence, channel building, and so on, while others are attributable to ancillary activities associated with oil and gas activity (navigational traffic, climate change, etc.). Many people made their careers in the oil and gas industry, and the region leans right in terms of political orientation; both these facts hinder the embrace of an 'environmentalist' perspective. There is also a sense that technology will lead to advancements to further mitigate environmental risks (elevated homes and highways, new building technologies, advanced levee construction), and that it is only a matter of time before some advancement in technology renders all the "hand wringing about climate and hurricanes and the coast moot," as one research participant observed. This increases the dependence on technological intervention. There will come a point when these interventions are either not sustainable in terms of pragmatic concerns over cost-effectiveness, or quality-of-life issues over what degree of risk is tenable and acceptable, or both.

A second barrier to developing a holistic approach to landscape alteration and degradation is the way in which disasters are managed at the level of state and federal government. Disaster declarations are a complex process with political and economic implications that dictate when and how they are made. But a key feature is that they refer to an isolated event or disaster, and not to the chronic long-term conditions of Gulf Coast living. They can only be made for an acute event, not for the accumulative impacts of decades of gradual change. The cumulative effects of landscape changes, environmental degradation, social disruptions, and skyrocketing insurance costs may all come together to form the 'real' disaster facing Gulf Coast residents. This is a slow-moving series of linked events, not a discrete disaster easy to declare. For an acute event, you can pose for pictures as you write the checks, and move on to the next crisis. For a long-term chronic crisis, it is much more difficult to strategically position or frame a response. Recovery funding is similarly dependent upon acute storm events: a number of NGOs that expanded their operations in Texas and Louisiana, post–Katrina and Rita, were forced to scale back their operations as the funding dried up.

There are drawbacks to focusing on the layered effects when problems are complicated; they are rarely easily or elegantly solved. One reason that the operations of FEMA (Federal Emergency

Management Agency) are focused on acute events is the relative simplicity in management that goes along with writing a check for losses sustained in a specific disaster, rather than a complex analysis of the social and environmental factors amplifying the risk of hurricanes on the Gulf Coast. This ruthless pragmatism of government agencies has left a void, and two primary forces are working to address these complex problems, albeit from radically different perspectives. First, existing and novel non-governmental organizations are taking up some slack in addressing the effects of environmental risk along the coast. Second, the insurance system governing the Gulf Coast operates on a continuum between collective and actuarially monetized risk. On one end, case management and careful consideration of the experience of individuals and households helps NGOs better address local problems, while on the other, actuarial tables and rational choice models are deployed to determine what is cost-effective, and therefore permitted.

"The industries were there because of the river," John McPhee (1987) famously wrote in his essay in *The New Yorker*. "They had come for its navigational convenience and its fresh water. They would not, and could not, linger beside a tidal creek. For nature to take its course was simply unthinkable. The Sixth World War would do less damage to southern Louisiana. Nature, in this place, had become an enemy of the state."

## Note

1. There is a long history of this economic expansion, beginning with agriculture (especially sugarcane) and timber (cypress extraction), continuing into hunting, trapping and seafood, and culminating in the current state of oilfield work. See Austin and Woodson (2012) and McMahan (2014) for details.

## References

Austin D., and D. Woodson, eds. 2012 *Gulf Coast Communities and the Fabrication and Shipbuilding Industry: A Comparative Community Study, OCS Study*. New Orleans, LA: U.S. Department of the Interior, Minerals Management Service, Gulf of Mexico OCS Region.

Bürgi, M., A.M. Hersperger, and N. Schneeberger 2004 Driving forces of landscape change—current and new directions. *Landscape Ecology* 19(8):857–868.

Jackson, S. 2005 *Un/Natural Disasters, Here and There. Understanding Katrina: Perspectives from the Social Sciences*. New York: Social Science Research Council. Accessed from http://understandingkatrina.www.ssrc.org/Jackson.

McMahan, B. 2014 Environments of Risk in a Dynamic Social Landscape: Hurricanes and Disaster on the United States Gulf Coast. University of Arizona (USA), ProQuest, UMI Dissertations Publishing 13483.

McPhee, J. 1987 Atchafalaya. *The New Yorker*. Accessed February 23 from http://www.newyorker.com/magazine/1987/02/23/atchafalaya

Thorne, S.E. 1993 *Negotiating Health Care: The Social Context of Chronic Illness*. Thousand Oaks, CA: Sage Publications.

---

The ways in which cultures construct notions of risk are linked to their perceptions of dangers, the technologies that they use to detect and make visible what is dangerous, and the sociopolitical systems that set up the rules about how to respond to what is defined as a 'risk' (Cartwright 2013; Douglas and Wildavsky 1983). Risk becomes a concept that is multi-level, culturally specific and contextually enacted in particular socio-legal systems. Cartwright (2016), in her work on the health dangers associated with oil and gas fracking (hydraulic fracturing) and other occupational and environmental health issues, uses the term 'eco-risk' to describe the interactive process of constructing risks as they are embedded in and part and parcel of ecological systems. The environment is an integral component of the risk-milieu; the cultural appraisals of dangers are not created in an intellectual vacuum, but rather are responding to and modified by natural (and in the case of fracking, unnatural) forces that are key to how the risk is defined, understood and deployed. Eco-risks are often legally defined, substantiated and contested via the many existing socio-legal codes across the globe, further illustrating their sociopolitical presence and import.

The long-term chronic environmental crises described by McMahan, above, and Adams, below, can also be parsed through Singer's expanded notion of the term 'syndemic,' one that includes "climate change, social collapses and anthropogenic disruptions of environmental systems, and environmentally mediated class boundaries and distinctions" (Singer and Bulled 2014: 4). In this iteration of the concept of syndemic, there is an emphasis on the larger social and environmental disruptions that influence and create a plethora of new co- and multi-morbidities. This emphasis on various levels of pathogenicities is reflected in Adams's case study. Being poor, African American, and at the mercy of a variety of unscrupulous 'aid' organizations created a situation of never-ending despair in the aftermath of Hurricane Katrina.

## 11.4 Disastrous Recovery

*Vincanne Adams*

In the wake of one of America's greatest disasters, Hurricane Katrina and the breaching of levees and subsequent flooding of 80 percent of the city, New Orleans residents struggled to rebuild their lives, their homes and their city. That this was a task of extraordinary proportions was not surprising. That even after four and five years many residents were still struggling to rebuild and get out of their temporary housing was somewhat more surprising. Most surprising of all, though, was how all of this occurred despite enormous federal and state outlays of resources to 'speed up' recovery.

The story of how vast amounts of public money were spent to fuel profit-driven recovery industries that left many New Orleanians 'high and dry' is one that can teach us much about contemporary forms of market-driven governance or as some would call it "neoliberalism," and about how the machineries of disaster capitalism ultimately fail to ensure effective recovery for those who fall victim to disasters. In my book, *Markets of Sorrow, Labors of Faith: New Orleans in the Wake of Katrina*, I describe what happens when we allow the private sector to take charge of the public safety net. I argue that New Orleans' experience with disaster was not exceptional but rather exemplary for many communities experiencing disaster in the US today.

Take the case of Henry and Gladys Bradlieu, an elderly African American couple who owned their home in the middle-income neighborhood of Gentilly. Henry was a three-time Purple Heart recipient (for enduring injuries as a soldier in a US war), and Gladys was a retired clerk from city hall. Their house was flooded with up to 10 feet of water. When they returned from their evacuation to Texas three weeks after the floodwaters receded, they began living in a small Federal Emergency Management Administration (FEMA) trailer and tried to figure out how to rebuild. Temporary housing in a FEMA trailer and a check for several thousand dollars was the only effective disaster recovery help they would get from the government.

Henry and Gladys wondered how their historically safe neighborhood ended up nearly entirely underwater. Newspapers called it a 'natural disaster,' but most residents knew otherwise. The impact of the hurricane and the floods in New Orleans were hardly natural. They could be traced to both the destruction of protective wetlands in Southern Louisiana *and* the unrepaired levee system of the city, which in turn could be traced to revolving-door relationships between the Army Corps of Engineers (responsible for repairs to the levees) and large for-profit military subcontractors (Haliburton, Bechtel, the Shaw Group) who received large funds to do the government work of repairing them. For years, these companies ignored warnings about weakened levees and turned a blind eye to the environmental decline that came along with oil industries in the Gulf from which they were profiting. The wetlands south of the city had been disappearing at a rate of 13 square miles per year because of the sea channels, and by 2005 the natural landmass that would have protected the city from the harshest impact of the hurricane at landfall was gone. Whatever safety the levee system once insured for the city was long gone by 2005 as executives in these for-profit companies invested more federal resources in foreign oil wars than infrastructure at home.

The destruction from storm and floods was only the first disaster to come for residents of New Orleans. The second disaster would begin as soon as people like Henry and Gladys returned to rebuild. The second-order disaster was a result of recovery being slowed by agencies that found ways to profit on the publicly funded engines of recovery.

The first line of support for the Bradlieus, for instance, might have been insurance. They didn't have insurance in part because they had no mortgage, but even if they had, it is unlikely they would have gotten any payout. Most people only had hurricane insurance, and insurers refused to pay for flood damage. Ironically, the government supported insurance companies' claims that floods were not caused by the hurricane but from broken levees, even while the Army Corps claimed that the floods were not caused by unrepaired levees but from the hurricane. Market-driven governance works like this: the interests of corporations are placed above those of ordinary citizens who are victims of disasters.

A second option for the Bradlieus might have been to take a loan from the Small Business Administration program, a program designed during the Roosevelt years (1933–1936) to leverage Federal resources for victims of disaster. Here homeowners were invited to borrow money against their income and remaining property assets—to essentially turn their lives into a business investment. The Bradlieus didn't qualify for this. But even if they did, how could they take out a loan for a home they could no longer live in and that they would then not be able to afford? This would be "adding insult to injury," they said. Many of their neighbors agreed. Still, this was the only option for many residents, especially renters who did not own their homes. Once again, the banks that were offered these federally guaranteed loans were able to make money on the victims of disaster.

A third option was The Road Home Program. Here was the help the Bradlieus and so many others needed. The Road Home program provided federal funds to the state-run Louisiana Recovery Authority to give homeowners financial help to bridge the gap between what their insurance paid and what it cost to rebuild, based on assessed values of their homes. Henry and Gladys applied for Road Home funds, but this proved not to be an easy process. Complaints of the poor performance of the Road Home were rampant and uniform: uneven and slow distribution, they lost paperwork, demanded documents that didn't exist, and consistently undervalued homes, especially in African American neighborhoods. After nearly two years of waiting, the Bradlieus were denied funds on the grounds they could not show title to their home. This is because they bought their home through a bond-for-deed sale, directly from the owner—a strategy used by many African American families in the 1970s when most banks refused to give them mortgages. But, when Gladys went to the Road Home office to explain this, the officer they spoke to said he had never heard of bond-for-deed sales. "How could they work in New Orleans and not know about this?" Gladys asked.

People waited years for help, living in their trailers and blaming the government for the slow bureaucracy of the Road Home program. Yet the program was not actually run by the government at all, but by a for-profit company called ICF International. ICF had designed the program for the federal government, before being awarded its management in a no-bid contract. A month prior to winning this contract, ICF held an IPO enabling stockholders to buy ICF stock at around $12 per share. One month after being awarded the contract for the Road Home, their stock skyrocketed to over $25 per share and continued to grow throughout their three-year contract. They repeatedly gave out millions of dollars in bonuses to their executives, and despite massive complaints from homeowners that they were not getting the help needed, ICF managed to get even more funding as they neared the end of their contract. Since for-profit companies are rewarded for their stock portfolios and their ability to retain operating capital, accountability to recipients of aid was easy to neglect. This was true even after congressional oversight hearings that exposed the failures of ICF.

By year six after Katrina, only about half of the residents who applied for Road Home funds had received any money, and most of those who did complained that it was not nearly enough to rebuild. In 2008, the Bradlieus were still trying to figure out how to rebuild, still stuck in their trailer in the front yard of their gutted home. When, that year, FEMA told them they could buy their trailer for $25,000, they wondered why they should be asked to pay for it. Then they learn that the for-profit companies who built these trailers (Haliburton and Bechtel) and shipped them to New Orleans were paid roughly $229,000 per trailer. Henry and Gladys Bradlieu, like others in New Orleans, felt that had they been given that amount of money instead of Haliburton or Bechtel, they would have not only been able to buy a trailer of their own, but also to rebuild their home by then.

In year four after Hurricane Katrina, people were suffering. The stress of living in a state of waiting, of being in perpetual striving toward recovery but not making progress, of being uncertain about their financial future, took a toll on most residents. Mortality doubled in the first two years after the floods, with a three-fold increase in heart attacks. By year four, people were suffering from high rates of stress disorders, including eczema, asthma, hypertension and depression. People talked about being in a "never ending funeral," or being "on a hamster's wheel" that they could not get off, no matter how fast they ran. Suicides were rampant. We have referred to this "chronic disaster syndrome" (Adams et al. 2009).

The biggest source of frustration was with the Road Home program. Henry and Gladys petitioned over a four-year period to obtain Road Home funds. They would travel week after week to the Road Home offices, over unrepaired roads, through their still-devastated neighborhoods where there was no mail service, no streetlights, and where endless gutted or ungutted moldy homes still sat. Once at the Road Home office, they would be turned away. In 2009 Henry learned that they would be denied Road Home funds for a second time, even after obtaining affidavits from the previous owner. He told Gladys he was going to take a nap and when he laid down, he suffered a stroke that left him paralyzed and bedridden for the rest of his life. Gladys took care of him after that. Wiping his body in the sweat-soaked bed at the end of their trailer, trying to navigate the system of applications, arbitrations, complaints and denials, she was desperate, and completely on her own. There were days when she thought she could not go on.

Gladys's only hope now was with the volunteers. In the absence of effective help from federally funded for-profit companies that were making money on the disaster, many residents in New Orleans started to rebuild on their own. Coming to their aid in the first two years post-disaster were over 17,000 volunteers from all over the world. Henry and Gladys would only recover because of this help. Many of these volunteers were organized by their churches, many were simply good Samaritans. Returning residents started community organizations and pooled their resources to help one another rebuild.

Caroline, for instance, was a housewife and mother from Lakeview who returned and organized her neighborhood to dig out and help others rebuild on their own. In late 2009, she got a call from Gladys asking if she could help paint her home. Caroline said she could help her, but when she probed a bit further about the condition of Gladys's home, she said: "But she didn't even have any walls on her home. There she was, living in this trailer taking care of Henry. No walls on her home . . . Can you imagine, [Henry] a three-time Purple Heart recipient, and this was how he was treated?"

Caroline tapped into the enormous outpouring of volunteers who showed up at City Park, the Good News Camp, which hosted thousands of volunteers who camped out in tents and cooked in a communal kitchen. Every day, she would pick up the volunteers and bring them back to homes in her community. By 2009, the volunteers came from all over the country directly to her organization, a small non-profit rebuilding group that got funding from the Episcopal Diocese. She took these volunteers to the home of Henry and Gladys. She had some volunteers nailing drywall to the open studs, others cleaned up old windows so they could be put back in. One volunteer knew how to do electrical wiring, another, how to putty windows. In between these efforts, she went fundraising. Overwhelmed by the story of Henry and Gladys at the Episcopal diocese in Seattle, one parishioner shouted out, "Screw the government, I'll pay for their roof." And he did. He wrote a check for $4,000 right there on the spot. By June 2010, Caroline said: "800 volunteers and 2 years later, we got the Bradlieus back into their home."

★ ★ ★ ★

Time after time, we heard amazing stories of human compassion, of how a can-do spirit and an act of financial generosity got one family, then another, then another back into their home. I volunteered, and every year brought family members, young and old, to help out. This flow of volunteers, we should recall, was not simply a spontaneous response to a humanitarian crisis. The use of volunteers, charity, and faith-based institutions to fill in the gaps in the safety net was by design. During his presidency, George Bush Sr. called upon Americans to let a thousand points of light rise up and take care of those in need. By 2005, when Katrina hit, this idea of letting private sector charity fill in the gaps in the safety net had become institutionalized. In part as a means of deflecting attention from conservative efforts to reduce government spending on safety net programs and in part an effort to shift responsibility for these programs to the for-profit and non-profit private sector, the privatization of public institutions was nowhere more visible than in post-Katrina New Orleans. Public–private partnerships like the Points of Light Institute or HandsOn Network had become institutional responses to need in America. These organizations mobilize large numbers of volunteers and service workers by leveraging federal money in order to get private and corporate philanthropy into safety net activities. The government's Corporation for National and Community Service (also set up during the Bush years) helped ensure that these public–private infrastructures received federal money but also turned humanitarian work into profit-making opportunities.

People in New Orleans who volunteered or got help from volunteers talked about "putting hands and feet to the gospel" and "doing God's work." But the effort exceeded that of the churches and of religion. Volunteers described their experience as creating bridges between rich and poor, Black and White, religious and non-religious, Democrat and Republican. Among those who received volunteer help and those who gave it, one could witness a revitalization of sentiments of American exceptionalism. Volunteering enabled people to form authentic communities, egalitarian and presentist, giving them a feeling of being part of a community that was larger than themselves and that spread as far and wide as the nation itself.

The sense of emotional urgency aroused by the need to help that was seen in the volunteer sector post-Katrina New Orleans was powerful. In fact, we might think of the engines of charity-based recovery as forming an *affect economy* in which we increasingly rely on volunteer and charity workers to fill in the gaps in the safety net. It is important to remember that the labor in the affect economy is motivated by an emotional sense of obligation to help. It is work that usually is unpaid, underpaid or even paid for out of pocket by the volunteers themselves.

The growth of the charity/volunteer sector in the case of post-Katrina New Orleans is exemplary of the growth of this sector in the US, as neoliberal policies shift more and more of the safety net to the private sector, including churches, NGOs and philanthropy. But the New Orleans' experience also offers an opportunity to witness the corporatization of charity and volunteerism. Growing at a pace that exceeds faith-based charity, the new secular institutions, including philanthropy-based corporations, NGOs for volunteer and service work, are perceived as the newest success of neoliberal reforms. In the volunteer sector, we see the blurring of for-profit and non-profit, public and private, volunteerism and underpaid labor—blurrings that are seen as innovative solutions to age-old problems of socioeconomic poverty, and not just responses to disaster relief. But, as philanthrocapitalism creates opportunities to merge business strategies with the work of humanitarianism, even non-profit charities are now asked to run themselves like, and merge with, for profit businesses.

Take, for instance, the return of ICF international that so tragically bungled the dispersal of federal resources to returning homeowners post-Katrina. After its contract with the Road Home program ended, ICF turned its attention toward opportunities that would enable it to get government subcontracts to help oversee and manage support for—you guessed it—faith-based volunteer community organizations. The company that had failed to put market strategies to work for the recovery of returning residents was now going to obtain federal funding to marketize the world of grassroots, non-profit volunteer services. When Caroline heard about this, she said, "It's enough to make me sick." Considering that most volunteers do not get paid (and often pay to do this work), and that others are usually grossly underpaid to do service work, one wonders why companies like ICF should earn any profit at all from this.

Henry Bradlieu died in 2011, three months after he moved back into his home. Gladys held the funeral and wore a creamy white dress that Henry loved to see her in and she rented a white Cadillac for his second line, a musical funerary procession that follows after the formal burial procession. She was happy to let me tell her story so the whole world would know what went on down in New Orleans. Her story is not just about what happened in New Orleans, but about a more pervasive crisis in America as we turn our safety net responsibilities over to the private sector and leave humanitarian interventions to the demands of the for-profit market. As climate change brings more disasters like Hurricane Katrina into our midst, let us only hope that the engines of neoliberal governance are revved up to respond in ways that can be improved upon.

### *References*

Adams, V. 2012 *Markets of Sorrow, Labors of Faith: New Orleans in the Wake of Katrina.* Durham, NC: Duke University Press.

Adams, V., T. VanHattum, and D. English 2009 Chronic disaster syndrome: Displacement, disaster capitalism and the eviction of the poor from New Orleans. *American Ethnologist* 26(4):615–636.

## The Pluralea of the Anthropocene

In a recent article, Merrill Singer (2009) discusses the need for an 'engaged' medical anthropology. In elaborating on this, he writes of the role of anthropologists working in collaboration with the communities they study, and with activist movements, so applying anthropological insights to amelioration. While this might apply to any of the work we do, he relates this specifically to environmental research, noting that "in light of recognition of pluralea interactions as representing grave threats to all human futures, there is a critical need for a new applied narrative of environmental health equity and action" (Singer 2009: 815–816).

In her case study above, Vincanne Adams highlights how new forms of organizations emerged from the environmental and social devastation produced by Hurricane Katrina. In the aftermath of the hurricane, new non-profit charities took over when governmental agencies failed; for-profit businesses came about to regulate the new non-profits. Comparisons with other regions and countries that have experienced similar large-scale environmental disasters shows the cracks in the particular social infrastructures of affected areas. Who responds? Who lends a hand and why? Anthropologists have provided rich analysis of the social outcomes of disaster in relation, for instance, to the earthquake in Haiti (Farmer 2012), the nuclear pollution of Chernobyl (Petryna 2003), and the tsunami in Sri Lanka (Hastrup 2011). Large-scale environmental disasters and their human sequelae are increasingly important topics for medical anthropologists to study as we move through the Anthropocene, where the health of vast numbers of humans is negatively affected by an increasingly sullied and unnatural world.

In concluding this chapter, we return to two First Nation tribes in Alberta, Canada: the Mikisew Cree First Nation and the Athabasca Chipewyan First Nation. These two groups are located below the third-largest deposit of bitumen (crude oil) in the world—the Athabasca Oil Sands. The area, along the banks of the Athabasca River, was once teaming with wild game and abundant harvests of edible plants. It is now in close proximity to the gargantuan open pit mines in the Oil Sands, a site of near-unimaginable environmental devastation. In a community-based, participatory research study that combined Indigenous tribal knowledge and state-of-the-art toxicology, these two tribes, in conjunction with researchers from the University of Manitoba and the University of Saskatchewan, carried out interviews and environmental testing and reviewed medical records of tribal members living in the area. The results of the study indicated that eating traditional, wild foods and/or working in the Oil Sands resulted in elevated cancer and other illness rates, especially among women. The researchers found high concentrations of arsenic, mercury, cadmium and selenium in animals traditionally hunted by the tribes. The moose, ducks, muskrats and beavers living in the area are now considered to be unfit for consumption (Tyas 2014). Less wild meat and much less wild fish is consumed by tribal members now than a generation ago. Like the Sami, the Inuit and others living closely to nature, consuming wild foods is part of the cultural expression of identity and a validation of one's place in the environment; when these foods are no longer fit for human consumption part of the culture dies. Preserving the environmental knowledge of these Indigenous groups gives us a vision not only of what life on Earth was like in the past, but, it is hoped, will allow us to imagine and create a planet that is healthy once again.

## References

Basso, Keith H. 1996 *Wisdom Sits in Places: Landscape and Language among the Western Apache*. Albuquerque, NM: University of New Mexico Press.

Cartwright, Elizabeth 2001 *Espacios de Enfermedad y Sanacion: Los Amuzgos de Oaxaca, Entre la Sierra Sur y Los Campos Agricolas de Sonora*. Hermosillo, Sonora, Mexico: El Colegio de Sonora.

——— 2007 Bodily remembering: Memory, place, and understanding Latino folk illnesses among the Amuzgos Indians of Oaxaca, Mexico. *Culture, Medicine and Psychiatry* 31(4):527–45.

——— 2013 Eco-risk and the case of fracking. In *Cultures of Energy*. S. Strauss, S. Rupp, and T. Love, eds. Pp. 201–212. Walnut Creek, CA: Left Coast Press.

——— 2016 Mining and its health consequences: From Matewan to fracking. In *A Companion to Environmental Health: Anthropological Perspectives*. M. Singer, ed. Pp. 417–434. Hoboken, NJ: Wiley.

Cushing, Frank Hamilton 1981 *Zuni: Selected Writings of Frank Hamilton Cushing*. Lincoln, NE: Bison Books, University of Nebraska Press.
Desjarlais, Robert 2003 *Sensory Biographies: Lives and Deaths among Nepal's Yolmo Buddhists*. Berkeley, CA: University of California Press.
Douglas, Mary, and Aaron Wildavsky 1983 *Risk and Culture: An Essay on the Selection of Technological and Environmental Dangers*. Berkeley, CA: University of California Press.
Farmer, Paul 2012 *Haiti after the Earthquake*. New York: Public Affairs.
Feld, Steven 1990 *Sound and Sentiment: Birds, Weeping, Poetics and Song in Kaluli Expression*. Philadelphia, PA: University of Pennsylvania Press.
Fortun, Kim 2001 *Advocacy after Bhopal: Environmentalism, Disaster, New Global Orders*. Chicago, IL: University of Chicago Press.
Furukawa, Hiroko, and Rayna Denison 2015 Disaster and relief: The 3.11 Tohoku and Fukushima disasters and Japan's media industries. *International Journal of Cultural Studies* 18(2):225–241.
Haraway, Donna 2015a Anthropocene, Capitalocene, Chthulocene: Donna Haraway in conversation with Martha Kenney. In *Art in the Anthropocene: Encounters among Aesthetics, Politics, Environments and Epistemologies*. Heather Davis and Etienne Turpin, eds. Pp. 255–270. London: Open Humanities Press.
——— 2015b Anthropocene, Capitalocene, Plantationocene, Chthulucene: Making Kin. *Environmental Humanities* 6:159–165.
Hastrup, Frida 2011 *Weathering the World: Recovery in the Wake of the Tsunami in a Tamil Fishing Village*. New York: Berghahn Books.
Latour, Bruno 2014 Anthropology at the time of the Anthropocene—a pesonal view of what is to be studied. *Distinguished Lecture, Meeting of the American Anthropological Association*. 16 pp. Washington, DC. Accessed from http://www.bruno-latour.fr/sites/default/files/139-AAA-Washington.pdf.
Lowe, Celia 2010 Viral clouds: Becoming H5N1 in Indonesia. *Cultural Anthropology* 25(4):625–649.
MacPhee, Marybeth 2012 *Vulnerability and the Art of Protection: Embodiment and Health Care in Moroccan Households*. Durham, NC: Carolina Academic Press.
Manderson, Lenore, Jens Aagaard-Hansen, Pascale Allotey, Margaret Gyapong, and Johannes Sommerfeld 2009 Social research on neglected diseases of poverty: Continuing and emerging themes. *PLOS Neglected Tropical Diseases* 3(2): e332.
Olson, Valerie A. 2010 The ecobiopolitics of space biomedicine. *Medical Anthropology* 29(2):170–193.
Petryna, Adriana 2003 *Life Exposed: Biological Citizens after Chernobyl*. Princeton, NJ: Princeton University Press.
Seremetakis, C. Nadia 1993 The memory of the senses: Hisotrical perceptions, commensal exchange and modernity. *Visual Anthropology Review* 9(2):2–18.
Singer, Merrill 2009 Beyond global warming: Interacting ecocrises and the critical anthropology of health. *Anthropological Quarterly* 82(3):795–819.
Singer, Merrill, and Nicola Bulled 2014 Ectoparasitic syndemics: Polymicrobial tick-borne disease interactions in a changing anthropogenic landscape. *Medical Anthropology Quarterly*. DOI: 10.1111/maq.12163.
Stephens, Sharon 1995 Physical and cultural reproduction in a post-Chernobyl Norwegian Sami community. In *Conceiving the New World Order: The Global Politics of Reproduction*. Faye Ginsburg and Rayna Rapp, eds. Pp. 270–287. Berkeley, CA: University of California Press.
Tyas, Michael 2014 Health Study in Fort Chipewyan 2014-Full Report, Accessed July 31, 2015 from http://onerivernews.ca/health-study-press-release-2014/.
Wolf, Meike 2015 Is there really such a thing as "one health"? Thinking about a more than human world from the perspective of cultural anthropology. *Social Science & Medicine* 129:5–11.

Mobile Showers, 2015. St. Johns, Oregon, USA.
© 2015, Mary Anne Funk. Printed with permission.

## *About the photograph*

*Diane, who with her husband, is currently homeless in St. Johns, Oregon, exits one of the mobile shower units provided by St. Johns Showers for the Homeless. The mobile shower trailer is equipped with two showers; it comes to the town once a week. "It's a blessing to have a place to get clean," said Diane, "and it's the only one in our area." Jim, back left, and Shannon, back right, volunteer their time driving the mobile shower unit to various locations around Portland, Oregon. They scrub each shower stall after each person uses it, so it is clean and ready for the next person.*

*This photo is part of an ongoing ethnographic photo documentary focusing on redefining our perceptions of homelessness and poverty, in order to advocate for policy change, social awareness, and to develop community-based participatory programs.*

—Mary Anne Funk

# 12

# Global Quests for Care

*Elizabeth Cartwright, Lenore Manderson and Anita Hardon*

Worldwide, growing numbers of people seek health care beyond their home settings and national borders. For some people, this is volitional, as they move to take up new opportunities or to improve their own or their children's life circumstances. Far larger numbers of people move internally and across borders a result of economic and environmental crises, civil war and human rights abuses, or because basic human needs cannot be met in the post-conflict settings in which they find themselves. People move across borders in search of health care and treatments that, in different settings, might be less expensive, more sophisticated, more accessible, or subject to different legislation and control. Increasingly, medical institutions in high- and middle-income countries promote surgical procedures and care to an international market, while pharmaceutical companies and practitioners test out new drugs and procedures in different country settings. In health care, it seems, borders can be especially porous. In this chapter, we explore how and why people seek medical care, treatments and cures across borders, between countries and health systems, and we examine the various implications of this medical travel.

Janzen's (1978) seminal work in medical anthropology, *The Quest for Therapy: Medical Pluralism in Lower Zaire*, focuses on understanding local logics of illness causation and treatment and, once described, on understanding the social relations around treatment decision making and the search for a cure. This foundational description of the treatment-seeking path serves as a jumping-off point for understanding how people navigate the global health care arena. We interrogate how moving one or more parts of the treatment-seeking process to a geographically different location changes individual and family expectations, the resources that people require and might access, the possibility of a cure or life-changing intervention, and eventual outcomes. Medical pluralism within and between countries thrives because of differences in modalities of care, technologies and settings, and what these offer to people who are ill, injured or distressed. We then move to questions of citizenship and belonging, and consider how civil status and legality influence life experiences and the choices available for health and medical care. We end this chapter with a discussion of people who are semi- and permanently displaced—both individuals and larger social and ethnic groups—and the implications of this for access to and quality of care.

The 'quest' is an apt metaphor for how people sometimes need to engage in long and arduous searches across time and space for treatment, a cure or relief from their symptoms. In his discussion of therapeutic decision making and the quest for a cure, Janzen notes the significance of "the composition of the therapy managing group within its social field, the role of the therapist or group enacting therapy; the technique; and the total cost of therapy" (1978: 156). He goes on to discuss what happens when there is disagreement over the diagnosis and/or over the course of treatment to be pursued, and considers how conflicts are settled at multiple levels. He then traces the multiple options that individuals have with respect to different healing traditions present in their geographical locations. Janzen's work is as applicable now as it was four decades ago; it offers us a framework to examine how the individuals and their families in this chapter use a variety of strategies in conjunction with new medical and communication technologies (as discussed in Chapter 8), in a world interconnected via many forms of physical and virtual transportation.

The idea of volition is central to our examination of the global movements of people. How and why do people move between different regions or countries and what effect does that have on their health care–seeking behaviors? What motivates people to leave their own country specifically to receive health or medical care? Certain movements are made by choice, as is broadly described in the literature on medical travel, but many others travel for economic reasons or because of access to better or different technologies and professional expertise. Others seek health care in unfamiliar settings because of their own migration, necessitated by the larger forces of economic needs, violence or natural disasters. We address this spectrum of possibilities below.

## Globalized Care

The global quest for treatments has been referred to by anthropologists and other social researchers variously as medial tourism, medical travel, and transnational and cross-border care—depending on motivation and context (Kangas 2002). Personal care services like health spas and cosmetic surgery in particular have been likened to 'tourism,' and these are often marketed in ways that mirror recreational travel: the opulent settings of some internationally renowned hospitals recall a vacation; medical services may be bundled with holiday activities; often the tourism infrastructure brings individuals to a place, arranging international and in-country transportation, hotels and visas, and brokering the medical services required. Many times this kind of travel takes place at great expense to the individuals involved given their unfamiliarity with the medical setting, including in relation to language, institutional and legal environment, and provisions of care. Sometimes, this type of medical travel is conceptualized as an act of faith or a pilgrimage (Song 2010), the last step, or the last hope, in resolving a particular health problem. When strategizing the quest for therapy, patients and their families use the ambiguities and grey zones in between and around various international spaces to maximize their ability to access what they need, or think they need, to maintain or improve their health and wellbeing.

Many different constellations of people and resources are involved in seeking care in different places in the long and the short term. People with sufficient resources may travel from low- and middle-income countries to international medical research centers and hospitals, wherever they believe they have access to the best care available; oftentimes, they incur serious personal debt to seek care abroad. People may travel from one high-income setting to another, too, in search of the newest medical therapies, those that are most likely to promise extended survival or cure to a life-threatening condition. People may seek care that is unavailable in their home countries for political or religious reasons, or because of the local limitations related to biomedical ethics; this is often the case for people seeking gender reassignment (Aizura 2010), stem cell therapy (Song 2010), in-vitro fertilization or surrogacy (Bergmann 2011; Inhorn and Patrizio 2012; Whittaker and Speier 2010). Conversely, individuals may travel for care to avoid personal debt, both for major procedures and for routine services such as for dental surgery or optometry, hip replacements or coronary stents. Meanwhile, less well-off people in rich countries are at times able to seek medical advice and/or purchase pharmaceuticals that are less expensive in other settings, as has occurred in the cross-border medical travel between the US and Mexico for decades

(Dalstrom 2012). Some medical travelers are migrants who return to their countries of origin for care. In these cases, their capacity to communicate with health professionals in their own language; their understanding of the health system and relations between patients and clinicians; and the availability of family support are all compelling reasons to travel for care. More generally, global treatment seeking and global care occurs because what individuals need or want is not always available or affordable where they live. Transnational medical travel is a way to plug the holes in an inadequate personal safety net.

People who move between countries for medical travel are vulnerable to the legal and institutional constraints, to the immigration and medical structures and securities of different countries; they can be targets of scams, deportations and incarcerations. The rules of a new country, and of what is and is not possible, can be challenging to learn in the best situations. However, it is also true that some people exploit these jurisdictional imprecisions and the gaps in ethics guidelines, surveillance and control, as indicated by the grey areas in relation to experimental biological treatments (Tiwari and Raman 2014) and commercial gestational surrogacy (Deonandan et al. 2012; Pande 2011).

In the following case study, Andrea Whittaker and Chee Heng Leng describe their on-going research on medical travel in Thailand and Malaysia; they highlight that much medical travel occurs for prosaic procedures, like the treatment of chronic infection, for heart surgery, hip and knee replacements, and the like, as well as for conditions such as quadriplegia that may never be resolved.

---

## 12.1 Medical Travel

*Andrea Whittaker and Chee Heng Leng*

People have long traveled across national borders to access health care. In the past such travel typically involved travel by patients from developed countries or elites from developing countries travelling to wealthy countries to access care in high-tech specialized clinics in the US or Europe. Although presently the largest movements of patients are between European countries, there is a growing trade in patients from wealthy countries travelling to low- and middle-income countries such as Thailand, Malaysia, Mexico and India for care. Travel by patients between lower-middle-income countries for services unavailable in their home countries or higher-quality care has become more common. For example, Thailand is an important regional hub for medical care, especially among the growing upper middle classes of Cambodia, Vietnam and Myanmar. Their income allows them to travel for a standard of care not accessible back home. Similarly, Malaysia is an important medical destination for Indonesian patients who constitute over 90 per cent of the foreign patient trade (Toyota et al. 2013).

Medical travel is an assemblage of medical technologies, staff, global air travel, Internet marketing, international accreditation and health insurance. It relies upon pre-existing infrastructure and human resources and draws upon local service and tourism industries. Many governments such as Thailand and Malaysia also support the trade through schemes such as land tax exemptions for the hospitals, special medical visas, incentives for the importation of medical equipment and government-supported marketing campaigns. It is seen as an important source of export revenue for lower- to middle-income countries.

The following case studies were collected between 2008 and 2013 in hospitals in Thailand and Malaysia. Over sixty patients were interviewed in four private hospitals. The stories complicate our views of medical travel—who travels and why. As these stories reveal, people travel for acute and chronic health conditions as well as rehabilitative care; their stay may be transient or very long term. Some people travel across borders regularly for all their health care.

Travel by Indonesians reflects growing dissatisfaction with the quality of care in Indonesian hospitals. For example, Ibu Siti was interviewed in a Penang Hospital in Malaysia accompanied by two of her daughters. She is sixty-two years old from Aceh, with four daughters and two sons. Her husband is a high school teacher. This is the fifth time she has visited this hospital in Penang. She came to seek treatment for her diabetes and on this trip has also been diagnosed as having a cardiac problem. She first came to this hospital in 2012, to seek treatment for her abscessed leg. She had been to a private specialist in Aceh for a check-up, but did not seek treatment there, as her family heard a lot of stories about poor results of operations in Aceh. Given a long history of insurgency against the Indonesian state, the health infrastructure is poor and Acehnese mistrust government institutions.

> There are stories that leg operations in Aceh causes paralysis, but here, we hear Indonesians get good treatment results, they won't get wounded, so we choose to come here . . . [medical service] better here. Many Acehnese people seek medical treatment here, we have heard about people's experience back in Aceh.
>
> *(Ibu Siti)*

There are many private hospitals in Penang marketing to Indonesian patients. Acehnese started coming to Penang for health care in the 1990s when they traveled by ferry. Now a regular direct flight between Peneng and Aceh has facilitated the movement of patients. For Ibu Siti and her family, the other private hospitals in Penang are too expensive so they continue to come to this hospital. They are paying out of pocket; Ibu Siti and her daughters calculate that the cost of seeking medical treatment in Penang is lower than that in Aceh, even when their air tickets and accommodation are taken into account. They stayed at the same hotel for each of their five visits. They came to know about their hotel through the air ticket agent, who gave them the contact number.

A number of Thai private hospitals specifically cater for their foreign patients with luxurious hotel-like furnishings, translators and culturally diverse cuisines. Unlike the Malaysian hospital catering for Indonesians, these hospitals cater for upper-middle-class patients from the region as well as patients from Europe, Australia and the US. For example, Anh is thirty-nine years old and works as an education consultant in Vietnam. She previously lived for a while in the US. She had flown to Thailand with her daughter who had been suffering from a fever. The family kept her home from school and gave her Tylenol and Advil to stop the fever and took her after two days to a clinic in Hanoi, one frequented by foreigners. The doctor took a blood test: "I talked to the doctor and told her that I was thinking—at that time I was thinking of taking her to Bangkok because we always come here for any kind of medical and then she said 'No, she's fine' and gave her medicines for the fever and the cough." After a few more days Anh and her husband became more worried and so flew her to Bangkok to the hospital. Within forty-five minutes of arriving they had X-ray results and her daughter was admitted with pneumonia. "We didn't plan to stay here this long, we thought it would only be a couple of days. We flew in Thursday night, Friday morning we brought her here. The doctor checked on her and sent her to do an X-ray and we found that she has pneumonia, and really bad. There was a lot of virus throughout and around the lung and that's why we're still here." They intend to stay a further three days until their daughter improves.

Anh joked that she was a 'frequent patient.' Her family always comes to this hospital for care. Her brother came here for care after a stroke and Anh had a sinus operation in the past for which she returns for follow-up care. During the time they lived in the US, she traveled back here to undertake in-vitro fertilization. Her husband had laser eye surgery the previous year and plans to undertake knee surgery later in the year. He comes regularly for health checks and an anti-aging program, as he explained:

> In Vietnam we don't have very good health care system. . . . I think Vietnam has good doctors but the system doesn't allow them to really care for other patients because they work all day long and even after they go home, they have a private clinic at home. You cannot serve hundreds of different people all day, you get tired and you cannot focus on what you do. Facilities in Vietnam are not very good. I don't know if you've been to Vietnam or not but if you go to hospitals there I'm sure you'd be scared. People are just everywhere, on the floor—100% of hospitals in Vietnam are overloaded right now. They don't have a good private hospital like here. They're starting to have some but that's clinics, not full facilities like at a hospital.

For other patients, the trip to Thailand is to seek expertise simply unavailable back home. Not all are wealthy; many undertake loans and debt to pay for their care. For example, in another ward of the same Thai hospital I met with Hagos from Ethiopia. Hagos is thirty-one years old, and a university lecturer. He was involved in a car accident resulting in neck and spinal cord injuries and quadriplegia. Four days after his accident he and his family decided to come to Thailand for treatment:

> We explored the hospitals in Ethiopia, but the damage is very severe. It was very severe and I need further treatment. So we decided to come to Thailand because we had prior knowledge about this hospital, that they will offer better treatment, and better services. So right after, on the third or on the fourth day that the accident happened to me. . . . I was here. I was very early, and it has been like a month and something since I am here.

A friend who had medical training initially accompanied him on the seven-and-a-half hour direct flight to Thailand. Later his father traveled over to care for him staying in the hospital room with him. He shared a four-bed room with three other patients as they could not afford the private single-bed rooms.

Although Hagos is wealthy by Ethiopian standards he had no health insurance. The family had pooled their savings together to pay for the medical expenses. "We don't have health insurance, medical insurances, so we have to cover the bill by ourselves. So we have decided to cover the bill partly by my family, and partly by my university." Hagos cannot afford to stay for much longer. So far he estimates that he has spent USD 45 000 on his treatment and stay:

> But the only thing now, when you are talking of the price, or whether something is expensive or not, it just depends on where you are, okay? Well, I heard that one of the major customers of this hospital are also Europeans. Why Europeans? Probably in Europe there might be hospitals better than this, but the price is very much higher. So they came here to get quality service with better price. But for us, for Ethiopians, this is quality service with a very expensive price. So whether to tell that something is expensive or not, well it depends where you are, so for us, for Ethiopians, it is expensive, it's a very expensive one.

With his money running out he hopes he will be able to pull himself up so he can get into a wheelchair by the time he has to return to Ethiopia.

> So long as my spinal cord has damage, it will take a very long time for rehabilitation and physical therapies. . . . So after I am able to use the wheelchair successfully I'll go back to Ethiopia, and I will restart my physical therapy then. So we are supposing that just after a week from now, I might be discharged from the hospital. . . . They might not have good rehabilitation services, but the choice is—now the choice is between whether you can or you cannot. . . . In Ethiopia there are some rehabilitation centres, and also we have rehabilitation centres might not be to the standard, but they have at least the minimum materials, and the minimum human resource to run it. In terms of advice for other people coming to the Thai hospital, my advice will be only one [thing]; they need to take it out of their pockets, and their pockets should be full.

Health insurance plays an important role in facilitating medical travel. Many patients have 'portable' insurance which allows them to be reimbursed for treatment undertaken in other countries. Some countries send citizens to other countries to receive care unavailable at home or because it costs less to send them than to provide similar services within their national health systems, in effect, outsourcing their care (Whittaker 2015). Likewise, a number of insurers send their clients to other countries for care to lower costs.

Stories of medical traveled such as these reveal a diverse range of motivations and circumstances surrounding people's therapeutic itineraries. Other studies likewise document the diversity: from cosmetic surgery patients travelling on group tours combining surgery with visits to exotic beaches; patients travelling to circumvent home country restrictions on forms of treatment such as stem cell treatments or commercial surrogacy or commercial ova donation; members of diasporic communities travelling home for care; to patients from countries which lack particular expertise or equipment travelling to undertake treatment in a more medically sophisticated location.

* * * *

There are concerns about the effects of medical travel upon health equity in lower- and middle-income countries such as Thailand and Malaysia. At one level it could be argued that medical travel allows individual patients such as Ibu Siti and Hagos access to health care they could not obtain at home. Inequitable home health systems lacking universal health coverage and having high health costs or long waiting lists encourage the movement of patients across borders. For example, many of those travelling from countries such as the United States are those who are uninsured or cannot afford their treatment or medications back home. Patients are 'outsourced' by their countries health systems or insurers because of the cost saving such outsourcing represents to the government or insurer. However, medical travel remains inaccessible for those in the poorest health who are not able to travel. Access remains stratified across lines of health status and mobility.

Medical travel also has implications for national health systems. The effects of the trade upon local health systems depend upon the degree of privatization already existing, whether there is excess capacity within the private sector, and the degree of government control, subsidization and regulation exercised over the private health sector. A well-financed public system is an important buttress against an erosion in local health equity for lower-middle-income countries. But even then, experience in Thailand suggests that a two-tiered health system develops with major implications for a brain drain for the health workforce away from public hospitals towards the private hospitals catering for foreign patients (Kanchanachitra et al. 2011). Hazarika (2010) reports that India suffers drastic shortages of 600 000 doctors, 1 million nurses and 200 000 dental surgeons and a shortage of medical specialists in local community health centres while over 75 percent of human resources and medical technologies are in the private sector. Further growth in the private sector due to the growing medical travel trade in India could exacerbate the shortages in the public system as health care professionals move to the better pay and conditions in the private sector.

Finally, the growth of medical travel has implications for patients as citizens. Provision of health care through private or public mechanisms forms an important part of a state's relationship with its citizens. As people travel across borders to receive care, the expectations and relationship with their national health system changes and attenuates and they no longer have the same stake in nor benefits from their citizenship. As the informants in these case histories reveal, home health systems are viewed with suspicion, mistrust and frustration that the same quality of care and services is perceived to be unavailable back home. Rather than being a public good, health care is traded as a commodity, accessible to a mobile few who can summon the financial means to afford it.

## *Acknowledgments*

This research was supported under Australian Research Council's *Discovery Projects* funding scheme (project number DP 1094895). We wish to thank Por Heong Hong who acted as a research assistant for the Malaysian part of this study.

## *References*

Hazarika, I. 2010 Medical tourism: Its potential impact on the health workforce and health systems in India. *Health Policy and Planning* 25:248–251.

Kanchanachitra, C., M. Lindelow, T. Johnston, P. Hanvorvongchai, F. Lorenzo, L. Nguyen, S. Wilopo, and J. Frances de la Rosa 2011 Human resources for health in Southeast Asia: Shortages, distributional challenges, and international trade in health services. *The Lancet.* DOI: 10.1016/s0140–6736(10)62035–1.

Toyota, M., H.L. Chee, and B. Xiang 2013 Global track, national vehicle: Transnationalism in medical tourism in Asia. *EJOTS: European Journal of Transnational Studies* 5:27–53.

Whittaker, A. 2015 'Outsourced' patients and their companions: Stories from forced medical travellers. *Global Public Health* 10(4):485–500.

---

Above, as Whittaker and Chee describe, people who have the means are able to seek out care in other places across the globe, often where cultural preferences as well as technical needs can be met—hence Malaysia's market advantage in providing care to Muslims from Indonesia, South Asia and the Middle East. Sometimes, specific procedures are sought that are not available to individuals in their home places, but in addition, people's perceptions of *quality* play an important role in the medical choices that they make. Quality is an ephemeral concept, a mix of information, illusion, deception and hope, and understandings of the strength of professional training, and the technology, diagnostics and medicines that are available in different places. Commercial hospitals and specific surgeries and treatments are circulated and promoted through a variety of communication platforms, via word of mouth, and, in some cases, as covered by government programs (Kangas 2002). 'Doctor-shopping' is a simple way to talk about peoples' search for care and cure, and the forms that this search takes in various situations are telling of locally available technologies and health care systems as well as reflective of cultural notions of healing, ethnophysiology, and possible alternative treatment regimen.

## Structural Vulnerability

Although in the medical tourism literature, there has tended to be a focus on wealthier (although not necessarily wealthy) medical travelers, many people engage in the global quest for cures from a very disadvantaged position. Because of their location economically, ethnically and legally, poor people are sometimes forced into seeking care outside of their home terrains. Structural vulnerability is a useful concept for thinking through this situation (Cartwright 2011; Quesada et al. 2011). The term 'structural vulnerability' makes precise the ways in which whole groups of people fall through established safety nets. They are structurally vulnerable because of the economic and political forces and institutions (educational, medical and legal) present in their societies, in conjunction with the ways in which they are discredited, discriminated and devalued by the larger society in comparison to others around them. Importantly, structural vulnerability attends to how individuals have internalized those negative evaluations; this internalization is necessarily partial and is dependent upon the social context and the individuals themselves. The dual descriptive nature of structural vulnerability at both the social group and the individual level makes it a particularly interesting conceptual tool to use in exploring trans-border quests for care. Structural vulnerability in one's home place may lead to medical travel or, conversely, it may be the result of immigrating to another place in search of a better life. Both of these situations point to precarious states of being that are well suited for exploration by medical anthropologists.

While the notion of structural vulnerability has been explored among immigrant groups in the US and elsewhere, another more global aspect to structural vulnerability is represented in this chapter (see also Cartwright and Manderson 2011). Global economic and social hierarchies are constructed from various vantage points of power, and these hierarchies result in overlapping mosaics of difficulties and opportunities for an increasingly 'globile' (mobile and global) world. There are many variations in how nation states react to immigrant groups within their borders. For example, the Roma, Romani, Sinti travelers and gypsies have moved throughout Europe for over a thousand years, and their history is characterized by their need to exist on the fringes of societies, stigmatized and negatively targeted because of their ethnic affiliation and the stereotypes that derive from this. Lorenzo Alunni (2015), in his work with the Roma, describes the effect on their health and wellbeing following their move to Italy from Yugoslavia and the Balkan states predominantly in the 1960s and 1970s. Far from being assimilated into Italian culture, the Roma have kept to themselves, both by choice and discriminatory practices, and live in impoverished camps on the outskirts of urban centers, necessarily relying on limited and at times sub-standard medical care. Those who are especially vulnerable are newer immigrants, still trying to make sense of the organization of social life in Italy and still striving to create social networks with more established Romas. Participants in Alunni's research sought medical help across borders, going to France and back to Romania, when they needed procedures such as dental care or abortions, respectively unaffordable or unavailable to them in Italy. While in France, Roma have access to more material things than they might in Italy—apartments and more government assistance, for instance—but as described by Alunni, these medical travelers returned home to the camps in Italy because of their affective ties to family and, despite the dismal conditions in which they lived, to the familiarity of their everyday lives.

What health rights does citizenship confer? Within a particular country, how can people access better care for their families? Kate Goldade (2011) illustrates how Nicaraguan women immigrants strategize to have their babies in Costa Rica while they are there working in the coffee fields. Although in the country illegally, the Nicaraguan women are able to receive low-cost or free medical care during their pregnancies, post-partum care and tubal ligations in the Costa Rican hospitals. But their children have an even greater advantage because of the principle of *jus soli* that operates in Costa Rica: they obtain Costa Rican citizenship by virtue of their birthplace. Strategizing care, giving birth, and quests for citizenship highlight some of the biosocial complexities of cross-border approaches used by individuals around the globe (see also Castañeda 2008).

Below, Heide Castañeda describes the heterogeneity of families along the US–Mexico border, and the significance of immigration status to their entitlements to health care services. In a family, one child may be a US citizen, one a Mexican citizen, with parents who may be of different immigration statuses. Families go back and forth in a strategic manner, seeking care, buying pharmaceuticals, and obtaining hospital, medical and other health services. Simultaneously using two health care systems puts the families described by Castañeda into the position of having to endure a double dose of bureaucratic tedium, but this is balanced by the flexibility and a sense of control that comes from working the various constellations of binational medical resources.

## 12.2 Health Care along the US/Mexico Border

*Heide Castañeda*

More than 11.5 million people live in economically and socially interdependent communities along the two-thousand-mile-long US/Mexico border, where transnational strategies have always been integral to daily life. Borderlands are ideal sites in which to examine issues of inclusion, exclusion, and the various forms of health citizenship these boundaries imply. Inequalities of access and delivery are particularly sharp in communities along the US/Mexico border, where, in addition to underfunded local public health infrastructures, there are high numbers of unauthorized or undocumented persons, who cannot obtain affordable, quality health care. They remain ineligible for all publicly funded health services except perinatal and emergency care, despite recent health care reform in the US. This un- and underinsurance has broad spillover effects on families and entire communities.

One unique feature of the border region that highlights transnational connections is the heavy presence of mixed-status families, in which members are stratified by juridical categorizations related to immigration status (Castañeda and Melo 2014). Some 2.3 million mixed-status families live in the US (Passel 2011), consisting of variable constellations of citizens, permanent legal residents, undocumented immigrants, and individuals in legal limbo such as recipients of Temporary Protected Status (TPS) or Deferred Action for Childhood Arrivals (DACA). About three-quarters of all children of unauthorized immigrants, and most children in mixed-status families—about 4.5 million—are US citizens by birth. The composition of mixed-status families is not static, as members may move in and out of households and between statuses over time. This complexity and fluidity impacts access to resources for individual members in relation to public institutions especially health care. In addition, the 'illegality' of some family members influences opportunities for all, including those who are recognized as citizens. For instance, despite US citizen children's eligibility for benefits such as Medicaid and State Children's Health Insurance Program (SCHIP), those with undocumented parents access benefits at a lower rate, as parents avoid institutions and limit or delay services for children due to fear of deportation or that enrollment will affect future chances at regularization.

The Rio Grande Valley in the southernmost part of Texas is unique among border communities because of the physical separation created by the river, which splits sister cities with historical interconnectedness. With a population estimated at 1.3 million and rapidly growing, it is a blending site of communities geographically, socially and economically closely integrated with Mexico. Latinos of Mexican descent account for 89% of the population and the preferred language for 90% of area residents is Spanish. The highly mobile population includes binational extended families and large numbers of people who cross the border daily for work and recreation. Many Mexican citizens can apply for a Border Crossing Card (B1/B2 visitor's visa issued for 10 years at a time), allowing them to cross frequently. However, this mobility only applies to some people. Many are undocumented—having entered the US on an unauthorized basis or overstaying a visa—and so are relegated to life within a small strip along the border. Unable to re-enter the US if they cross back into Mexico, they are also unable to travel to other parts of the US, since this requires inspection at one of the Customs and Border Patrol traffic checkpoints along major roads and highways. As a result, undocumented persons describe being 'trapped' in the region, despite the fact that their Mexican town of origin may be only a 10-minute drive away. Many families on the US side live in *colonias*, unincorporated neighborhoods often lacking city services such as water, electricity and sewage. Nonetheless, despite these disadvantages, *colonias* provide access to low-cost housing and facilitate the maintenance of multi-generational households.

Below, I focus on the experiences of mixed-status families in the US/Mexico borderlands to emphasize the interconnectedness of immigration and health care policies. Because of this convergence in their lives, families and individuals employ specific health care strategies, access informal medical and dental services at a high rate, and engage with transnational opportunities to address illness.

## *Health Care Strategies of Mixed-status Families*

Veronica is 34 years old and has lived in the US for eight years since moving from Reynosa, Mexico, about a 30-minute drive from her current home in Texas. Her husband works in construction, and she is a stay-at-home mother of five children, three born in the US and two in Mexico. Like other parents, she frequently experienced dilemmas when her children, who have different forms of access to the health care system, became ill at the same time, as is often the case with respiratory and other common childhood infections. She and other parents often had to spend the greater part of a day or two visiting first a pediatrician for a citizen child with Medicaid, followed by long waits at charity or community clinics hoping to have any uninsured children seen for the same condition. This results in time off from work, loss of income, and is logistically difficult if no transportation is available. Veronica described the difficult decisions she faces when an undocumented child becomes ill:

> It's very difficult, because those children who have Medicaid and were born here have more privileges, like going to the doctor. I struggle a lot when they get sick because you have to pay for the doctor and sometimes you don't have money. A consultation is very expensive here. Those that have Medicaid, you immediately go to the doctor if they get sick or have an accident. It's not the same for those without, if they fall or get hit. Like one time my girl, who doesn't have Medicaid, fell. She was playing, running, and she banged into a chair and got a bump that swelled up and almost burst open. You know it's something that is worth going to the doctor for, but then you don't have money for it. All you can do is try to get the swelling to go away. It's better to put Vick's [VaporRub ointment] or something like that on it to bring down the swelling. For a heavy blow like that, you have to get X-rays. If that happened to someone with Medicaid you would immediately go to the doctor, because you know your insurance will cover it.

Like Veronica, many parents shared their regret at being unable to take their children to the doctor for an illness or accident for which they would have immediately taken a child with Medicaid. Instead, treatment consisted of home remedies, over-the-counter medications, or, as described below, 'leftover' medications from others. This creates distinctions within the family that even the children notice, as stratified access based on legal status may lead to preferential treatment, resentment, and hierarchal relations within the family.

Maria Elena is a 33-year-old mother of three children, two born in Mexico and one in the US. She and her husband Everardo, a bricklayer, arrived 12 years ago, also from nearby Reynosa. They moved to the US primarily to provide a better education for their children, as "over there, school is very expensive but not very good." Despite the fact that all her children go to school together and are being educated in the same way, Maria Elena recognized that there are distinct advantages for her US-born daughter:

> She's six years old and was born here. She doesn't know anything, she doesn't say, like, 'I'm from here, and you're not,' to her brothers. But the kids do notice: 'Mami, why does she get benefits and we don't?' They see the difference. I tell them, 'son, because you weren't born here. You were born in Reynosa. And she is from here.' The same thing at the doctor's, they see the difference, because I take her since she has Medicaid, but not them. At the doctor's they just ask, 'Do the boys have Medicaid too?' 'No.' 'Ok, then just give them this over-the-counter medicine.'

For undocumented parents and children, sources of formal care include low-cost community health centers, charity care, or the emergency room. Maria Elena explained, "for the adults, well we wait until, honestly, until we're really bad, and they we go to [a community health center] or if we're really, really bad, to the hospital." While it was anticipated that the burden placed on emergency rooms by the uninsured would be remedied with the recent US health care reform (Affordable Care Act of 2010), the large undocumented population in border communities are not among those who can acquire coverage. Despite some structural and institutional changes, the burden of filling gaps in health care will continue to fall on governments and organizations at the state and local levels. This is a unique challenge in many border communities, where public health infrastructure is significantly underfunded. Community health centers continue to be a vital source of medical care for immigrants without coverage; however, there are not enough to serve the entire population.

## *Informal Practices*

As formal systems fail to meet the needs of a large segment of the population, alternative and informal channels of care proliferate. One common practice is the sharing of medications prescribed to citizen children to treat undocumented siblings and parents. As Alan, a 22-year-old undocumented student at the local community college, recalled, "When people prescribed something to my brothers who had Medicaid, they didn't use all of them, we would use them. That is how we would do it." The mutual assistance evident in sharing prescription medications sometimes extended beyond the immediate family. Lisa is a 20-year-old US citizen with one undocumented and two citizen siblings and undocumented parents. She recalled how 'leftover' medication circulated not only within the family, but also in the wider community:

> There's always leftovers. Even the neighbors would call us and be like, 'Oh, my son is coughing,' or 'we have a cough, do you have anything?' 'Oh, yeah, I took her to the doctor and aquí está la medicina que me sobró [here's the medicine that was left over].' So it's always counting down medicines to see who needs it. . . . So it's always been like, handing down medicines, or seeing what we have in the cabinet and always trying to save anything because we can't afford that type of medicine in case someone gets sick.

Medication use is a socially embedded practices, and saving, sharing, and re-using medicines is broadly practiced. Sharing not only serves the immediate need of treating illness but also creates the obligation of reciprocity between individuals (within a family) and households (within a community), an advantage in conditions with limited resources. However, sharing is problematic when, for example, a course of antibiotics is cut in half, rendering it less effective for both people who take them. Thus, a socially valued and pragmatic act of sharing may lead to twice the negative outcomes; half of an antibiotic regimen may be worse than none at all. The wellbeing of US citizen children is thereby directly affected by a family's mixed status.

Another common practice rooted in the transnational reality of daily life is traveling to Mexico to visit health professionals or purchase medications. As Israel, a 21-year-old undocumented college student, noted, "We have people bring medicine over. We'll get flu medicine, penicillin, just regular stuff. Injections, you know." Rather than crossing the border themselves, many people rely on others to bring or send medications. Traveling to Mexico for medical and dental services has been a common practice for several decades, but increased border militarization since 2007 and amplified scrutiny of papers has decreased people's ability to do so. Amanda—who has three children, the eldest of whom was born in Mexico—pointed out, "We can't even go for treatment to Mexico, not even to Reynosa, not even that . . . you can't leave here." As Mexican citizens, they are no longer able to obtain medical care in Mexico because they would be unable to return to their home in the US. Additionally, the availability and affordability of services has been impacted by the 'brain drain' of physicians and dentists away from northern Mexican border towns due to violence in recent years, coupled with elevated costs as a result of the local narcoeconomy.

As a result of the inability to travel across borders in recent years, there has been an increase in strategies such as purchasing prescription medications offered—unlawfully and in an unregulated manner—by vendors at local flea markets. Based on observations at booths and discussions with vendors, these include antibiotics of various classes, steroids to treat inflammation, insulin, birth control pills, and emergency contraception. Angela, a 39-year-old woman from Zacatecas, sells homemade tamales. She noted that, "If people don't have insurance, they go to the flea market when they have an infection or something and take pills that they buy there. They may not even know what it is. I don't go, I don't do that, but I know people who go there to get injections or some pills." In addition, some practitioners operate out of homes or at flea markets, including dentists licensed in Mexico (but not the US) or nursing assistants who provide injections for a fee (although not legally permitted to do so). Some dentists operate out of their own or patients' homes. Lisa, introduced earlier, noted that, "they do house visits, and I'm sure they're not supposed to, but they do it to help out the communities so we're very thankful for that." Marina, a 42-year-old woman, added:

> There are no dentists here for us. Some come from Mexico and we seek them out in homes, but it is risky for them because they could get caught, could get in trouble. But we need them. My mother has gone to such dentists who are not licensed here. Actually I need to go, too, but I am afraid. It's not the same, like being in a clinic where you know exactly how everything was cleaned, and that they have everything they need. To work out of a house, it could be that they have cleaned everything, but it's better not to do it that way.

These informal practices are the direct result of stricter border policing in previous years and the inability to travel for transnational care-seeking. Due to these limited opportunities to obtain services, undocumented immigrants are forced to seek health care that is improvisational and may pose additional risks.

## *The Convergence of Immigration and Health Care Policy*

The dilemmas and practices described here highlight the need to examine the intersections of immigration and health care policies, as each has direct and indirect impacts upon the other. A number of historical and geographical factors have resulted in the prevalence of transnational connections and identities along the US/Mexico border. This includes mixed-status households, which are a major feature of the contemporary migration landscape (Castañeda and Melo 2014). Experiences of these families—including some 4.5 million US citizen children—have significant implications for the future of health policy and consequences for future immigration reform. In late 2014, President Obama announced a plan that provides administrative relief to up to 5 million undocumented immigrants who have lived in the US for at least five years. The program significantly impacts undocumented members of mixed-status families, protecting them from deportation and allowing them to work legally. What it does not do, however, is provide these individuals with any form of public health benefits or affordable health insurance through the mechanisms of the 2010 reform; in regards to health care access, therefore, their situation remains the same. Other features unique to the region I have highlighted here include an established pattern of transnational care-seeking that has recently been severely curbed as border enforcement has increased, along with high poverty and inequality, which produce and reproduce informal health care practices. Despite limitations, families and communities develop ways to cope with the lack of accessible and affordable health care.

However, the consequences of disparate access to quality care on lifelong health outcomes are well established. Since undocumented immigrants will account for up to 25% of all uninsured in coming years, their exclusion implies serious limitations for the future of health equity in the US. There are stark cultural, political and health implications in the continued existence of a medical underclass comprising over 11 million people. Limited access to health care creates widening disparities for an already impoverished and marginalized group. Specific shifts at the policy level will be a necessity to address this burgeoning public health issue created through the convergence of immigration and health care policy.

### *References*

Castañeda, H. 2015. Mixed-status families in the Rio Grande Valley of Texas: Health disparities along the US/ Mexico border. In *In Between the Shadows of Citizenship: Mixed Status Families*. A. Schueths and J. Lawston, eds. Pp. 106–117. Seattle, WA: University of Washington Press.

Castañeda, H., and M.A. Melo 2014 Health care access for Latino mixed-status families: Barriers, strategies, and implications for reform. *American Behavioral Scientist* 58(14):1891–1909.

Passel, J.S. 2011 Demography of immigrant youth: Past, present, future. *The Future of Children* 21(1):19–41.

---

There are many reasons for moving back and forth across the border of the US and Mexico (Cartwright 2011). Poverty and lack of the possibility of economic and educational advancement have led many families to emigrate from Mexico to the US, either formally or informally, over the last fifty years. More recently, intractable violence in Mexico and Central America has led to new waves of people crossing the border, often at great personal risk. Upon arrival and long after that, the life possibilities that are open to new residents are constrained by the legal status that they are able to attain (Cartwright 2011; Holmes 2013; Quesada et al. 2011). Individuals coming from Mexico and Central America move through the immigration system in the US slowly, at best. Many people reside without citizenship or residency papers for years, vulnerable to incarceration and deportation if they try to seek medical care. Others obtain Lawful Permanent Residency (LPR) status that comes with some benefits, the obligation to purchase health insurance and the annual increasing penalties if one does not, and mandatory waiting periods for receiving government benefits of Medicaid and health insurance for children of up to five years in some states (www.healthcare.gov).

Biology, in this case one's state of health, often plays a role in whether or not one can make a move to another country and be allowed to stay there for a length of time. Miriam Ticktin (2011) makes an interesting comparison between the role that biology plays in the treatment of immigrants in France and in the US with respect to the quest to gain citizenship. In France, the official approach over the last twenty years has been to use ill health favorably in the equation of whether or not an individual can stay in the country when he or she arrives there illegally or overstays a visa. Under the 1998 'illness clause,' French immigration law tended towards granting immigrants permission to stay in France if they required medical treatment and if returning to their countries of origin would put them in a situation of not being able to obtain the necessary care. Originally a very humane document, the illness clause has become increasingly restrictive since its implementation with increased agitation from right-wing nationalists and as French mainstream sentiment towards immigrants has deteriorated. By the early 2000s, France had moved away from using illness as the basis of allowing individuals to obtain legal papers to stay in France. With the violent attacks in January 2015 on the French newspaper Charlie Hebdo by French-born Muslims, and the immediate and vicious push-back against immigrants in that country, especially in the suburbs of the large urban centers like Paris (Ellick 2015), the power of the illness clause may be further eroded. In the US, in contrast with France, a 'security-state' ethos prohibits individuals from entering the country or applying for visas if they have communicable diseases of public health significance, severe mental illness or drug addiction, although since 2009 individuals testing positive for HIV virus have not been excluded from entry or from seeking residency.[1]

Below, Susann Huschke explores the precarious nature of living as an undocumented immigrant from another angle. The social services provided by the German state are inaccessible to Eduardo, the Colombian immigrant at the center of her account. Huschke details the human exploitation and deprivation that borders on slavery, experienced by many immigrants, and through this sheds light on how people without official status find meaning in their lives in such a way that makes it possible for them to endure many hardships.

## 12.3 "I Haven't Paid This Karma Yet"

*Susann Huschke*

Eduardo and I first met in the spring of 2008. Another Colombian research participant, Mercedes, had suggested that I meet him and briefly told me his story—a story of what some might call modern slavery. Eduardo, she explained, had lived in Germany illegally for over 20 years. He had worked for a wealthy couple in one of Berlin's affluent neighborhoods without ever being paid, and was deprived of basic rights and mistreated in numerous ways. For most people from countries outside the European Union, the only way to legally live and work in Germany would be as a spouse or child of a German national. Consequently, labor migrants enter the country illegally or, as in the case of most undocumented Latin American migrants, come as tourists and stay even after their visa expired. If they find work—and most do—they work illegally.

Eduardo's story sounded like the worst case scenario: no pay, no rights, and very little contact with other people for two decades, trapped in a villa as an obedient servant. Mercedes arranged for me to interview him. I subsequently met him several times over the next three years, and continue to receive updates about his wellbeing via email until today. His story touched and upset me but he also taught me a great deal about the complexity and ever-changing nature of lived experience, the many faces of social suffering, and about human resilience in the face of structural and interpersonal violence.

I first interviewed Eduardo in his tiny, cluttered room in a state-run hostel for migrants and asylum seekers. He had been put there after the police arrested him in the summer of 2007. It was a rather unlikely coincidence (he would argue that it was fate): the police came knocking on his door looking for a neighbor. Due to his poor spoken German, they identified him as a 'foreigner' and asked him for his passport which he couldn't produce: his employers had taken it from him many years ago. The police suspected that he was undocumented and arrested him. In custody, he was diagnosed with tuberculosis and transferred to a hospital. After treatment, his physical and psychological state remained fragile, and his deportation was temporarily suspended. He was allocated a room in the refugee hostel and given access to basic medical care and social benefits. Meanwhile, a lawyer working for the Jesuit Refugee Council took on his case to try and legalize his status.

Eduardo told me that he had come to West Berlin in 1986 when he was already 40 years old in the hope of finding work. Unlike many other labor migrants, he did not leave behind a wife or children, only his mother. A few months after he arrived in Berlin, he found work as a house servant for an Argentinean-Italian couple, employees of the Argentinean embassy. He moved in with them and lived with them for the next 20 years in their four-story villa, cleaning their house and garden. He lived in the cellar, in a room without heating, kept more or less warm by the central furnace providing heat for the house. He had to leave the door open because of the fumes coming from the furnace. About his work situation and the relationship with his female employer, he said:

> They didn't hit me, but verbally, verbally they treated me really bad. . . . I told her at one point that I would prefer if she hit me than being treated in a bad way verbally, I would prefer that. Because getting beaten, there are things you can take for that, right? There are pills and creams to make the pain go away, right? The pain goes away fast. The moral pain on the other hand does not go away, that pain stays. But yes, my God, it was difficult.

His employers ordered him to stay away from the neighbors and the visitors who frequented the place—often international guests and employees of other embassies: "I was afraid of them. They did not allow that I brought anyone to my room, and I could not speak to the people who lived upstairs either. They would always check on me." Eduardo wanted to study German but he was not permitted to leave the house for a language course. His boss threatened to send him back to Colombia any time he disagreed with her—that is, to denunciate him to the police and have him deported.

I got to know Eduardo as a shy and quiet person, a small, thin and sickly looking man with pale skin, blurry eyes and flat cheek bones. He whispered more than he spoke, and avoided eye contact. However, once he realized that I was not going to judge any of his views on life, fate, and spirituality, Eduardo opened up and passionately shared his story with me. His tiny room at the hostel was crammed with old newspapers, on the walls he put up pictures from magazines, mostly of pretty young women. There was no room to sit down, but there was a sense of order to the seemingly chaotic assemblage of collected things: everything had its place.

During our conversations, which lasted hours, I learned about Eduardo's views on what had happened. I discovered that there were loopholes in the otherwise tight regime his employers subjected him to. He told me how he used to ride his bike to the nearby forest, he went shopping at the grocery store, and he also sporadically and secretly worked for other families as a paid domestic worker, earning a little bit of money most of which he sent home to his mother. These contradictions led me to wonder about individual agency in the face of severe political, spatial, economic, or social constraints.

When I first asked him why he had not left, he told me a story of how he had been thrown out of the house by his employers once after an argument, and had considered looking for help from a Christian support organization, but then his employers 'forgave' him and he decided to stay. When we returned to this incident in a second interview, he added the following explanation:

> What happened was that I did not want to leave the neighbors, the other neighbors who were really nice and who I loved very much, I used to visit them. They provided me company, and so I said to myself, now that they [the employers] have forgiven me, if I go [to live elsewhere] I will feel alone. Here, on the other hand, I have those neighbors who I know and whom I greet every day.

Eduardo therefore had good reasons to stay: he appreciated the security and his acquaintances in the house and in the neighborhood. He preferred the known evil, with a measure of comfort and security, to uncertainty.

Apart from the constant abuse by his employers, Eduardo also suffered from his failed attempts to find a partner. He was well aware that marriage was the only way to legalize his status, obtain a permanent visa in Germany, and so earn more money to send home. He explained: "The only thing that could save me was a girl." He was reminded of this by his mother in the letters she sent him, in which she complained: "How long have you been there now? Why have you not sought the visa yet? So much time has passed!" In order to uncover the cause of his misfortune, his mother consulted a diviner back home in Colombia. It was revealed that Eduardo's misfortune with women and his experiences of interpersonal violence and exploitation were linked. They had the same cause: black magic. From then on, Eduardo began to understand that all of his suffering was mainly a spiritual matter. He explained to me that his employer's aggression against him was the work of a *mago negro*, a sorcerer, from Colombia: "He leaves his body and comes to this house [the villa where he lived] and bothers the woman so that she then bothers me. This happened for 20 years and I put up with it." The sorcerer also affects his ability to love: "This black magician with his bad energies does not allow that I love a girl. My feelings are totally blocked. My heart, my feelings are like a rock. It's impossible." Moreover, the sorcerer bewitched Eduardo directly, causing for example itching, raw sores on both his legs, and he made his eyes burn and his stomach hurt: "The sorcerer is sending me acid."

When we talked about how he could protect himself against sorcery, Eduardo pulled out a magazine from one of the piles in his room and showed me an advertisement of a Dutch healer who claimed she can cure a multitude of physical and spiritual ailments. I asked him whether he wanted to get in touch with this healer:

*Eduardo:* Why don't I do it? Because time hasn't come yet. I have done bad things in other lives. There are no causes without consequences. I was a dirty sorcerer in other lives, too, and I caused the same harm.

*Susann:* So she can't cure you?

*Eduardo:* She can cure me!

*Susann:* Aha, ok.

*Eduardo:* But it is not the time for her to cure me.

*Susann:* How do you know it's not the right time?

*Eduardo:* Because . . . God is just. I have paid a karma in this house. Two karmas, there are two karmas. The one I have now; and the one that I already paid, for 20 years. The police arrested me and I stopped paying the karma. I hope that they give me the visa and I can begin anew to live a life without offending others. There is justice on Earth . . . and she can cure me, but not yet.

In addition to the explanations his mother offered him in her letters, his understanding of suffering was shaped by the teachings of a master, his *maestro*, as he calls him. He had met him years earlier when he was experiencing economic misfortune in Colombia and was looking for spiritual support. The maestro introduced him to the concept of karma, and gave him one of his main sources of comfort which he still reads: a book by Ramatis, a Brazilian prophet, whose teachings involve extra-terrestrial life and the coming of judgment day.

When I met Eduardo once again in June 2010 to catch up and to see if I could help with his legal case, I was still struggling to understand why in 20 years, he had never attempted to leave the abusive relationship. After the meeting, I wrote in my fieldnotes:

> After repeatedly avoiding a direct answer to my question, he finally replied rather impatiently: "I didn't leave because first of all, I didn't want to, and secondly, because I understood that I was paying a karma." He went on to explain that he had realized early on that he would not be able to leave that place. Even suicide would not have been an option, he explained, because one cannot flee karma by dying; it would simply start all over again in the next life. One cannot avoid or heal suffering, one can only bear it. "Suffering is purification," he said, and "suffering is not gratuitous." Finally starting to feel that I understood his perspective, I commented that it probably became easier to bear the pain once one understood why one is suffering, and he nodded in agreement. Earlier, he had said something similar: "Yes, this gives me strength to continue living."

Eduardo's experience exemplifies the complexity of suffering: there are no simple answers. The ways in which we tell a story matter (cf. Abu-Lughod 2005). Emphasizing the structural violence (cf. Farmer 1997) that shapes the lives of undocumented migrants like Eduardo renders claims of justice and inclusion more powerful. At the same time, I wanted to avoid the pitfalls of representing research participants as 'suffering strangers' (Butt 2002), as faceless, voiceless, and helpless puppets, made to dance (or rather, suffer) in a way that suits the researcher's own political agenda (cf. Huschke 2015a).

Eduardo, one could argue, was caught in a web of abuse and dependency, made up of restrictive migration regimes and woven by those in power to support the global demand for cheap labor in a capitalist economy, in a deeply unequal society. Eduardo had very little agency; he felt powerless in the face of the very tangible effects of structural violence. From his point of view, though, things look somewhat different. He is well aware that he could have left, but this would only have perpetuated the circle of suffering he was caught in. Eduardo's spiritual understanding led him to see his suffering as unavoidable, as a necessary and ultimately beneficial catharsis. He was preoccupied not so much with his immediate physical and emotional wellbeing, but with the long-term effects of his decisions. Further, he did not perceive his living situation as entirely bad: he appreciated the safety and comfort of knowing his way around and being able to draw on a—albeit limited—social network, mainly his neighbors. To understand Eduardo's experiences from his angle helps to challenge the dominant perspective of what it means to be well and healthy: "[The study of individual subjectivity] holds the potential to disturb and enlarge presumed understandings of what is socially possible and desirable" (Biehl and Moran-Thomas 2009: 270).

Eduardo found a way out of the dilemma many of my research participants experienced: their suffering as criminalized, marginalized, and excluded migrants seemed incomprehensible and profoundly unjust. Eduardo saw his suffering as unavoidable bad karma, to be endured, with light at the end of the tunnel: the suffering would end once the karma has been paid. It helped him to achieve a fragile equilibrium, a state of acceptance—although not a state of wellbeing.

At the time of writing, Eduardo lived in a state-sponsored home for the elderly in Berlin. After six years of legal struggle, he was granted an extremely rare residence permit on humanitarian grounds, based on psychological assessments that certified 'a mental illness,' his lawyer told me. The decision also took into account that Eduardo was now nearly 70 years old and had no relatives in Colombia able or willing to take care of him. Obtaining a legal status allowed Eduardo to finally rest, with some peace of mind. When I asked him how he felt now compared to before, he replied: "calmer, livelier." He still feels the pain of being lonely: "I never had the chance to create a home, the sorcerer did not let me." But he has found a friend in a middle-aged woman from Colombia who sees him as a father figure, and he appreciates his new-found stability and safety (cf. Huschke 2015b). Undocumented migrants like Eduardo—many of whom experience anxiety, sleeplessness or depression, social isolation, economic deprivation and exploitation, and physical ailments such as chronic pains and acute infections—ultimately find relief in gaining legal status.

### *References*

Abu-Lughod, L. 2005 Writing against culture. In *Anthropology in Theory. Issues in Epistemology*. H. Moore and T. Sanders, eds. Pp. 466–479. Malden, UK: Blackwell.
Biehl, J., and A. Moran-Thomas 2009 Symptom. Subjectivities, social ills, technologies. *Annual Review of Anthropology* 38:267–288.
Butt, L. 2002 The suffering stranger. Medical anthropology and international morality. *Medical Anthropology* 21(1):1–24.
Farmer, P. 1997 On suffering and structural violence. A view from below. In *Social Suffering*. A. Kleinman, V. Das, and M. Lock, eds. Pp. 261–283. Berkeley, CA: University of California Press.
Huschke, S. 2015a Giving Back. Activist Research with Undocumented Migrants in Berlin. *Medical Anthropology*. Available from http://dx.doi.org/10.1080/01459740.2014.949375
——— 2015b Fragile Fabric: Illegality knowledge, social capital, and health-seeking of undocumented Latin American migrants in Berlin. *Journal of Ethnic and Migration Studies*. Available from http://dx.doi.org/10.1080/1369183X.2014.90774

---

In some instances, such as those described by Huschke, the tension of not belonging is only partially resolved, even after many years. The ways in which Eduardo makes sense of his suffering are a good example of how people come to understand and give meaning to their plights. Similarly, in their work with Ecuadorian migrants to Italy, Raffaetà and Duff illustrate that these individuals created attachment to place through assemblages of "social, material and affective resonances, experiences and resources" (2013: 342). Over time, these migrants created social spaces and participated in activities that enabled them to interact in pleasurable ways similar to those in which they might have participated at home in Ecuador. Raffaetà's and Duff's emphasis is on the process of 'practicing' place in a way that acknowledges its physicality as well as the affective feelings that migrants develop for it. The deprivations and discrimination against immigrants are such that it is almost inevitable to focus on the negative aspects of their lives, and to overlook the resilience of people, regardless of their legal status, that enables them to care for themselves and others.

## Survival Migrants and Global Violence

We opened this chapter with comments on population mobility in the twenty-first century, and above, we have drawn attention to the vulnerability of people who lack legal documents that allow them to stay and that provide them with access to health care and other human services. Far more vulnerable, and increasingly characteristic of global population movement, are those who Alexander Betts (2013) has characterized as 'survival migrants.' In his study of failed and fragile states—countries that often have a prior history of civil war and/or other regional wars, and that lack a viable and competent system of governance—people may move across borders out of desperation. The impetus to move is tied, too, to environmental change and natural disasters, food insecurity, and continued generalized violence that creates precarity both on an everyday basis and in relation to a future (Hedman 2008). Such reasons do not accord with the international criteria for humanitarian migration and resettlement; these require people seeking asylum to show cause, primarily by demonstrating the risks of persecution for reasons of race, nationality, membership of a particular social group, or political opinion. These criteria do not necessarily support requests for asylum and resettlement on the grounds of living under conditions of sustained local violence, economic duress, environmental devastation or protracted civil conflict. Current definitions of asylum seekers and refugees therefore systematically exclude many people from claiming protection. Further, how people are categorized can vary from one country to another, as not all countries accept this international definition, and often the validity of claims for sanctuary is determined at a national level, as we discuss further in Chapter 16. The United Nations High Commissioner for Refugees (UNHCR) classifies many of these people as 'irregular' migrants or 'people of concern.' Their legal status, the risks associated with their travel, the conditions in temporary and long-term detention camps, some of which have existed for decades, with generations born into the camps, seriously impacts on the health and wellbeing, directly and indirectly, of these seeking sanctuary (Holzer 2015).

As we have shown in this chapter, there are many kinds of global quests for care. Motivations, resources and results differ according to such things as civil status, laws around particular procedures, monetary resources and the larger sociopolitical realities of different places. The complexities of decision making, managing conflicts between family members and others as well as the ever-changing arrays of possible treatments and procedures makes these quests a vital, ongoing field of investigation within our discipline.

## Note

1. US Code 1182—Inadmissible aliens http://www.law.cornell.edu/uscode/text/8/1182.

## References


Aizura, Aren Z. 2010 Feminine transformations: Gender reassignment surgical tourism in Thailand. *Medical Anthropology* 29(4):424–443.

Alunni, Lorenzo 2015 Securitarian healing: Roma mobility and health care in Rome. *Medical Anthropology* 34(2):139–149.

Bergmann, Sven 2011 Fertility tourism: Circumventive routes that enable access to reproductive technologies and substances. *Signs* 36(2):280–289.

Betts, Alexander 2013 *Survival Migration: Failed Governance and the Crisis of Displacement.* Ithaca, NY: Cornell University Press.

Cartwright, Elizabeth 2011 Immigrant dreams: Legal pathologies and structural vulnerabilities along the immigration continuum. *Medical Anthropology* 30(5):475–495.

Cartwright, Elizabeth, and Lenore Manderson 2011 Diagnosing the structure: Immigrant vulnerabilities in global perspective. *Medical Anthropology* 30(5):451–453.

Castañeda, Heide 2008 Paternity for sale: Anxieties over "demographic theft" and undocumented migrant reproduction in Germany. *Medical Anthropology Quarterly* 22(4):340–359.

Dalstrom, Matthew D. 2012 Winter Texans and the re-creation of the American medical experience in Mexico. *Medical Anthropology* 31(2):162–177.

Deonandan, Raywat, Samantha Green, and Amanda van Beinum 2012 Ethical concerns for maternal surrogacy and reproductive tourism. *Journal of Medical Ethics* 38(12):742–745.

Ellick, Adam B., and Alderman, Liz 2015 Crisis in France is seen as sign of chronic ills. *New York Times*, January 15. Accessed from http://www.nytimes.com/2015/01/15/world/europe/crisis-in-france-is-seen-as-sign-of-chronic-ills.html?_r=0.

Goldade, Kate 2011 Babies and belonging: Reproduction, citizenship, and undocumented Nicaraguan labor migrant women in Costa Rica. *Medical Anthropology* 30(5):545–568.

Hedman, Eva-Lotta E. 2008 *Conflict, Violence, and Displacement in Indonesia.* Ithaca, NY: Cornell University Press.

Holmes, Seth 2013 *Fresh Fruit, Broken Bodies: Migrant Farmworkers in the United States.* Berkeley, CA: University of California Press.

Holzer, Elizabeth 2015 *The Concerned Women of Buduburam: Refugee Activists and Humanitarian Dilemmas.* Ithaca, NY: Cornell University Press.

Inhorn, Marcia C., and Pasquale Patrizio 2012 The global landscape of cross-border reproductive care: Twenty key findings for the new millennium. *Current Opinion in Obstetrics & Gynecology* 24(3):158–163.

Janzen, John M. 1978 *The Quest for Therapy: Medical Pluralism in Lower Zaire.* Berkeley, CA: University of California Press.

Kangas, Beth 2002 Therapeutic itineraries in a global world: Yemenis and their search for biomedical treatment abroad. *Medical Anthropology* 21(1):35–78.
Pande, Amrita 2011 Transnational commercial surrogacy in India: Gifts for global sisters? *Reproductive Biomedicine Online* 23(5):618–625.
Quesada, James, Laurie Kaine Hart, and Philippe Bourgois 2011 Structural vulnerability and health: Latino migrant laborers in the United States. *Medical Anthropology* 30(4):339–362.
Raffaetà, Roberta, and Cameron Duff 2013 Putting belonging into place: Place, experience and sense of belonging among Ecuadorian migrants in an Italian Alpine region. *City & Society* 25(3):328–347.
Song, Priscilla 2010 Biotech pilgrims and the transnational quest for stem cell cures. *Medical Anthropology* 29(4):384–402.
Ticktin, Miriam 2011 *Casualties of Care: Immigration and the Politics of Humanitarianism in France*. Berkeley, CA: University of California Press.
Tiwari, Shashank S., and Sujatha Raman 2014 Governing stem cell therapy in India: Regulatory vacuum or jurisdictional ambiguity? *New Genetics and Society* 33(4):413–433.
Whittaker, Andrea, and Amy Speier 2010 "Cycling overseas": Care, commodification, and stratification in cross-border reproductive travel. *Medical Anthropology* 29(4):363–383.

Cooperation for Peace and Unity: Rebuilding Health Infrastructure, 2012. Kabul, Afghanistan.

## *About the photograph*

*One of my Afghan colleagues, Fatimah, leads a discussion about peace-building. This conference was initiated by a local NGO where I worked as a visiting senior researcher in 2012–2013. Fatimah's session spoke to male and female doctors, elders, midwives, psychologists, teachers, and youth leaders in the health and education sectors. The conference was attended by people of mixed ages, gender, and tribal groups; this particular session was facilitated by an unmarried, professional local woman. The conference and the session were the first of its kind in Afghanistan. "All voices and perspectives are important for our future," Fatimah says.*

—Athena Madan

# 13

# War, Violence and Social Repair

*Lenore Manderson, Elizabeth Cartwright and Anita Hardon*

War leaves an indelible imprint on social life, shaking the confidence and trust of those who survive. Civil wars perhaps especially have substantial effects on interpersonal relationships, since there are few options to externalize the responsibility for the perpetration of violence. Soldiers and others who support war efforts routinely construct the enemy as the Other, but in civil war, those who instigate terror, raze villages, rape, torture, and wage war may be neighbors or family members, driven by ideological differences, political and economic grievances, structural inequalities, greed for power or anger at their powerlessness. Internationally and within nations, war and other civil violence and terrorism have profound effects on population health and medical care, food supplies, disease transmission, and the health of health providers, civilians, and soldiers. This has become a growing field of interest and urgency for medical anthropologists.

This urgency is underscored by the continued scale of violence and terror worldwide. The two world wars of the twentieth century resulted in vast loss of life and massive social and economic disruption, and both the League of Nations post–World War One and the United Nations post–World War Two were established to work towards world order and social justice and to find ways to defuse and contain political tensions and address the social and economic forces implicated in them.[1] But war has not abated. At time of this writing—mid-2015—there were around 50 serious conflicts worldwide, of which a number, such as in Palestine (from 1948), Southern Philippines (from 1969 to the present), and Afghanistan (from 1978) have continued for decades: everyday life for people living in these areas is characterized by chronic militarization. Further, some wars—in Iraq, Syria, and Nigeria (Boko Haram)—have escalated in recent years, now contributing to the highest death tolls per annum. Other wars, shocking in their brutality, have moved the setting—the 'theater' of war—while continuing to maintain a regime of terror. One example is the war waged by the Lord's Resistance Army, which operated in northern Uganda and South Sudan in the 1990s and is now active primarily in the DRC (Democratic Republic of Congo) and ACR (African Central Republic).

The collateral damage of these wars, and of state-sanctioned structural violence, militarization, systematic terror, and torture in various other countries, is reflected in the staggering numbers of displaced persons globally. The UNHCR (United Nations High Commissioner for Refugees) has estimated that one in every 122 people in the world is displaced (http://www.unhcr.org/558193896.html). At the end of 2014, 59.5 million people had been forcibly displaced, of whom two-thirds were internally displaced and were living within national borders.

Those living in camps are visible as refugees, but the majority of people relocate as individuals or in larger groups to other safer places, moving to live on the edge of cities or to share houses with kin in other regions. Using these UNHCR estimates, in 2014 some 20 million people had fled across borders, but of these, only a proportion are officially registered as asylum seekers or refugees, living in second- or third-country displacement camps, detention centers, and refugee processing centers. Others escape to and reside in new countries without registration as refugees, without papers and without rights, living in constant fear of deportation. Many have lived under these conditions for decades, often—where conflict has been sustained—for generations. The largest refugee camps, including in Kenya, Ethiopia, and Jordan, now function as major, established towns, where medical, psychological, and social problems are compounded by environmental and economic stressors, restrictions on food and water, aid dependency, trauma, and uncertainty.

## The Health Effects of War

Under the conditions of civil war and under regimes of terror, and in environments in which there is high risk of personal and property violence, people's health is affected in multiple ways. This creates challenges for the ruling government (and alternative governments) in meeting the health and other needs of civilian populations and the military. The most immediate demand on health services during times of war is the provision of medical and surgical care of people directly affected, depending on the site of war and the modes of violence, through bombing, landmines, road vehicle and aircraft accidents, gunshot wounds, rape and torture, and other specific acts of brutality and execution. In major conflicts, multiple logistic, financial, personnel, and personal challenges interfere with the delivery of emergency and continuing care. Such care is disrupted by systems challenges—the infrastructure and supplies necessary for emergencies and the maintenance of the delivery of basic primary care—and by the difficulties associated with providing rehabilitation, counseling and mental health care for those directly affected, and supporting health care providers dealing with the high mortality rates of soldiers and civilians. These challenges persist despite the translation from ground wars to 'smart' wars, with drones targeting strategic structures rather than people, that purportedly limit the destruction of civilian lives. Even so, most wars fought in the early twenty-first century are not 'smart'; they involve nasty direct assaults on populations and their places of residence.

Given these challenges, international and civil wars and other forms of extreme violence force governments to meet emergency medical needs at considerable cost to population health (see further Chapter 15). Fighting and peace-keeping are expensive. Funds are diverted that might have been used to maintain local health facilities, support primary health care workers, and undertake illness prevention and health promotional activities. Low- and middle-income countries already face endemic problems of poorly maintained facilities, lack of and broken equipment, stock-outs of essential drugs, shortages of doctors and other health workers, and poor quality of care; these problems are magnified in war and discourage people with acute infections and chronic conditions from seeking care.

Frontline health care workers toil under dangerous conditions, often in the line of fire. In her account of a public hospital in Honduras, Adrienne Pine illustrates the pressures on under-resourced nurses as they try to meet the medical and surgical needs of people who have survived stray gunshots, domestic violence, muggings, intentional attempts at homicide, and quotidian threats to their lives from vehicle injuries, infectious disease, and structural violence. All of these things threaten the lives of health care workers as well as their patients, and the families of both, as they struggle in an environment of unparalleled violence.

## 13.1 Honduras: Practicing Wartime Healing

*Adrienne Pine*

In 2013, Honduras's homicide rate was the highest in the world at 90.4 per 100,000, approximately 18 times higher than the murder rate in the United States, and 113 times the average of countries like Italy, the Netherlands, Japan and Australia. These statistics are often used to frame discussions of Honduras. But murder rates can obscure the embodied truths of wartime survival, especially in a country not officially at war. If you ask Hondurans today, most of them will tell you they live in a war zone, usually in reference to the ever-present threat of bodily violence. But metaphorical wars declared against drugs, terror and crime in Honduras are also real wars against the poor. Perhaps nowhere is the current state of war in Honduras more evident than in public hospitals. There, health care workers, suffering the violent impact of neoliberal economic policies in the form of lack of supplies, infrastructure and staff, must attend to the survivors of the war on 'the street.' Employees of Honduras' national teaching hospital, Hospital Escuela, refer to their workplace as a 'war hospital.' Underfunded public hospital buildings are in various states of decay, and police and soldiers roam the hallways, sometimes removing patients from their beds for 'interrogation.'

Since 2008—a year before the military coup that overthrew democratically elected president Manuel Zelaya and precipitated a dramatic increase in all forms of violence in Honduras—I have carried out multi-sited fieldwork among nurses on numerous fieldwork trips to the country. Between July 2013 and June 2014, I worked as a visiting professor at the National Autonomous University of Honduras (UNAH) in the capital city Tegucigalpa, where I taught medical anthropology to nursing students and carried out participant-observation fieldwork in the maternal-child section of Hospital Escuela, administered since early 2013 by UNAH. During this period, health care workers carried out numerous labor stoppages. A variety of scandals plagued Hospital Escuela, including violence carried out against family members by private security guards (themselves former soldiers and police officers, products of the Honduran war machine), assaults and kidnappings. At the same time, the hospital was undergoing a dramatic UNAH-led labor restructuring, allegedly in order to address poor workplace discipline. Unions were under attack, and nurses and other workers were subjected to biometric screening on a daily basis, like the newly installed finger-scanning machines that they had to use at the beginning and end of every 8-hour shift.

With each new crisis, a new mechanism of labor control was implemented as a solution. Following the highly publicized kidnapping and 'miraculous' return of a newborn baby girl (following which newspapers announced the baby's name had been changed to 'Milagro'—miracle) from the maternity ward in May 2014, it was announced that security cameras would be installed throughout the entire hospital. While some nurses welcomed the security measure, others theorized that the whole kidnapping had been orchestrated as an excuse for hospital management to control their every move, weakening their ability to care for patients for fear of being sanctioned by an administration that did not understand the nature of war hospital care.

Between crises, nurses and other health care workers still have to contend with the day-to-day reality of working in an (unofficial) war-zone hospital. The implications of this began to become clear to me on my first full day of fieldwork in the Hospital Escuela's pediatric orthopedic ward. Just after the morning shift began at 6 AM, Edith, the charge nurse, instructed me to interview patients and their families. In the main hall, seven children, ages 4 to 16, lay in cage-like iron beds—none of which had functioning guard rails. Most children had casts on. I began to hesitantly ask each one why they were there. Below, I draw on my fieldnotes describing my first interactions with the young patients and their family members.

> I went to talk to the girl . . . who looked to be at least sixteen. I stood next to her, awkward.
> "What happened to you? Did you break a bone?" I asked.
> "(Inaudible)" is what I scribbled in my notes as her response.
> "What did you say?"
> "*Un tiro*" [a bullet] she said with exhaustion, sadness or something similar.

She wore a full leg cast. I asked if the shooting was recent, saying also that her toenail polish, visible through the toe-hole in the cast, was pretty. It was different colors and had designs, looked recently applied but likely before the cast. She had been shot in the knee last Friday. A poor-looking, frail woman, whom I guessed to be her mother, came and began attending to her.

The frail woman asked me something, and I didn't understand right away. She was asking about a wheelchair so she could bathe the girl with a bullet wound to the knee. Useless, I directed my question to a woman who was with a 14-year old boy in traction, with 'Arthrogryposis multiplex congenita' scribbled on the paper at the foot of his bed. She was washing him and I assumed she was an auxiliary nurse but it turned out she was his mother. As Edith had explained to me the previous day, due to staffing shortages (i.e., firings and hiring freezes), jobs that nurses should do were frequently passed on to family members, and mothers were all in charge of washing their own kids now. In either case, she knew the scene better than I did. "You are going to wash her?" she asked the frail lady, who indicated she was indeed planning on washing the girl. The experienced mum said that there was only one wheelchair on the ward and it was too small for the girl.

Also, it was being used by the little boy, Carlitos. Carlitos, dependent on a wheelchair because of his condition (Edith used the term 'eggshell' in explaining it to me), wheeled himself around in his tiny wheelchair, joking and flirting with all the nurses like he owned the place. Edith told me that almost as soon as he left, he'd break something again and be back in.

So, no wheelchair for the quiet bullet-wound girl with pretty toes. "Do you have soap?" the hospital-experienced mother of the boy in traction asked the frail confused mom. The frail mom held up her little bar of soap. I later realized one of the auxiliary nurses sometimes handed out tiny bars from storage to the parents, though really they were expected to bring their own. What you have to do, said the experienced mum, is wash her on the bed, like I'm washing him (demonstrating a basic sponge-bath technique; she was clearly a pro). The frail mom tried washing her daughter for a couple minutes then came back to the experienced mum, asking her directly (instead of addressing me this time) if she had a robe with which she could cover the girl for privacy. "No," she said kindly, but with a tone that indicated to me that it seemed obvious to her, a regular, what this woman—certainly not a regular—should do. "You have to ask for a patient robe to cover her."

A young boy lay in another bed with his left arm and leg in casts. His mother and father, who looked particularly poor and rural, were loving with him. They told me he had been hit by a taxi in Esperanza, Intibucá. They gently moved his limbs about trying to make him more comfortable.

Another cute little girl (under 10, I'd guess) with one arm in a cast was being gently led in by her father to her mother who waited by her bedside. They had hung a pink princess mosquito net over her otherwise prison/military-looking cot. Later, during rounds, Edith explained to me that the girl had been shot at by thieves, who had held up her family while they were in their car in traffic. They taken everything from them, and then shot her in the elbow, as she sat between a parent and her little brother. Just for good measure. The page taped to the foot of her bed read, "Grade III A open fracture left radius." "We are all exposed here," explained Edith. . . .

In the room that was designated for quarantine, with only one bed, sat a boy with a cast on. Edith explained to me that the doctors had been planning to amputate, but God had saved his limb. It was a shotgun accident. The shotgun belonged to his father's friend and it went off accidentally. The same kid, months earlier, had spent five days in a coma, she told me, after getting run over by a car. Edith explained to me that with an *escopetazo* [shotgun wound/attack] the bullets are dirty, and you have to put the victim on antibiotics so they don't get necrosis. "This is the hospital of the poor," she said, "We have to pray to God to take care of them, we have no other choice." This was partially in reference to the fact that the hospital never had the necessary supplies; the *escopetazo* kid's mother was off buying some bandages and medicine. Families almost always need to purchase meds and supplies on their own, and nearly everything is available within walking distance, from drugstores to the funeral homes surrounding the hospital. But often the drugs in question aren't stocked, whereas there always seems to be a good supply of coffins on display. . . .

Edith called me to join the residents (two tall young men with a combined air of importance) and her on rounds. They started with the quiet teenage girl who had been shot in the knee, with the full leg cast on. Edith told me it was a *bala perdida*—a stray gunshot. I felt then, and also later, that this may not have been the whole truth. Or at least that there was much more going on. She seemed profoundly traumatized (in a way that, in this context, a stray bullet didn't sufficiently explain for me). . . .

A handwritten sign on the wall over the boy in the back corner of the 'contaminated' room read "Children are the most important thing in the world; the problem is we can't even take care of the world." Pointing to the boy under the sign, Edith half-whispered to me that he had "RM," hoping I'd understand. Then seeing I didn't, she went ahead and said *retraso mental* [mental retardation]. An Aguazul bottled water truck had run him over.

In the small kids' room, a mother was sitting with her two-year-old child, who had one arm in a cast. Edith asked what had happened to him and the mother replied that he had fallen from a chair. Edith looked at me knowingly and said, "[h]e didn't fall from a chair. That doesn't happen from a fall off a chair."

In subsequent weeks, it became clear that the child injury patterns I observed on my first day of fieldwork were representative and symptomatic of war in Honduras. While there were injuries like those that regularly put me into casts as a child, such as falls from trees, poorly executed acrobatics, and so on, they were a small minority compared to gunshot wounds and car accidents. In Honduras, the frequency of car accidents is tied to neoliberal structural violence; a lack of auto safety regulation and investment in public infrastructure contribute to the increased deadliness of transportation, especially for the poor. And as Edith implied, the results of intimate and domestic violence (which cannot be easily disentangled from the suffering of war) were ever-present.

The Honduran murder rate has increased in a context of impunity for the perpetrators of violent crime (both overtly political and 'general'). So too have survivors. It is rare to meet a Honduran who has not lost a loved one to homicide. Trauma that would likely be diagnosed (if psychiatric care were available to the general population) as PTSD has become a generalizable Honduran societal fact. But PTSD also medicalizes as a *disorder*—as if there should be a normal, orderly way to assimilate the murder of a loved one—forms of violent subjectivation that are much more complex and culturally rooted than the diagnosis implies.

Nurses in every ward of Hospital Escuela provide care to people who are impacted by violence that goes well beyond what might be knowable from simple murder statistics or easily readable symptoms. Most hospital patients (and Hondurans in general) are survivors in multiple literal and metaphorical senses. They survive violent physical attacks. They survive their deceased family members—victims of murder, car accidents, untreated illnesses and other forms of structural violence. They survive hunger, intimate violence and the fear of living with the 'insecurity' that maims their neighbors, family members and friends, and sometimes themselves.

Whatever the immediate symptoms that bring Hondurans to seek care, this context is their reality. It is the daily reality for Honduran nurses as well. As is the case throughout Honduras, an hour in Hospital Escuela does not pass without a conversation about the dangers on the street. And how could it? The primary focus of most Honduran nurses' jobs is caring for people whose bodies are damaged and destroyed by that violence. But the undeclared war in Honduras also impacts nurses' own bodies. They too are robbed at gunpoint; they too lose family members and friends to violence. They too live every day with fear. Indeed, many nurses became leaders in the resistance movement that opposed the 2009 coup, precisely because of the moral authority that their gendered position as healers accorded them and their intimate understanding of the human costs of the post-coup repression and usurpation of democracy.

In 2013–2014, however, the post-coup resistance movement had reverted to a politic of individualized survival and fear in a militarized state. One morning in the pediatric orthopedics ward, Edith sighed and told me that the previous day, two of her nurses had been robbed on the way to work. She had had her wallet stolen on the bus on the way to her shift a few weeks earlier, but she said she was assaulted less frequently than the others, a fact she attributed to her strong Catholic faith. It used to be worse, though, she said. She used to be afraid to walk down the halls of the hospital, because of armed thieves who would attack hospital workers and family members. "Now with the security guards things have gotten a little better," she told me.

The presence of armed security forces in the hospital was a controversial topic among nurses and patient family members. Police and soldiers also regularly patrolled the halls. In an election riddled with fraud, Edith's candidate Xiomara Castro (of the resistance-affiliated LIBRE political party) lost to Juan Orlando Hernández, whose 2013 campaign promise was to further militarize the country in the name of 'security.' Despite her satisfaction with reduced hallway robbery, Edith was aware of the inherent contradictions of the politic of workplace security, and of the huge problems that private armed guards caused for patients and their families—another regular topic of conversation among nurses and family members. Guards arbitrarily refused entry to some family members, and frequently demanded bribes (*pa'l fresco*—a little something for a soda) for the privilege of visiting patients. This led to emotional stress and physical altercations between family members and guards. It also complicated patient care for nurses. Since family members are responsible for providing nearly all supplies, medication and food for patients, there is little that nurses can do to ensure proper care if family members are not permitted to enter.

In April 2014, UNAH administration unveiled a campaign within the hospital to stem corruption. Posters were placed in numerous locations on all floors of the hospital and at the entry gates to the parking lot, controlled by the private security guards. On one version, the word NO hovered over a cartoon image of a guard receiving a bribe. The other contained the same message with blocky cartoons of a series of workers meant to represent a nurse (female), doctor, security guard and office worker (all male).

Despite the violence and corruption, nurses express a determination to care. In my fieldnotes, I recorded a conversation with an auxiliary nurse:

> "If I win the Diario," Leticia began telling me, then backed up. "I mean, I don't really buy tickets, but if I were to win the lottery, I would spend a million on this ward so it would have the latest equipment. The *cunas* [cribs, but applies to the beds used by the bigger kids too] don't have guard rails. They had guard rails but they're all broken," she said. She then told me (in response to my question about it) that sometimes kids have indeed fallen from the beds, "and imagine—their bones are already broken."

Nurses in Honduran public hospitals face a series of interconnected challenges in providing care. In every ward, patients, family members and nurses alike suffer from the war around them. Hondurans' visible physical symptoms as well as other trauma (PTSD) symptoms are so ubiquitous that they are largely invisible. Nurses provide care without supplies or sufficient personnel, in a work context that limits their job security and scope of practice. This results in further violence toward patients that nurses struggle to mitigate. Although they hold a wide range of political views, they uniformly express affection and empathy for—and solidarity with—their patients. They build creative strategies with family members for providing care that neither party can afford. They build relationships with individuals and organizations (often without the knowledge of administrators) to solicit donations of supplies and informational materials. They hustle construction materials and paint to keep their wards functional and even beautiful, thus creating a symbolic refuge from the violence that surrounds and to such a great extent defines their work. And they dream of achieving their ideal practice, through collective struggle or other means.

## Studying War and Terror

War undermines the basic conditions of everyday life: that is its purpose. Bombs intentionally target infrastructure essential for communication and commerce—roads, bridges, dams, markets, buildings of cultural significance (temples, churches)—in order to prevent health care and other services from reaching populations, cutting off people's access to markets, and destroying crops, water and sanitation facilities. 'Scorched earth' has a long history as a strategy of war by government military forces, with the aim of destroying assets potentially available to enemy forces, including food sources, industry, communication and transport. The strategy was used against Napoleon in Russia, in the American Civil War, in the Boer War in South Africa, by Stalin against German troops during World War Two, in with toxins as part of its weaponry throughout the so-called Cold Wars of the 1950s–1970s in Southeast Asia. During the Vietnam War, the highly toxic herbicide Agent Orange was used to destroy crops and foliage to expose possible enemy hideouts, and Agent Blue was used on rice fields to deny food to the Viet Cong and to those with whom they were living. Both herbicides had immediate effects on food supplies, leading – as crops were poisoned and destroyed – to hunger and under-nutrition. But in addition, Agent Orange especially has had long-lasting (if contested) effects on population health, with continued high incidences of birth deformity, skin disease and cancer in areas that were saturated with the chemical during the Vietnam War (Uesugi 2015). The health effects of these chemicals persist over generations (Lykes 1994), yet despite public outrage, they have continued to be part of the armament of the military in more recent civil wars in Central America and elsewhere.

Few anthropologists have conducted fieldwork when war has been at its most fierce. Ethics committees are as reluctant as families to support individual anthropologists for reasons of their personal safety, and because of the ethical issues that arise when researchers work in settings where everyday life is precarious and populations are under acute stress. But anthropologists have certainly been present during some wars, and have provided us with nuanced accounts of the impact of violence on health and wellbeing. Our understanding of terror and torture as a vehicle of control, and the threats of greater violence that are embodied in such technologies, owes much to the work of Allen Feldman (1991) in northern Ireland. Carolyn Nordstrom (1977) conducted innovative, multi-sited ethnographic research in Mozambique from 1988 to 1996, documenting the brutality and profiteering but also the resilience and vision of people determined to resolve everyday challenges and to work creatively to rise above the social fissures of society to bring conflict to an end and to work towards a future. Tim Allen, Sverker Finnström and their colleagues (Allen and Vlassenroot 2010; Finnström 2015) all worked during the years of terror of the Lord's Resistance Army in northern Uganda and contiguous territories, and documented everyday life in times of war. Carlos Iván Degregori (2012) provided a chilling description of life lived during the revolutionary war of the Shining Path in Peru, and Kimberley Theidon's discussion of that same conflict highlights the particular toll of this war on women and children (Theidon 2012; 2015). In their efforts to make sense of the continuation of genocidal wars in the second half of the twentieth century, Alex Hinton and colleagues (2002; 2004) have drawn attention to how cultural notions of difference and purification have been used by armies, colonizing governments, dictatorships and civilian populations to justify atrocities against large populations. The accounts on which these anthropologists draw derive from narrative accounts from survivors, and from fieldwork in the aftermath of war, to highlight the corrosivity of violence on the bodies, minds and physical landscapes of survivors, impacting at individual, community and national levels across generations. In doing so, they remind us of the moral obligations of anthropology. By exposing the motivations of genocide, and the indifference of others as it unfolds, Hinton argues that "anthropology can be a spur to action . . . the first step towards combating it" and of helping societies to "determine how to move on" (2002: x–xi). As we demonstrate in this chapter, this same imperative for research applies to all aspects of war and violence, and the particular work of medical anthropologists in this domain.

## PTSD and the Medicalization of Horror

Civil wars of the late twentieth and early twenty-first century have characteristically used technologies of terror to force unsympathetic civilians to submit, and, depending on standpoint, to defeat the contested government or to undermine the efforts of insurgents. Civil wars fought in countries such as Angola (officially 1975–2002), Mozambique (1977–1992) and Sierra Leone (1991–2002), for example, which often followed extended wars of independence, have been marked by the common use of summary executions and mass graves, vicious amputations, beheadings and disembowelment, violent rape, and the kidnapping and deployment of young children, invoking indelibly ugly images globally as well as locally. In an article published in *Daedelus*, anthropologists Arthur and Joan Kleinman (1996) argued that the use of photography and particularly of video recordings had commercialized suffering and misfortune: the appropriation of images of suffering for mass entertainment left viewers insensitive to the pain and horror of war and other tragedies, creating a sense of indifference to the causes of such suffering. They were prescient of the extent now that media is deployed, although they failed to anticipate—how could any of us have done so?—how social media has enabled warring factions to extend the reach of terror in contemporary major wars. Consider from the early twenty-first century especially the video recordings of beheadings, posted on various web pages, by the Taliban in Afghanistan, the Islamic State in Syria, Boko Haram in Nigeria, and other terrorist organizations. The fear is not of the violence generated by these posts alone, but that we all might become inured to them, as we might become inured to drone technology, as 'simply' aspects of 'modern' warfare.

The close exposure to such acts of terror by civilian populations, the management of maimed bodies and corpses by health workers, and the involvement of soldiers in witnessing and sometimes enacting such violence, takes a massive toll on people's mental health. Awareness of the mental health costs of war are not recent, of course. Shell shock, later also referred to as combat stress, was already well described early in the First World War (Jones and Wessely 2014; Myers 1915), as a reaction to horror by men at the scale of carnage and the immediacy of killing and death; soldiers engaged directly with enemy soldiers on battlefields as well as with prisoners of war. Witnessing violence and its concomitant feelings of horror and powerlessness left many people—men, primarily, since men vastly outnumbered women in direct warfare—unable to function; because of this, they faced accusations of malingering, cowardice, weakness and supreme disloyalty.

The psychiatric category of post-traumatic stress disorder (PTSD), as published in *DSM-III* (1980), was developed to diagnose the psychological experiences of people shocked by the trauma of war or other distressing events such as interpersonal violence or its witness, or a traumatic accident. The symptoms were specified to include the persistent recollection of particular events, including intrusive flashbacks, vivid memories, recurring dreams, hyperawareness, sleep disorders, irritability and difficulty concentrating, such that individuals were overwhelmed to the point that they had difficulty managing everyday life (American Psychiatric Association 1980). This classification of PTSD, updated in subsequent versions of the *DSM*, was strongly criticized, partly because of its questionable appropriateness in different contexts (Solomon and Canino 1990), but also because it medicalized and individualized experiences of war and other deeply disturbing events; that is, the application of the label of PTSD treated the embodied reactions to horror and brutality as individual pathology, while the distanciation of others from these same events was represented as normal and functional. From a critical perspective, it seems extraordinary that a functioning human is one who can rationalize and so dismiss something as disturbing as watching a body blown apart or a gang rape, while the responses of those who struggle to find meaning in witnessing such violence are considered pathological. As medical anthropologist Allan Young (1995) argued for Vietnam veterans, PTSD needs to be considered as a cultural construct, rather than as a reflection of the efforts of individuals to find moral meaning in an ugly and deeply contested war.

On the other hand, medicalizing trauma allows people to replace personal psychological distress with a health problem that can be addressed. In writing of war in Sierra Leone, Doug Henry (2006) illustrates how Sierra Leoneans speak of sustained trauma, associated with fears of being murdered and stories of cannibalism that destroyed their ideas of what it means to be human. In his account, existential and physical unease is medicalized and managed as *haypatɛnsi*, a local illness category that translates as 'spoiled heart.' This condition—the bodily expression of constant fear and anxiety, chronic illness and disability, and economic insecurity—is distinct, despite that it borrows its name from hypertension and is treated as if it were hypertension, with people prescribed Inderal (propranolol, a beta-blocker) when the drug is available. The stigma associated with what people do and see in war, their difficulties in making sense of this, and the links between masculinity and soldiering, mean that soldiers are advantaged by not revealing war experiences and by disguising their struggles on return. Alexander Edmonds illustrates what might be at stake were a soldier to admit to PTSD.

## 13.2 Does Sgt Pearson Have PTSD?

*Alexander Edmonds*

When I first met Sgt Pearson at a Starbucks in the Army town of Fayetteville, North Carolina, in 2012, he ordered a latte with soy milk.[1] At age 29, with 10 years of military service and four tours of duty in Iraq and Afghanistan behind him, he is what infantrymen sometimes call a 'war dog,' a seasoned soldier. In addition to lactose intolerance—a condition he blames on drinking long life Army milk—he's got 70 percent hearing loss, stress fractures in his feet, and a mostly healed broken shoulder. After work he has to lie down for an hour "just to be able to do anything." He is worried by the prospect of being away from his seven-year-old son during an impending deployment to Afghanistan. He's got "a bit" of agoraphobia, a lot of insomnia, and he said, "nightmares." Then he added thoughtfully: "Not really nightmares because they aren't fictional, just memory replaying."

Does Sgt Pearson also have PTSD, an acronym that has become so widely known that in many countries it is not necessary to write it out? The American Psychiatric Association introduced the term 'post-traumatic stress disorder' in 1980. It has a range of symptoms, including notably the flashbacks or 'memory replaying' that Sgt Pearson mentioned, as well as avoidance and hyperarousal. Though PTSD is also diagnosed in civilians, it has become the most significant mental health problem in combat veterans in the West. It is estimated that around a fifth of the two million American veterans of Iraq and Afghanistan have PTSD. But it is not known how many ultimately will get the disorder since symptoms can develop months or years after exposure to a traumatic event.

When I met Sgt Pearson again in 2013, he had just returned from Afghanistan—his fifth tour of duty. He told me he had been ordered to have a 'PTSD test' by a superior but had not yet done so. The outcome of that test, if he ever gets it, could have major consequences. A 'service-connected' diagnosis of PTSD—a category that recognizes that illness resulted from military service—can confer a substantial disability pension. The US Veterans Affairs Administration (VA) spent around 36 billion dollars on disability compensation in 2010. Yet despite the potential benefits a PTSD diagnosis confers to veterans, many active duty soldiers fear that it would land them a despised job as what Sgt Pearson calls a 'desk jockey' or end their career.

In this section, I reflect on what happened before that 'PTSD test'—why he was ordered to have it and why he did not want to follow this order. I draw on pilot anthropological fieldwork with soldiers who have been in combat and are now stationed back home. Below, I explore soldiers' perspectives on the military and health care institutions that play an important part in their lives post-deployment, and how they come to accept—or reject—clinical interpretations of their problems.

Given all that is at stake with PTSD, not surprisingly the disorder has sparked major controversy. One issue is its prevalence. Humanitarian responses to war and disaster in the developing world now often include mental health services to prevent or treat PTSD. Derek Summerfield (1999: 1460) has argued that such efforts make disaster into a 'mental health emergency writ large' and can weaken collective forms of coping and healing.[2] Others decry the widening range of people being diagnosed with PTSD: victims, perpetrators, and witnesses of violence as well as those who give care to the traumatized and even those who observe traumatic events in the media.

Fassin and Rechtman (2009) counter that the PTSD illness concept is not 'good' or 'bad' in itself, but reflects an altered moral attitude towards the ill or injured person that goes beyond clinical issues. They argue that previously those who suffered from medically unexplained symptoms caused by violence or accident were often suspected of malingering, or else of unconsciously seeking 'secondary gain' (i.e. the benefits that can be gained through illness, such as sympathy, care or disability pensions). The PTSD concept in a sense 'exonerates' the ill person and shifts the 'blame' for illness onto an external event. As a result of the social and material benefits that can follow from its diagnosis, PTSD is—in Rechtman's words—the only kind of psychological disorder "you want to have" (2004: 914).

However, like many active duty soldiers in the US military, Sgt Pearson did not seem to 'want' this diagnosis. Criticism of the overdiagnosis of PTSD has largely focused on civilians or veterans who have left military service. The moral significance and material effects of a PTSD diagnosis are often quite different for soldiers still in the military. In the American Army, there has been rising concern that soldiers with PTSD are not getting expert help. Some studies have found that less than a quarter of soldiers who are "positive for a mental disorder" (as determined by an anonymous survey) ever see a provider (Hoge et al. 2004). Suicide rates have been rising, and outpaced combat deaths for the first time in 2012. In response, clinicians and military leaders have launched major suicide research studies and an ambitious resilience training program. They have also conducted quantitative studies of 'barriers to care' that seek to understand why so few soldiers seek mental health treatments. One study that found that "negative attitudes towards treatment inversely predict treatment seeking" and concluded, logically enough, that policy should aim "at reducing negative attitudes towards mental health treatment" (Kim et al. 2011: 65)

Ethnographic research can complement such quantitative research by exploring how such negative attitudes are generated or sustained by daily life and institutions. Sgt Pearson has to date never received a mental health diagnosis, but was admitted into an alcohol abuse program some years ago: "Someone in the 25th in Hawaii decided all these guys just needed counselors. So we would meet in a coffee house, or for lunch, like here." Today he says he only has one to two drinks a day—but then scoffs that this is the Army's "official definition of an alcoholic." He was given antidepressants by an Army doctor, but stopped taking them as they made him feel worse, and had "male" side effects. He also saw a social worker "around three times" after his first deployment. He said "she was educated in talking to people, but we had no common experiences." He added, "Things you did there would be unforgiveable here. That weighed on me. I went to talk to a Baptist preacher back home. He was a Vietnam vet." He was also ordered to see a psychologist in Afghanistan, but he stopped seeing him after a couple of sessions. And most recently he received that "command referral" to get a "PTSD test."

It is not entirely clear whether Sgt Pearson has ever voluntarily *sought* or even *received* mental health care. While he did *choose* to see the social worker, preacher, and GP, he was *ordered* to see the psychologist and to have the PTSD test. This mix of choice and coercion in a therapeutic trajectory that took him from medical to psychotherapeutic to pastoral care makes it hard to determine whether he encountered a 'barrier' to mental health care. Recently, the military has tried to de-stigmatize PTSD, partly to make it unnecessary to *order* soldiers such as Sgt Pearson to see a clinician. For example, military leaders have been using new language to discuss mental health—or what is often now called 'behavioral health.' Some leaders have dropped the D from PTSD since 'disorder' sounds more serious than 'stress.' Others refer to PTSD an 'injury' to emphasize that it was honorably 'earned' during combat.

Sgt Pearson seemed aware of such efforts but was skeptical: "It's like there are two levels [of leadership]. At the higher level there is the liberal voice of the Army that says meet up, help each other. It cares about high suicide rates. They started treating PTSD as if you're, like, a rape victim, using the same treatment. Hopefully it works. But then on the lower level [of the Army] PTSD is really stigmatized."

The soldiers I spoke to mostly belonged to this 'lower level': enlisted men, NCOs, and a few lower-ranking officers. This group—while by no means representative of the enormously diverse Army—seemed to have precisely those 'negative attitudes' identified by quantitative research. It was not that PTSD was a taboo topic for them; it came up frequently in conversation and often in a joking manner. But more serious talk about PTSD often mentioned soldiers 'who get paid for PTSD.' Sgt Pearson said: "These guys on the big bases, who never saw combat and did paperwork. Some of these guys get paid for PTSD. It really bothers me. Some guys I knew I had to stop talking to them, people who faked PTSD." Others went further, claiming that *anyone* who 'got paid for PTSD' didn't really have it. Although several soldiers openly talked about having some of the symptoms that have now become recognizable signs of PTSD in American popular culture—such as hitting the floor in response to a sudden noise—they thought that most of their comrades who 'get paid for PTSD' do not really have the disorder.

The comments of Sgt Pearson and his comrades might conceivably be changed by training and education programs. Yet, I think these soldiers are not uninformed about mental health problems. Rather they also possess some insight into the current institutional and moral climate in which PTSD is diagnosed and lived. While Sgt Pearson did have a few contacts with caregivers, he was largely unhappy with what happened: "Army doctors are biased. If you start saying anything [about a work dispute], they might side with your commander." Civilian norms around patient confidentiality often do not apply to soldiers. For example, a clinician may be obligated to reveal information about clients to their commanding officers. Of course patient confidentiality is never an 'absolute' right and the limits to that right in the military are based partly on common sense concerns around giving weapons to someone with a disorder or who is on medication. Yet military clinicians sometimes have fundamentally competing obligations: to heal patients and to support military operations.

It has been said that the 'true patient' of the military psychiatrist is the Army itself. This professional position can create major ethical dilemmas. During World War I a psychiatrist who found a case of war neurosis to be false might send his patient back to the front. Today a soldier judged to have 'fake PTSD' would not be sent to a war zone; more likely, steps would be taken to *remove* him from a war zone. Yet the clinician still has unusual power over the soldier-client. 'Withholding' a service-connected diagnosis of PTSD can deny a soldier disability benefits. The high moral and material stakes of PTSD were made evident in recent scandals about the 'downgrading' of PTSD to a pre-existing condition such as a personality disorder. As Kenneth Macleish (2013: 127) points out, questions about overdiagnosis are inevitably bound up with the enormous economic stakes of disability compensation as well as the "weightier moral economy of who bears responsibility for the effects of violence." Many active duty soldiers are simply concerned that a diagnosis of PTSD can harm their career and status as a 'good' soldier who stoically 'sucks up' pain and suffering.

Capt Mulhern said: "Like anyone who has seen a lot of combat, I have a little bit of PTSD." What is a little bit of PTSD? In epidemiology, clinical trials, and disability assessments, disorders are present or absent; they must be counted. Of course there are more or less severe cases. But what Capt Mulhern meant I think was *not* that he had a mild case of PTSD. Rather, he seemed to be getting at ambiguity in the PTSD concept itself. This captain—who seemed to be highly respected by his subordinates—did not see his symptoms as evidence of mental disorder but rather as evidence of being a good soldier. Symptoms such as a violent temper or jumpiness or a tendency to brood were testimony to having served in combat. They reflected an unspoken sentiment that has perhaps taken root in the US since the Vietnam War: combat messes you up a bit. After war it is normal to be a bit abnormal. What defined PTSD as a mental health disorder for these soldiers was not the presence or absence of PTSD symptoms, but rather the official diagnosis and, paradoxically, the disability pension that might come in its wake.

The logic for them seemed to go something like this. Real soldiers—those who've been in combat—have PTSD symptoms by virtue of being real soldiers. But real soldiers don't get *diagnosed* with the disorder because they know that such symptoms are one of many risks of the job. And those who are diagnosed with PTSD cannot be real soldiers because they violate a soldierly ethos by seeking benefits for simply doing what they're paid to do.

This reasoning about illness and malingering was different from that used by clinicians. For clinicians, illness stigma and malingering are problems that are 'external' to the illness itself. Stigma prevents the person who truly has PTSD from getting treatment or benefits he or she deserves. Malingering is a related, but almost inverse problem. Soldiers who fake or exaggerate symptoms, one neuropsychologist told me, hinder his ability to properly measure clinical outcomes, a problem he resolves by administering effort tests to patients. This position of clinicians—logical as it is—is different from that of soldiers who speak with what Sgt Pearson called the second voice of the Army. For these soldiers symptoms are less important than the issue of disability pension. It seems they could not, or would not, divest PTSD of its material and moral significance. Their attitude echoes the generally suspicious stance taken by military clinicians themselves towards soldiers in earlier eras when the discipline was heavily influenced by psychoanalysis and its concept of secondary gain.

Sgt Pearson said he had done "unforgiveable things there." Later he mentioned one incident in Iraq, when he had mistakenly killed civilians by firing a grenade launcher at a farmhouse he thought was occupied by insurgents. He explained, "Killing does affect you. If it doesn't affect you then you are a sociopath. I mean it doesn't affect some people that much, but if it doesn't affect you at all then you're a sociopath."

What he seems to say here is that it is normal for killing to affect you, so why should those affected by killing be seen as mentally ill? Isn't the soldier who is *not* affected at all by killing the one who is ill, a 'sociopath'? These questions are perhaps one reason why he remains ambivalent about clinical care. The clinical encounter can seem to exclude the moral significance of violence—who did what to whom and whether it was justified, honorable, courageous, wrong, or cruel? In some forms of psychotherapy, a goal is to process emotions such as guilt. Yet the therapeutic attitude—it's OK: I sympathize with you now as a suffering patient with a right to heal—might seem to the soldier to fly in the face of what he knows, which is that what happened was terribly wrong.

Paradoxically, the reverse kind of moral dissonance can also happen in treatment: what was normal during combat becomes immoral when confessed to a clinician. Either way, the difference in how violence is morally valued in clinical as opposed to military situations may contribute to the pervasive feeling among soldiers that "you can't understand if you weren't there." It might also be one reason Sgt Pearson preferred to talk with a Baptist preacher who was a Vietnam veteran rather than with a social worker.

I don't know if Sgt Pearson has PTSD and he probably didn't either the last time I spoke with him. He seemed open to the possibility yet was also deeply skeptical: "I can function. I'm truly not sure if I have it. I have changed over the years, but I'm not sure it's PTSD. I don't want to be on meds. I'm not interested in taking a pill because the Army tells me I need to function."

What might appear to be a negative attitude on his part perhaps indicates his uncertainty as to whether the clinician's 'true patient' would be himself—or the Army. But Sgt Pearson seems to also have a kind of disquiet, a more fundamental doubt as to whether intense stress, killing, seeing others die—and other horrors that he views as a normal part of the job—could actually make him ill in the first place.

Sgt. Pearson's experiences are not easily encompassed by the diagnosis of PTSD. And why should they be? No one's life can be reduced to a mental illness category. But PTSD is currently made to do a lot of explanatory 'work.' It can explain why a soldier is having life problems, but explain *away* bad behavior such as stony silence around loved ones. It can determine entitlements to disability pensions, or end a valued career as a professional soldier. It can signify the heroism of self-sacrifice, or the horrors of combat, or simply weakness.

Yet the PTSD diagnosis also leaves unexplained questions that most plague Sgt. Pearson. "Why are (some of) my comrades ill or homeless, and I'm functioning when we both had the same experiences? Isn't it after all normal to feel this way after all I've been through?" And at moments when he is prone to darker thoughts about the war, he also wonders simply "why am I alive and (some of) my enemies dead?"

Perhaps Sgt Pearson has a borderline case. He has some symptoms, but he also functions. Ultimately, whether he has PTSD will be determined not only by past violent events, but by the interpretation of affliction, including his own interpretation. As he moves through different military and health care environments these interpretations will change—and bring new consequences. One task of the anthropologist is to study such a journey: to try to understand what the PTSD description means and what it does for people differently situated in the world.

## Notes

1. I use pseudonyms in this article. I draw on pilot fieldwork conducted in Fayetteville, North Carolina. This research is part of a multi-country study of soldiers' reintegration and psychological wellbeing and health after combat, which is funded by the European Research Council.
2. There is also a growing anthropological literature on PTSD and soldiers' combat experiences. See Macleish (2013), and Wool (2013) for excellent ethnographies of US soldiers' bodily experiences, and Finley's (2011) nuanced analysis of PTSD among veterans in the VA system. Allan Young's (1997) now-classic work critically discusses the notion of traumatic memory at the core of PTSD.

## References

Fassin, D., and R. Rechtman 2009 *The Empire of Trauma: An Inquiry into the Condition of Victimhood*. Princeton, NJ: Princeton University Press.

Finley, E. 2011 *Fields of Combat: Understanding PTSD among Veterans of Iraq and Afghanistan*. Ithaca, NY: Cornell University Press.

Hoge, C.W., C.A. Castro, S.C. Messer, D. McGurk, D.I. Cotting, and R.L. Koffman 2004 Combat duty in Iraq and Afghanistan, mental health problems, and barriers to care. *New England Journal of Medicine* 351(July 1):1, 13–22.

Kim, P., T.W. Britt, R.P. Klocko, L.A. Riviere, and A. Adler 2011 Stigma, negative attitudes about treatment, and utilization of mental health care among soldiers. *Military Psychology* 23:65–81.

Macleish, K. 2013 *Making War at Fort Hood: Life and Uncertainty in a Military Community*. Princeton, NJ: Princeton University Press.

Rechtman, R. 2004 The rebirth of PTSD: The rise of a new paradigm in psychiatry. *Social Psychiatry and Psychiatric Epidemiology* 39(11):913–915.

Summerfield, D. 1999 A critique of seven assumptions behind psychological trauma programmes in war-affected areas. *Social Science & Medicine* 48(10):1449–1462.

Wool, Z.H. 2013 On movement: The matter of soldiers' being after combat. *Ethnos* 78(3):403–433.

Young, A. 1997 *The Harmony of Illusions: Inventing Post-Traumatic Stress Disorder*. Princeton, NJ: Princeton University Press.

As we have suggested, framing PTSD as a psychiatric condition caused by the inability of a person to make sense of life after a particularly traumatizing event or extended circumstances minimizes the scale of war, displacement or flight, and the multiplicity of events that constitute these. It overlooks too how people relive terror, fear, and insecurity, and how this shapes everyday behaviors and practices, not only among those directly affected, but also among their descendants. In exploring this, Julia Dickson-Gomez (2002) writes of the hyperawareness and persistent social fears that frame how people live their lives in El Salvador, where the memories of the brutality of its most recent war, and the chronicity of poverty and structural violence, have resulted in people being chronically distrustful of neighbors, politicians, and the police. Like Doug Henry (2006) for Sierra Leone, Dickson-Gomez is concerned with the transgenerational effects of this—in terms of social engagement, trust, parenting skills, and children's interactions. PTSD, as it unfolds through generations, is collective, not merely individual. Her image of "the sound of barking dogs," and villagers' rapid retreat into the jungle at this warning, echo with other accounts of a habitus of vigilance among people in places where security is tenuous. Sverker Finnström's poignant example (2015) of the scent of soap in northern Uganda is another example of how people internalize their experiences of living with violence. Structural and physical violence, and the constant threat of reprisal, retaliation, and terror, force people to flee, hide, and modify everyday life to avoid exposure. They are constantly vigilant.

Below, Meagan Wilson illustrates the internalization of risk among people from minority populations who had left Burma (Myanmar) and were living illegally in Thailand. She focuses on Shan women, who have crossed the border into northern Thailand for refuge from violent militarization and economic deprivation. But, without papers, despised and distrusted by local Thai, these women are stripped of autonomy, mobility, agency, and their health. They live and work illegally, suppressing any possible behaviors, speech, or habitus that might render them as vulnerable in these sites as they were in their homelands.

## 13.3 Life in a State of Fear

*Meagan Wilson*

Militarization, and the surveillance in and of everyday life that comes with it, force individuals and communities to adopt strategies of self-surveillance to survive physically and emotionally. 'Burmese' (people living in what is known as Myanmar) have lived under the oppressive conditions of militarization for decades. The Department of Psychological Warfare (now closed but long operating), the Military Intelligence spies in tea shops, the conspicuous number of people wearing government uniforms and weaponry in public spaces: these converging actors established an environment of fear and control, particularly in urban and central Burma, such that ordinary citizens were consistently on guard (Skidmore 2004). In rural Burma, however, particularly in the border homelands of ethnic minority groups, the control was direct rather than implied—routine military action against villagers, and the occupation of their lands, led to the regular violation of spaces and bodies. As a result, self-surveillance—being on guard, always—became integral to their everyday life. When Burma moved from military rule to a 'civilian democracy' in 2011, the changing political climate offered optimism and hope. However, armed conflict and economic hardship has continued in the rural border regions, and ethnic minority groups continue to live without stability, safety or security. Because of these conditions, large numbers of people from these various ethnic groups—including the Karen, Karenni, Kachin, Rakhine, Rohingya and Shan—leave for 'a better life' in Thailand. But, as I describe below for Shan women, 'illegal migrants' in northern Thailand must continue to use the same strategies of self-surveillance to ensure their survival.

The migrant women from rural Shan State, with whom I worked in Thailand, had all experienced militarization both directly and indirectly. Their land was confiscated and appropriated for use as a military base; they were forced to work as human porters and laborers; they were raped and forced to marry members of the Burmese Army; they had inadequate food and health care. To survive the structural and direct violences of militarization, women devised various strategies of self-surveillance and survival action. For instance, families made regular 'runs' to the forest when the Shan State Army (SSA) intelligence advised them that the Burmese Army was approaching their village, taking with them their source of livelihood (buffalo) and enough rice for a few days. Women would retreat to known landmine-free spaces, feeding their families on frogs, rats and bamboo until it was safe to return. If the Burmese Army was too close, strategies such as 'no smoke' (cooking fire) and silence were essential to ensure their invisibility and so their safety. These strategies were often difficult, with hungry young children, screaming babies, the pain of childbirth, and the travel of sounds across long distances in the still forest night.

Self-surveillance strategies were gendered in particular ways. Women mapped out sites of sexual violence. Roads that lead into Shan villages are associated with rape by members of the patrolling Burmese Army, and women always avoided these routes. Shan families sometimes encouraged their sons to join the Buddhist monastery at an early age to avoid 'being taken' by the Burmese or Shan State Armies as soldiers or porters, but for women to escape recruitment as army nurses or wives, their surest strategy was often to leave their homeland. At the same time, many young Shan people envisaged a future in Thailand that was freer, with better economic prospects and enough cash to send remittances home to 'feed the family.' On arrival, they often found that they had lost not gained freedom, as they faced control and surveillance of a different kind. Harassed by Thai government officials, police, civil society, employers and other migrant workers, migrants from Burma needed to re-invoke strategies of self-surveillance in order to survive.

* * * *

Around three million migrant workers live in Thailand, most from Burma and many in Thailand illegally. Thai government policies concerning migrant registration are complicated, time-consuming, expensive for migrants, and vulnerable to fraud. Migrants usually arrive in Thailand with little or no money, or with a debt owed to a broker, and Thai employers often initially pay the worker's permit. This process creates a relationship of debt-bondage, whereby migrants work without pay, often for extended periods, to reimburse their employer what they owe them (and often much more).

Migrants who do not register, who travel outside of the province in which they are permitted to work, or whose employer withholds their registration documents or work permit, are considered by the Thai government to be 'illegal.' They are vulnerable to arrest, detention and deportation. Only two of the women I interviewed, when I worked in north Thailand from 2011 to 2013, were legal migrant workers; the majority had to monitor where they traveled (avoiding known police checkpoints), decide whether or not to leave an abusive employer, to speak always in Thai rather than Shan, to reduce the possibility of their arrest and forced return to Burma. The majority cross the border without papers, and in an unfamiliar environment, with a new language, no understanding of how Thai migration policy works, and vulnerable to arrest and abuse, they go into hiding. Out of hiding, they need to find a way to stay in Thailand without the papers that might help them do so. Yan (pseudonym) explained to me that "one year after you have arrived you need to renew your permit, but if you can't read or write, you do not know that, and the government says you did not renew so you are illegal."

* * * *

In May 2014, the Royal Thai Army launched a coup d'état against the caretaker government in Bangkok, and the Thai military junta established a National Council for Peace and Order (NCPO), which formed the Committee on Solving Migrant Problems (CSMP). 'Problem solving,' in operational terms, means that migrant communities have been more intensely surveyed and controlled. Raids on known 'urban migrant villages' and growing numbers of people from entire villages are being deported back to Burma, injecting a deeper level of fear into the migrant community, increasing a sense of desperateness among them to remain invisible to the Thai Government and to those who act on its behalf (any Thai person in uniform). The strategies they adopt draw upon their home experiences—the negotiation of safe geographical space and the use of silence.

Migrant workers' fears in reaction to Thai police are visceral. As a researcher, I became afraid and nervous around the police, merely from observing migrants' bodily reactions to them. I recalled in my fieldnotes the experience of going through several checkpoints over the course of a 4-hour bus trip to the Thai–Burma border with some Shan migrants—students—with valid papers:

> I was the only person other than the driver to not manage a single moment of sleep on the bus. I had only been in the country a couple of weeks and it was the first trip I had taken to Mae Hong Son Province. The winding roads tracing the jungle-dense mountains induced nausea, and I wondered how the person sitting in front of me managed to nap so easily in between periods of heaving into a plastic bag. But bodies around me, heavy with sleep, were conditioned to rest for periods, and then to awaken with the slowing motion of the bus coming into a checkpoint, quickly shifting from a state of deep exhausted sleep to one of nervousness and panic. As the bus slowed and stationed at the checkpoint, I noticed that my friend Pee Mai was now awake and alert, her legs rapidly twitching up and down, next to mine, the way *soi* dogs move when they are scratching their parasitic wounds. She sat forward on her seat and began biting her nails as the police entered the bus and walked up the narrow aisle toward us. How did they know who was Thai or Shan . . . to miss the first few rows of (Thai) people and come straight to us? My heart was pounding as I made eye contact with one of them. I knew they were wondering what I was doing there, the only *farang* on the bus, sitting with Shan. In my mind I knew I was safe, but somehow my body didn't. I was sweating and my mouth was dry. When the police officer asked for my passport, I fumbled around trying to find it; my panicked hands were not working. When I eventually handed it over, I was repeatedly asked something in Thai that I did not understand and so I could not respond. I was told by friends to always play the 'no Thai language' card with police. My friends answered for me. I was handed back the passport and then it was their turn for an inquisition. The three of us were legally permitted to be in Thailand and to be on that bus. But the fear of the uniform, the gun, the power differential, overwhelmed us all. How on earth would someone without papers—an 'illegal'—feel in that context?

The conditioned response of fear to the military and police among those with whom I worked is so deep and palpable, it entered my world. However, migrant fears are not merely conditioned responses that originate in their home state. Their fears around militarization and surveillance in Thailand are well founded. Human rights abuses by police, military officers and immigration officials against migrants have been well documented, and include rape, murder, extortion, and other forms of violence. The perpetrators of these abuses are rarely held to account by Thai law (Human Rights Watch 2012; Raks Thai Foundation 2011).

Migrant workers devise behaviors that allow them to survive in their 'new' militarized state. Self-imprisonment is a common strategy to avoid police surveillance, especially during the first months of arriving in Thailand, while getting their bearings. At first, my friends told me, "all (official) uniforms look the same" and new arrivals are conditioned to fear uniform: it has the potential to "make us very hungry." When Mae One first arrived in Thailand, she stayed inside the factory where she was employed for two days without eating, because she mistook the factory security guard for a policeman.

Another Shan woman, Mee Mee Naing, was 17 years old when, on reaching the Thai border, she was taken directly to a garment factory. She imprisoned herself within the factory compound for seven years to avoid the police and possible jail, because she was not registered: "I did not dare go outside the factory. I have never seen what a jail in Thailand looks like but I was scared to go. Even if I needed something from outside the factory I asked someone to get it for me."

The 'business' of illegal migration (Andersson 2014) is an important source of income for Thai police and their families, especially when a 'high quota' of fines from 'illegal aliens' is consistently reached. Police corruption is widespread along the Thai–Burma border regions, and migrant workers regularly need to adapt their self-surveillance behaviors according to changing police practices. One woman explained that "because we cannot have bank accounts, we women used to keep our savings in our underwear (bra) but then the police, they realized this and started to do a body search, and take all of our money, so we had to hide it somewhere else." Another participant feared having her savings stolen and explained to me that she would never keep her savings on her body. She worried especially that, if she were in a motorbike accident, police or hospital staff would take it:

"I thought of a place in my room where no one would ever find my money, a very dirty place, no one would ever touch."

Thai police devise ways to increase their illegal migrant 'quotas,' such as by surveying the bodily markers of migrant otherness and apprehending people on that basis. For Burmese women, waist-length hair, pulled back into a ponytail, is a symbol of cultural identity. When Thai police arrest unregistered migrant women, they often reportedly cut women's hair very short to make them easy to identify as 'illegal' should they re-cross the border into Thailand, so that they are more easily re-arrested and fined (Raks Thai 2011). Migrant men are also assessed for physical markers of cultural identity, and so their physical appearance is monitored in public spaces: "When Thai police come and they see the (Shan) men, they will check their arm to see if they have a tattoo. If they have Burmese or Sanskrit letters, they will ask them for money" (Thong Kham).

Government policy, police surveillance and corruption lie at the heart of migrant workers' fear. Shan migrants devise strategies to reduce their contact with authorities and to remain invisible and silent. Migrant workers have less control when it comes to employer surveillance and abuse, often committed with the intention of maintaining control over employees and increasing employee productivity through fear. The emotional impact of employer control on the women I interviewed was disturbing. Win Win described her five years of work at Golden House as ' a troubled life.' She rarely left the worksite and was surveyed whenever she did, in case she tried to leave. She lived in a room with 15 other women in a space that was only about 3 by 4 metres in size, which made negotiating everyday living such as eating, sleeping and showering difficult; it created fear and anxiety for all of them. The living quarters at the factory were extremely dangerous, and Win Win feared she would die of electrocution:

> Sometimes when it was raining, the water would come in. Nobody could sleep. We would have to hug our clothes and pillow and stand up, we could not sit down. Later when the rain stopped we would have to clean and then sleep. Somebody died from electrocution. I was very afraid (of dying) so I went to sleep on the table.

Listening to Win Win's story, I felt claustrophobic and anxious; I wondered how I would ever have the strength to survive five years of living the way she had been forced to live. The everyday suffering she communicated was deeply troubling, but I sensed there was more to her story. Her narrative spun round and round an intrusive character—the boss's son—until she finally explained this to me. He was a rapist. He regularly raped the 'beautiful' women who worked and lived in the factory. "I got this too," said Win Win, "but some women, they got a pregnancy." I witnessed the terror from her past being embodied in the present. She was shaking with fear, while sharing the terror. It was as though her trauma was indelible. But with determination, she articulated her feelings: "Sometimes when the boss's son hurt me, other people saw it (because of the cramped living quarters). I felt ashamed. I thought if I get stressed, it only hurt me more. I cannot do anything to him. I just had to live like that and let things go." Win Win knew of two women who became pregnant following rape, and who had had abortions:

> For one of them, a big stick was brought into the factory to do the abortion there. The other woman, the boss took her to the hospital to do the abortion. They were afraid other people would know about them and what they were doing to people if the women had the babies.

Win Win eventually devised a strategy to escape factory life. She found a husband. She did not love him or even like him, but she could survive on the outside with him. However, the 'troubled feeling' of life at Golden House traveled with her when she finally left. Like many other women I spoke to, Win Win had to negotiate ways to monitor her own mental health and implement strategies to keep emotions in check. All of the women I interviewed had survival strategies that revolved around the notion of *mee sa thi*: to not think too much. *Mee sa thi* requires that a person 'get out of their head' and into their body, to connect with the wider physical world: singing, playing with children, praying to Buddha, smoking a cigarette, chewing betel, laughing at someone or something, admiring a beautiful flower, preparing Shan food. All of these can help a person to *mee sa thi*, avoid reflecting on their 'troubled life' as a migrant worker and their compromised freedom, even more tightly surveyed and controlled in Thailand than in their homeland. Win Win explained: "In Shan State, when there is not civil war, I can go wherever I like all of the time. Civil war does not happen all of the time. (But) here, when I was working in the factory, if I went outside the boss would send someone to follow us. I think I was happier in Shan State." Kyi also felt happier in Shan State and explained that she needed to monitor her state of mind regularly in Thailand: "I often think, oh I am thinking too much; if I keep thinking back to my life in Burma I cannot be happy. So I try to stop thinking and I calm down a lot."

Migrant women strongly believe that working in Thailand and the suffering that comes with it is essential for the survival of their families back in Burma, and they believe that one day they will be able to return home with enough money for a decent life. In reality, migrant workers often become trapped in a liminal world in Thailand, with no economic means to leave, where dreams of a free future are just dreams. The emotional self-surveillance strategy of *mee sa thi* prevents migrant workers from thinking their dreams through to the point where they actually map them out in the world in concrete terms. If they did, they would realized their dreams are unattainable. Emotional stability would be lost, along with the will to survive.

### *Acknowledgments*

This case study is based on ethnographic fieldwork conducted in Chiang Mai and Tak Provinces, northern Thailand, from 2011 to 2013. A 2011 Prime Ministers Australia Asia Award and a Monash University Australian Postgraduate Award Scholarship supported this research, which was conducted for a PhD. It included 26 in-depth interviews with women, mostly of Shan ethnicity, who had migrated from Burma. All women referred to their homeland as Burma, not Myanmar. I use the term 'Burmese migrants' to refer to all migrants from Burma, not Shan specifically.

### *References*

Andersson, R. 2014 *Illegality, INC.: Clandestine Migration and the Business of Bordering Europe.* Berkeley, CA: University of California Press.

Human Rights Watch 2012 Ad Hoc and Inadequate: Thailand's Treatment of Refugees and Asylum Seekers, 13 September 2012, ISBN: 1-56432-931-3. Accessed from http://www.refworld.org/docid/5052e51b2.html.

Raks Thai Foundation 2011 A Joint Submission on Migrant Workers and Their Families for the 12th Session of the Universal Periodic Review, October 2011. Accessed from http://lib.ohchr.org/HRBodies/UPR/Documents/session12/TH/JS4-JointSubmission4-eng.pdf.

Skidmore, M. 2004 *Karaoke Fascism: Burma and the Politics of Fear.* Philadelphia, PA: University of Pennsylvania Press.

---

The lives of the women who Wilson describes are circumscribed by structural vulnerability, as we discussed also in Chapter 12. People who lack official residential status of any kind have no rights to health. They are subject to exploitation as workers; they lack secure housing, state education, and social support; and their poverty and the secrecy in which they live lead to health problems. Yet gaining papers comes with a cost. In Chapter 12, we discussed Miriam Ticktin's (2011) work in France, where people can claim ill health as a reason for a continued authorized temporary visa, with the option of legal residency after five years. Cristiana Giordano (2014) provides us with a contrasting example in Italy, where residence is secured from claims of abuse. 'Credible' and recognizable narratives are essential for this, and so women provide the bare facts of their transit from home country to Europe, elaborated by their brokers to authorities as accounts of forced migration, violence, and abuse to ensure new residential status.

## Reconciliation and Repair

The aftermath of war necessarily involves not only physical reconstruction but also nation building. War memorials and museums acknowledge the costs of conflict to human life, but in so doing, war often appears to be glorified, the contributions of soldiers and civilians represented as sacrifices to nation. On the other hand, many museums are built to reinforce peace and human rights. The Holocaust museums in Germany and worldwide have arguably led the way in supporting resistance to war, racialized terror, and assaults on human rights, with a growing number of museums now established, globally, to honor the memories of other people subject to genocide or killed either on the frontline or as 'collateral damage' in other internal and international conflicts. In Berlin the Jewish Museum, the Topography of Terror Museum, the Memorial to the Murdered Jews of Europe, the Roma Holocaust Memorial Pool, the Stolperstein or 'stumbling stones' set into the cobblestone streets of the inner city, all reflect how one city has worked to make sense of its history of terror and brutality. The city has literalized the commitment of its majority population to avoid repetitions of crimes against humanity by reminding its citizens (and visitors) of the past.

Memorials of regret have increasingly been only one way by which countries have sought to redress the wrongs of militarization and terror. Truth and Reconciliation Commissions have been a powerful technology of restorative justice, through which people have been encouraged to speak of wrongs against them, or their role in perpetrating such wrongs, in war and in other violent circumstances, despite the potential risks they take in making public their own subject status (Hayner 2001; Manderson et al. 2015). In Argentina, following the restoration of democracy after seven years of military dictatorship, the National Commission on the Disappearance of Persons (CONADEP) was established with the task of exhuming unmarked graves. In 1986, an NGO, the Argentine Forensic Anthropology Team *(Equipo Argentino de Antropología Forense)* was established to continue this work; it has subsequently worked in many countries to train nationals to exhume bodies and identify human remains as a component of restorative justice. Below, we turn to the aftermath of the Spanish Civil War and the dictatorship that followed, a period of repression that ended only in 1975. Rachel Ceasar considers the symbolic importance to families of the exhumation of bodily remains from mass graves, some 70 years after people disappeared and were presumed to have been lost forever.

## 13.4 Exhuming the Disappeared

*Rachel Carmen Ceasar*

New advances in science and technology are increasingly used in post-conflict countries to protect heritage, promote reconciliation, and bring justice to victims of human rights violations. The current exhumation of mass graves in Spain is one example of how everyday citizens negotiate knowledge and memory in the aftermath of conflict, and through exhumation, people seek a sense of justice and peace. But what happens when bodies are exhumed?

For 17 months, I conducted ethnographic fieldwork and explored the archives related to the current exhumation of mass graves that date from the Spanish Civil War (1936–1939) and the Francisco Franco dictatorship (1939–1975). During and after the Civil War, the institutions of Church and political regime went hand in hand: "Religion and the army," explained Luisa of San Pedro, Spain, "were the two pillars of Francoism." People who were suspected of resistance were arrested, interned in concentration camps, and executed; over the course of the war, 130,000 civilians and Republican partisans were killed. Most were buried anonymously in unmarked mass graves, their deaths denied or suppressed by those who lived nearby. Because the Catholic Church had legitimized the Francoist regime, these mass graves of Republican partisans were often located in or around Church cemeteries. For this reason, I spent much of my time in Spain at graveyards in churches.

One night, on her patio, Luisa told me that her father's body was located in an unmarked mass grave in the San Pedro cemetery. Now in her 70s, Luisa grew up knowing that her father had been killed 'for his ideas,' a common way of referring to people with leftist leanings in Spain. Luisa knew why, but she did not know how or where her father Martín was killed. Exhuming his remains would be testament to his death and evidence of the violence committed against civilians like him, during the Spanish Civil War and subsequently—for a total of 40 years.

"If we ourselves don't know where he is—there's nothing to justify [his death], just bones. Just to think this . . . " Luisa gasped, and let her voice drift into the muggy summer night. "For all that they excavate, I can't say that this [person] is mine. The only thing we have that makes him distinct [from other bodies in the grave] is that in the pueblo, there were few people who wore boots. It's what we have." She showed me a photograph of her father. "It's the only one [photo] we have, and that's because I stole it!"

Luisa, like many daughters of the Civil War period, awaits the discovery of her father. "*Yo tenía mucha falta de mi padre*, I missed my dad a lot—I don't know if that happens to all girls." The boots and the single, spectral photo are provisional placeholders for the missing body—until the exhumation.

Unlike the refugee bodies of asylum seekers in France today, whose identity is authorized by medical expertise (Fassin with d'Halluin 2005), the exhumation of the disappeared in Spain is a collaborative process driven by local and descendant communities, together with archaeologists and local historical memory associations. In Martín's case, Luisa searches for evidence and truth to the specifics of a body—a body that they may never find. In this manner, the corporeality of Martín's absent body is paid homage to by his family and community, because to *not* do so would be to refute the killings of and violence against Republicans during the war and dictatorship.

"All this began because of a genealogical error, because I didn't know how many brothers my grandfather had or what their names were," Ana, Luisa's daughter, explained. Ana was speaking of a common taboo reflected in the genealogies of Spanish families from either side of the war. "What we would like, I suppose, as families of the disappeared, is to find his remains."

Like many Spaniards who grew up at the end of the dictatorship, and so were born in the 1970s, Ana uses the human rights term *desaparecido* or 'disappeared' to refer to her grandfather and other victims killed by the Nationalist Army during the Civil War or under the Francoist dictatorship. This much Ana and Luisa knew: Martín was killed and left in an unmarked grave in the cemetery of San Pedro, an earth so cracked and dry that bones exhumed here pulverized when touched.

When a grave is opened, the context is contingent on the political climate surrounding the exhumations, as well as the sentiments circulating among families hopeful that they will find a body they can claim. Many people were complicit in the oppression under Franco, and even after the war and dictatorship had ended, men like Martín could not be exhumed; the government, the Church and families were silent of their fate.

Until 2000, families were prevented from exhuming and reburying Republicans. The circumstances of the dictatorship and democracy prevented further investigation of the conflict from taking place, eliminating the possibility of a national truth and reconciliation commission. This enforced silence by the government extended to the living as well as the dead. Communities wanting to locate and rebury Republican soldiers and civilians, who had been killed by the Nationalists and their supporters, were discouraged from doing so by the ruling Francoist regime. But in October 2000, over 60 years after the end of the war and 20 years into democracy, the first exhumation of a Republican mass grave took place in Priaranza del Bierzo, Spain, and is considered the first to be conducted by a technical team of archaeological experts. Through the recovery of Republican remains, and the conduct of proper burials for these remains, the exhumations provided Spaniards with an historical alternative to the recent past.

My exchanges with Luisa and Ana inside and outside the graveyard revealed feelings and interests that might seem irrelevant to the larger goals of history that privilege objective knowledge of the past. Yet by approaching the exhumations in San Pedro as another perspective of the past—that of the disappeareds and their descendants—Luisa, Ana, and other descendants of the dead contributed to the exhumation process and played an important role in reframing contemporary Spanish history.

The desire to produce knowledge about the bodies of the disappeared, and for Luisa and Ana to prove Martín's existence, suggests varying (and sometimes conflicting) interests of stakeholders involved in exhuming over 70 years after the war: proper reburial, recognition of victims, a multivocal perspective of the past, evidence of human rights violations. In addition to the work of archaeologists at the San Pedro exhumation, Ana's genealogical work and Luisa's photos of her father also formed part of exhumation ethos.

"There were two classes of citizens: those who had won the war, and those who had lost. It was like in India—do you know the caste system? It was the same here," Ana explained. "Because of your religion!" added Luisa. Luisa explained that her family were forced to go to church, as was then expected in San Pedro and throughout Spain during the dictatorship. "The losers were segregated [from the war victors]—they didn't even go to the same dances together," Ana added. "It was a Nationalist–Catholic regime, this Catholicism, so everyone went to mass." Catholicism in Spain, the women explained, was not just a religion or part of the regime; it was part of everyday life. During the war, acts of religiosity were an important part of Spanish society—and politics.

The importance to both Luisa and Ana to know Martín by obtaining his body also reflected a Catholic sensibility. Their desire for his physical remains was a kind of *sacramental symbolism*, what anthropologist Joseba Zulaika has described as a "concern with certain limiting concepts having to do with life as a whole, the notion of death included" (2000[1988]: xxv).

In *Basque Violence: Metaphor and Sacrament*, Zulaika examined sacramental symbolism as an analytic to understand the subjective expression of Basque political violence and terrorism in his hometown of Itziar. Zulaika contextualized Basque political violence within the broader cultural and moral framework of Basque nationalism and its sacramental aspects. In the same manner, Luisa's and Ana's desire to find Martín reflected their desire to reclaim his body from a regime, religious institution and ideology that contributed to his death. By doing so, Luisa and Ana aimed to "resacramentalize" (2000[1988]: 48) or reclaim Martín according to their own values and practices.

★ ★ ★ ★

In August 2012, together with members of a local historical memory association, an archaeological team, and other families of the killed, we spent one month searching for Martín and another 200 persons killed in the San Pedro cemetery. Could one of these bodies be Martín? Ana visited the exhumation site where we were working, in a section of the cemetery. This was a section that, during the war, was separated from the sacred Catholic grounds; this was where Republicans were killed and left in unmarked graves.

When one of the graves was finally opened, Ana stepped down into it. Speaking from inside the grave with a cigarette cast to her side, she glanced down at the six exhumed bodies, tied together at the wrists with electric cable. "For me, this *is* the homage," she said without looking up, referring to the ritual devotion of the exhumation process itself, and not necessarily the identification of the individual bones to particular bodies. I followed Ana's gaze at the exposed bones before me; I could not even begin to understand what the bones and their exhumation must have meant to her and her family.

Luisa and Ana knew that they might never locate Martín's bones. At the most, they could only hope that the exhumation process itself—not the exhumation of his individual bones but of *someone's* bones—might purge Martín from his current place of violent death and obscurity, and purge them of the pain they felt that this was so.

Exhumation began with their desire to know where Martín was, and to retrieve his remains. Luisa and Ana hoped to recover some aspect of Martín—material or otherwise—for, as Luisa explained, "*Creo que todavia no me ha salido del cuerpo.* I believe it [the tragedy of my father's death] still has not left my body."

The desire to know and feel who her father was, via the exhumation, animated (and sometimes depressed) Luisa and her daughter during the exhumation process. The ethos motivating Spanish practices and rituals embodies what it means to be Spanish—and to have lost a family member, often more than one, during the war and the dictatorship.

Although the war and dictatorship were now in the past, and the Catholic Church no longer possessed the political power it had once enjoyed, Luisa was doubtful about the possibility of any formal form of reconciliation taking place in Spain in the near future. "*No se cae ni la puerta de la iglesia*—not even the door of the church falls," she told me. The institution of the Church cannot fall so quickly; the vestiges of its role in political oppression linger in the Spanish present.

### *Acknowledgments*

This case study is drawn from a larger study on the science of exhuming mass graves in contemporary Spain. I thank the editors of *Somatosphere*, Eugene Raikhel and Elle Nurmi, and Neda Atanasoki, Mariam Lam, and the other members of the University of California Humanities Studio Workshop on "Humanitarian Ethics, Religious Affinities, and the Politics of Dissent," for their warm encouragement and support on aspects of this research. I also acknowledge the local and descendant communities at the San Pedro exhumation, in particular, Luisa and Ana, as well as Alfredo González Ruibal and Xurxo Ayán Vila at the Institute of Heritage Sciences in Santiago de Compostela, Galicia (INCIPIT-CSIC). Their narratives and guidance on the Spanish exhumations were an invaluable source of knowledge that informed this paper.

### *References*

Fassin, D., and E. D'Halluin 2005 The truth from the body: Medical certificates as ultimate evidence for asylum seekers. *American Anthropologist* 107(4):597–608.

Zulaika, J. 2000 [1988] *Basque Violence: Metaphor and Sacrament.* Reno and Las Vegas, NV: University of Nevada Press.

## Militarizing Emergencies

With violent disruptions, or as a result of other unanticipated emergencies, including environmental disasters, states are often unable to provide basic services. Many countries also face chronic problems reflected in administrative incompetence, corruption, and failure of the rule of law. In the face of administrative ineptitude, the absence of judicial processes, or simply frail systems of government and broken infrastructure, external assistance may be inevitable in ways that bypass political partisanship. The Red Cross, established in 1863 and restructured in 1919 as the International Federation of Red Cross and Red Crescent Societies, has worked for over 150 years to provide emergency medical assistance and public health programs. Médecins Sans Frontières, discussed in Chapter 15, was established in 1971 at the time of the Biafran war in Nigeria, and in 2015, it was working worldwide to provide health and medical assistance to people impacted by epidemics, disasters, and civil disruption. Other independent organizations and multilateral agencies, private foundations and NGOs, operating with support from individual donors and donor countries, assist in providing essential food, medicine, water, sanitation, and emergency medical aid. There has been continuing debate about the approaches taken by some of these organizations, and the extent to which they bypass and so fail to address the politics behind the chronic incapacity of certain countries to meet the health and medical needs of their populations. Hence, the important point that Doug Henry and Susan Shepler have made of "chronically acute crises," alluding to the recurrent difficulties that many countries have to respond at times of crises, as rooted in problems of poverty, poor health infrastructure, and challenges in governance (2015: 20). War, we have noted, exacerbates poverty as its armies internationally destroy infrastructure and undermine governance.

The case studies in this chapter have emphasized the sorrow, suffering, and fear that saturate everyday life under conditions of conflict. Yet in the face of such tragedy, humans show remarkable resilience. Much of this resilience is at an individual level, whereby people work to make sense of seemingly impossible odds, but resilience also occurs at the community level. Even in societies where trust has been badly broken—in Rwanda, for instance, or in Palestine and Israel—people continue to build affective ties and sustain families, to support each other whether they stay or flee, and to take a public stand either on specific occasions—as in truth and reconciliation commissions—or through their life's work. While many people, individually, are shattered by catastrophe, remarkably communities continue. One role for medical anthropologists is to further better understand how we, with health and other professionals, might repair the social fabric as well as individual bodies and minds.

## Note

1. Some 17 million soldiers and citizens were killed during World War One. Over 20 million were killed just 50 years before in the Taiping Rebellion (1850–1854) in southern China (http://www.britannica.com/event/Taiping-Rebellion).

## References


Allen, Tim, and Koen Vlassenroot 2010 *The Lord's Resistance Army: Myth and Reality*. London: Zed Books.

American Psychiatric Association 1980 *Diagnostic and Statistical Manual of Mental Disorders, 3rd ed. (DSM-III)*. Washington, DC: American Psychiatric Association.

Degregori, Carlos Iván 2012 *How Difficult It Is to Be God: Shining Path's Politics of War in Peru, 1980–1999*. Madison, WI: University of Wisconsin Press.

Dickson-Gomez, Julia 2002 The sound of barking dogs: Violence and terror among Salvadoran families in the postwar. *Medical Anthropology Quarterly* 16(4):415–438.

Feldman, Allen 1991 *Formations of Violence: The Narrative of the Body and Political Terror in Northern Ireland*. Chicago, IL: University of Chicago Press.

Finnström, Sverker 2015 War stories and troubled peace: Revisiting some secrets of northern Uganda. *Current Anthropology* 56(S12):S222–S230.

Giordano, Cristiana 2014 *Migrants in Translation: Caring and the Logics of Difference in Contemporary Italy*. Berkeley, CA: University of California Press.

Hayner, Priscilla B. 2001 *Unspeakable Truths: Transitional Justice and the Challenge of Truth Commissions*. Abingdon, UK: Routledge.

Henry, Doug 2006 Violence and the body: Somatic expressions of trauma and vulnerability during conflict. *Medical Anthropology Quarterly* 20(3):379–398.

Henry, Doug, and Susan Shepler 2015 AAA 2014: Ebola in focus. *Anthropology Today* 31(1):20–21.

Hinton, Alexander Laban, ed. 2002 *Annihilating Difference: The Anthropology of Genocide*. Berkeley and Los Angeles, CA: University of California Press.

——— 2004 *Why Did They Kill? Cambodia in the Shadow of Genocide*. Berkeley and Los Angeles, CA: University of Califonia Press.

Jones, Edgar, and Simon Wessely 2014 Battle for the mind: World War 1 and the birth of military psychiatry. *Lancet* 384(9955):1708–1714.

Kleinman, Arthur, and Joan Kleinman 1996 The appeal of experience; the dismay of images: Cultural appropriations of suffering in our times. *Daedalus* 125(1):1–23.

Lykes, M. Brinton 1994 Terror, silencing and children: International, multidisciplinary collaboration with Guatemalan Maya communities. *Social Science & Medicine* 38(4):543–552.

Manderson, Lenore, Mark Davis, Chip Colwell, and Tanja Ahlin 2015 On secrecy, disclosure, the public and the private in anthropology. *Current Anthropology* 56(S12):S183–S190 .

Myers, Charles S. 1915 A contribution to the study of shell shock. Being an account of three cases of loss of memory, vision smell and taste, admitted into the Duchess of Westminster's war hospital, le tocquet. *Lancet* 1:316–320.

Nordstrom, Carolyn 1977 *A Different Kind of War Story*. Philadelphia, PA: University of Pennsylvania Press.

Solomon, S.D., and G.J. Canino 1990 Appropriateness of *DSM-III-R* criteria for posttraumatic stress disorder. *Comprehensive Psychiatry* 31(3):227–237.

Theidon, Kimberly 2012 *Intimate Enemies: Violence and Reconciliation in Peru (Studies in Human Rights*. Philadelphia, PA: University of Pennsylvania Press.

——— 2015 Hidden in plain sight: Children born of wartime sexual violence. *Current Anthropology* 56(S12):S191–S200.

Ticktin, Miriam 2011 *Casualties of Care: Immigration and the Politics of Humanitarianism in France.* Berkeley, CA: University of California Press.

Uesugi, Tak 2015 How "Agent Orange Sickness" emerged in the United States and stayed hidden in Vietnam. *Medical Anthropology* DOI: 10.1080/01459740.2015.1089438.

Young, Allan 1995 *The Harmony of Illusions: Inventing Posttraumatic Stress Disorder.* Princeton, NJ: Princeton Unversity Press.

Anthropometry in the Guatemalan Highlands, 2014. Sololá, Guatemala.

## About the photograph

*Medical anthropologist Caitlin Baird measures the growth of children at an NGO-sponsored preschool in the Guatemalan highlands. Baird works with Wuqu' Kawoq: Maya Health Alliance, as part of a team of anthropologists, physicians, nutritionists and NGO workers who are studying child growth in rural areas of Guatemala.*

*Guatemala has the highest rate of chronic malnutrition in the Western Hemisphere, and one of the highest rates in the world. Chronic undernutrition in early childhood, a condition commonly referred to as 'stunting,' causes permanent delays in cognition and in the development of multiple organ systems, leading to both a loss of productivity in adulthood and a hugely increased disease burden. The combined burdens of greatly increased incidences of chronic disease and lower economic productivity as adults are mutually reinforcing and act as a drain on already at-risk households, re-establishing the cycle of poverty (and undernutrition) in subsequent generations.*

—Caitlin Baird and Amber Urquhart

# 14

# Genes, Kinship and Risk

*Anita Hardon, Lenore Manderson
and Elizabeth Cartwright*

In her fascinating book *The Century of the Gene*, sociologist Evelyn Fox Keller (2001) argues that we need to understand the multiple histories and understanding of the word 'gene' to make sense of contemporary 'gene talk.' More than a century after the word 'gene' was coined, biologists still do not agree on what a gene is nor what it does. Keller describes how the term 'genetics' emerged in the early twentieth century, when Hugo de Vries, a Dutch botanist, and his contemporaries rediscovered the rules of inheritance that Gregor Mendel, a solitary Austrian monk, had found 40 years earlier in his investigations of pea plants (Keller 2003). A new and well-developed branch of plant breeding emerged with a focus on the material basis of inheritance patterns; in this context, the English scientist William Bateson (1906), speaking to a congress of botanists, coined the term 'genetics.' The term 'gene' was introduced three years later by the Danish plant physiologist Wilhelm Johannsen, who wanted a new word to replace earlier concepts such as 'genmules,' the Darwinist unit of pangenesis, and de Vries's term 'pangens.' Johannsen argued "it appears simplest to isolate the last syllable, 'gene,' which alone is of interest to us" (1909: 124).

These developments occurred long before James D. Watson and Francis Crick unraveled the structure of DNA in the 1950s (Olby 1974), after which genes became defined as segments of DNA that carry coded information, and geneticists began to study hereditary conditions and traits regulated by specific DNA sequences. The late twentieth century was a period of optimism for geneticists, who, through international collaborations and generous public and private funding, set out to map the human genome. At the time, scientists expected that the genetic information uncovered through such large-scale mapping exercises would result in new knowledge of the regulation of bodily processes, which could be instrumental in diagnostics and therapeutics. Knowing which nucleotide sequences produce which proteins would enable genetic engineering to shape the biological basis of life. A rapidly expanding biotechnology industry invested in medical genomics, with increasingly strong ties between geneticists and commerce, all promising a better future in which both disease prevention and treatment could be personalized. Celera Genomics, a private company, aimed to speed up the mapping of the genome in order to gain intellectual property protection for the genes that it identified. By 2000, Celera Genomics had applied for thousands of patents on the genes it had discovered (Happe 2013).

The potential patenting of human life led to much concern among scientists and the public in the United States and Europe. On March 14, 2000, at the US National Medals of Science and Technology award ceremony, President Bill Clinton announced a joint agreement between the United States and the United Kingdom to lead the way in opening access to genomic research, declaring that the raw data generated from the human genome project should be accessible to all. Clinton stated, "we must ensure the profits of human genome research are measured not in dollars but in the betterment of human life" (Venter 2007: 299). The US Patent and Trademark Office decided that 'raw data' on the sequence of human genes could not be patented.

## Genetic Testing

These developments followed the growing use of genetic testing, beginning in the 1980s in antenatal care, to identify abnormalities in the fetus, as described in Rayna Rapp's ethnography entitled *Testing Women, Testing the Fetus* (2000). The range of conditions for which the fetus could be tested was limited, however, and testing in pregnancy was generally accompanied by genetic counseling that provided the pregnant woman or couple with some basic understanding of the test results. As described in the first case study in this chapter, in some places companies only provide results of genetic testing in writing, and clients are left to decide for themselves how to make sense of them. Without accompanying information, these tests may be given more weight than they should, and even with counseling, the meaning of any risks may be poorly understood. Being tested can lead to worries about future health, including unnecessary investments in all kinds of preventive procedures and products, as we discuss below.

A major breakthrough in understanding the relationship between heritability, risk, and genomic medicine was the identification of the BRCA (breast cancer) gene sequence; the gene is used to identify women with a high BRCA-related susceptibility for getting breast and/or ovarian cancer. To prevent these cancers, women who have this gene may be advised to have their breasts and ovaries removed. Since testing for BRCA has become common, many women, including actor Angelina Jolie, have shared their experiences through social media and news reports. Women diagnosed with BRCA are also advised to tell female family members about the diagnosis, and they in turn need to decide whether they want to be tested, and what to do if they test positive. BRCA diagnostic tests can have significant consequences for women and their families, including, when ovaries are removed, surgical menopause. Even so, the general public does not seem to see testing for the BRCA gene and subsequent preventive medical interventions as an extreme measure. Rather, breast cancer advocacy programs and women who have been tested for the gene argue that the BRCA test empowers women to make lifesaving decisions. Social studies of science and technology have described how women with the BRCA gene mobilize through the Internet and breast cancer support groups based on this shared genetic disposition (Klawiter 2008).

As the Human Genome Project identified more and more genes, it became clear that the association between gene sequences and the incidence of disease is not as straightforward as first thought. While a connection has been made between breast cancer and BRCA, further epidemiological research has shown that the genetic mutation in BRCA1 or BRCA2 accounts for only a small proportion of all breast and ovarian cancers, and not all women with the genetic mutation will develop either disease or both diseases. Men with BRCA gene mutations are also at higher risk of breast cancer, as well as prostate cancer, but preventive surgery has not been proposed as an option for men. Thus, genetic researchers increasingly point to the complex interactions between genetic makeup, biological processes, 'lifestyle' (or personal habits), and the environment that determine the expression of genes.

Given this complexity, how should people make sense of positive test results, especially when geneticists can only give rather rough predictions about the chances of becoming ill? Margaret

Lock (2013) conducted a study on the biological understandings of one such ambiguous condition, late-onset Alzheimer's disease (AD), which genomic researchers associate with the so-called APOE gene. This gene comes in four forms, but only one of them, the APOEε4 allele, has been found to put people at risk for AD. In her extensive review of the biological studies, Lock points out that getting AD is not necessarily related to a positive test for the APOEε4 allele. At least 50 percent of people with this gene do not get AD, and 30 to 60 percent of people with AD do not have the gene. Meanwhile, more and more genes are being identified as having a positive association with AD. Lock's work reveals that biological insights into late-onset AD are subject to continual revision, and are still far from conclusive. As a consequence, Lock suggests that learning about one's APOE status "does not provide information about a highly probable future: it only raises a possible scenario, that everyone living in a family where AD is present has already entertained at some point in their life" (2013: 63).

Lock's research is illustrative of a growing number of social science studies that have followed the developments in the mapping of the human genome and medical genomics with some skepticism. Paul Rabinow (1996), writing about biosociality, and Nikolas Rose and Carlos Novas (2005), offering the idea of biological citizenship, have both anticipated that increasingly social life would be structured around our genetic identities. Lock (2013) has countered that when the relation between genes and illness conditions is ambiguous, genetic makeup fails to be a basis of our identity. Social scientists have also emphasized that disease is not only a biological condition located in the body, but also the outcome of socioeconomic and environmental conditions. In some cases, genetic testing may resolve for an individual the uncertainty around whether or not he or she will develop a disease, and so enable the person to plan for this possibility, as Flaherty and colleagues (2014) have illustrated for people with family histories of Huntington's disease. But others warn that genomic medicine could decrease the quality of life for people who interpret risk as disease, and who thought they were healthy until they received the positive results of a genetic test. Uncertainty remains, however, even with positive test results: a person may or may not develop a particular disease, and if they do, there may be no treatment for it.

In the first case study in this chapter, Suli Sui and Margaret Sleeboom-Faulkner discuss how genetic testing is promoted in China, with promises to clients that such tests 'decode the mystery of life' and 'predict future health,' even if the results of the tests have limited predictive value about future health and wellbeing. Sui and Sleeboom-Faulkner describe how private genetic testing companies send clients reports on their predisposition to get a large number of diseases as 'high,' 'medium,' or 'low' as compared to 'average' individuals without a given gene. If, for example, the results identify a medium-level risk of developing Alzheimer's disease, the reports indicate that the client's probability of getting the disease is five times higher than a person without the gene. The companies usually offer lifestyle advice, including the use of vitamins and minerals. Thus, the tests work to medicalize people's lives: not only are they defined as "pre-symptomatically ill" (Rose and Novas 2005: 445), but also they are expected to buy preventive nutritional supplements to avoid becoming ill.

Besides testing for susceptibility to disease, Sui and Sleeboom-Faulkner reveal that some companies offer genetic tests personality traits such as optimism and shyness, the likelihood of becoming addicted, and intelligence. The prices of such tests are quite high for ordinary people, varying from 400 RMB to 10,000 RMB (approximately US$650–1,600). Still clients seem interested in having the tests done. In a country with a one-child policy, parents want to find out what their child is worth. Sui and Sleeboom-Faulkner suggest that being diagnosed with a particular gene not only determines one's own sociality, but also parents' investment in their children's future.

## 14.1 Direct-to-Consumer Genetic Testing in China

*Suli Sui and Margaret Sleeboom-Faulkner*

Commercial genetic testing (CGT) is usually promoted among individuals with an increased genetic susceptibility conferring a predisposition to future disease symptoms (Fulda and Lyken 2006). In China, the application of predictive genetic testing and its commercialization are becoming increasingly popular. Many biotech companies offer genetic testing to customers, including individuals or organizations. Customers buy tests direct from the company through company agents or online. The scope of testing services is widening, and the modes of business operation are becoming simpler, with some companies solely selling tests online, as is the case with Genetic Testing Net and Zhong-Ren Gene-Net.[1]

In February 2014, the Ministry of Health (MOH)[2] and the China Food and Drug Administration (CFDA) jointly issued the *Notice on Strengthening the Management of the Clinical Use of Gene Sequencing Products and Technology*. This notice points out that gene-sequencing products and technologies, including prenatal genetic testing, belong to the field of advanced product and technology research. They involve issues of ethics, privacy, protection of human genetic resources, and biological safety, and issues related to technology management, price, and quality control of the medical institutions that offer genetic testing services. The purpose of this notice is to guarantee the safety and efficacy of genetic products, and to strengthen the supervision of clinical applications of new genetic technologies. The notice has officially stopped all clinical application of genetic testing in medical institutions until new licensing criteria and relevant regulation are enacted. However, the management and operation of genetic testing offered by companies is outside the supervisory radar of the MOH or the CFDA, so that companies are continuing to offer genetic tests for clinical purposes (Sui and Sleeboom-Faulkner 2007).

As with other commercial companies in China, biotech companies planning to set up a genetic testing business need only to apply for a business license from the local Industrial and Commercial Bureau. They are neither limited in how they market their testing products, nor do they need any special medical qualifications or permissions from the MOH and the Ministry of Science and Technology; the staff of these companies do not need any medical qualifications. CGT in China was incentivized by the availability of genetic testing technologies and a large pool of potential customers. Currently, the testing price varies from 4,000 to 10,000 RMB (US$650–1,600; in November 2014, the exchange rate was 1,000 RMB = US$164) depending on the number of testing items.

Some biotech companies declare on their websites that they collaborate with a state university or research institution, even when only one or two technical advisers are employed there. Such declarations aim to obtain the trust of the public, as people are presumed to trust state educational institutions more than commercial enterprises. In the advertisements, the companies offer genetic testing services for a wide range of multifactorial diseases. For example, Chongqing Xiehe Gene Center declares that it offers genetic tests for the genetic predisposition to 110 diseases. One principal staff said that technically their genetic testing could test for more than 1,000 diseases, but most of them were rare. Thus, considering low detection efficiency and high testing cost, 110 tests (related to the three main kinds of human diseases—cardiovascular diseases, cellular immunity, and cancer) were finally selected for their testing services portfolio. A test requires one or several drops of blood or a few mucous membrane cells from the client to test whether he or she is a carrier of genes associated with certain diseases, so to determine his or her predisposition status.

To gain market share, some biotech companies have agents in more than one large city. The first author visited one biotech company in Shanghai with an office and special agent in Beijing. This company offers 'door-to-door' services (other online companies do not involve any intermediary agent). After payment, the company sends sample equipment to clients by express mail for the collection and return of oral mucosa. After the tests are analyzed, the client receives the report of her or his predisposition status. Usually, the results are stratified into three categories of risk—'high,' 'medium,' and 'low'—but no exact percentage is attached to these. The probability of contracting a disease is explained in terms of likelihood in comparison with the 'average' individual without such a predisposition. If, for instance, the results identify a medium-level risk of developing Alzheimer's disease, the explanation is that the client has a five-times increased probability of developing the disease compared to an average person. The report usually provides advice on a healthy diet and lifestyle, allegedly to prevent and avoid the disease in question.

In promotion, biotech companies use attractive and striking phrases, such as "decode the mystery of life, predict your future health," "genetic testing—the most fashionable healthy lifestyle in the twenty-first century," "personal treatment, decode health," and "create a healthy life based on genetic technology." These sensationalist advertisements aim to convince potential customers that genetic testing can provide predictable health and a healthy future. Some companies claim that genetic tests are suitable for anyone, and encourage healthy people to purchase tests for themselves, their partner, their children, and their parents. They also encourage employers to purchase tests for their employees. To deal with commercial competition, some companies offer sets of services at discount prices, exemplified by the test price of Chongqing Xiehe Genomics Center being 3,995 RMB for 33 diseases, 6,995 RMB for 60 diseases, and 9,995 RMB for 110 diseases.

Besides the genetic testing for susceptibility genes, some companies, such as Chongqing Xiehe Genomics Center, Zhong-Ren Gene-Net (Beijing), and Henan Yujing Bio-Technology Company, offer genetic tests for "talent genes," "rational drug use," and "safe drug use." These kinds of tests mainly target children, usually between 4 and 13 years of age, a huge group of potential customers in China. Talent genes divide into several major items, including personality, emotion, art, sport, and IQ, and 40 to 50 sub-items including genes for optimism, shyness, passion, depression, puppy love, and alcohol addiction. The various gene items all include numerous potential abilities and imperfections. The price of testing for one talent gene item is approximately 600 RMB, and 4,000 RMB for a set. Additionally, some companies sell a test for the "beauty gene," targeting women longing for beauty, a large group of consumers. For example, Henan Yujing Company sells genetic tests for "skin color," "no wrinkles," "antioxidation," and "cell renewal." It declares that such tests can assess the most suitable nutrients and beauty determinants for skin.

## *Reliability of Test Results and Advice*

The results of commercial genetic tests, as noted, are usually expressed in terms of high, medium, and low susceptibility, partly because commercial genetic testing is not formally regulated, and because there is still no universal standard for describing genetic risk. Companies claim that the criteria used in test reports draw on state-of-the-art scientific papers in the fields of medicine, genetics, biology, and epidemiology. Until today, regulatory bodies have not decided which institutions should have the authority to approve and control such criteria. Several biotech companies ask their clients to answer a questionnaire about their everyday life. Shanghai Rongjian Bio-Technologies Co., Ltd tell their clients that precision and honest responses will help the company confirm the relationship between the client's genetic heredity and his or her lifestyle. The company also asks for reports from recent health check-ups to help the company provide health advice. To some extent, then, the clients themselves provide the testing results for the report.

Usually, the reports contain a section on the actual predictive result of the test and a section on prophylactic measures, aimed to enhance the significance of the outcome. The advice usually concerns lifestyle, diet, and intake of vitamins and minerals. For example, the Beijing Huada Gene Research Center gives the following advice to prevent senile dementia in a sample report of the genetic test results: "Take physical exercise for half an hour each day; do mental exercises at least for two hours a day, by for instance playing chess, playing cards and reading; do not smoke; eat five pieces of fresh fruit and vegetables every day; drink eight cups of water a day; and, take enough rest and sleep."[3] Although this advice may be sound, one need not take a genetic test to obtain such knowledge. In fact, the positive result of a test may have a negative effect on the behavior of the person concerned. For instance, Mr Chen, a Beijing citizen in his 40s, now feels free to smoke more after taking a predictive test that did not detect any gene that predisposes him to developing lung cancer.

Doubts arise about the existences of such items as the so-called talent gene, and whether they can be standardized for genetic testing and diagnosis. Without reliable proof of the existence of such genes, the genetic tests for 'talents' may misinform education choices for the children with or without a certain talent gene. In some cities the genetic test for talent has become popular. In 2009, the city of Chongqing carried out a 'talented baby project,' using advanced genetic testing to select 50 children with special talents, and helped cultivate them according to their talent gene. Given the reliability of these tests, an emphasis on the talent gene might create the impression that inborn intelligence is more important than postnatal education and diligence. This may unduly influence a child's motivation and the parents' willingness to invest in his or her future. Furthermore, parents' high expectation of their children might lead to disappointment, and undermine the relationship between parents and children.

### *No Genetic Counseling Available*

According to Article 11 of the UNESCO *International Declaration of Human Genetic Data*, it is an ethical imperative that where genetic testing has likely significant implications for a person's health, appropriate genetic counseling should be made available. Genetic counseling should be non-directive, culturally adapted, and consistent with the interests of the person (UNESCO 2003). However, questions arise as to who is the appropriate expert to counsel the patient and/or the family, and how to counsel them (Fulda and Lyken 2006).

While companies claim to offer counseling, in practice this involves a promotional session used to introduce the price of the genetic test, the test procedure, and the benefits of the tests on offer. No professional genetic counseling is provided. The perceived risk of developing a hereditary disease, especially those for which there are no cures or for those that are severe, is usually accompanied by considerable psychological distress (Friedrich 2002). As environmental factors complicate the interpretation of genetic test results, even with well-trained genetic counselors this is difficult. In China, without appropriate support from professional genetic counselors, test results that are unreliable, and important genetic abnormalities that remain undetected, no responsible provisions can be made for the potential patient (Sui 2009).

### *Genetic Information and Privacy*

Genetic information is different from other personal information because it concerns the privacy of family members who usually share the same or similar genetic information (O'Neill 2002). The confidentiality of genetic information is regarded as highly important both for individuals who undertake tests and those who may be affected by the test results. The failure to protect privacy and confidentiality could lead to genetic discrimination. Biotech companies with access to genetic information on their clients and their families have the duty not to reveal the data to third parties. This is crucial as such information is of potential interest to employers, insurance companies, and even governmental agencies. Especially genetic information of children or other vulnerable persons should be carefully protected, as they are dependent, and not fully competent or able to make free and informed decisions. For example, the sale of predictive genetic tests for talent or potential might harm these children. If such a child is branded for carrying, say, a gene for being 'prone to violence' or 'prone to depression,' it may have a deep psychological impact on the child, and could result in social stigma and genetic discrimination. Similarly, if a child is regarded as a 'genius' on the basis of a genetic test, the high expectations of parents and society regarding gifted children might also put a heavy burden on the child.

### *Personal Autonomy and Free Decision Making*

Personal autonomy is regarded as a basic ethical principle, and refers to the individual's capacity for self-determination. People have a right to make their own decision to undergo a test, independent from the views of others. The situation in which parents buy genetic tests for their child is more complex. If a child's adverse test results are received at a very early age, the child may have to live for a long period with the prospect of a later onset of the disease in question, and parents will fail to achieve their original well-meaning intention. The child, in turn, might regret knowing about the diagnosis, the consequences of which might continually worry them. To some, not knowing may mean a less worrisome life.

The decision to take a talent gene test is made by the parents, who usually ignore the personal feelings of their children about the matter. Parents in China are especially concerned with the education of their children, and are willing to invest heavily in this (Kipnis 2011). The association between talent, education, and the future of the family household is important, leading to decisions in which the children themselves sometimes have little say. In fact, a similar situation exists when adult children buy tests for parents or when a spouse buys a test for his or her partner. For example, one company offers a special set of tests for elderly people, named 'filial piety.' It encourages people to buy their set of tests as a gift to express their filial piety to their parents. These issues, concerning personal autonomy and free decision making, should be taken into account and should be part of any attempt to regulate commercial genetic testing and enhance its supervision.

### Conclusion

In China, genetic tests by biotech companies are increasingly available. Partly due to the lack of regulation and supervision of commercial genetic testing, socioeconomic and ethical governance issues have emerged. The issues brought about by the application of commercial genetic testing, including the effects of misleading advertising practices, the suitability of the groups targeted for potential customers, the reliability of test results, and the unavailability of genetic counseling for clients, deserve more attention from the public and the authorities.

### Notes

1. The website of Genetic Testing Net is: www.jiyinjiancewang.org. The website of Zhong Ren Gene Net is: http://yiyaowang.org/index.html.
2. On March 17, 2013, the Ministry of Health had changed its name to National Health and Family Planning Commission. For better understanding, this case study still uses MOH.
3. The template concerned is downloaded from: http://south.genomics.org.cn/genetest/template.doc.

### References

Friedrich, M.J. 2002 Preserving privacy, preventing disclosure becomes the province of genetic experts. *Journal of the American Medical Association* 21:815–821.

Fulda, L.G., and K. Lyken 2006 Ethical issues in predictive genetic testing: A public health perspective. *Journal of Medical Ethics* 32:143–147.

Kipnis, A.B. 2011 *Governing Educational Desire: Culture, Politics, and Schooling in China.* Chicago, IL and London: Chicago University Press.

O'Neill, O. 2002 *Autonomy and Trust in Bioethics.* Cambridge, UK: Cambridge University Press.

Sui, S. 2009 The practice of genetic counselling—A comparative approach to understand genetic counselling in China. *BioSocieties* 4:391–405.

Sui, S., and M. Sleeboom-Faulkner 2007 Commercial genetic testing in mainland China: Social, financial and ethical issues. *Journal of Bioethical Inquiry* 4(3):229–237.

UNESCO 2003 International Declaration on Human Genetic Data. Article 11. Adopted unanimously and by acclamation on 16 October 2003 by the 32nd session of the General Conference of UNESCO.

---

The marketing of and consumer desire for testing such a broad range of genetic traits is a new development in China, as is the case elsewhere in the world. Biotech companies increasingly see a profitable market for these tests, even if they cannot offer much in terms of prevention or cure. However, as Sui and Sleeboom-Faulkner point out, questions arise about the reliability of such tests. Tests can have false-positive and false-negative results, and gene–environment interactions and lifestyles also determine both the expression of a gene and life outcomes regardless of genetic profile.

## Kinship and Genes

Kelly Happe (2013) argues that the recent advances in genomics have contributed to a discourse in which disease is evidence of inherited defects, rather than embodied life. It is, she argues, a political worldview that ignores the embedding of bodies in historically specific environments, in particular socioeconomic circumstances. The fact that exposure to toxins and living in polluted environments can cause cancer is increasingly ignored. Pollution is seen to be an inevitable characteristic of modern life and cannot be avoided. Rather than toxins, "inherited genetic mutations are the polluting agents, a conceptual displacement that is part and parcel of a neoliberal logic that valorizes personalized, market-driven medical and public health interventions" (Happe 2013: 139).

Genetic testing anticipates the likelihood of the development of a particular inheritable condition, as we have noted, but while genetic work-ups may illuminate the risk of disease for a particular patient, tests provide little hope for its treatment and cure. However, biomedical research continues to explore the potential for treatment, and so sustains hope among families affected by congenital disease for the future treatment and cure of conditions that are presently fatal. People with family histories of incurable inherited conditions are especially attracted to the promise of future medical discovery and success, and in this context various biomedical companies build on medical knowledge for profit. Hence the lure of banking genetic material on the basis of the promise, or possibility, that discoveries downstream will provide desirable treatments and cures. In the following case study from Simonetta Cengarle, we see how human umbilical cord blood has been commodified, with commercial banks employing notions of potential risks for and anticipated cure of disease for the baby, its siblings, and even its parents. According to the American Academy of Pediatrics (AAP), an infant's cord blood should not be banked for future use by that infant because:

> Most conditions that might be helped by cord blood stem cells already exist in the infant's cord blood (i.e., premalignant changes in stem cells). Physicians should be aware of the unsubstantiated claims of private cord blood banks made to future parents that promise to insure infants or family members against serious illnesses in the future by use of the stem cells contained in cord blood. Although not standard of care, directed cord blood banking should be encouraged when there is knowledge of a full sibling in the family with a medical condition (malignant or genetic) that could potentially benefit from cord blood transplantation.
>
> *(American Academy of Pediatrics Section on Hematology/Oncology et al. 2007: 167)*

The AAP goes on to encourage the donation of cord stem cells to accredited national cord blood banks, but it cautions that private storage of umbilical cord stem cells should not be viewed as health insurance. Given the extremely high prices of cord blood banking, and the ways in which parents' fears are manipulated through vague claims of efficacy or the promises that medical science might deliver in the longer term (Petrini 2014), the following case study makes an important contribution to our knowledge about how this practice has been taken up in Southeast Asia and elsewhere.

---

## 14.2 Harvesting Umbilical Cord Blood

*Simonetta Cengarle*

Parents attending private antenatal medical clinics are often surrounded by information on umbilical cord blood banking. In a brochure, THAI StemLife, an umbilical cord blood bank, warns parents to "Protect your family's future by preserving its origins!" The text continues, "one thing we cannot do is to predict what will happen to our loved ones in the future. Unfortunately, accidents or disease can strike down even the fittest of us. But now doctors have a new weapon in the ongoing fight against disease in the form of stem cells: microscopic miracles, which are found in each and every one of us." The company's video on YouTube (in Thai) takes the viewer through a similar message (https://www.youtube.com/watch?v=EWIZanUvf04&feature=youtube) ending with the tagline, "Stem Cells for a Safer Feature." On its website, mention is made that "medical advances now allow you to choose a 'true life insurance' which can be used to save or prolong the life of the policyholder. This true life insurance enables individuals to store their own stem cells for their own future health needs" (http://www.thaistemlife.co.th/content/?p=104&c=54).

The language used in the sales pitch and marketing materials of umbilical cord blood banks makes it difficult for parents to ignore. It characterizes a notion of good parenting and strikes at the deepest fear of every parent: something may happen to their child. Cord blood banks are essentially assurance of the future health of a child. In their marketing, these biobanks use ideas and language affiliated with the private banking sector and investment companies. Clients are seen as investors who need to accumulate some security for the future. As one of my research participants, Esther, noted, "if something happens, I mean to these girls, they get a disease or whatever, and it might help and I would really regret that I haven't done it."

Since the first successful transplant in 1988, in which stem cells from a sibling were used as an alternative to bone marrow to treat Fanconi's anemia, umbilical cord blood banks have sprung up everywhere (Gunning 2004). Now banking storage facility companies are growing and expanding in all parts of the world. Asia is particularly prolific. Although in this case study I draw on research I conducted in Thailand and Singapore, the business of biobanking is global. Many parents are interested in covering their children's future by buying insurance that may save their lives. What most parents are not told, and may not understand clearly, is that private banks count on future technology to reap the benefits of umbilical cord blood cells rather than being grounded in technology currently in use. Although many conditions now can be treated with stem cells, there is an expectation too that future medical breakthroughs will provide the necessary medical fix for others (Brown and Kraft 2006: 314). At the moment, the amount of cells collected would be enough only to treat a child and not an adult, but in marketing their services, the banks use vague language to assure potential clients that the technology to grow cells in culture will soon be available. Also, the cord blood is rarely used for autologous transplant; more often it is used for siblings.

Stem cell research is dynamic and fast moving. Whilst results have not been transferred so rapidly as to change people's lives, commercial companies have been quick at attracting an ever-expanding audience which lives in hope and fear: fear of what the future may hold in terms of diseases and hope in a new technology that, will, one day, miraculously cure all chronic and deadly diseases. In the span of just a couple of years, private stem cell banking companies have gone from offering only umbilical cord blood storage facilities to banking almost every possible source of stem cells. As Siew Lian commented, "It's like, you buy it not knowing whether you will use it or not. So to me I feel like, if you look at this point of view, it's more like an insurance. You may use it, you may not use it. At the end of the day your money for the 21 years may just go down to waste but at the end of the day you could still use it . . . Of course we hope that our money will just go to waste and we will never use it again at all. So to me I feel like that is why it is like an insurance, you do not know whether you will need it." Her husband David added, "Ultimately it's just an insurance for the kid. Our parents, none of our parents did it before, it's just that nowadays a lot of parents are a little bit receptive to any idea that is good for their kids. Be it their education, be it the food, be it whatever, you know, as long as the best is for their kids."

Two different types of cord blood storage are available in most countries including Singapore and Thailand: public cord blood banking and private cord blood banking. The umbilical cord blood of the newborn, which until recently was discarded at birth, can be donated to public cord blood banks for use by anyone in need. The umbilical cord blood units are held separately and registries are available for patients throughout the world. Although most countries have public banks, women do not routinely receive information about donation when they go for antenatal checkups. During my observations in public or, as they are called in Singapore, restructured hospitals, women in the late stage of their pregnancy are approached by the staff of the public bank and asked to donate. This is done discreetly whilst women, often accompanied by a family member, wait for their antenatal appointment. It often takes more than one session for women to agree to staff sitting down with them to explain what cord blood donation involves. However not all women are approached, due to limitations of time and sometimes a decision by a bank representative about whether or not the woman looks 'healthy' enough to donate.

Although public banks are increasingly available, donating umbilical cord blood is not always straightforward. Some national banks only accept blood from families from under-represented ethnic minorities or from families with a known genetic risk curable with blood transplant. This is done to capture as wide a sample as possible. In addition, not all hospitals or birth centers in a given country have the expertise to collect umbilical cord blood. Therefore, although a family may be willing to donate, the hospital may not be in a position to collect. In addition, confusion is common, especially regarding payments and benefits.

Private banks, of course, are more flexible; they collect in any hospitals or birth centers, and a representative is available at any time to collect blood after delivery. Private cord blood banking refers to the collection of umbilical cord blood stored in private banks, or in storage space in public stem cell bank facilities, to families who wish to preserve the umbilical cord blood of their newborn. Private cord blood banking facilities have increased dramatically since 2000 with the emergence of regenerative medicine, whereby the potential of cord blood stem cells has gone beyond the use to substitute bone marrow in transplant and has expanded the potential therapy for a number of degenerative diseases in both adults and children.

Employees of private biobanks target women who present for antenatal checks at hospitals or clinics through promotional material and face-to-face conversations. Depending on the internal policy of the hospital and clinics, women and couples are approached on a regular basis and given information. Companies also reach their clients through marketing techniques, including social media campaigns, baby fairs, printed media, and television advertisements, depending on the context in which they operate. In Singapore, most biobanks sponsor prenatal talks with an array of neonatal experts, including doctors, breastfeeding consultants, and baby massage therapists, and run contests for parents with the possibility of a private consultation with the bank going on concurrently during the event. Once clients sign up, they are offered various payment plans, including a one-time fee that will cover payments until the baby is 21 years old, a 10-year plan with annual payments until the baby is aged 11 and then annual payments, and an initial fee with subsequent annual payments.

In Singapore, private banks offer deals to couples that sign up at any of the fairs or workshops and talks organized by them, with the initial payment at around SGD 1250 (c. US$1,000) and an annual fee of SGD 250 (US$200). There is the added advantage that families can pay for stem cell storage with the child development fund, an amount of money that the government gives to each child born in Singapore from Singaporean parents. With this incentive, the initial payment is reduced to SGD 625 and the difference is paid through the baby bonus. The annual payments remain the same. Once parents sign up, the mother is asked to have a blood test and to complete a form collecting demographic and health history information, and they are then given a cord blood collection kit to bring to the delivery.

Private banks are more aggressive in their campaigns than are public banks. Meiling told me she had just settled into her room at a private hospital and was getting ready to go into the operating theater to have a caesarian section for her twins, when the nurse who was helping her asked her if she wanted to buy additional benefits for her baby, including insurance for the ICU, baby photographs, baby hands and feet impressions, and stem cell collection.

> I think most mothers have heard of stem cell banking but they do not know so much until they get to the day of delivery. Because on the day of delivery and after the day of delivery, you will get like a few people visiting you trying to sell you things. I think the typical ones are if you want to have the feet or the hands, and then you will also be asked if you would like a photograph taken by a professional of the baby. And they were asking me for a photograph and the stem cells.

Meiling was very nervous and left it for her husband to decide whether they would sign on for the collection of umbilical cord blood or not. The nurse showed them two brochures, the husband chose one bank, and within a few minutes the representative of that bank was in their room. The process was quick: it only took a few minutes. Meiling told me that she bought the ICU insurance policy straight away, and that her husband bought the stem cell collection and storage with their baby bonus money.

In the context of stem cells, new meanings of kinship and social relationships emerge. For example, in an article in *The Straits Times* (5 May 2008), it was reported that a woman had stored the stem cells of her two children in 2005 and 2006, and although the youngest child had died after seven weeks due to a congenital heart defect, the mother had continued to store his cord blood. Stem cells are imbued with meaning beyond a common understanding such as medical insurance and regeneration, and bring new affections into play, new hopes, and the realization that a part of the child is still with the family (albeit in a laboratory). The lost child lives on in its stored blood.

What we are experiencing is the medicalization of kinship in which "family and kin relationships are being drawn into the biomedical domain" (Finkler 2000: 3), especially when related to the explanation of genetic transmission of diseases. Finkler's notions of kinship and transmissions are applicable to stem cells. Families decide to bank not only as an investment (whether they will use the stem cells or not) but also because there are perceived or real risks of recurrent diseases, often on the basis of family history: "My husband and I both have a history of cancer on both sides of the family. We decided to bank our youngest's cord blood. Hopefully we will never need it" (FB message board on Viacord and Cryo-Cell group, 9 September 2009). Or as one woman I interviewed put it:

> Okay, I thought a little bit before I did it, I maybe was like when I was three to four months pregnant . . . after reading through it (some articles she was sent from friends), I decided to go for it. For two reasons actually, one is because of the child, in case they get, actually if they get sick and partly it is because of myself because my mum actually passed away from cancer. Her whole family actually died of cancer. So I have a very high risk of inheriting the genes.
>
> *(Nalina)*

These new forms of kinship may play a limited role in society, yet despite this, blood still has an important role in modern societies where biogenetic kinship is valued (Finkler 2000: 35).

Public and private stem cell banks have unique characteristics within tissue economies. Public banks create clinical value for cord blood by storing a wide range of tissue types from various ethnic groups. In contrast, because private banks store the stem cells into a private account, these cells are withdrawn from public market circulation. Tissues and body fragments have social lives. Umbilical cord blood has changed status in the last 10 years, from waste, regularly incinerated after birth, to a highly valuable therapeutic and remunerative material. However, in hospitals, waste tissue is used in two ways (Annas 1999: 1523). One is in medical research; the second is its use for commercial purpose. The waste or abandoned material is easy to use and is not subject to the ethical considerations tied up with informed consent procedures. The recent Human Tissue Act (2004) in Britain mentions that it is good practice but not mandatory to obtain consent for research on non-fetal products of conception (like the placenta or umbilical cord), and such types of tissue will continue to be used without consent requirements providing the proposed research using those tissues has ethics approval (Human Tissue Authority, Code of Practice, 2009, www.hta.gov.uk/guidance/codes_of_practice.cfm).

Given that, increasingly, diseases are given a genetic origin, it is not surprising that, reading various stem cell blogs on the Internet, parents decide to bank the umbilical cord blood when they have a family history of diseases such as cancer, Parkinson's, Alzheimer's, and so forth. Social groups form around a specific and shared characteristic of biomedical origin. These collective groups open new spaces for public debate using public spaces on the net. Community groups, NGOs, and individuals who want to bring clarifications, personal experiences, and advice in relation to banking stem cells have emerged as well. For example, the Parent's Guide to Cord Blood is a website set up by the parents of a girl who died of leukemia (http://parentsguidecordblood.org). The aim of the website and its work is "to educate parents with accurate and current information about cord blood medical research and cord blood storage options. . . . The second mission of the Parent's Guide to Cord Blood is to conduct and publish statistical analyses on medical research or policy developments which could expand the likelihood of cord blood usage."

Within the process of becoming parents and exploring the possibilities of using the umbilical cord blood, couples assign new meanings to blood whereby its biological, social, and even economic values cannot be separated but rather become connected and reconnected in the process of banking. With umbilical cord blood banking, decisions to donate or store stem cells with health implications for the family or the community are imbued with hopes for the future.

## *References*


Annas, G.J. 1999 Waste and longing—The legal status of placental-blood banking. *New England Journal of Medicine* 340(19):1521–1524.

Brown, N., and A. Kraft 2006 Blood ties: Banking the stem cell promise. *Technology Analysis & Strategic Management* 18(3–4):313–327.

Finkler, K. 2000 *Experiencing the New Genetics: Family and Kinship on the Medical Frontier.* Philadelphia, PA: University of Pennsylvania Press.

Gunning, J. 2004 A worldwide study of umbilical cord blood banking. European Group on Ethics in Science and New Technologies. Opinion 19. Ethical aspects of umbilical cord blood banking. Luxembourg: Office for Official Publications of the European Communities.

Rucinski, D., R. Jones, B. Reyes, L. Tidwell, R. Phillips, and D. Delves 2010 Exploring opinions and beliefs about cord blood donation among Hispanic and non-Hispanic black women. *Transfusion* 50(5):1057–1063.

Genetic testing not only affects the individuals who receive reports on their genetic makeup. It also affects the siblings and children of the person who is identified as being susceptible to a specific disease. For example, when diagnosed with the BRCA gene, indicating a high risk of getting breast cancer and an elevated risk of other cancers, women have to decide who to tell and how. Like other stigmatized conditions, having the BRCA gene puts the individual into the position of revealing her diagnosis or not. Living with the social repercussions of the diagnosis may be very difficult, especially for people who have Ashkenazi Jewish heritage, among whom especially high rates of breast and ovarian cancer are due to this mutation. This risk profile is particularly disturbing where many people have lost friends and loved ones to cancers with these genetic origins, and where universal screening is encouraged, so pushing such women to take prophylactic action. While in the 1960s and 1970s, anthropologists emphasized that how people act is determined by culture, in the present genomic medicine has led to a renewed emphasis on heredity. Accordingly, anthropologists are studying how developments in the field of medical genomics lead to a rethinking of kinship. How does genomic medicine affect the way we think about family ties? Marilyn Strathern (1992) points out that the basic idea that offspring should resemble their parents, while they may also differ from them, has been extremely durable in Western culture, as evidenced in expressions like 'like father, like son.' How do people incorporate genetic test results into their discourses on family (non)resemblance?

The third case study, by Janice McLaughlin, shows how ideas about family resemblance play a role in consultations between parents of children with unexplained developmental abnormalities and pediatric geneticists, who nowadays can make use of 'whole genome sequencing' to find the genetic variations causing the disorder. She shows how clinicians adhere to a 'genotype first' process of diagnosis to identify unexplained developmental abnormalities. Although whole genome sequencing allows for complex biogenetic analysis of genetic variation, and defines relatedness at a molecular level, families continue to make their own assessments of similarity in everyday life. As an example, she describes how one grandfather saw nothing unusual about his grandson's height: "I just think it's the family," he remarked. "We're all tiny." McLaughlin shows how family members can use genetic test results to characterize and validate some relations as closer than others. People are surprised when, in some cases, genetic tests do not confirm the inheritance of a trait. The identification of a shared genetic disposition for learning difficulties, for example, creates a link between a mother and her daughter, as the mother finds parallels in her own childhood to the challenges faced by her daughter. McLaughlin suggests that the results of genetic technologies will not replace narratives of resemblance, but rather, will be incorporated into the social and intimate lives of families.

## 14.3 Genetics, Childhood Development and Kinship

*Janice McLaughlin*

Children with unexplained development problems and unusual (dysmorphic) characteristics are often referred to a genetic service or clinic to see if chromosomal variation may explain their issues. Patterns of referrals have increased in much of the global North as advances in genetic testing have expanded the genetic variations linked to a range of childhood conditions. Until now diagnosis has been a combination of family history, close examination of the child's features and biochemical analyses of the child's DNA from blood samples. Pediatric genetics is set to see significant changes in its diagnostic practices and possibilities over the next few years as we move into the era of whole genome sequencing (WGS).

Unlike current diagnostic practice, WGS, through complex biochemical manipulation of blood samples and bioinformatic analysis, enables (theoretically) all genetic variations within a person's genome to be identified. What in the Human Genome Project took billions of dollars and many years to do very soon could take thousands of dollars and days to produce. The most difficult element is the complex work required to separate out variations that are clinically meaningless from those that are significant. This is the key challenge (and current barrier) to bringing WGS into clinical practice. Considerable research activity is underway in multiple research studies across the globe to establish which variations that can now be 'seen' are clinically relevant. One such study in the field of childhood disorders is the UK Deciphering Developmental Disorders (DDD) study (http://www.ddduk.org/). The hope is that such studies will both explain the genetic underpinning of a range of developmental issues in children and have the potential to offer greater predictive power and therapeutic application.

While WGS remains primarily a research endeavor, elements of sequencing technology are entering clinical practice. For example, sequencing is being used to search for particular traits with already known clinical significance in areas such as immune disorders. These changes in clinical practice are already leading some commentators to suggest that in the future the child's body will become increasingly irrelevant to the diagnostic process. An editorial in 2008 in the *New England Journal of Medicine* proposed that "(c)linicians, like researchers, can now shift to a 'genotype first' model of diagnosis for children with unexplained developmental abnormalities" (Ledbetter 2008: 1729–1730). Medical sociologists and anthropologists have been quick to point out the dangers of a move away from an interest in symptoms displayed on the body to variations found in a computer simulation (Navon 2011). They warn of a future where variations found in DNA will be used to create "ontologically disputed borderline forms of disease" (Buchbinder and Timmermans 2011: 57).

Drawing from a study undertaken in pediatric genetics in the UK,[1] I explore the social and cultural significance of the body within childhood and kinship. Can the body be replaced by WGS findings in family practices of meaning making? And what happens when children help shape these processes?

## *Following the Body*

The presence of the child's body during clinical consultations is significant to the implications genetics has to how the child is understood by family members (McLaughlin and Clavering 2012). However, if the detailed physical examination of the child's body becomes much less important in the future era of WGS, will that mean that the body will become less important to how the child is understood? In clinical observations parents were often impressed by the complex diagrams geneticists produced (all names given below are fictional):

*Dad:* It was fascinating 'cos you think, how can you do that with something, you know, that's so little? And just blast it apart.

*(Brown Family, Second Interview with Mum and Dad)*

Below, I suggest that lack of clinical interest in the body in WGS will not necessarily mean that the body will drift into the background and that biomedical processes will dominate. Assuming this will happen minimizes the social and cultural significance of how identities form for children via the important mediating influence of those close to them, particularly parents and other intimate family members. Exploring the significance of these broader processes of meaning making with and through bodies requires pulling back from the clinic to a broader landscape. While the clinic is a place of authority, it is not the only space within which children's bodies are represented, visualized, analyzed and made sense of. Multiple interactions can occur in multiple spaces, and the child's own enactment of her body may have a role in how others interpret it and give it meaning. One value of ethnography is that it allows the researcher to observe such interactions and to follow people through multiple places that play a part in forming their identity.

## *Resemblance*

The way people in a family interpret a child is influenced by the way they look at and read their body, often in comparison to others, searching for resemblance. People trace resemblance and give meaning to it through stories of similarities between family members. Existing familial understandings of shared inheritance may either support the possibility that a genetic variation of significance, or more commonly, that the factor in development, framed as a 'clue' to genetic peculiarity, was instead a shared family trait and appeared as perfectly normal to them. For example, a grandfather at a first consultation explained why he did not think there was anything significant or unusual about his grandson's height:

*Geneticist to grandfather:* And what about you?
*Grandfather:* No. I just think it's the family.
*Geneticist:* So, it's the family?
*Grandfather:* We're all tiny.
*Geneticist:* Hm. Let's just do the family tree.

*(Dougherty Family, Observation of First Consultation, Mother, Maternal Grandfather, Son, Geneticist present)*

Appeals to resemblance can also be used to emphasize connections and relatedness that genetics may not see or consider important. In one family in the study, the father's line was not reflected in the family tree the geneticist took in the consultation. The reason for the exclusion was that the trait being explored was only inherited via the maternal line. Biomedically, the father was irrelevant. However, the exclusion clearly pained him, as he felt he was being erased in some form as a component in the making of his daughter's life. In response he always spoke in interviews of ways she took after him both physically and in character:

*Dad:* She's got my temper!
*Gran:* Yes, I was going to say that.
*Mum:* Yes, yes.
*Dad:* Yeah, attitude and stubbornness.

*(Henderson Family, Third Interview with Dad, Mum and Maternal Gran)*

Through claiming resemblance, validated by the maternal side of the family agreeing, the father found a way back into accepted narratives that he is important in his child's present and future.

Another route to finding resemblance as meaningful is seeing it in others said to have similar variations in their genetic makeup. In the study, such resemblances were often found online as families searched for information about the types of variations or syndromes they had been told their child had. What they found were pictures of dysmorphology (unusual looking children). While one response was to express fear and discomfort that their child might be one of those children, another reaction was to find connection and community through interpretations of physical resemblance. At times, the virtual similarities found online and understandings of familial resemblance came together, reflecting ways in which bodily similarities were participating in practices of familial forms of recognition. The Todd/Richardson family were told on the phone that their child had a variation on a particular chromosome and were invited to a second consultation to discuss the finding. As with the others, they went on the Internet and saw pictures of children with the same chromosome error. They described what they saw as like "looking at their son."

*Mum:* We went through lots of websites, all about Chromosome X. We found a family down south. They have a four-year-old girl and we sent them a photo of Harry. We all agreed they looked exactly the same, only she had long hair. They have the same chubby cheeks and that look in their eyes.

We look across to Harry, who is playing with plastic shapes near the play house.

*Mum:* He does have slightly chubby cheeks but then, I think, he is only just over two years old, and his dad has quite a round face.

*(Todd/Richardson Family, Observation Notes of Waiting for Second Consultation to Begin, Mum, Dad, Child present).*

During the consultation the geneticist suggested that what they had seen on the Internet was not relevant as the variation was different from what they had understood from the telephone call. However, reflecting after the consultation, the father commented:

*Dad:* We could see how all these children look the same. Just having something different about your genes brings them all together, whatever the deletion-point-this-that-or-the-other is.

*(Todd/Richardson Family, Observation Notes of Second Consultation)*

Seeing resemblance is an important component to making connections with others, something the body is culturally positioned to provide.

## *Incorporating Genetics*

It is more useful to think about the information and understandings that emerge from genetics as being incorporated into, rather than determining, kinship connections. One purpose of family stories is in shaping the boundaries of who is in and who is out. One maternal grandmother had a strong narrative that her family history was one of good moral character and physical health, passed on by the women in the family:

*Gran:* Very strong women in this family. When I think about every woman in this family, we're very strong women. You know I'm talking about like *my* blood line, my mam, grandmother, great gran, very strong women. Always been like that.

*(Brown Family, Second Interview with Maternal Gran)*

Her granddaughter was diagnosed with 'developmental delay' and was being seen by the genetics service. From early on, the grandmother had assumed the biological father must be the person she had inherited her problems from. Her justification was several examples of illness and learning disability found in his family and not in hers (she claimed). To her surprise, the genetics service said the genetic trait they had found came neither from the father nor mother (it was a de novo variation). Because genetics could not validate her separation between her family and that of the father's side, she switched her focus to emphasizing the inheritance of social and moral character rather than genetic matter:

*Gran:* So I bring this line up of going back. And you know, my granddaughter will say, "what about your mam? And her mam?" And I'll get pictures, and I'll say that was my mam, that's my nana. . . . And I say, look at her lovely hair, and she's got a lovely clean apron on, because obviously in them days they always wore the aprons. And I always seem to push, *clean* and *manners*. . . . They're not going to be, they're not going to be *dragged* up. . . . Because I was brought up, and so was my, this line, as you call it, these women behind me.

*(Brown Family, Second Interview with Maternal Grandmother)*

The grandmother moved between genetic and social versions of inheritance as she worked best to maintain a narrative that the family was of moral character, and some version of that inheritance would be passed on to her granddaughter.

The creative use of different versions of what families share, including genetics, was also evident in other families. Alice, who was nine years old, had been identified as having a genetic variation that the geneticists linked with learning difficulties. Her diagnosis also led to the identification of the same variation in her mum. In an observation and informal interview with Alice at her home, she commented:

*Researcher:* Did your mam say how you have the same as her?
*Daughter:* Yes.
*Researcher:* What do you think about that?
*Daughter:* Hm [long pause, looks up and smiles] Happy.
*Researcher:* Happy, in what way?
*Daughter:* We're the same.

*(Collins Family, Notes of Single Interview with Daughter)*

Likewise, the mother spoke positively about sharing this trait with her daughter—a connection she felt meant they were closer than she was to her biological son:

*Mum:* Everything now is focused on looking ahead for Alice. Just having this information now for her will make a difference. . . . It helps make sense of things for me, just as it does for my mam, and my nanna, and my granddad. . . . All my focus is on both my kids. I'm really proud of them. I'm close to them both, but I'd say I have a special relationship with Alice. I think that's because I can see what she is going through. It is hard to think of the future for her. She is in mainstream school, the local school they both go to. There are great teachers there, really friendly, one lives just over the fields behind us. I went to a special school that was horrible, though they've changed now, but Alice is fine in mainstream. I don't see any problems for her with carrying on with that—she loves school.

*(Collins Family, Single Interview with Mum)*

It is possible to read the mother and daughter's special connection as being made through the narrative of genetics. This would imply that genetics can be used to validate some relationships as being closer than others. However, this narrative is about more than genetics; it is important to read it within its full social history. The mother's childhood had been difficult, she had struggled in school and in social interactions, she had been bullied, marginalized and had terrible experiences of schools for 'retarded' children. The mother's story is of a shared connection to her daughter that is about being similarly different in body and mind (now with a genetic explanation), which will be lived (she hopes) in a very different way by her daughter, because she lives in a very different world than the one her mother grew up in.

## Conclusion

WGS will produce new ways of capturing and analyzing genetic variation; however what will this mean to families, or indeed children themselves? Will complex algorithms and displays of biochemical analyses become dominant in interpreting children, particularly as "other"? For now I would say while this is possible, it appears unlikely. Children's bodies, in particular how they look, who they look like and who they don't look like, are integrated into the social and intimate lives of their families. The importance of embodiment, as practice and narrative, to kinship making means that it is unlikely to be simply replaced by other sources of narrative, instead it is more likely that WGS, like current genetic technologies and stories, will become part of how children's bodies are understood.

## Acknowledgment

Funded by the Economic and Social Research Council, between 2008 and 2011, as part of a larger team, I undertook ethnographic research in the UK examining the experiences of children and their families who had been seen by a genetics service. The fieldwork, over an 18-month period with 26 families, consisted of a mix of non-participant clinical and non-clinical observations, narrative longitudinal interviews and creative practices (used with the children). Within families we worked with parents, siblings, other significant family members and the children referred (with the very young children that work was limited to the observations). Families also shared the material they were sent from the genetic service detailing what diagnosis (if any) had been made. The project obtained ethical approval through the NHS National Research Ethics Service (NRES).

## Note

1. For a more detailed account of the research project I draw from here, see McLaughlin (2014).

## References

Buchbinder, M., and S. Timmermans 2011 Medical technologies and the dream of the perfect newborn. *Medical Anthropology* 30(1):56–80.

Ledbetter, D.H. 2008 Cytogenetic technology—genotype and phenotype. *New England Journal of Medicine* 359(16):1728–1730.
McLaughlin, J. 2014 Digital imagery and child embodiment in paediatric genetics: Sources and relationships of meaning. *Sociology* 48(2):216–232.
McLaughlin, J., and E.K. Clavering 2012 Visualising difference, similarity and belonging in paediatric genetics. *Sociology of Health and Illness* 34(3):459–474.
Navon, D. 2011 Genomic designation: How genetics can delineate new, phenotypically diffuse medical categories. *Social Studies of Science* 41(2):203–226.

## The Turn to Complexity

Two decades of mapping the human genome have led to the identification of 23,000 genes—which, it turns out, is equivalent to the number of genes found in a roundworm. Faced with these figures, observers have increasingly acknowledged that "nucleotide sequences themselves were neither the 'holy grail' nor the 'code of conduct' that the proponents of the projects hoped they would be" (Rabinow and Bennett 2012: 13). Clearly the identified genes could not account for the complexity of the human organism (Rheinberger 2004). Perhaps, as Evelyne Fox Keller (2001) argues, genes have outlived their role in understanding biological phenomena and heredity. It has become very difficult to define what genes actually do, and there is a substantial gap between the overflow of information and its "promised transformation into ameliorative and lucrative applications" (Rabinow and Bennett 2012: 14). As biologists continue to examine the complex regulatory systems through which organisms function, they realize that the phenomena they study are more complex than they imagined. In surveying this situation, Hans-Jorg Rheinberger (1997) argues that biologists need to build a middle ground between complexity and simplicity into their epistemic practices.

Such complexity is built into the framework of the field of environmental epigenetics, which examines the multiple, overlapping relationships between the environment and genes (Landecker and Panofsky 2013). In this subdiscipline, genes are not thought to determine behavior unilaterally, rather, diverse factors, such as violence, family life, and toxins, can affect our genetic makeup. In this context, the environment refers to both the physical and the social environment in which people live, and the uterine environment, which in turn is influenced by the external and internal (health and nutrition) environment of the mother. Using the notion of biological plasticity, researchers argue that biology and behavior are co-produced (Mol and Law 2007).

In the fourth case study in this chapter, Stephanie Lloyd presents the results of an ethnographic study conducted in an epigenetics laboratory where scientists have been exploring whether early childhood abuse is linked genetically to suicide later in life. More specifically, epigenetic researchers are examining how negative experiences in early life are embodied, physically changing genomic expressions and ultimately increasing people's chances of becoming, in her example, 'suicide completers.'

## 14.4 Suicide and the Epigenetic Turn

*Stephanie Lloyd*

Though elevated in particular eras, regions and subpopulations, suicide is an act deeply ingrained the past and present of human history. Those analyzing suicide from an anthropological perspective have described it as a type of sociality, a way of living as much as dying, intimately connected to time, place and people's personal lives (Staples and Widger 2012). Seen in this way, suicide has complex meanings and relationships with innumerable life factors related not only to people lost to suicide but also those around them.

The scope of suicide is now regularly documented alongside other 'health topics.' The WHO suggests that globally one person commits suicide every 40 seconds. Every three seconds someone attempts suicide. Given these rates, suicide is of great concern as a clinical problem. Clinical researchers link suicide to developmental factors, pathological personality traits, psychopathology, and biological and genetic predispositions. They are most concerned with suicide attempters, as these people are statistically most likely to become what they refer to as 'suicide completers.'

My focus is on emerging epigenetic explanations of suicide that attempt to both overthrow and integrate the disparate lines of reasoning emerging from sociohistorical and clinical research. Epigenetic researchers suggest that negative early life experiences are literally embodied, physically changing individual genomic expression and ultimately increasing people's chances of becoming suicide completers. Though the science is in its infancy, the perspective is gaining currency and influence in mental health research and well beyond.

*Environmental epigenetics* is a subfield of epigenetics focused on how environmental factors—from socioeconomic disparities, to individual life experiences, to environmental toxins—affect the body. The rapidly evolving interdisciplinary field of research uses the notion of biological plasticity as a base upon which to construct theories of how nature and nurture co-produce biology and behavior. In a 2011 public lecture Michael Meaney, a leading environmental epigeneticist, identifies the foundations of contemporary environmental epigenetics in the work of biochemists over the past 10–15 years. This research, he contends, made it evident that "the activity of a gene depends upon the context in which it operates." Through the process of this research, he says:

> (I)t was found that some of those structural modifications that occur to the DNA [as a result of the environment] . . . they are not so dynamic, they can actually be long term. They can actually stably alter the structure of the DNA . . . we have all known that in the course of development or even in adult life, there are occasions when people are exposed to environmental events or to biochemical events that are so profound that they stably alter the activity of genes. In other words, the effect isn't simply there during the period of the event, it endures well beyond. . . .

This brings Meaney to observe that one can then ask how the social or educational environment, or nutrition in early life, could stably alter the genome. He compares the orientation of this research to studies of gene activity near the turn of the century. In that era, "everything seemed to start from the gene." Epigenetics research, he contends, displaces attention from the gene by underscoring the extent to which "the environment can influence the DNA which then influences particular traits." He concludes, "now the DNA . . . actually becomes integrated within the environment and we start to have a more complete understanding of how the environment literally produces biochemical structural changes to the DNA, that then produces changes in the phenotype [a behavioral or physical outcome]" (Meaney 2011).

Meaney provides a digestible, if widely encompassing, explanation of epigenetics and the critical role of the science in repositioning of our understandings of humans and how we become who we are at any given moment in our lives, from diseases to personality traits. However, the science he describes is an incredibly complex field, with competing theories and models muddled by difficulties in operationalizing the questions epigeneticists wish to pose in day-to-day lab work. For instance, how are environmental factors embodied and what is the impact of this embodiment?

A group of Montreal epigenetics researchers are trying to answer this question. Specifically, they ask, how do you look inside someone for evidence of early childhood abuse and then identify how this is linked to suicide later in life? Following Meaney's assertions about the epigenetic effects of early maternal care and documented links in psychiatric literature between early childhood abuse and later psychopathology and suicide, these researchers search for biological traces of abuse that occurred in childhood during what are referred to as 'critical moments' of neuroplasticity, when the brain is considered particularly sensitive to environmental factors.

Conducting this research entails many challenges. Epigenetic changes resulting from environmental exposure is dynamic, as Meaney noted. Further, they are tissue specific: if you want to know what the effects of abuse are on the brain, you need brain tissue. As a result, the Montreal researchers are dependent on their affiliated brain bank which houses thousands of brains, including those of suicide completers. In these brains, they search for epigenetic changes such as methylation patterns they consider to be the durable imprints of early childhood adversity. They then attempt to identify if these methylation patterns are distinctive from those in the brains of normal controls or even people lost to suicide with no history of child abuse. Thus, the work of these researchers focuses on the remains of suicide completers, samples of fixed or frozen brain tissues, in the hopes of identifying the particularities of the methylome that predisposes people to suicidal behavior.

In this case study, I focus on the incipient models that underlie and guide their brain-based research: models of how the environment—an event in the form of early childhood abuse—'gets into' people's brains and leads to a behavior—the enactment of suicide later in life. Constructing these models remains a distinct challenge. The lead scientist of the research team, Gustavo Turecki,[1] described how he conceived of the embodiment of adversity:

> We don't know how it gets in, we don't understand how it is that, let's say . . . a social event, whatever it is, or psychological event . . . gets into the actual genome. Basically we know associations. . . . So from the work in animals, that suggests a *causal* relationship between . . . events in the [early] life of the rats . . . or whatever model that's used: that negative environments induce epigenetic changes. So it's clear from the animal work that there's a causal relationship. Yet, even from the animal work, it's not clear how it gets to the genome. So how it gets from the actual experience to the genome, that is not clear. The hypothesis, primarily based on Michael's [Meaney] work. You know Michael's work is about tactile stimulation, so it's how the mom [rat] licks and grooms. . . . So it is really the stimulation of the skin that induces the changes. So in that case there are a few, let's say, putative mechanisms that are related . . . this tactile stimulation releases one neurotransmitter that then [affects] levels of one particular hormone, and that leads to a cascade of events that leads to these methylation changes. But it's not yet . . . 100 percent. It's not [yet] a robust hypothesis. . . . So that's what we know from the animal work, from the human we don't know anything. . . .

When I asked whether he believed a similar mechanism would nonetheless be implicated in humans, he replied:

> No, it wouldn't because it is a different mechanism . . . because in the rat . . . it is really related to the stimulation of the skin, which has nothing to do with the human experience, right? The human experience is more about psychological impact, so we still don't understand, that is a problem. We don't understand how is that emotions are processed let alone what the link is between emotions and the genome. So these are parts that have big question marks.

A geneticist (Carl Ernst) and postdoc (Paul) in Turecki's research group delved somewhat deeper into the challenge of ascertaining how an event is embodied:

*Paul:* It's a very large question and I guess every part of the neuroscience field is trying to understand this question using different techniques and tools. So we are using post-mortem brain tissues and we are making correlations, while people who are working with mice, for example, are trying to look more closely at causality, but then it is only a model [of mice, with unknown implications for humans], so every approach has its limitations . . . if you wanted to speak more precisely about child abuse, I mean, I guess there are many systems in the brain that are dedicated to the neuronal coding of life experiences and its different components, so there are cognitive components, there are emotional components, there are memory components, and so every component is going to be modified and impacted by life experiences.

When I asked how they saw the neuronal coding leading to durable changes, Paul responded:

*Paul:* So then it's a very long chain of events . . . you could imagine that some brain regions are encoding the emotional aspect of pain, okay? And so when you are subjected to repeated child abuse these regions are repeatedly activated, and so this triggers repeated release of some neuropeptides, and some neurotransmitters in this specific region are going to trigger specific downstream signalling pathways that are going to be modified in the long term and this may be long-lasting . . . up until adulthood. I compared [the abused research subjects, i.e., suicide completers] to subjects who were not exposed to child abuse and this, this [difference] . . . is reflected at the epigenomic level.

In terms of how they conceive of the 'signal' ultimately leading to the changes at the epigenetic level, they contend:

*Carl:* Well, they could be linked, I mean . . . I wouldn't phrase this just in term of methylation, but just like dynamic changes to the genome, to modulate gene expression and the protein translation, it is all that. But that's a good example, you have some neuronal network that's responsible for coding, pain, it releases some . . . [Paul continues.]

*Paul:* Or reward, or I mean . . . emotions, memory, because you have to remember the adverse experience, you have to, I mean, to develop strategies to avoid them, so there is a cognitive part because you are trying to implement behavioral strategies to avoid your mother when you are feeling that she is becoming aggressive, I mean, there are so many components, it's amazingly complex, right?

*Carl:* I agree with all that, but then you have the release of, let's say, some peptide, it binds to its receptor on the outside of the cell, that triggers the second messenger system in the cell, which ends up interacting with the genome. This is factual, this happens all the time, it happens with a lot of peptides we know about. Some of those genomic changes may inform [its] DNA methylation state, and if that neuronal network is repeatedly activated by a constant stimuli, in this case abuse to the child, the constant release of that hormone, let's just say, or peptide or whatever, the constant stimulation in that second messenger pathway, and the constant interaction of the genome with the molecular factors end up having a long-lasting change, which could be DNA methylation, which could be histone changes [another epigenetic mechanism], but just global change in the genome, that are gonna result in up, down, changed regulation. I mean, these things *do* happen, if you eat broccoli every day for a year, you are gonna change the epigenetic pattern of your stomach cells, y'know, you are gonna change the bacteria in your gut, you know, there's tons of stuff that happens because of some external stimulus. We're just using a complicated one, like a behavior, like child abuse . . . but if you just change your diet, your epigenetic pattern is gonna change in your food related cells.

Carl's statement gets at the heart of the challenge in their work: trying to link a complex life event (early childhood abuse) to what they see as a pathological behavior (suicide) later in life. Social and personal environmental factors such as child abuse are complex and their putative effects are not as easily traced as compared to relatively more straightforward environmental factors such as toxins, whose effects are nonetheless complex. Trying to link biological changes to suicide later in life represents an additional challenge, as it is not a well-defined disease state, but a complex behavior.

Added to this uncertainty of how to link an event to a behavior at a biological level, there are questions about the fidelity of the epigenetic traces they seek to characterize, given that their research methods involve looking for relatively distant history imprinted on the brain. When asked how he knows with any certainty that he is seeing embodied childhood abuse in the brain, Turecki responded:

> We don't. This is a *clear* problem of the approach that we have. So ideally what you would like is, you know, to have access to the brain before it [abuse] happens and after it happens, and then look at the consequences. But you cannot. So, in humans, because what we study has to be based on the study of the brains, you can only have access to the brain after people die. So any other ways of looking at the brain would be limited. So these individuals were abused, a lot of things happened after they were abused in childhood, and they died many years later. We don't know if what we are looking at is directly related to the abuse or as a consequence of everything that happened after, or a combination of both. We don't know.

Their beliefs about plasticity, induced durable changes in the brain and how this leads to behavior are constantly shifting as new findings are produced. Each piece of research provides footholds for them to move forward, with a plethora of new questions.

Correlating a methylation pattern with a behavior is at least as difficult as tracking its embodiment, as Turecki notes:

> Every time we find a methylation pattern, so the first question we always ask is: what is its functional implication? Does . . . differential methylation in this area, does it lead to different function somewhere or not? . . . But we don't [yet] understand the function implication of these genes [where differential methylation patterns are identified]. . . . So, then what you can do? So there is this one study that we are doing. . . . [linking differential methylation to regulatory outcomes on a specific gene] So this is a big story that took a lot of work . . . then after all this work, we injected viruses containing the [genetic] sequence that we wanted to test in the brain of mice. . . . And when we do that we induce the expression of this thing, the mice become more aggressive, significantly more aggressive. So that is the way you can follow up [on] . . . the *behavioral* consequences of this differential methylation that you find.

By complementing human studies with animal models, they are able to begin painting a portrait of the process whereby early childhood abuse is linked to a methylation change and a behavior, in this case aggression, a trait associated with suicide. Gustavo considers this a considerable success.

Carl remains less certain about the solidity of behavioral epigenetic theories in general:

> You know what I want? I want one example of a behavior . . . I don't care what it's in. Where some event, some stimulus is given to the animal and you can repeatedly see an obvious, clear, clean epigenetic mark . . . I just want . . . some external environmental factor that is actually causing some molecular epigenetic change. That's all I want . . . just *something*, and I've yet to see anything that shows a causal link [to his satisfaction] between an outside behavior and DNA methylation, or whatever, pattern. . . . I just think, we just need a *model*. If it's true [that child abuse can lead to methylation that is still visible in the brains of suicide completers] then we should be able to find it. . . .

Given their professed uncertainty and the challenges inherent in their research, one might be tempted to ask why they bother asking such questions. Doubt that plagues their nascent theories notwithstanding, Carl noted that they *do* know that experience affects biology and these findings are enough to push them to continue looking for causal evidence of their hypotheses. As they search for this proof, they focus on mapping as robustly as possible the molecular traces of abuse they are able to render visible in the brain by virtue of their lab work, moving forward pragmatically and casting aside their doubt and uncertainty in their day-to-day practices.

Their molecularized view of suicide risk, in which the risk profile of individuals who die by suicide is constructed in their absence through epigenetic research on their remains, results in the origin of suicide risk and ultimate suicidal behavior being shorn of its links to proximate factors such as pain, personal loss or even imitation, and equally, of other environmental circumstances whereby suicide might be seen as an act of defiance against a political system or state of injustice. Instead, the time frame of interest is shifted to distal factors earlier in life, such as abuse suffered during childhood at critical moments of neural plasticity. The induced risk profile in the brain is ultimately seen as leading to suicide completion. Simultaneously, diagnostic attention and early therapeutic intervention shift temporally leading to earlier clinical surveillance and enrollment in risk management. Epigenetic research is lending support to molecular characterizations of mental illness characterization overall.

The epigenetic view of suicide risk follows on other epigenetic research in which the environment is increasingly viewed as a set of molecularized risk factors to approach with trepidation, with limited attention to molecular 'protective factors.' This research has profound implications beyond mental illness including understandings of humans as 'biosocial becomings' (Ingold and Palsson 2013: 9). In this view, humans are constantly in formation with biological and social forces interacting fluidly. While the researchers cited in this text adopt something similar to this biosocial view of suicide and human development, the difficulty of operationalizing complex biosocial developmental pathways evidently remains a significant challenge. As a result, complex fluidity becomes stunted, as researchers reductively focus on the few risk factors their current tools and practices can identify and measure, diminishing the complexity of experience and behavior in the process.

## *Note*

1. Principal investigators of the team have waived anonymity as the distinctiveness of their research makes them readily identifiable. Trainees and staff in the research team have been anonymized.

## *References*

Ingold, T., and G. Palsson 2013 *Biosocial Becomings: Integrating Social and Biological Anthropology*. Cambridge, UK: Cambridge University Press.

Meaney, M. 2011 Panel member in Beyond your genes: Epigenetics, and how early experiences may affect your health later in life. November 30, McCord Museum, McGill University, Montreal, Canada. http://publications.mcgill.ca/medenews/2011/10/03/beyond-your-genes-epigenetics-and-how-early-experiences-may-affect-your-health-later-in-life/

Staples, J., and T. Widger 2012 Situating suicide as an anthropological problem: Ethnographic approaches to understanding self-harm and self-inflicted death. *Culture Medicine and Psychiatry* 36(2):183–203.

Because the genes unraveled in the human genome project have failed to provide clear-cut codes to many of the biological processes in our bodies, and because biological processes are the result of complex interactions between environment, genetic makeup, and lifestyle, it has proven much more difficult than expected to develop preventive and curative interventions based on the results of the human genome project alone.

## Synthetic Biology

In the new millennium, synthetic biology has emerged as a new field of research in which new therapeutics are being developed. Synthetic biologists aim to construct completely novel biological entities and redesign already existing ones, and to re-engineer complex biological systems and make targeted interventions possible. They have succeeded in making synthetic insulin from the bacterium *Escherichia coli,* are developing a precursor to the antimalarial artemisinin from *E. coli* and yeast, and are developing a semisynthetic derivative of artemisinin to enhance its bioavailability. They have also created more precise ways of diagnosing disease, based on people's specific genetic makeup, allowing for more targeted therapeutic interventions.

Calvert (2010: 97) suggests that synthetic biology aspires to "draw inspiration from the technological achievements of other branches of engineering, such as aircraft and computers, and conclude that it is 'economically and socially important that we improve the efficiency, reliability and predictability of our biological designs' (Arkin 2008: 774);" synthetic biologists aspire to replicate the successes of engineering. This cutting-edge science, however, is surrounded by a great deal of uncertainty. What kinds of 'interventions' will be possible? What are the risks? What are the ethical questions related to engineering biology?

The fifth and final case study in this chapter is concerned with the way in which people in the United States and the United Kingdom view the benefits, risks, and possible future applications of bio-engineered products, including cognitive/memory enhancements via chips implanted in the brain, and newer, stronger, lighter-weight and longer-lasting prostheses. These innovations both involve nanotechnologies—tiny engineered particles and structures that range in size from 1 to 100 nanometers (one nanometer is one-billionth of one meter). Synthetic biologists are especially interested in nano-enabled drug-delivery systems. The small size of the particles makes it possible to better target delivery, including across the blood-brain barrier. Worldwide, nanotecholoogy development is the focus of significant public and private investment, and nanotechnology is being used in pharmaceuticals, diagnostics, tissue regeneration, and cancer treatment, offering glimpses of bionic/cyborg futures.

---

## 14.5 Techno-Benefits and Social Risks

*Barbara Herr Harthorn*

US and UK publics hold diverse, complex, and often conflicted views as they co-imagine and debate potential futures for medical and enhancement nanotechnologies in deliberative workshops. In contrast to fears their governments express about potential public backlash spilling over from past concerns about genetically engineered organisms, food, and other new technologies, both US and UK publics overwhelmingly considered a future embellished by new nanoscale medical technologies to be beneficial, demonstrating a surprisingly pro-technology acceptance of these largely unknown technologies, including those with radically new and different properties, functionalities, and capacities (Pidgeon et al. 2009). People did raise significant concerns, however, related to the unlikelihood of equitable distribution of these goods, and the trustworthy management and wise governance of their safety and potential unintended consequences by corporations and governments.

Nanotechnologies are a large class of engineered particles and structures ranging in size from 1–100 nanometers, where 1 nanometer measures 1 billionth of a meter, with exciting new optical, chemical, and electrical properties at this molecular scale. Scientists and engineers synthesize these nanoscale particles and devices from an array of materials, some of which, like carbon nanotubes, have no corresponding forms at larger scale. In the US and now over 40 other countries around the globe, nanotech development has been a focus of significant public and private investment; in the US, public investment takes place through the National Nanotechnology Initiative (NNI), founded in 2000 (see nano.gov). Nanomaterials are currently being incorporated into an ever-increasing range of products and devices, including an estimated 247 nanomedicine applications as of late 2013 (Etheridge et al. 2013). Yet despite approximately $20 billion investment in US R&D so far, public awareness of nanotechnologies is low (see Satterfield et al. 2009).

How we form trajectories for the future is a critical one in anthropology broadly and in studies of public imaginaries about techno-medical futures specifically. Since Margaret Mead's anticipatory anthropology, anthropologists have asked probing questions about how people make sense of the future across cultural and other boundaries, and how new technologies foster fundamental changes in contemporary culture and society. Cultural and social landscapes of inequality play an important role in such processes, and ideas about risk and the ethnography of contemporary fractured, ruptured experience are a notable part of this story.

## *The Nanomedicine Case*

Nanotechnologies are being developed for use in products such as pharmaceuticals, medical imaging and diagnosis, implantable materials, tissue regeneration, multi-functional systems that incorporate several such capabilities, and especially cancer treatment (Etheridge et al. 2013). The promise of targeted drug delivery at the molecular site of incipient tumors offers therapeutic drug delivery, but far more potentially 'transformative' (and disruptive) futures are anticipated with the introduction of these technologies into personalized medicine and point-of-care diagnostics. For example, biomarkers of a huge array of diseases will be instantly discernable by 'lab on a chip' or 'lab on a pill' devices much as home pregnancy tests and implanted diabetes insulin pumps currently provide local diagnostics and treatment. The unusually strong and lightweight characteristics of nanoenabled materials make them attractive for prostheses, and when combined with nano tissue regeneration, offer glimpses of imminent bionic/cyborg futures.

In 2007 in the US and UK, and again in the US in 2009, my colleagues and I convened a number of small quasi-representative groups of everyday citizens to engage in extended (4.5 hours) facilitated dialogue about the benefits, risks, and imagined future implications of these novel medical technologies. Groups ranged in size from 9–15 participants. We selected across what were then judged to be near-, middle-, and long-term technologies as the main nanomedical topics for discussion: lab on a chip (remote, low cost, low energy, rapid diagnostics at point of care); lab on a pill (multi-function molecular device with capacity to travel undetected within the body to the site of tumors which it can then diagnose and treat in situ with 'nano bombs'); new, stronger, lighter-weight, longer-lasting prostheses (e.g. for hip and other joint replacements); and, the idea of cognitive/memory enhancement via implanted chips in the brain.

In general, people were deeply ambivalent. They had gained sufficient knowledge about these esoteric technologies to enter into informed, reasoned, and in some cases passionate debate about the benefits and risks they potentially posed to individuals and society (Pidgeon et al. 2009). Medical technologies, in comparison with energy technologies, produced a far more individualized response, and risk concerns centered primarily not on the technical risks posed by uncertainties about physical/technical hazards of new materials, but on social risks or societal hazards—particularly the potential for inequitable distribution of both benefits and harms, for invasion of privacy, for loss of care through remote sensing, self-management systems, and for irresponsible or incompetent governance of what many saw as unruly and greedy corporations. Most participants also drew quite sharp distinctions between technologies for restoring 'normal' functioning and those that might enhance abilities beyond the normal, although a minority of participants fully embraced more extreme aspects of potential bodily and cognitive enhancements and, thus, the end of death as we know it.

### *Targeted Drug Delivery*

The cancer context is, unsurprisingly, one with which many participants had had some direct or indirect experience, and perceived urgency for new, more effective and less debilitating treatments tended to reduce perceived risk. For example, in a nanomedicine session in the US in 2009, participants readily agreed that using a risky application to save someone's life was an entirely different thing from using it in 'a cosmetic' or 'another salt shaker.' The touted benefits of earlier tumor detection and molecular-level treatment as a future replacement for full-body chemotherapy and radiation were intuitively appealing to most participants:

> As a healthy individual I think I would be leery of it at first but I know people that have cancer right now and I know that they are just fighting for survival and that their views might be a lot different than mine. They might not worry so much about it because they are not healthy.
>
> *(Jacquelyn)*

Yet technology also evoked many concerns. Some related to the specifics of how this embodied technology would work. For example, issues of controllability focused in particular on information and privacy, including how imperfect or intercepted transmission of information from mobile-embodied devices to medical practitioners could result in loss of privacy (and potentially, loss of rights to medical insurance if the information disclosed a disease):

> I would be okay with the information being used you know on like a general level but if it is used as a specific way to you know discriminate against someone you know with employment or something like that, that is definitely something that is very troubling.
>
> *(Ross)*

> There is one thing I was thinking about, the way we sort of tested Internet technology was to put it out there, and then also when we discovered we had spam, and then we built spam filters. But, I don't know that I, if I had a lab on a pill, I don't know that I would want to use that as a test, figure out how people are gonna hack into that, and then you know, I mean if it's transmitting it could be receiving, it's kind of an interesting thing.
>
> *(Jordan)*

> I think that that [technological controls] is essential, that yeah, because these devices are invisible it is something that the individual it comes prepackaged you cannot manipulate, it is not like a radio that you can turn it on and off. Be willing to develop devices that individuals will have to block information to see if maybe the particular pill is not genuine or . . .
>
> *(Simone)*

Remote sensors, which are likely to become widely used applications of nanotechnologies and include those in lab on a pill/targeted drug delivery, raised heartfelt issues about self-management of care and technological (robotic) substitution for direct, human, hands-on care. Other concerns centered on more general social hazards and suggested underlying ambivalences about the desirability of advancing technological solutions to societal problems in the face of many apparent contradictions posed by increasing inequalities, perceived loss of care, dehumanizing practices, and human foibles and weakness. Trust was a main issue and included questions of trust in technological R&D as well as in medical professionals' knowledge and dissemination. Also, many people questioned whether the new technology was really needed and whether it really would do more than current technology, whether technological fixes could really solve problems, and how cost entered into it. People often wanted the technology but resented the ever-escalating costs associated with this:

> I glanced at this one here [an article] and the thing that really bothered me and I think the whole thing is, it's from the concept of profit. Do we need cosmetics or personal care things which are going too expensive now, more than what, you know, what they really are. To put nano materials in that to further enhance it at double the price now where it leaves. *It's where you use it for making the most money rather than for doing the most good.*
>
> *(Gary)*

> I want to discuss the fact that what we are doing is, we are really looking at the end result and not the cause and that what we are doing is fixing something but without saying why. And I think that there's moral implications with saying why do we need all of these things? If we have bad water or we have pollutants, putting that little pill in or a microchip will get you an end result, but we are still not dealing with the fact that we have these pollutants, that we have issues that we are simply skating over and I think there are areas that still need to be looked at. . . .
>
> *(Melanie)*

In comparative discussions in the US and UK, another difference that emerged was that US participants more consistently adopted a consumerist stance to discuss the meanings of nanomedical technologies. They tended to understand the technology as a commodity, as 'goods' to be acquired, and that stance drove a sense of desirability, of consumer choice as the same thing as personal choice, and of competition for desirable resources. US participants seemed to impute perceived benefit more uncritically than those from the UK, who had a deeper understanding of risk controversies and deliberation as outcomes of past governmental risk management mistakes.

One aspect of this consumerist stance included a 'trickle down' idea about benefit distribution among US but not UK participants—they thought that access would be restricted to the wealthy initially, but would eventually become more universally affordable and available. In the UK, the benefits were seen as remaining with industry and the wealthy.

> These developments will be expensive, that means they could very well be limited only to the monied classes and they won't be available throughout society or the world.
>
> *(Sadie)*

> I think in that sense it might be in a lifetime or two but in the future it's like vaccines, like getting a flu shot. It will be available for future generations, but we have to start somewhere.
>
> *(India)*

Concerns about over-reliance on technology and laziness were recurrent throughout the discussions, and reflected deeper moral issues about the seductions of technologies and human susceptibility to poor choice making when presented with such easy (but costly) options.

> You know, *it's just a quick fix*. You know, instead of me, . . . right now my girlfriend said I'm a little fat, so I could get a pill, a fat burner pill, but I'm riding my bike to work every day and I'm losing weight so it's, you know, what we always want—that pill—we always want to give me something to fix so that I can continue my unhealthiness.
>
> *(Tate)*

## Intelligence Enhancement

US President Bill Clinton introduced the National Nanotechnology Initiative in 2000, when he conveyed the excitement of this nanotechnology 'revolution' by inviting us to "(j)ust imagine . . . shrinking all the information at the Library of Congress into a device the size of a sugar cube" (http://www.nature.com/news/2010/100901/full/467018a.html). In exploring possible future uses of nanotechnologies to enhance intelligence and memory, discussion splintered between imagined enhancement of intellectual capacities in general or to increase specific knowledge. Participants were concerned about what was 'natural' in enhancement technologies, but also what was necessary or unnecessary, and how enhancement could unfairly give advantage to some more than others.

> I think that living forever, it's—I don't think it should, we would become very overpopulated. . . . So we live forever, when people come and then when people born and then it gets more populated and more populated and we just keep living forever. And I think that, I mean it's good to have it, in the sense that it helps you with, you know, give strength. That's good, but to be able to use over- over strength yourself, that's just wrong because you're pretty much, if you want to become a basketball player like he said or baseball player you could, you could become

maybe because of the little nanos that make you stronger, and so people that don't have it, that are trying their best to do it, that's kind of wrong.

(*Yolanda*)

Discussions readily moved in the direction of seeing the potential for increasing inequality through unfair access and possible misuse, while discussions about enhancement versus therapy repeatedly raised issues about whether or not living longer was a social good.

There's moral questions throughout . . . this whole idea is, is ethically a mine field isn't it, for every facet of it. You know will it be like the reparative medical stuff, the enhancement stuff, the fact that if it could be used for this good it can be used to making people clever, it can made, used for making dumb; if it could be used for making people superpowerful, it would be able to make people, you know, fall apart at the seams, and you can't see it, you know. So it's the sort of base terror, isn't it, you're being haunted by something you don't actually know you're being haunted by, because you can't see it, you can't feel it unless you've got those crystals somewhere embedded in your arm or something.

(*Lance*)

The speed of technological change also seemed to challenge "our ethical foundation," as Sam described it: "I'm not sure we have the luxury of time, nanotechnology is changing so fast, the capabilities are increasing so rapidly, that maybe our ethical foundation isn't sufficiently developed to observe, analyze, and make recommendations, on what's happening" (US 2007, male).

## Conclusion

Public narratives about nanomedicines that emerge in deliberative dialogue reflect a distinctly post-modern and fractured set of ambivalences about risk, benefit, and inequality; distributive justice, technological desires, deep distrust of government and/or corporate actors; and uncertainty about public preparedness for full participation in the emergent international project of 'responsible development.' In this description of a series of discussions with diverse groups, social location and identity played a major role as participants debated and staked out positions on the multiple meanings of health and harm, risks and benefits, and the deeper ethical challenges posed by new technologies for therapy and human enhancement. As illustrated in other anthropological work, medical technologies are not stand-alone 'objects' bearing only technical attributes, as often conceptualized and described in the science and engineering world, but rather are inextricably bound to and entangled with the social worlds that they inhabit and through which they are constructed.

## Acknowledgment

This work was conducted in the US National Science Foundation–funded Center for Nanotechnology in Society at the University of California at Santa Barbara (CNS-UCSB), which I direct. This work was funded by NSF through cooperative agreements # SES 0531184 and # SES 0938099 to the Center for Nanotechnology in Society at UCSB and grant # SES 0824024. Views expressed here are my own and do not reflect the views of the NSF. My lead collaborator in this project is Nick Pidgeon, Cardiff University, UK; colleagues Karl Bryant, State University of New York at New Paltz, Jennifer Rogers Brown, Long Island University, and Christine Shearer, University of California at Irvine are the 'we' in this case study. All names cited in quotes are pseudonyms.

## References

Etheridge, M.L., S.A. Campbell, A.G. Erdman, C.L. Haynes, S.M. Wolf, and J. McCullough 2013 The big picture of nanomedicine: The state of investigational and approved nanomedicine products. *Nanomedicine: Nanotechnology, Biology, & Medicine* (9):1–14. Accessed from http://dx.doi.org/10.1016/j.nano.2012.05.013

Harthorn, B., C. Shearer, and J. Rogers 2011 Exploring ambivalence: Techno-enthusiasm and scepticism in US nanotech deliberations. In *Quantum Engagements: Social Reflections of Nanoscience and Emerging Technologies.*

T. Zulsdorf, C. Coenen, A. Ferrari, U. Fiedeler, C. Milburn, and M. Wienroth, eds. Pp. 75–89. Heidelberg, Germany: IOS/AKA Press.
Pidgeon, N., B. Harthorn, K. Bryant, and T. Rogers-Hayden 2009 Deliberating the risks of nanotechnologies for energy and health applications in the United States and United Kingdom. *Nature Nanotechnology* 4(2):95–98.
Satterfield, T., M. Kandlikar, C. Beaudrie, J. Conti, and B.H. Harthorn 2009 Anticipating the perceived risk of nanotechnologies. *Nature Nanotechnology* 4:752–758.

---

Interest in the cyborg in anthropology emerged in the 1990s, in response to developments in the field of assisted reproduction, including techniques such as in-vitro fertilization. Robbie Davis-Floyd and Joseph Dumit (1998), and the authors who contributed to their edited volume, point to the utility of an ethnographic inquiry that does not reject technologically assisted reproduction as unnatural. Instead, they argue, ethnographers should use the metaphor of the cyborg to better understand how reproduction is mediated by different kinds of technoscientific interventions, and the implications of this in terms of biological, cultural, and psychological evolution. More recently, Lenore Manderson (2011: 63) has observed that with the rapid development of bio-engineered options, cyborg bodies have become increasingly ordinary, even expected, just another item on a menu of medical interventions for people seeking treatment or repair.

In this chapter, we have illustrated how developments in the field of genomic medicine and synthetic biology are rapidly changing how we understand our bodies, as well as what we expect from physicians and surgeons, their procedures, and medications. We are offered many techniques through which we can manage our lives, including rapid diagnostic tests that tell us our genetic makeup and that allow for more targeted and tailored therapies. Our cyborg bodies are increasingly commonplace. While geneticists were initially optimistic about finding the biological basis of life, today, however, they acknowledge that much of life is not explained by genetic sequences. Both our genetic makeup and our risk of disease are affected by the circumstances in which we live our lives, and gene expressions are mediated by environment and lifestyle, pointing the direction for new medical research. In personalized medicine, information on our genetic makeup is combined with other biomarkers, bodily traits, and 'lifestyle' data, and combined with up-to-date data on therapeutic efficacies to make treatment decisions. Such advances open up as many social and ethical questions as they resolve biomedical ones, creating a new demand and urgency for medical anthropological research.

## References

American Academy of Pediatrics Section on Hematology/Oncology, B.H. Lubin and W.T. Shearer 2007 Cord blood banking for potential future transplantation. *Pediatrics* 119(1):165–170.
Arkin, Adam 2008 Setting the standard in synthetic biology. *Nature Biotechnology* 26(7):771–774.
Bateson, W. 1906 The progress of genetic research. Third Conference of Hybridization and Plantbreeding, London: 90–97.
Calvert, Jane 2010 Synthetic biology: Constructing nature. *Sociological Review* 58(Supp.):95–112.
Davis-Floyd, Robbie, and Joseph Dumit, eds. 1998 *Cyborg Babies: From Techno-sex to Techno-tots.* New York: Routledge.
Flaherty, Devin, H. Mabel Preloran, and Carole H. Browner 2014 Is It 'Disclosure'? Rethinking tellings of genetic diagnosis. In *Disclosure in Health and Illness.* M. Davis and L. Manderson, eds. Pp. 89–103. London: Routledge.
Happe, Kelly E. 2013 *The Material Gene: Gender, Race, and Heredity after the Human Genome Project.* New York: New York University Press.
Johanssen, W. 1909 *Elemente der Exakten Erblichkeitslehre.* Jena, Germany: Gustav Fischer.

Keller, Evelyn Fox 2001 *The Century of the Gene.* Cambridge, MA: Harvard University Press.
——— 2003 *Making Sense of Life: Explaining Biological Development with Models, Metaphors and Machines.* Cambridge, MA: Harvard University Press.
Klawiter, Maren 2008 *The Biopolitics of Breast Cancer: Changing Cultures of Disease and Activism.* Minneapolis, MN: University of Minnesota Press.
Landecker, Hannah, and Aaron Panofsky 2013 From social structure to gene regulation, and back: A critical introduction to environmental epigenetics for sociology. *Annual Review of Sociology* 39:333–357.
Lock, Margaret 2013 *The Alzheimer Conundrum: Entanglements of Dementia and Aging.* Princeton, NJ: Princeton University Press.
Manderson, Lenore 2011 *Surface Tensions: Surgery, Bodily Boundaries, and the Social Self.* Walnut Creek, CA: Left Coast Press.
Meaney, M. 2011 Environmental epigenetics: How childhood experience regulates your genes. Lecture, Center for Studies and Behavioural Neurobiology, Concordia University, Montreal, Canada.
Mol, Annemarie, and John Law 2007 Embodied action, enacted bodies: The example of hypoglycaemia. In *Biomedicine as Culture: Instrumental Practices, Technoscientific Knowledge, and New Modes of Life.* Regula Valérie Burri and Joseph Dumit, eds. Pp. 87–107. London: Routledge.
Olby, Robert Cecil 1974 *The Path to the Double Helix: The Discovery of DNA.* Seattle, WA: University of Washington Press.
Petrini, Carlo 2014 Umbilical cord blood banking: From personal donation to international public registries to global bioeconomy. *Journal of Blood Medicine* 5:87–97.
Rabinow, Paul 1996 Artificiality and enlightenment: From sociobiology to biosociality. *Essays on the Anthropology of Reason.* Pp. 91–111. Princeton, NJ: Princeton University Press.
Rabinow, Paul, and Gaymon Bennett 2012 *Designing Human Practices: An Experiment with Synthetic Biology.* Chicago, IL: University of Chicago Press.
Rapp, Rayna 2000 *Testing Women, Testing the Fetus: The Social Impact of Amniocentesis in America.* New York: Routledge.
Rheinberger, Hans-Jörg 1997 Experimental complexity in biology: Some epistemological and historical remarks. *Philosophy of Science* 64(Supp.):S245–S254.
——— 2004 Introduction. In *Classical Genetic Research and Its Legacy: The Mapping Cultures of Twentieth-century Genetics.* Hans-Jörg Rheinberger and Jean-Paul Gaudilliere, eds. Pp. 1–5. New York: Routledge.
Rose, Nikolas, and Carlos Novas 2005 Biological citizenship. In *Global Assemblages: Technology, Politics, and Ethics as Anthropological Problems.* Aihwa Ong and Stephen J. Collier, eds. Pp. 439–463. Malden, MA: Blackwell.
Strathern, Marilyn 1992 *After Nature: English Kinship in the Late Twentieth Century.* Cambridge, UK: Cambridge University Press.
Venter, J. Craig 2007 *A Life Decoded: My Genome, My Life.* London: Penguin.

Dr. Rahim and Nepalese Boy after Earthquake, 2015. Yangri, Nepal.

## *About the photograph*

*Dr Fahim Rahim, a nephrologist, helps a sick child in Yangri, Nepal, following the 7.8 magnitude earthquake that struck the area in May 2015. Nearly one billion people living in remote parts of the world lack access to adequate biomedical care, and in times of natural disaster, they become significantly much more difficult to reach to provide aid. An unknown observer took the photograph for him.*

—Beenish Mannan

# 15

# How the Logics of Biomedical Practice Travel

*Elizabeth Cartwright, Anita Hardon and Lenore Manderson*

Biomedicine travels; it is a malleable and mobile set of practices and ideas. It travels with practitioners and is used in far-flung places. It travels via educational strategies and via the use of accepted and codified texts that set the 'standards' of diagnostics and treatment. Educational approaches travel out from large, well-funded teaching institutions to rural areas, then to more remote and less well-resourced areas of the globe. Biomedicine travels via books and training sessions and via the Internet in and out of clinical spaces, weaving itself into homes, public buildings, and the most remote nooks and crannies of the world. Practitioners and patients, who in the past had much less access to biomedical technologies, use it in ways unique to local situations. Many times, in low-resource settings, biomedical protocols are enacted using only parts of the system for which they were originally designed. Inadequate hospital infrastructure, lack of medications, difficulties in the procurement of medical supplies, and lack of fully trained personnel all create situations that result in incomplete diagnostic and treatment regimens and less than optimal outcomes, as we have noted already.

In this chapter, we trace some of the many ways that the logics of biomedical practice travel. The case studies we include show how biomedicine travels via changes in medical education curricula (Georges and Davis-Floyd), changes in market demand (Gerrits and Hörbst), via medical education experiences (Benton), and with humanitarian relief efforts—with their ambiguous and sometimes contested results (see the cases from Frankfurter and Redfield). In each of these case studies, we see that biomedicine is not the unassailably dominant force that we might think. Rather, it is a way of being and doing the work of healing that, when confronted with different local contexts, results in a plethora of hybrid ideas among both patients and practitioners. Biomedical elements are grafted onto partial clinical infrastructures, pharmacopeia, and various pre-existing systems of healing (see Karchmer 2010).

One response to the perceived 'need' for biomedical 'help' in resource low settings is for biomedical practitioners from the global North to bring their skills directly to the region in question. This occurs in two ways: by offering care to people during natural disasters, wars, and epidemics, and by providing intermittent medical and surgical interventions where such care is not locally available—fistula repairs, cataract surgery, oral and maxillofacial surgery, for example. Local health care systems and public health programs may be underdeveloped or under-resourced, and medical staff lack specialist training, hence the role of external providers to work with locals to strengthen capacity, as occurs through development assistance with international NGOs such as Save the

Children and World Vision. Other international organizations provide more specific services: the Fred Hollows Foundation, for instance, focuses on treatable and preventable vision problems; Hamlin Fistula Ethopia operates primarily to treat obstetric fistula for women and girls in that one country; Mercy Ships provides surgical and medical care off the shores of Africa. In other emergency contexts, including where health systems are damaged or destroyed, groups such as Médecins Sans Frontières (MSF) and other nongovernmental organizations (NGOs), and the militaries of donor nations, necessarily care for large numbers of people under these circumstances.

The complexities of humanitarian interventions have recently come under study by medical anthropologists (Bornstein and Redfield 2010); the meanings of donating time, money, and expertise differ culturally, as do expectations of the continuation of aid and the actual uses of donated money and goods. Often, local practitioners work side by side with humanitarian volunteers, learning new aspects of biomedicine within the context of the unfolding crisis or in managing particularly challenging surgical and other procedures. The details of how biomedical knowledge is passed between practitioners in situations, like the Ebola outbreak of 2014 that took place in Sierra Leone, Guinea, and Liberia, illustrate how biomedical knowledge is disseminated, understood, and retained over time in settings where there are multiple, often competing forms of ethnophysiological understandings. We return to the topic of Ebola towards the end of this chapter.

## Biomedical Interruptus

While doctors, surgeons, and students training to be health care professionals have taken biomedical treatments to people in need of medical assistance—sharpening their own skills in the process—there has been limited progress in providing adequate medical training for local residents of resource-poor areas to become fully trained in medicine, nursing, dentistry, and midwifery, or in public health (Ezeh et al. 2010). Education is expensive, often requiring those who want to become health care professionals to relocate to urban areas, far from their rural or otherwise marginalized homes. The expense, time lost from family obligations, and inadequate educational preparation during primary and secondary school create situations where academic success, especially at higher levels of education, is very difficult to achieve even for the most intelligent and dedicated of students. Health care professionals in large urban centers often don't want to or can't leave their families and jobs to relocate to more rural/underserved areas; thus, extensive regions of the world are 'served' by clinics that are understaffed and under-resourced, and which are only minimally effective at best.

The difficulty of enticing health care professionals to work in rural areas has been addressed in part by transporting patients to the nearest urban center with some semblance of hospital facilities. Transporting patients to urban hospitals can be difficult, dangerous, and expensive; untrained taxi and bus drivers often provide patient transport because there are no medically outfitted ambulances—let alone functioning emergency medical systems (EMS) that include medics, communication, triage and supportive medical decision making and treatment during the long hours of transport to the nearest, best-equipped hospital. Even those urban hospitals are often incapable of treating the patients that arrive at their doorsteps, due to lack of facilities, general medical staff, specialist skills, medications, and supplies. Moreover, poor conditions of employment and limited opportunities for on-the-job training make it difficult for personnel to do their jobs well. These deficiencies create clinical landscapes that have meager assemblages of possibilities—both physical and intellectual.

Livingston (2012) ethnographically describes one of these partially equipped, resource-poor hospitals in her book *Improvising Medicine: An African Oncology Ward in an Emerging Cancer Epidemic*. She takes us into a barely functioning ward in Botswana where patients arrive with advanced cancers: they have tumors protruding from their bodies, they are cachexic and in unbearable pain. At the hospital they receive the most basic, and oftentimes futile, care. The practitioners in her book, the doctors and nurses and aides, work in conditions where they are called upon to personally sacrifice a great deal for their patients, and they do. The scenes she describes are beyond imagining for the lucky minority of the people on this planet who live in well-resourced settings. For the majority of humanity, these scenes are all too familiar.

Non-standard clinical practices emerge where there is less institutional and/or administrative oversight; clinicians who dwell in the electronic periphery of Internet connectivity are operating at and with a disadvantage. Many of the case studies in this chapter, and indeed, in the entire volume, are located in this Internet-outback. Practitioners have incomplete tool sets and experience difficulty upgrading their skills and knowledge compared with those who reside in core connectivity spots and have easy access to such information, as we considered in Chapter 8. These landscapes—physical and virtual—are superimposed upon one another. Physical spaces are accessed by roads, cars, planes, mini-buses, and foot, resulting in a particular geographic distribution of medicines and diagnostic tools that define how biomedicine is experienced by practitioners and their patients. This physical landscape is now overlaid by an Internet landscape. The quality, reliability, and cost of Internet services create a filter—one that can change quickly for better or for worse depending on the quality of the telecommunications systems upon which they depend. These two landscapes—the physical and the virtual—when viewed simultaneously, can work together to create more access to biomedical understandings for those on the periphery, as we have described in Chapter 8, or they can cut off communities of practitioners and their patients from many aspects of biomedicine that could improve their practices and their outcomes.

## Exchanging Contesting Paradigms

The cases in this chapter illustrate the ways that the techniques and bodily movements of clinical practice travel and are shared between practitioners from very different places and clinical spaces. These new ways of enacting biomedicine are taught and re-taught in settings that have very different access to the tools of the trade—pharmaceuticals, medical instrumentation, and biomedically trained staff. Nia Georges and Robbie Davis-Floyd, in the case study that follows, illustrate that the transferring of medical paradigms through education and training is not a one-time event. Contesting biomedical (or slightly less-biomedical) perspectives on the manner in which practitioners should interact with their patients can be entertained and sometimes taken on as the definitive clinical protocol in light of the arrival of new biomedical ideas and information. Georges and Davis-Floyd illustrate how a critique of biomedicine travels from midwives and alternative birth practitioners in the US into standard, hospital birthing protocols in Brazil.

---

## 15.1 Humanistic Obstetrics in Brazil

*Eugenia Georges and Robbie Davis-Floyd*

*Question:* Where does your passion come from?
*João Batista:* From my heart. I believe in justice and rights, and women in Brazil have neither, and their rights are most violated during birth, even the rich ones. It's ironic that the rich women have the worst treatment and they believe they have the best. . . . I want every ob professor to teach this to his residents: The woman is the center of care!

Throughout Latin America, 'humanization' is the unifying concept around which social movements to demedicalize and transform childbirth care have coalesced. In the 1980s, feminist, environmentalist, human rights and other oppositional movements began to proliferate in the democratizing aftermath of authoritarian rule. Since then, Brazilian humanization activists, like their counterparts in other countries, have engaged in ongoing political struggles to produce new meanings and cultural categories with which to challenge dominant medical discourses (Alvarez et al. 1998). Although a diverse project, humanization at its core is grounded in the principle that safe, respectful and supportive maternity care, free of unnecessary medical interventions, is a woman's human right as a citizen of the nation. Following this logic, unnecessary obstetrical interventions have come to be redefined as a category of violence against women—'obstetrical violence.'[1] The humanization movement insists that care of mother and infant consist exclusively of 'best practices' that are supported by up-to-date, evidence-based medicine. The objectives of humanization are simultaneously ethical (promoting women's rights to respect and autonomy and committing doctors to protecting those rights) and instrumental (improving outcomes for mother and infant).

Here, we describe our research with 32 Brazilian obstetricians (half men, half women) who have chosen to replace the conventional obstetrics in which they were trained with a humanistic approach.[2] Humorously, they call themselves 'the good guys and girls.' For some, the preponderance of the scientific evidence in favor of supporting normal, physiologic birth was most important in their decision to transform their practices; for others, it was allegiance to humanistic values rooted in their intersecting affiliations with feminism, left-wing politics, liberation theology, environmentalism, traditional medicine and midwifery, and New Age or older spiritual beliefs. Despite their differences, however, they typically shared an ethos that blended, or at least honored, all these dimensions to varying degrees. Some had private practices that catered to highly motivated, middle-class women—those who, in one doctor's words, were "100 percent to 10,000 percent committed to normal birth." These interviewees attended homebirths almost exclusively and had cesarean rates of 7 to 10 percent. Others in private practice had cesarean rates of 14 to 30 percent, because they chose to care for all who came to them, believing that every woman has the right to a humanized birth, whether a natural birth at home or in hospital or a scheduled cesarean section. Others worked in the public health system, providing care mainly for poor women. Some interviewees occupied influential positions in the Ministry of Health; others held faculty positions at some of Brazil's most prestigious universities.

For example, Ivo Lopez and João Batista worked as clinical directors of Hospital Sofia Feldman in Belo Horizonte, Minas Gerais, a high-risk center providing tertiary-level care available to all women in the state. More than 50 nurse-midwives practice there along with 20 obstetricians; 14 doulas are provided by the hospital for women who bring no companion. They have very low rates of forceps, vacuum extraction, and episiotomy (around 8 percent for each); electronic fetal monitoring is used only intermittently; delayed cord clamping and vertical births are the norm. Sofia Feldman is widely regarded as an exemplar of humanistic practice in the international childbirth movement. In December 2012, Director João Batista stated:

> Our cesarean rate is 24 percent . . . one of the lowest in Brazil—but last year it was only 20 percent—it went up four percent in one year [when] there was a huge increase in high-risk women sent to us from all over the state. We are the biggest high-risk hospital in the state—so our PNMR is 14.4/1000. Ten percent of our babies are preterm. More than 50 percent of the babies here are from other cities. Maybe we are doing the best we can. [Preterm labor and premature rupture of membranes are common—hypertension is the really big one]. If they come to us early, we can stop it, but most are sent late.

When we asked him, "Are you afraid of birth?" he responded, "No. I am afraid of not doing all I can for women."

## *Guidance from Books and Films*

Working in relative isolation because they did not know of each other when they began their new trajectories (mostly in the late 1980s and the 1990s), these humanistic obstetricians undertook a process of self-education to effect their personal and professional transformations. In almost all instances, this process was profoundly influenced and guided by a common corpus of resources, including texts by international advocates of childbirth reform. These included the works of French doctors and reformers Frederick Leboyer and Michel Odent, the Uruguayan obstetrician Roberto Caldeyro-Barcia, British authors and activists Sheila Kitzinger and Janet Balaskas, and US writers

Marsden Wagner, Penny Simkin, Marshall Klaus, Henci Goer, Elizabeth Davis, and Robbie Davis-Floyd. Melania Amorim explained:

> I started my Master of Science Program [in 1993]. And I was introduced to evidence-based medicine. I also had to read philosophy of science and I was introduced to Thomas Kuhn and research on scientific revolutions. And I began to think, "Oh! It can be different." And I started to read about the [standard obstetrical] procedures . . . and I could realize that they were not necessary! That there was no evidence to support my practice. But I couldn't realize, why, with so much evidence against these procedures, they always have been performed by other obstetricians. . . . And so finally I encountered Robbie's book *Birth as an American Rite of Passage*, and I could understand what evidence-based medicine didn't explain—why these procedures have been maintained for decades with no scientific evidence. And that's why her book changed my life.

There were domestic critiques of conventional childbirth too. Moysés Paciornik's book *Aprenda a Nascer e a Viver com os Indios* (*Learn How to Live and Give Birth Like the Indians*) and film *Birth in the Squatting Position* (1979), as well as the birthing chairs he designed to simulate the benefits of the squatting position (*parto salvagem*, 'savage birth'), had a significant impact on alternative birth activists not only in Brazil but also internationally. Hugo Sabatino, professor of obstetrics at UNICAMP, was also an early pioneer whose teachings and writings on squatting birth were highly influential in stimulating the movement for humanized birth. And Dr. Jose Galba de Araujo, in the state of Ceará in the 1970s and 1980s, worked to humanize hospital birth and provide support to the rural midwives in his catchment area; his work remains influential today. Not only did these texts and individuals serve as exemplars—some of their international authors, such as Odent, Wagner, and Davis-Floyd, themselves developed close ties to the humanization movement, visiting Brazil on numerous occasions, offering workshops and delivering keynote speeches at conferences.

## Guidance from Midwives

Another source of guidance was the revaluation of midwives. The contributions of traditional midwives were recognized, and occasionally romanticized, as part of the more general valorization of the experiences of subaltern and Indigenous women (Tornquist 2004: 62). In some rural areas, especially the Amazon region, Indigenous midwives still play important roles in childbirth. Some of the older obstetricians we interviewed expressed respect and appreciation for the skills of the midwives who had assisted their mothers and other relatives. Two of our interviewees, Claudio Paciornik (son of Moysés) and Bernadette Boussada, had learned directly from traditional midwives. Claudio and his father spent years observing the birth practices of the Kaigang and Guarani tribes on Indian reserves in Ibirama and Xanxare in South Brazil during the 1970s. Bernadette, in contrast, met traditional midwives for the first time when she was assigned after residency as the sole obstetrician in the only maternity hospital in the remote city of Coraçao de Jesus, Minas Gerais, for one year (1997–1998). Three *parteiras tradicionais* working in the hospital as auxiliary nurses became both her assistants and her guides. The two most important things they taught her were respect for the woman and respect for the physiology of birth. She watched them treating women with kindness, attending labor without interfering, and using movement and upright positions. She slowly started to practice more and more like they did. The first intervention she gave up was Kristeller (pushing on the abdomen), then forced pushing (yelling at the woman to "Push, push!"), then episiotomy, "then all the rest of it all at once!"

In addition to supporting the remaining Indigenous midwives in Brazil, all of our interviewees believed that ideally, professional midwives should be the primary birth attendants in Brazil because they are or should be the experts in normal physiological birth. When there are professional midwives in their cities, our interviewees are generally happy to work with them and support them however they can. They lament their scarcity (professional midwives attend less than 10 percent of Brazilian births) and some actively work to create and support midwifery programs and to develop birth centers in which the 'midwifery model of care' can be actively implemented. They try to practice the midwifery model, based on keeping the woman at the center and on understanding and facilitating normal physiological birth with a minimum of interventions. Some even call themselves "midwife-obstetricians," yet Carla noted, "I would like to say I'm a midwife, but I can't say that because I have this little monster in my head saying, 'Oh, shouldn't I do an episiotomy?' I'm an obstetrician, I'm a teacher." In turn, given their general ostracism by their technocratic colleagues, these obstetricians usually receive their primary emotional and psychological support from their local midwives, doulas, and birth activists.

## *Social Movements and Networks*

In 1993, ReHuNa (Rede pela Humanização do parto e Nacimento), the Network for the Humanization of Childbirth, was officially formed as an NGO to advocate for a woman's right to a humane and respectful childbirth experience grounded in scientifically sound practices. ReHuNa is composed predominantly of obstetricians, gynecologists, public health doctors, nurses, professional midwives, and other highly educated professionals. Their dual identities as professional experts and movement activists have enabled many members to move between the state sector and civil society and participate in the democratic institutions constructed to bridge them. Our interviewees credited ReHuNa (which some of them were instrumental in creating) and other activist networks as critical sources of knowledge, support and guidance. Many met each other for the first time at ReHuNa conferences, and were delighted to discover that they were part of a community of like-minded alternative practitioners. Today humanistic obstetricians keep in touch via Facebook and other social media, where they engage in vigorous debates over such issues as whether it is wise and safe to assist a woman with two or more previous cesareans to give birth vaginally, how best to turn a breech baby, what constitutes optimal nutrition during labor, and so on. Melania Amorim described her discovery of the Internet chat group *Parto Nosso* (Our Birth) as "magic. . . . It was a revolution in our life because you could find persons whose thoughts were similar to ours. We were not crazy. We were accepted, we were stimulated." When they experience ostracism or outright persecution from their technocratic colleagues (often called *cesaristas* for their routine use of cesareans, as opposed to *vaginalistas*—another term by which our 'good guy' interviewees may identify themselves), they turn to these activist networks for both advice and support.

## *The Process of Transformation*

Even for those whose narratives of transformation highlight a pivot point—an 'aha moment' in which their perspectives abruptly shifted—changing their day-to-day practice was in almost all cases a slow and painful process that required the conscious choice to dissociate themselves from their habitual forms of thinking and acting. As Ricardo Jones (2009: 292) eloquently put it, "between rationality and fear, there was a considerable space still to cross." Changing deeply embodied knowledge and habits typically occurred "step by step, step by step" (Paulo Batistuta) and could take years to complete. For many, the episiotomy was one of the hardest routines to give up. Carla Polido noted:

> [As a student at UNICAMP] I saw squatting births with Dr. Hugo [Sabatino] but it didn't influence me—the other professors said he was a lunatic, "That's for savages!" At the time I also thought that those women were too demanding. Today I *want* women to have demands, but at the time, I was Dr. House—I was the boss. We didn't value the woman, we thought that we were the stars of the show, not her. From there was a whole process, one birth after another. The toughest thing for me to stop doing was the episiotomy. I had this fear of the perineum blowing up!

Even once convinced that a procedure was unnecessary, finally relinquishing it was sometimes still a struggle, as Marilena Pereira described regarding a homebirth she attended:

> Many times, I still couldn't believe it would be OK not to cut. When the birth would start, I'd say to myself, "It's going to tear, it's going to tear," and I would cut. Sometimes I would succeed in not cutting, but I was afraid, really afraid, so many times I would cut. [It took me] the next two years to give up doing episiotomies.

Yet give them up they did, soon discovering that upright positions were much more likely to prevent vaginal tears. The routine use of Pitocin to speed labor was quick to go; hard to learn was the patience it takes to attend births without rushing the process. Helio Bergo noted:

> When I started with squatting birth . . . I saw that there was no need any more for forceps and other aggressive and invasive methods. . . . Little by little I was learning to watch the rhythms and this is what enchanted me. Do not intervene. [Just] watch . . . I learned from attending homebirths. To respect the rhythm was the hardest thing to learn. And it's what brought me the most benefits. When I learned how to do that, things became easy. But it was hard to let go . . . to not intervene. Because there are moments when, without any doubt, it is necessary to intervene. To know, to decide which moment it is, without intervening prematurely, is difficult.

During residency, Carla Daher was shocked and horrified that thirsty laboring women were not allowed to drink. She began to offer them water and to encourage them to move around. Like Helio Bergo, she learned from a sympathetic colleague "to be patient, to just watch women work, to be nice." She gave up manual extraction of the placenta because she saw it as "unnecessary and cruel." As a result, her obstetric chief called her practice "messy" and made sure she was the only resident in her cohort not hired by the hospital from which she graduated. (She later went into private practice.)

For our interviewees, giving up routine interventions meant placing great emphasis on treating their patients as individuals. The five-minute standard question and answer sessions were replaced with prenatal visits lasting an hour or more, childbirth education classes which they often taught themselves, and discussion sessions in which their patients could come together in groups and ask all the questions they wished. Those in private practice made conscious decisions to limit the number of patients they took on so they could provide high-quality humanistic care. They accepted lower salaries (and often ostracism from other obstetricians) as the price to pay for the much greater levels of satisfaction they experienced.

## *Conclusion*

Although their numbers are small, these obstetricians and their colleagues have managed to provide options for humanistic maternity care, including homebirth, in nearly every major city in Brazil. They have succeeded in creating a few exemplary institutionalized humanistic practices such as Hospital Sofia Feldman and the Maternity Department of the University of Santa Catarina. Some of our interviewees are promoting large-scale policy change via influential positions in the Ministry of Health. Some are instrumental in Rede Cegonha (the Stork Network), inaugurated in 2011 to transform maternity care throughout the nation's Unified Health System (SUS), within which 76 percent of Brazilian births take place (Brazil SUS nd:10). This program is designed to lower maternal mortality and the national caesarean rate, which in 2012 averaged 55 percent nationally and reaches 80–95 percent in private hospitals,[3] by replacing the dominant technocratic model of maternity care with a humanized paradigm. Our research on the roles of our interlocutors in this new program and on their unfolding transformations in practice is ongoing and will be further presented in future publications.

## *Notes*

1. In 2010, for example, Venezuela became the first nation in the world to recognize obstetrical violence as a legal category and punishable offence.
2. We use the terms 'humanized' and 'humanistic' for the sake of clarity and simplicity. Many of our interviewees considered themselves to be not only humanistic (keeping the focus on the woman, treating her with kindness and respect), but also holistic (perceiving the body as energy and birth as a spiritual process) (see Davis-Floyd 2001).
3. Maria Esther Vilela, Women's Health Coordinator, Brazilian Ministry of Health, personal communication, March 26, 2014.

## *References*

Alvarez, S.E., E. Dagnino, and A. Escobar 1998 Introduction: The cultural and political in Latin American social movements. In *Cultures of Politics/Politics of Cultures: Re-Visioning Latin American Social Movements*. S.E. Alvarez, E. Dagnino, and A. Escobar, eds. Pp. 1–32. Boulder, CO: Westview Press.

Brazil, SUS (Unified Health System) n.d. Manual Pratico para Implementacao da Rede Cegonha. Accessed from www.saude/mt/gov.br/arquivo/3062

Davis-Floyd, R. 2001 The technocratic, humanistic, and holistic models of birth. *International Journal of Gynecology and Obstetrics* 75(Supp 1):S5–S23.

Jones, R. 2009 Teamwork: An obstetrician, a midwife, and a doula. In *Birth Models That Work*. R. Davis-Floyd, L. Barclay, B.A. Daviss, and J. Tritten, eds. Pp. 271–304. Berkeley, CA: University of California Press.

Tornquist, C. 2004 Parto e Poder: el movimento pela humanização do parto no Brasil. Ph.D. dissertation, Universidade Federal de Santa Catarina, Florianópolis, Brazil.

In the case study above, we have seen how a *critique* of biomedical obstetrics travels. Through books, seminars, and writing in scientific journals and through social media, the dissident voices of midwives and other alternative birth practitioners and advocates from the US, Canada, and Europe have influenced colleagues the world over. These authors show how the highly intrusive medical techniques routinely practiced by mainstream biomedical obstetricians are being challenged and how a critical consciousness is developing among some obstetricians in Latin America with respect to unneeded interventions in the birthing process.

Elizabeth Cartwright has been working with a group of US doctors who are carrying out a series of teaching seminars on the implementation of diagnostic sonography in rural Peru. Using a sonogram machine that was donated to the small clinic a few years ago, and never put into clinical practice because none of the Peruvian practitioners knew how to operate it, the medical doctors associated with Cartwright's projects have shown Peruvian doctors how to do obstetrical examinations with their sonogram machine. The curiosity of local Quechua-speaking women about the images and what can or cannot be seen about their babies importantly informs how the medical practitioners use the technology within their clinical setting. The discourses surrounding childbirth are changed with the introduction of this machine, notions of risk are different with the sonogram, and expectations and blame now include the 'machine.' This is similar to the process that occurs in clinical settings where monitoring equipment like the electronic fetal monitor is used (Gammeltoft 2007; Georges 2008). Practitioners become dependent on the output of the machine to assess maternal and fetal wellbeing (Cartwright 1998), while at the same time losing the more embodied and experiential knowledge and ways of seeing and assessing a birthing woman (van der Sijpt 2014).

Technologies that fundamentally change biological processes, such as the following case study on the transfer of the technology of assisted reproduction to sub-Saharan Africa, are also moving on an increasingly large scale to areas of the globe that formerly did not have access to them. The innovators that establish clinics and hospitals based on new technologies stand to earn a great deal of money, as they change local practices in related fields and create a medical workforce that has new skill sets, educational needs, and future aspirations. In the following case study, Gerrits and Hörbst highlight how ARTs have opened up a lucrative entrepreneurial space in sub-Saharan Africa.

## 15.2 Entrepreneuring Barren Grounds

*Trudie Gerrits and Viola Hörbst*

Assisted reproductive technologies (ARTs) are increasingly becoming available in the global South. In sub-Saharan African countries, they are predominantly offered in private clinics. Private clinics providing these and other new technologies have dramatically changed the landscape of health care in the region, forming an integral component of African health care in the twenty-first century.

In Ghana, private clinics offering ARTs have become a 'booming' business over the last decade.[1] These clinics have mainly been founded by gynecologists trained abroad in assisted reproduction, particularly in Europe, before returning to Ghana to start their own clinics.[2] This phenomenon suggests that the 'brain drain' is not a straightforward phenomenon; while some well-trained African professionals leave their countries to work elsewhere, leaving the local health care system without their expertise, others return as medical entrepreneurs.

The rise of the private health sector in sub-Saharan African countries (including Ghana) occurred as part of structural adjustment programs in the 1980s and 1990s. This increase of free health markets in the global South is often criticized for boosting inequalities in health care. In the field of ARTs, private clinics are sometimes seen as exploiting poor and desperate women and men in need of fertility treatment. The high costs of these treatments, and the (ab)use of gamete donors and surrogates, are usually discussed in terms of "stratified reproduction" (Inhorn and Birenbaum-Carmeli 2008). Here, we show how the initiatives of these entrepreneurial doctors trigger professional opportunities and aspirations and new forms of entrepreneurship. The cases we present also provide a glimpse of the cultural changes induced by these entrepreneurial initiatives. Thus, we revive the anthropological—transactionalist—notion that conceptualizes entrepreneurs as individuals acting out of self-interest and by doing so, forming an impetus for cultural change (Barth 1967). We suggest, too, that the various entrepreneurs involved in the business of ARTs hold structurally different positions, which make them unequal actors in this field.

Below, we draw on the life stories of four professionals involved in the market of ARTs, and show how this created professional opportunities and aspirations in life. These stories highlight the prospects of different actors in the ART business: they concern a gynecologist; an embryologist; the director of an agency intermediating between 'wish parents,' donors, and surrogates; and a woman acting as a surrogate. Pseudonyms are used for the professionals and the clinic.

## *From Gynecologist to Clinic Director*

Dr. Aidoo founded the LeLena Clinic in 1985. Before that he worked for 19 years in Germany where he studied medicine and specialized in gynecology. He decided to return home and to invest his savings to start a gynecological clinic in Ghana. He felt that "as a medical doctor he could be more useful in Ghana than in Germany." At that time, Ghana experienced dramatic economic decline and political unrest: "The conditions were very, very bad: hunger, and there was a *coupe d'état*. The economic situation was very, very bad. . . . Apart from that, it is not by chance that God had me born here . . . I have a duty here."

Back in Ghana, Dr. Aidoo found that infertility was a major problem, and he returned to Germany again to further specialize in in-vitro fertilization (IVF). In 1995, he carried out the first successful IVF in Ghana in his own clinic. His successes received a lot of media attention—in particular the births of twins, triplets, and higher-order multiples. His reputation grew rapidly in Ghana and beyond. His patients now come from all over Ghana and from neighboring West African countries, such as Gabon, Cameroon, Ivory Coast, and Mali. Some of his clients were Ghanaians living in Europe and the US, returning 'home' for treatment.

Over the years, with bank loans, Dr. Aidoo gradually expanded his clinic. In 2012 the clinic performed a total of 630 IVF treatments, and by 2013, the clinic employed around 130 staff members, including health professionals, administrative and support staff. In addition to 'conventional' IVF, the clinic offered treatments using donor gametes and surrogacy. The number of treatments per annum decreased in recent years, probably due to the recent presence of other fertility clinics: Dr. Aidoo believes that his success encouraged others to get involved in the IVF business.

This growing competition did not prevent Dr. Aidoo from investing in the future of the clinic; rather it made him into a flexible entrepreneur. In 2012 he started constructing a new building to expand the fertility clinic. Given his new competition, he decided to turn the building into an intensive care unit, another area of health care in Ghana that, he claims, needed urgent attention.

While he presents himself—again—as a pioneer, addressing the exigent health needs of the Ghanaian population, most Ghanaians cannot afford the treatments in his clinic. Most of his clients were middle class, highly educated, and/or own or work in private companies. LeLena Clinic had served the fertility needs of the happy few and generated employment for many people (within and beyond the clinic); and the clinic brought prosperity to Dr. Aidoo, as a taxi driver commented by pointing to his Mercedes Benz, parked in front of the clinic: "He has a huge car park—like the pop stars in America—and many properties too. Dr. [Aidoo] is a very, very wealthy man."

## *From Laboratory Technician to Clinical Embryologist and More*

Embryological laboratory work is an indispensable part of IVF treatments, but a scarce specialization in Ghana (and the entire African continent). Clinic directors have to invest heavily to recruit international embryologists and/or in training local laboratory staff. In 2013, Richard was in charge of the IVF lab work at LeLena Clinic. He started working there in 2001, initially in the general laboratory, but soon Dr. Aidoo gave him the opportunity to work in the IVF lab. The senior lab technician was supposed to train him on-the-job, but this did not work out well, as Richard complained: "For 3 years I was only observing." When the senior technician suddenly left the clinic in 2004, Richard had to take over his job, but felt ill prepared to do so. Dr. Aidoo hired a German embryologist to train him, both on-the-job in the clinic and in Germany. Additionally, Dr. Aidoo sent Richard to international conferences and training courses abroad. Richard repeatedly stated: "I am very grateful to all people that helped me achieve this, and I owe it to God as well. I am very grateful to Dr. Aidoo."

Recently Richard obtained his Master's in Clinical Embryology from a UK university (that he financed himself), as a result of which he is one of the best-educated ART professionals in Ghana. Upon his return, new opportunities and aspirations soon developed. He started teaching at university level and is now dreaming about "picking the excellent students and training them to work in clinics." He plans to do his PhD and envisions an academic future as a university professor in human embryology. Reflecting on his prospects, he confirms that he is "thinking of the money too," as do others in this field. Some months after Gerrits's last fieldwork visit, Richard explained on the phone that he had been dismissed from the clinic for unclear reasons. Yet, with his scarce embryological expertise, he had full confidence in his future employability.

## *From Infertile Woman to Director of an Intermediary Agency*

Agencies acting as intermediaries between wish parents and gamete donors and surrogates form a recent spin-off of the introduction of ARTs in Ghana. Both clinics Gerrits studied offered treatments with donor materials and surrogacy, and made use of intermediary agencies to treat non-medical issues.

The director of one such agency is Janet, a Ghanaian woman. For a long time she resided in the USA, where she underwent two failed IVF treatments. In 2001 she returned to Ghana to do further IVF, but again the treatment failed. When she asked her Ghanaian gynecologist about the possibility of surrogacy in Ghana, he replied that technically he could do it, but he added: "Where in Ghana are you going to find a surrogate? Because of our culture, it is so difficult." However, Janet managed to find a woman who was willing to carry her embryos to term and Janet became the happy mother of twins. Her lawyer settled the legal issues, which was not easy as Janet explained: "In Ghana, of course, it was new. So, we didn't have any laws that say you couldn't do it, at the same time we did not have any laws that said you could do it."

At the request of her gynecologist, Janet started to think of ways to help other couples find a surrogate. A year later, she started her own agency, but in 2004 she did so "with big media attention," as she did not want it "to be a secret thing." In 2013 the agency had intermediated the birth of 42 children (most of them twins or triplets) using surrogacy and the birth of approximately 300 children using egg and sperm donation. The donors were university students who were in need of money to pay for their fees; the surrogates were single mothers, minimally 25 years old, who wanted "to improve their life conditions." Janet characterizes the major part of her clients as "career people, like lawyers and bankers. People, who . . . have spent all their life in school or in their career. By the time they realize they are forty, or forty plus, then it dawns on them that 'this is my problem.'" While initially acting out of self-interest, Janet introduced new (cultural) practices into the Ghanaian society, which in turn enabled her to run a profitable business.

### *From Single Mother to Surrogate to the Owner of a 'Chop Bar'*

Akuba is one of the women whose temporary job consisted of carrying the child(ren) for so-called wish parents or owners. When Gerrits first met her, she was pregnant with twins and hospitalized in LeLena Clinic, where she was supposed to stay until delivery. Akuba explained that a taxi driver had informed her about the work of donors. She had been desperate: her husband had passed away in a car accident when she was carrying her third child and his family had left her without any financial support. Her job at a filling station provided her with insufficient income to care for her children. After the conversation with the taxi driver, she visited LeLena Clinic, where she learned that she was qualified to be a surrogate—she was single and had given birth to at least one child. After passing the intake interview and a series of physical examinations, she was accepted. She received some information about the process (from embryo transfer to C-section) and about the babies: "They let you know that the babies are not for you. They give somebody's eggs to you. You know that. The eggs are not from you. The sperm is not my husband's. It is not my brood. I just carry them in my body." Moreover, Akuba was told that she was not to be in contact with the 'owners,' and would not see the children after birth.

Akuba only informed one person about her stay in this clinic in Ghana; others were told that she was traveling to South Africa to work. During her stay in the clinic—where she felt very well treated—a relative took care of her children, who she missed enormously. A fortnight after she delivered the twins, she was thrilled to return home, taking presents "from South Africa" for her children.

Akuba did not regret having been a surrogate. When Gerrits met her again (four months after the delivery), Akuba proudly showed her the 'chop bar'—a small street restaurant—she had set up. She had also bought a plot to build her own house and had rented an apartment; and she was proud that she could now pay school fees, buy better food, and meet the health expenses for her children. Akuba had been able to sustainably invest the money earned through surrogacy; some other surrogates, though, were less fortunate: they had to spend their money paying debts, and meeting the costs of funerals and relatives' health bills.

Akuba, though, was concerned that her chop bar would not bring her enough money to actually build a house on her plot. Therefore, she considered working as surrogate once again. Her hesitations clearly illustrate the two sides of being a surrogate: it is very attractive, as it gives a woman (of her socioeconomic stratum) access to a sum of money that she would not be able to earn with any other job; yet, it also requires a huge investment of her body and life—including missing her children—turning this new form of entrepreneurship into a demanding and ambiguous enterprise.

★ ★ ★ ★

Our cases confirm insights from other studies that the introduction of ARTs in the global South rather privileges the rich to access these treatments and exclude the poor, who may even get more frustrated about their fertility problem as the treatments are available in the country, yet not accessible to them (Hörbst 2012: 184). Our study also underlines that certain bodies (of poor women) are more 'bio-available' (Cohen 2008) than others to serve infertile couple's wish to have children. Both forms of stratified reproduction clearly increase inequalities in (access to) health.

In our contribution we have particularly drawn attention to the inequalities that exist between various types of entrepreneurs, resulting from the structurally different positions they hold. The degree to which they can profit from their entrepreneurship and the extent to which they can act as independent entrepreneurs varies hugely. The higher they stand on the societal and professional ladder, the more means to invest and safeguard their interests they have. Those at lower societal levels are more vulnerable, less autonomous, and for some, particularly surrogates and donors, the involvement in the ART market is only temporary and comes with serious (bodily) risks. While there may be economic stimuli resulting from the private health market in the contemporary neoliberalist era in the global South, this private market maintains and even reinforces power differentials and inequalities.

Finally, we have shown how therapeutic innovations to address infertility in the global South take place without the support of international donors or public money in the private health sector. This is in contrast to many other (reproductive) health issues where African states heavily depend on foreign aid (Hörbst and Wolf 2014). The autonomy from foreign money turns the ART clinic directors into independent decision makers regarding financial, medical-technical, and ethical questions. In a context where—to date—no ART regulations exist and effective quality control is hardly applied, the doctors' autonomous decision making highly shape the ways in which clinical practices and treatments are offered to patients. Subsequently questions emerge to which extent business interests and benefits impair medical practices (Hörbst and Gerrits 2015). As this autonomy is—presumably—a characteristic of private clinical entrepreneurships in Ghana and elsewhere in sub-Saharan Africa, our article provides a glimpse of future configurations of the health care system in Africa. With an increasing flow of various biomedical technologies to this continent our article thus hints at important new fields of anthropological research for future exploration.

### Notes

1. In 2013, a minimum of eight private clinics were known to offer ARTs in Ghana.
2. This case study draws on data collected as part of the comparative project 'Dynamics and Differences of Assisted Reproduction in Sub-Saharan Africa' (http://ssaart.wordpress.com). Here, we only use data from the Ghana study and mainly from one of the clinics. Ethical permission for the study was received from the Noguchi Memorial Institute for Medical Research-IRB.

### References

Barth, F. 1967 On the study of social change. *American Anthropologist* 69(6):661–669.

Cohen, L. 2008 Operability, bioavailability, and exception. In *Global Assemblages: Technology, Politics, and Ethics as Anthropological Problems*. A. Ong and S.J. Collier, eds. Pp. 79–90. Oxford, UK: Blackwell Publishing.

Ginsburg, F., and R. Rapp 1991 The politics of reproduction. *Annual Review of Anthropology* 20:311–343.

Hörbst, V. 2012 Assisted reproductive technologies in Mali and Togo: Circulating knowledge, mobile technology, transnational efforts. In *Medicine, Mobility, and Power in Global Africa. Transnational Health and Healing*. H. Dilger, S.A. Langwick, and A. Kane, eds. Pp. 163–189. Bloomington, IN: Indiana University Press.

Hörbst, V., and T. Gerrits 2015 Transnational connections of health professionals: Mobility, networks, and assisted reproduction in Sub-Saharan Africa (in Ghana and Uganda). *Special Issue, Ethnicity and Health.*

Hörbst, V., and A. Wolf 2014 ARVs and ARTs: Medicoscapes and the unequal place-making for biomedical treatments in Sub-Saharan Africa. *Medical Anthropology Quarterly* 28(2):182–202.

---

As described in this case study, women seeking to become pregnant have new options for trying to influence their biological processes—and those who can provide such services as IVF or surrogacy are in the position to profit from their work. The meanings associated with these 'new' kinds of pregnancies are embedded in the extant cultural meanings of conception, birthing, motherhood and fatherhood. These meanings play into culturally informed marketing techniques as well as the pricing of the commodity of fertility.

In his *Birth of the Clinic: An Archaeology of Medical Perception* (1973), Michel Foucault demonstrates how biomedicine developed in tandem with evolving scientific knowledge of the world as well as in step with changing social norms. The spread and uptake of biomedicine happens in a much less linear fashion in the twenty-first century. Information is exchanged rapidly and often incompletely. Implementation is enacted, not within institutions that have long histories of biomedical development, but rather that are new to the game of 'doing biomedicine.' As the case studies in this chapter show, the resulting configurations of technology and their social relational matrices are reflective of different historical traditions and different access to the accouterments of the hospital-clinic-industrial-complex. Medical anthropologists need to be sensitive to the way in which biomedicine is a rapidly changing assemblage (Lock and Nguyen 2010). In highly technological health systems electronic databases and oversight have led to the guidelines of medicine becoming more standardized and the judgment of clinical practice is changing from one based on a relationship that takes place in person to one that is mandated to be performed in a standardized manner and monitored closely by insurance companies who pay for health. 'Big data' will increasingly create opportunities to store, link, and collect patient data that will lead to different kinds of analytics, and these analytics will be used to change the way that hospitals are run and how care is given. But in the Internet-outback, older forms of practice may still be those based on human-to-human caring and thus may be superior in some ways. In less-resourced settings, clinicians, like other kinds of healers from different local and broader cultural traditions, necessarily rely on their capacity to interact with their patients and to 'read' their bodily and behavioral signs and symptoms.

Biopower, understood as the state-level control over populations through establishing norms related to the body, then enforced through mandating behaviors based on those norms, is changing rapidly as more and more medical knowledge is located on the Internet and in repositories of big data. As discussed in Chapter 8, the everyday habitus of many practitioners has changed, and they are becoming increasingly dependent on 'smart' technologies such as smartphones and tablets. The current medical doctors in resource-high countries are accustomed to not needing to recall information beyond where it can be immediately accessed on their mobile devices; practitioners in the Internet-outback still have to rely on traditional medical tomes, intuition, and only sometimes their electronic technologies.

## The Politics of Global Surgery

Global health is changing rapidly as progress is made in some areas, while other areas remain intractable or even worsening; conditions and resources change in response to shifts in economies, environmental transformations, and wars (see also Chapters 11 and 13). Global surgery is one aspect of global health that has lagged behind in its goal of expanding cadres of well-trained and well-resourced surgeons and facilities worldwide. As Mamta Swaroop, MD, FACS, writes on the Association for Academic Surgery official blog,

> Approximately 5 billion out of 7 billion people in the world have no access to basic surgical care. Lack of access to safe and timely surgical care results in more deaths and disability than HIV/AIDS, tuberculosis and malaria combined (Meara and Greenberg 2015). Although international priorities are starting to reflect the importance of non-communicable diseases, provision for essential services lags behind.
>
> *(Swaroop 2015)*

Below, Adia Benton follows the politics of the developing sub-specialty of global surgery within the larger field of global health. Benton focuses on the academic side of global surgery. In the academy, funding is competitive and the stakes are high for getting particular specialties recognized and on the agenda of all kinds of funding agencies.

---

### 15.3 Surgery and the Cultural Politics of Global Health

*Adia Benton*

In late January 2014, I attended the first of three commission meetings on global surgery, organized by *The Lancet*, one of the world's leading medical journals, and by Harvard Medical School, Lund University (Sweden), and King's College London, to address lack of access to surgical care in resource-poor countries. Attendance at the meeting in Boston was by invitation-only; invitations were distributed to a relatively small and select group of global health and surgical practitioners. After learning about the closed meeting through social media and professional contacts, I managed to secure an invitation to attend and observe in the second day's proceedings.

In a tightly packed program, speakers came to the podium to give presentations about issues that matter most in mainstream public health: cost-effectiveness; metrics of prevalence, disability, and mortality; institutional capacity building; global advocacy and policy stakeholder buy-in. Social justice and equity were notably muted in these discussions, but as in many mainstream public health projects, their relevance as an underlying motivation for action was nonetheless present. Surgeons, health and welfare ministers, prominent non-surgeon clinicians, and economists and health policy analysts from all over the world made brief presentations. The commissioners, a group of experts who would ultimately co-author the report of these meetings, sat near the front of the room in an area reserved for them. The commissioners represented a diverse swath of experts in health economics, engineering, surgery and public health, and hailed from each major region: Western Europe, southern and West Africa, and the Pacific.

Amidst concerns about transparency among the growing community of global surgery advocates, the presentations streamed live on the Internet for those who were not invited and unable to attend. Meeting conveners used social media platforms like Twitter and Facebook to foster a sense of openness and inclusion. Virtual and live attendees tweeted about the presentations, using the official meeting hashtag, #LGCoS1. Following Twitter chatter about the meeting provided me an opportunity to track interest in global surgery, gauge reactions to meeting proceedings, and visualize the ephemeral and shifting connections among global health actors—whether they were seated in the meeting room or observing with their smartphones halfway across the world. At times, it was difficult to determine where the (virtual) Twitter discussions began and live (meeting room) ones ended. Tweeted questions were read aloud by moderators during Q&A sessions; tweeters who recognized me in the room and were familiar with my research urged me to step up to the microphone. As the global surgery 'twitterati'—individuals who amassed a following on Twitter and frequently tweeted under the #globalsurgery hashtag—commented upon and posed questions about the presentations, audience members in the small meeting room quietly interacted with their counterparts in the room and in distance locales online.

When Paul Farmer, a well-regarded infectious disease physician and anthropologist, came to the podium, @bnwomeh, a surgeon tweeting from US Midwest, claimed to be #starstruck. He posted a picture of an article Farmer had co-authored with fellow physician-anthropologist and World Bank President, Jim Yong Kim, daring any onlooker to deny they had cited it in their work or referenced it on their presentation slides. The tweet started a chain of comments on the article's most referenced line, that surgery is 'the neglected stepchild of global health,' and garnered reverential recognition of Farmer's and Kim's commitment to improving access to surgical care among the world's poor.

The surgeon's tweet was both a joke and a provocation for its readers to contemplate—if only for a second—the strategies that global surgery advocates used for this newly delineated sub-discipline within global health. In particular, in his tweet, @bnwomeh called attention to the conventions (some might say, politics) of scholarly publication traditions, which implicitly require an author to situate one's work in relation to the existing literature. Not all articles are citation-worthy. As in other fields within global health, to become a well-respected scholar, one must also demonstrate facility with and knowledge about the state-of-art developments in their field; one must exhibit the ability to discern good science from rubbish—to cite the appropriate people and publications.

To reference the Farmer-Kim paper, specifically, was also to signal a strategic ideological alliance with a rich legacy of social medicine and anthropological critique of public health. Global surgery advocates' frequent reference to this paper for its evocative storybook-inspired kinship metaphor ('neglected stepchild') also suggested that the surgeons who are at the center of this movement had not yet found—and perhaps were not trying to find—their own powerful language to describe surgery's marginal and seemingly abandoned status within global health.

The tweet and the responses to it constituted a diagnostic event in the ebb and flow of 'transnational' communication about and within the global surgery movement. The event indexed several important issues related to global surgery's struggles to achieve legitimacy and to be fully adopted into an ever-expanding global health family, in relation to the 'cultural politics' of global health. Global surgery advocates have faced and continue to face multiple challenges in their quest to achieve legitimacy and acceptance in the field. Research and publication are one strategy they have used to reach their goals. While they have venues like *The World Journal of Surgery*, which started in 1977, for instance, to publish, their work has only recently received sustained attention in mainstream medical journals, as in the case of a special edition of *The Lancet* in 2014.

## Global Surgery's Imperial and Military Roots

'Global surgery' is relatively new, in the sense that surgeons, anesthesiologists, and their colleagues in wealthy countries have only recently began to carve out a specific place in mainstream public health initiatives that target low- and middle-income countries. Global health, generally associated with longer-term development programs, has also been subject to the same critiques as its development counterparts. The most frequent critique is that personnel, ideas, and funding from wealthy countries travel to and dictate health priorities of poorer ones. For the field's critics working within and outside its boundaries, this relationship has the potential to be neocolonial in its approach and aims. Thus, most global health institutions, including newly emerging ones aiming to address disparities in and unequal access to surgical care, attempt to confront this critique head-on and from the outset of projects.

Priorities in global health are largely oriented towards primary health care and low-cost preventive measures. These measures include prenatal care, 'attended' childbirths, childhood immunizations, management and treatment of common childhood illnesses, and infectious disease control. While surgery's importance for obstetric conditions is well articulated and accepted as global health orthodoxy, the significance of surgical care for treating victims of accidents and injuries, cancer, and other conditions has been marginalized in global health discourse, even as they constitute a significant health burden in global health metrics and indices.

However, transnational surgery projects have long existed. This is not simply because modern surgery travels and circulates to places under the umbrella of charitable biomedicine; it is also because surgery was uniquely implicated in the expansion of imperialism, and in the optimization of military forces, imperial and national, in combat. Many technological advances in surgery, moreover, resulted from need to treat a range of battlefield injuries. New weapons and war technologies have gone hand-in-hand with progress in surgical diagnostics, instruments, and techniques.

Surgery has also been a significant aspect of missionary and humanitarian interventions. From the colonial era through today, missionaries have set up hospitals in the most difficult-to-reach places; in some underserved areas, the only clinicians trained in surgery may be located in mission hospitals. Charities and non-governmental organizations that primarily perform surgeries, like Mercy Ships or Operation Smile, set up shop for weeks at a time, operating on patients who would otherwise have been unable to access surgical services. While many of these missions have relied on short-term international labor, they now appear to be operating under development principles—providing 'sustainable' and long-term solutions to lack of surgical care in certain places through education, training, and logistics management. And in humanitarian emergencies, whether coupled with military forces or not, surgery for conflict- and disaster-related injuries and conditions are foregrounded in a medical response.

## Radical Health Movements, Delayed Acceptance

The primary health care movement, which came to its peak with the WHO Declaration of Health for All in Alma Ata in 1978, featured contestations and debates about what kinds of care come under the rubric of 'primary health.' Primary health care proponents saw the model as revolutionary, central to liberation movements occurring in places like southern Africa at the time. Enlisting surgeons, specifically, who had largely been left out of the conversations about primary health care—particularly with the problem of 'appropriate technology' looming large—became an important goal for the program's architects. On June 29, 1980, Halfdan Mahler, then Director-General of the World Health Organization (WHO), addressed an audience at the Biennial World Congress of the International College of Surgeons in Mexico City. He urged surgeons to join a burgeoning primary health care movement:

> Surgery is a practical affair. It has to be by its very nature. . . . It cannot escape the political, social and economic factors that influence all human endeavours. Social injustice is socially unjust in any field of endeavour, and the world will not tolerate it for much longer. So the distribution of surgical resources in countries throughout the world must come under scrutiny in the same way as any other intellectual, scientific, technical, social or economic commodity. The era of only the best for the few and nothing for the many is drawing to a close. . . .
>
> Surgery clearly has an important role to play in primary health care and in the services supporting it . . . primary health care includes the appropriate treatment of common diseases and injuries. Surgical first aid is therefore an essential part of it. Without it, in spite of preventive measures aimed for example at preventing accidents, people will not have faith in primary health care. . . .

> This is the challenge I want to face with you. Conventional solutions are not likely to be very satisfactory. . . . There are tens of thousands of drugs on the world market, and only a few years ago the problem of getting essential drugs to the masses appeared just as intransigent to solution as the problem of providing essential surgery for the masses. So we set up a committee of experts on pharmacology and they arrived at a list of about 200 essential drugs that could cover most requirements. Could the international surgical community do the same for surgery?
> *(Mahler 1980)*

This official acknowledgment of the need for surgery to be a part of primary health care highlights surgery's 'practical' nature, its potential for realizing social and political justice, and, perhaps most notably, its potential for legitimizing primary health care as a model. Yet, despite the mounting enthusiasm about surgery in the immediate post–Alma Ata period, Mahler's plea to the surgeons largely went unaddressed by major international health institutions such as the WHO in the years that followed, leaving individual humanitarian service organizations to bridge the gap. Major institutions instead first focused on 'appropriate technologies' that could be delivered at low cost and with minimal training. The emancipatory potential of primary health care was further dulled by Cold War politics and international financial institutional responses to the 1982 Latin American debt crisis (Cueto 2004; Werner et al. 1997). Growing interest in health as a way to cultivate 'human capital' accompanied, and perhaps, encouraged greater involvement by World Bank in health issues. A focus by the World Bank on health also meant a fundamental shift in how health was conceived within health organizations, more broadly. The 'health for all' mandate of Alma Ata soon became 'investing in health,' with health being a function of human capital and oriented towards productivity, efficiency, and meaningful participation in a market economy.

Eventually, the list of essential surgical services that Mahler requested in 1980 was developed for use by the WHO, leading to the establishment of the Global Initiative for Essential and Emergency Surgical Care (GIEESC) in 2005. This initiative centers on promoting a set of training tools for district hospitals and 'surgical capacity' checklists. Even with this increasing recognition by mainstream global health institutions, the GIEESC faced almost certain dissolution. Plans were made to shut down the program. The input by a network of influential clinicians was able to secure support for the initiative's continuation, suggesting that expanded access to surgical care may be picking up support—precarious, though it may be—among global health power brokers.

Research funding agencies like the US National Institutes of Health (NIH) and among academic associations are beginning to pay attention to access to surgical care; the fact that the Global Commission mentioned earlier in this essay, exists, for example, is evidence of the growing recognition of surgery within these realms. And this is part of the strategy for legitimacy. So the logic goes: procuring research funding both enhances and reflects significance of the research problem; it also institutionalizes an issue and grants it validity and value within academic research disciplines. One surgeon, who had devoted much of his career to addressing surgical access, told me that although he had no interest in pursuing a career in academic surgery, he soon recognized that conducting and publishing research was the only way to bring attention to the need for improved access to surgical care. He began with the rudimentary data collection tools provided by the WHO, but soon branched out into more complicated and costly research involving population-based surveys. Within a short period of time, he and his co-authors—most of whom were developing country nationals—had become noted experts in the field.

## Conclusion

Access to surgical care has been peripheral to global health priorities but is becoming increasingly important as various forms of cancer, road accidents, and injuries—conditions that may require surgical interventions—gain prominence within the field of global health. The long tradition of scattered programs and interventions across the globe involving surgery appears to be shifting as global surgery advocates publish scholarly papers, build research portfolios at academic institutions, and push for better training, equipment, and supplies for performing safe surgeries in resource-poor settings. Professional associations are intensifying their global outreach, as they did in the 1960s and 1970s to the lead-up to Alma Ata: the Academic Association of Surgery (AAS) has a global affairs section and supports clinical training, and knowledge transfer between resource-poor settings in both the developed and developing world.

### *References*

Cueto, M. 2004 The origins of primary health care and selective primary health care. *American Journal of Public Health* 94(11):1864–1874.

Mahler, H. 1980 Surgery and Health for All. Address by Dr H. Mahler, Director-General of the World Health Organization to the XXII Biennial World Congress of the International College of Surgeons; Mexico City, Sunday 29 June 1980. L/81.7a. Geneva, Switzerland: World Health Organization.

Werner, D., D. Sanders, J. Weston, S. Babb, and B. Rodriguez 1997 *Questioning the Solution: The Politics of Primary Health Care and Child Survival, with an In-Depth Critique of Oral Rehydration Therapy*. Palo Alto, CA: HealthWrights.


---

Benton sheds light on some of the behind-the-scenes politics that either support or defeat the spread of international surgical training. Beyond the good intentions of medical practitioners—especially in this case, those in academe—there is always the question of who is paying for the expensive training missions, educational materials, medical equipment, and travel for teams of trainers; the resources needed to make surgical systems of care sustainable are often inadequate or non-existent. While surgeons dedicated to improving global health through disseminating their skills are addressing a slow-moving 'train wreck' of increasing trauma from more motorized vehicle travel and the recognized need for basic surgeries, too, the next section of this chapter deals with a faster 'train wreck'—we turn to consider the out-of-control, fast-moving, deadly Ebola epidemic of 2014 that took place primarily in Guinea, Sierra Leone, and Liberia.

## What Ebola Teaches Us

Death is dangerous from a cultural perspective: it underscores the unpredictability of life; unsettled souls and poses a threat to the living; and one death may be the harbinger of other deaths. This is literal when death is due to infection; hence the stigma commonly associated historically with the plague, smallpox, and tuberculosis (Goffman 1963) and in contemporary society, still with tuberculosis and now also with HIV. Ebola is the most recent virulent epidemic. By 1 April 2015, a year after the first cases of the current epidemic were diagnosed in West Africa, the World Health Organization had estimated a total of 25,000 cases and at least 10,000 deaths from the virus, many not reported. Medical anthropologists have played a major role in this epidemic for several reasons that highlight the social life of any disease, including by effective use of the Internet (http://www.ebola-anthropology.net/), and have drawn attention to the economic and political factors that have shaped transmission and control, the logics of public health, and the importance of working with, not despite, populations.

Ebola is transmitted through contact with blood and body fluids, including sweat, and so ordinary close human contact is a risk factor for infection. Consequently, close family members and health care providers caring for those who are sick are at high risk for infection. Practices around the care of the dead increase risk of infection, particularly in communities where bodies are washed and wrapped before internment. In the early months of the epidemic, in an atmosphere of heightened emergency and increased fear, there was little engagement with communities; consequently, communities resisted the demands placed on them to report cases of fever and death. During this period, Médecins Sans Frontières (see Redfield's case study, this chapter) played a major role in delivering medical care; until December 2014, almost all people with suspected Ebola infection were treated in MSF clinics primarily because of the fragile infrastructure, limited capacity of government hospitals, and the direct loss of the lives of local health providers. But communities themselves also drew on local knowledge of previous epidemics, of smallpox, for instance, to initiate quarantine and so to try to limit the spread of the virus.

Ebola impacted heavily on select communities in Guinea, Liberia, and especially Sierra Leone, already with fragile heath services and limited systems in place to manage medical emergencies. As Raphael Frankfurter describes, hospitals rapidly became overcrowded and cemeteries ran out of space; villages at times of necessity resorted to mass graves.

## 15.4 'Safe Burials' and the 2014–2015 Ebola Outbreak in Sierra Leone

*Raphael Frankfurter*

In late March 2014, the World Health Organization reported that the Guinean Ministry of Health had identified an outbreak of Ebola Virus Disease in the remote forests on the border with eastern Sierra Leone. An extremely contagious illness with no known cure, Ebola had been recognized and controlled in a series of outbreaks, almost exclusively in central Africa, over nearly four decades, but an outbreak had never been identified before in West Africa. Though it took months to recognize the hemorrhagic illness slowly spreading across rural Guinea, epidemiologists suspect that the outbreak began as early as December 2013.

Ebola is a zoonotic disease, endemic in a host species until it jumps to humans in a cross-species transmission event. This is thought to occur when a hunter or cook is exposed to the body fluids of an infected animal—often bats or monkeys, which are eaten as a major source of protein in many areas of remote West Africa. Those who are exposed undergo a 7–21 day incubation period during which they show no symptoms and are not infectious. Next, the early symptoms of Ebola are identical to far more common tropical illnesses like malaria and typhoid: sudden onset of a fever, fatigue, and abdominal pain. Patients begin to exhibit 'wet' symptoms like severe diarrhea and vomiting, and in the later stages of the disease some patients begin to bleed out of their gums, eyes, ears, and anus. The disease is transmitted solely via infected body fluids after the patient is symptomatic; as the disease progresses and the viral load increases, these fluids become increasingly contagious. Patients who only have 'dry' symptoms are hardly contagious; vomit and body fluids contain a larger amount of virus, comatose patients spread the disease with increasing intensity, and corpses are extremely contagious. Up to 90 percent of patients die with no rehydration or supportive care; with proper medical support that number can drop to well below 1 percent. The only means of controlling the disease from spreading is to isolate patients while they are sick, cared for by health workers wearing thick protective equipment obscuring their faces and bodies. Corpses must be wrapped in plastic bags, drenched with chlorine and buried by 'Safe Burial' teams wearing hazmat suits. In much of West Africa where families care for ill and dying family members in the home and intimately handle corpses before burying, communities and families may resist these restrictions on touching and caring for the ill and dead. For this reason, 'cultural practices' and 'burial beliefs' are often listed as primary modes of transmission and obstacles for responding health workers to overcome in order to contain the spread of the disease (Nielsen et al. 2015).

In late May 2014 the first cases of Ebola were publicly reported in a remote region of Sierra Leone. Seven people who had attended the funeral of a local healer in Guinea returned to Kailahun District, exhibiting the signs of the disease. In the time since the outbreak had been identified in Guinea, the Sierra Leone Ministry of Health had developed a strict protocol and response system for suspected cases: patients would be isolated in their community's clinic, blood samples would be rushed to a well-equipped hospital in Kenema with isolation facilities, and if positive the patients would be transported rapidly in designated Ebola ambulances to Kenema for basic care. What the response protocol did not account for, however, was a plan to communicate to families, empathetically and transparently, what was happening to patients after they entered the Ebola response system.

Within days of the first cases, things began to go awry. Families barged into the local clinic removing confirmed Ebola patients awaiting transport. Rumors began circulating within Kailahun and throughout the eastern regions of the country: "They're bringing people to Kenema to die! Once they take them away they don't come back!" Very quickly, those in affected villages came to know that Ebola was a near-certain terminal illness. Family members claimed they were not informed when, where, or even if their relatives died and were buried after they were whisked away from their community's clinic by men in hazmat suits. The families of the initial patients soon contracted the disease, and as the bodies of those who had succumbed were prepared for burial, scores of funeral-goers contracted Ebola. Some patients presented to clinics, and ill-prepared nurses found themselves at high risk. Within two months, the Kenema hospital isolation center had largely shut down: the doctor in charge and dozens of nurses had fallen ill and died. As the outbreak spread, so too did a flurry of shallow discourse and education campaigns organized by the Ministry of Health and responding non-governmental organizations (NGOs) aimed at halting 'harmful cultural practices' associated with the spread of the illness. "These people refuse to believe!" a local health authority told me in June. "Their cultural practices"—caring for and touching the ill in their homes— "are bringing it on themselves and the country."

By late October, treatment centers were overwhelmed, and nearly every hospital in the country had closed. A near-constant stream of possible Ebola patients weaved its way towards the capital city Freetown's hospitals, only to be turned away when facilities were at capacity. Meanwhile, health officials were faced with massive numbers of infected corpses. The Sierra Leone government identified large burial plots staffed by 'volunteer' grave-diggers across Freetown and hired more 'Safe Burial' teams. The Ministry of Health developed a sluggish but improving Alert Hotline to report sick and dead bodies in homes, but responding epidemiologists insisted that cataloguing and tracking the numbers of dead bodies was critical. "We're trying to create a system so that in any remote clinic in the country, the nurse will wake up in the morning, count their dead bodies, call it in and then get them in the ground," one expert explained at a coordination meeting in Freetown. The apparatus for processing the bodies involved creating an asocial system, where corpses would be taken promptly and far away from their families. But the emphasis on counting—in addition to burying, often slowing the response process—seemed to emerge from three broad goals: (1) to provide a means of determining where response resources should be concentrated, as a proxy measurement of the epidemiology of the disease; (2) as a way of enforcing the government-run 'safe-burial' system, and (3) to generate valued empirical public health data on how such an outbreak would run its unfettered course. In some districts, 'Safe Burial' teams apparently waited next to corpses for hours for separate teams of epidemiological officers to arrive and take a test swab to catalogue the body as 'Ebola-positive' or 'Ebola-negative.'

By November 2014, ambulances wailed all night answering the backlog of 'live alerts' and 'death alerts' in Freetown. The World Health Organization (2014) released recommendations for "Safe and *Dignified* Burials" (my emphasis), which mandated that bodies be buried in the vicinity of victims' families, communities be asked to participate in burying the coffin, and family members be informed of patients' outcomes, but these standards proved difficult to implement. Overwhelmed health workers and NGOs found it difficult to keep up with the number of corpses even without 'community assistance' and even more difficult to track where patients had come from when they entered large NGO Ebola Treatment Units. In the mornings in Freetown, amidst hundreds of billboards imploring citizens DO NOT TOUCH DEAD BODIES: CALL 117 FOR A SAFE BURIAL TEAM, clusters of people huddled in the street, each surrounding a corpse. Many victims seemed to have crawled from their homes at night, but the outbreak also seemed to have reached a certain threshold of destruction so that an increasing number of dead or dying victims were being dumped on the street by their families.

Kono District, a remote region in eastern Sierra Leone known both for its rich diamond flats and as the former epicenter of the country's 10-year civil war, was largely spared from the Ebola outbreak until late 2014. The region is extremely impoverished; the health care system in disrepair. There were two doctors for nearly 540,000 people, under-5 mortality rates often hovered above 2 percent, and most public clinics in rural villages were barely functional. In the context of this poverty, history of war, government reconstruction plans that had yet to materialize, and the whirlwind of humanitarian NGOs that had come and gone in the wake of the civil war, the population—largely from the Kono tribe—are known to navigate a particularly rich landscape of local and traditional healers rather than, or in tandem with, seeking biomedical care at their local clinics. Though the forms of care that these local healers practice vary—some specializing in herbal remedies, some in divining sorcery, some in extracting bullets shot by 'witch guns,' and some in creating amulets to protect from illness—the relationships that patients develop with local healers are profound, long-term, and intimate. Healers often treat patients in their homes, encouraging the participation of the patient's family in tactile rituals and processes of healing like washing the ill, massaging and applying herbal ointments, and hand-feeding herbal medicines. Other healers maintain 'healing compounds' where patients come to live alongside others for long-term treatment, creating bonds of solidarity among patients with similar conditions. Though differing beliefs in the causality of illness inform these practices, these local processes of caregiving serve as a rather pragmatic source of solidarity that enable patients to make-do through chronic and cruel poverty and disease.

Burials and funerals are similarly empathic and central events in Kono life. When a person dies in the home, immediate family members often communicate the death by wailing and sobbing as neighbors come to pay their respects. The male or female village elders then prepare the body for burial, by intricately washing the corpse and clothing it in a white sheet. Body washing is an especially elaborate ritual for the first-born in a family, including anointing the corpse so as to prevent the death from spreading to other children. The parents are also specially washed. For certain highly ranked members of the Kono male and female 'Secret Societies'—local organizations into which adolescents are initiated and then trained in the cultural knowledge, history, and day-to-day skills of Kono life—the *Sowei* (female society heads) or *Pomasu* (male society heads) are involved in the cleansing process. If there is a series of deaths in a family, diviners may also be called to investigate the corpse to ascertain the cause of death.

In many funerals, the corpse, shrouded in white, is displayed in an open coffin as large numbers of relatives gather to mourn. The immediate family often wails the loudest, comforted on all sides by relatives crying in sympathy. Before the coffin is brought to the burial plot, many of those at the funeral rush up to kneel at the foot of the coffin, touching the corpse through the white sheet one last time or kissing the coffin. Since the Kono tribe is half Muslim and half Christian, the final stages of the funeral are often directed by a pastor or imam according to religious tradition.

Though most of these rituals serve to ensure that the deceased is at rest and will not torment the living, many in Kono emphasize that the most critical aspect of a funeral is an emotional one. "People show you all kinds of concern—even people who never showed you concern when you were alive," a student from an interior village explained to me. Such expressions of empathy are particularly important for those in Kono who live in such proximity to serious illness and death. "People make sure that they wash you and the corpse fine, they go and cry with you—they make sure that everyone cries on the body, they dress them in white satin, follow the body to the cemetery. This is about showing concern." These funeral rituals bring families and relatives together for last-goodbyes and joint ritual, turning the uncertainty, chaos, and tragedy of a death into tangible, choreographed, and collective practice. When I asked the student if he sought to perform these burial rituals because he believed they were critical to prepare for the afterlife, he paused. "For some, this is about different beliefs," he said. "But for me, it is about tradition. I wish people to perform these rituals, for they are our tradition, but I do not *believe* 100 percent."

In November, a series of imported cases that had turned into small transmission clusters overwhelmed the Kono Government Hospital's isolation unit. The small, converted ward, intended to house only a few patients while they awaited transfer to a formal NGO Ebola Treatment Unit, filled with suspected Ebola patients, and many sent to the isolation unit as a precaution likely contracted the disease there. Over the next month, scores of bodies piled up as overworked 'Safe Burial' teams struggled to keep pace—by December, 40 infectious corpses were reported to be strewn in the isolation unit awaiting burial.

Without proper triage systems and frightened Ebola 'suspects' entering and leaving the isolation unit, nurses in the hospital were at great risk. Among those who contracted the disease in the hospital was Nurse Mamie Bangura, a nurse-aide in the women's ward. "My mother wanted to be where the patients struggled, that's what her job was," Mark, her son, explained to me. "She thought she contracted the disease taking care of patients at night during her shift, when the doctors aren't even there and she was the only nurse." One evening when she was at home, Nurse Mamie reported feeling cold and feverish. She checked into the woman's ward of the hospital that she had worked at, and the next evening she began to vomit. Her son continued:

> They thought it was malaria, but when she vomited they pulled her to the [isolation] ward. They said they would quarantine the whole area, and all the patients in the hospital ran away. She went to the hospital on Friday; on Sunday they told us she has Ebola. When I went to visit her, they told me at the hospital that she had gone to an 'Ebola place' in Kailahun District (an Ebola Treatment Unit operated by Doctors Without Borders).
>
> [A few days later] my uncle came to me and said: "Your mother died! Your mother has died of Ebola." They buried her in Kailahun. I don't know where, up to now nobody has told me. That's the thing that's been bothering me the most, I don't feel good. I haven't seen my mother's grave. Two things make me feel bad: (1) that I never got to speak to her, and (2) that it's my responsibility to bury her, I'm the first born child, but I don't know where she's buried. If I go

> to the grave I will be able to photograph it, just know that it's her grave. "This is my mother's grave!" I'll be able to show my children and say, "This is your Granny's grave!" But I don't know.

Mark returned again and again to the fact that he never saw any of his mother's belongings after her death. This was certainly a result of infection-control protocols—all clothes, cell phones, ID cards, and other items are burned as soon as they enter an Ebola isolation unit—but Mark fixated on this fact as if it were the most uncomfortable sign of disrespect. His mother seemed to have disappeared without a funeral and without a trace. "A lot is on me now. . . . I'm responsible for my family. This is all bothering me now."

Soon after Nurse Mamie's death, the outbreak exploded in Kono District, with up to 27 new cases per day. A large surge in cases occurred in Kombeh village, where a well-known taxi driver named James had fallen sick after driving a seriously ill friend to the hospital in his taxi. Aware of his risk, James went to the hospital for screening and isolation but was turned away at the triage gate by confused and overwhelmed medical officers. He returned home, and died several days later with his family nearby. James's family notified the Ebola hotline that they had a corpse in the house that needed to be buried, but the village leaders reported that it took 36 hours for the epidemiological officers and burial team to arrive. All the while, James's corpse lay in the house until it was brought to the mosque nearby for pre-burial rituals. Some villagers reported that one third of the community wept over and cleaned the body. Ten days later, dozens of villagers were reported ill in their homes, and Kombeh became the most severe hotspot in Kono District.

Though transmission rates greatly declined in Kono after the alert hotline was improved and a series of small Ebola 'Community Care Centers' were constructed in remote villages for patients to be isolated close to their families, in 2015 all burials are still outlawed except for those run by the 'Safe Burial' teams. Even people who die in vehicle accidents or of old age are to be tested for Ebola and buried far from family members by the teams. Though in practice this probably means that the vast majority of ('unsafe') burials are done in secret, pickup trucks with masked men in white suits sitting in the open back shuttle back and forth from the hospital to villages to the burial ground all day long. Occasionally a lone vehicle trails with family members who wish to watch the body go into the ground from a distance.

The Ebola outbreak and the 'Safe Burial' process has clearly ruptured rituals and funerary practices, but some families continue to wash bodies of relatives who have died in their homes, resisting the instruction of health workers and Sierra Leonean law. But as these cases illustrate, the reasons why 'unsafe' burials have been practiced throughout the 2014 Ebola outbreak are entangled and ambiguous. Cultural practices, intimate expressions of affect and empathy towards the deceased and mourners, a desire among family members to maintain 'traditions,' and the pragmatic consequences of an under-resourced and ineffective Ebola response system unable to remove corpses promptly, have all contributed to the propagation of the disease through corpses. Ultimately, the optics of the 'Safe Burial' system—every dead body wrapped in an anonymous white bag, handled like a disease-threat by a team of men in biohazard suits—may actually be the most disruptive aspect of the outbreak. Ebola containment is atomizing and inherently counter-social. Dying, death, funerals, and burials have been turned from central, tactile, and intimate social processes to grave public health hazards: bodies are no longer the remnants of lives lived to be mourned at funerals but disease-carriers to be processed and removed. "They say 'Safe and Dignified,'" a Sierra Leonean colleague told me in Kono as we watched the burial team weave its way with a bagged corpse to the mass burial ground. " 'Safe,' maybe, but certainly not dignified. Where are the relatives at the back crying? Where are the hundreds of people paying their sympathies? This is what we here in Sierra Leone imagine for ourselves. No—safe and *un*dignified burial is what this Ebola has done to us!"

## *References*

Nielsen, C.F., S. Kidd, A.R.M. Sillah, E. Davis, J. Mermin, and P.H. Kilmarx 2015 Improving Burial Practices and Cemetery Management During an Ebola Virus Disease Epidemic—Sierra Leone, 2014. *MMWR. Morbidity and Mortality Weekly Report* 64(1):20–27.

WHO 2014 How to Conduct Safe and Dignified Burial of a Patient Who Has Died from Suspected or Confirmed Ebola Virus Disease. Geneva, Switzerland: WHO. Accessed from http://who.int/csr/resources/publications/ebola/safe-burial-protocol/en/.

Frankfurter shows us how the epidemic played out on the ground. Practitioners relied on local networks of intelligence. Health workers in dedicated vehicles drove into villages where there were suspected cases of Ebola, entering houses to dispose of the clothing, bedding, and other items of those who had died, and swabbing corpses to confirm cause of death. Oftentimes they neither explained nor gained consent for these procedures and protocols from family members. They left with dead bodies, again without appropriate consent, disposing of the corpses in body bags—white for children, black for adults—ignoring the conventional use of white for all corpses. Ambulance drivers and body handlers suffered the most collateral damage. Colleagues from WHO working in Sierra Leone at the height of the epidemic in 2014 reported that paramedics were excluded from entering the hospitals to which they delivered those who were ill; instead, they were secluded in sheds in hospital yards, subjected to extreme stigma from other health workers and from their own families. Reduced transmission and the potential end of the present epidemic have occurred, and this has been facilitated through extensive community engagement, the provision of health information so that people understand how transmission occurs, and social mobilization to support communities where emergency responses might be required. The experience from Ebola emphasizes the need for consultation between epidemiologists, health workers, and communities and the significance of local traditions in the care of the dead both in relation to disease transmission and in regard for the dead and bereaved.

In the following case study, the final of this volume, Peter Redfield takes apart the logic of MSF, an NGO that was originally conceptualized in the 1970s as a non-political, quick-response team that would provide medical care in war zones and other crisis situations. MSF has grown over the ensuing decades to a very large, multinational player in the global health scene; MSF's impressive successes and now-established worldwide reach are also its greatest challenges.

---

## 15.5 Doctors Without Borders and the Global Emergency

*Peter Redfield*

For the contemporary aid world, the category of 'humanitarianism' commonly designates short-term relief work—actions intended to rescue people from immediate peril and promote their survival. Perhaps no organization exemplifies this emergency orientation more fully than Médecins Sans Frontières (Doctors Without Borders or MSF), which casts itself as a central actor of frontline medicine in crisis settings worldwide. Rushing breathlessly from site to site, the group has expanded from a ragtag French alternative Red Cross of the 1970s into a well-established multinational NGO known for both a combative tradition and excellent logistics. Moreover, it largely funds itself through private donations, with a budget now in the order of $1 billion a year.[1] Relatively rich and independent, it can chart its own course to a greater degree than most aid actors. In conceptual terms the group identifies more strongly with humanitarianism than development, defining its commitments through present states of crisis rather than future goals. While its projects encompass a wide range of medical activities, many well beyond urgent care, MSF's sheer existence extends an ambulance ethos on a worldwide scale. The group's trajectory thus outlines both the possibilities and the limits of emergency medicine as a response to economic inequities and political violence. To illustrate its mode of urgent action, below I sketch a day in the life of one project, combatting a minor epidemic among civilians caught in a conflict zone.

### *A Case of Cholera*

Near the end of 2004, I visited a cholera treatment center run by the Swiss branch of MSF in northern Uganda. The outbreak, focused in a displacement camp outside of Gulu, appeared to have run its course. Although the treatment center counted as an emergency project in the lexicon of the organization, the threat had subsided enough that there was little anxiety. Instead pressure came from an internal source: in a few days MSF planned to close the center and move the equipment back to its local headquarters, with an eye toward redeploying it elsewhere. Due to this self-imposed deadline on top of their other work, the staff found themselves quite busy. After a morning meeting in Gulu that ran late, I joined a Dutch doctor, a Portuguese nurse and a Ugandan driver to travel to the camp. We threaded our way through the center of town, packed with a mix of storefronts and signs for aid agencies, and set off on the dusty and bumpy road. Along the way the doctor briefed me on the project. He was by now quite familiar with the disease, having seen it many times before, and was thoroughly comfortable with the prescribed treatment. "It's quick, so OK," he noted, and then added, "With rapid response it's easy."

Indeed, for MSF cholera treatment had become something of a routine. Over the second half of the twentieth century, the disease proved a common scourge of human displacement, regularly appearing in refugee settings where people crowded together with contaminated water and poor sanitation. Cholera outbreaks had helped inspire the group's development of a kit-based logistics system in the 1980s, with depots full of prepackaged equipment ready for deployment in emergency. Once on site, the cholera kit provided any medical team with the essential means to set up a sanitized treatment zone, and confine and rehydrate patients within it. Most of the time the approach proved quite reliable in quelling an outbreak and dramatically reducing mortality. The combination of assertive public health and basic clinical care transformed an exceptionally deadly disease into a relatively ordinary problem.

This particular case appeared a success as well. The compound, when we finally reached it, stood largely empty. Surrounded by a tall reed fence, with chlorine sanitation stations guarding its entrance and a large MSF flag flapping overhead, it resembled nothing as much as a minor colonial fort. Once through the sanitation barrier, we entered a set of large tents with beds set up for patients. Only a few were still occupied, and the patients and caregivers looked more bored than distressed. As the doctor explained, diarrheal cases continued to trickle in, including a few cholera patients, but the records for the last few weeks indicated the epidemic was at its tail end. Since this structure was a temporary outpost, they would refer patients to the local hospital and dismantle the center before the equipment started to disappear.

A bit later in a nearby government health center, we attended an impromptu presentation by a Ugandan doctor from the nearest hospital. The MSF team and the health center staff (along with a curious patient or two) gathered around his laptop computer as he shared a slideshow he had just prepared on the outbreak for an upcoming workshop. The slides told a triumphal story. Over the past two months the area had seen well over 200 cases of cholera, most from the camp we had just visited. Testing showed all the springs and borehole wells were contaminated, and early projections suggested as many as 3,000 people potentially at risk. However, once MSF staff had set up their center and authorities held an emergency camp meeting, the infection rate plummeted. In the end there were only eight deaths, far below what might have been. We congratulated the doctor on a nice presentation. While cautioning that at a larger research meeting the audience might want larger numbers, our team leader agreed that this appeared a successful case of quick response and prevention. Of course, he mused, MSF had never told him to calculate the cost; in contrast to other health projects, with emergencies that was never a priority. The Ugandan doctor quickly responded with dark humor. Maybe he should just tell the government to send all the displaced people back home. After all, in camp cholera kills them, and in the country rebels kill them. Same result, but the rebels are cheaper! This off-color joke provoked general laughter. It also acknowledged the larger context within which this small epidemic occurred: the long-running, low-intensity war in the surrounding region, and the government policy of concentrating civilians into crowded displacement camps. Neither the Ugandan doctor nor the MSF team could do much about these background conditions. They might stop the spread of cholera and save lives. They might further work to secure a better and more reliable water source for camp residents. The larger risk factors behind contagion, however, would remain.

That evening, on the way back to MSF's compound, we passed a car smashed on the side of the road, a crowd around it. We stopped to investigate and picked up one man, who was bleeding and holding his arm and finger. He appeared to be in shock and kept apologizing to anyone who will listen. The team did not linger to inquire about the accident; as one of them warned me, crowds can grow dangerous if rumors spread. Short on bandages in the first aid kid, the MSF nurse used a wad of tissue as a compress as our impromptu ambulance sped to the hospital where the Ugandan doctor worked. We transferred our unexpected patient to the emergency room, and then, having effectively discharged him, we continued on home. By the next day the cholera treatment center had likewise folded its tents, its last occupants relocated to regular facilities, and its equipment dispersed to other projects.

## *The Medical Emergency*

Popular media attests to the degree of entanglement between contemporary health care and dramatic moments of life-saving intervention. The heroic doctor of television dramas is very much an action figure: a master not only of diagnosis, but also of defibrillation, delicate surgery, and the administration of wonder drugs, all provided at the last minute with hope nearly lost. Yet the conceptual frame of the medical emergency, like the capacity to respond to one, is actually a relatively recent innovation. Michael Nurok (2003) has convincingly shown how terms like 'accident,' 'reanimation,' 'resuscitation,' 'shock' and 'trauma' only combined into their now-familiar 'epistemological alignment' early in the twentieth century, partly catalyzed by the First World War. The component parts of emergency care, from first aid kits and ambulances to emergency rooms staffed by dedicated specialists, emerged onto the landscape of wealthier countries in stages between the late nineteenth century and the late twentieth, with a particular boost in the decades after the Second World War. Details vary by national context, but at a general level—the level at which anthropologists usually engage the past—one might say that the medical emergency appeared alongside industrial society and biomedicine itself. Older strands extend deeper into traditions of experimentation and surgery, particularly forms associated with war.

This background history matters for three reasons. First, it recalls the extent that medical emergency reflects a particular cosmology of time and etiology—one that assumes human actors can and should influence outcomes at an immediate material level. This cosmology is formally secular, at least in the sense that it prioritizes a technical rather than a divine set of nonhuman actors. It also assumes a world of machinery, risk assessment, and accounting, compounded pharmaceuticals and electricity. The balance between life and death has moved away from concerns about spiritual transgression or proper burial and toward reverence for biological existence, purifying the value of 'saving lives.'

Second, the conceptual lineage of the medical emergency underscores the importance of exceptional moments, a time outside of ordinary life when special equipment might be deployed and actions taken. This sense of exception is not identical to the state of exception in legal and political tradition, being located in human bodies rather than a collective body politic. Nonetheless, it remains intimately (if not always consciously) attached to claims of legal exception as well as the longer legacy of war.

Finally, the history of the medical emergency recalls that one of the many ways to distinguish maladies is by the temporality of their potential treatment. In some cases the sense of crisis is immediate and every moment counts. Others feature a slower rate of progression, whether positive or negative, and care becomes a long-term proposition. By definition emergency medicine—from its conceptual framework to its tools, traditions and attitudes—has never oriented itself toward extended care. Rather, it concentrates on the present, minimizing both history and future for the sake of current need. Indeed, a patient may be unconscious: an incapacitated body of unknown provenance, one that displays worrying signs and requires immediate treatment. In effect, the practical logic of emergency sorts through symptoms as much as it does desires. It thus can operate even in the absence of a speaking subject or known history. Action, rather than dialogue or contemplation, remains paramount; not acting may have deleterious consequences. All three of these points grow significant when extended to the realm of medical humanitarianism.

## *Life During Wartime*

Like all other human groups, nongovernmental organizations have their histories and habits. In the case of Médecins Sans Frontières, emergency plays a prominent role in both. Although not all forms of humanitarianism have emphasized immediate response, MSF descended directly from the Red Cross lineage of responding to war and disasters. By the end of the 1960s, its parent movement had expanded from its original concern with the battlefield suffering of wounded soldiers to encompass the plight of civilians. It had also moved well beyond its original focus on Christian Europe to engage in the fallout of decolonization and the Cold War. Thus the Red Cross found itself embroiled in the Biafran conflict in Nigeria, as well as the bloody birth of Bangladesh, two events that provided a catalyst for the formation of MSF. In this literal sense the organization was born from war. Although it would never limit itself to responding to conflict, conflict established its most defining norms. The group likewise appeared in the wake of fast transport, global communication and standardized emergency care for civilians, which in the French variation sent doctors straight to the scene of the accident. Imagining an organization of 'borderless' doctors, in other words, required more than humanitarian sentiment. It also required a particular configuration of possibilities, and a problem around which they might cohere. For MSF that problem was what its 1971 charter termed 'populations in distress'—or in a formulation used in later publications, 'populations in crisis.'

In operational terms MSF realized this classic métier of refugee work in camps on the border between Thailand and Cambodia in the late 1970s. For some key members, that experience stoked their ambitions to provide more efficacious care, as well as to develop a logistics system that would support moving rapidly from site to site. In rhetorical terms the organization's first publicity campaign had already provided a revealing slogan for such ambitions, suggesting that for MSF there were "two billion people in their waiting room." Given that the group was then but a tiny French initiative, hardly capable of delivering much to anyone, the slogan bore little relation to actual practice. Nonetheless, it both defined a problem, and established an expansive frame of potential response. If populations experienced distress worldwide, then care should adapt accordingly. The medical emergency had found a global scale.

By the time I began conducting research in the early 2000s, MSF had greatly expanded in size and scope beyond these modest beginnings. The name now encompassed a factious family of national sections, undertaking a shifting range of projects and initiatives around the world. Many of these extended well beyond the most immediate frame of emergency in medical terms, including such things as vaccination programs, psychosocial counseling, health education, pharmaceutical advocacy and the provision of AIDS drugs. In Uganda, where I did much of my fieldwork, the organization was involved in all of these pursuits in one fashion or another, as well as conducting epidemiological studies of drug protocols and responding to emergent diseases like Ebola (which surfaced in the same general setting a few years after the cholera outbreak described above). It launched a spinoff NGO involving traditional healers, and ran a garage facility to maintain vehicles for the wider region. In practice, the concept of crisis proved elastic. From the perspective of MSF, that was precisely one of its values in the formulation 'populations in crisis,' permitting a greater degree of latitude than was true with 'emergency.' Still, the group regularly fretted over the limits of its mission, alternately launching new experiments and drawing back from them. What should it try to do, and what should it leave to others?

## *A Global Band-aid*

A humanitarian organization like MSF often encounters the question of why it does not address root problems. Is not crisis response—particularly the theatrical, media-saturated international variety—like applying a band-aid rather than treating the underlying pathology? MSF's standard reply is staunchly realist. A crisis response remains limited by definition. Taking the medical metaphor seriously makes this clear: a band-aid, like an ambulance, seeks only to address an immediate problem, nothing less or more. For better or for worse that is precisely the temporal logic of emergency. To address chronic or future problems would require other equipment. It would also risk overlooking immediate needs, even as it might produce dependencies or new forms of domination, intended or not. For MSF lives cannot be exchanged, and a population should ultimately determine its own fate. Thus while the project of saving lives might have political implications and effects, it cannot substitute for a political plan or obscure political responsibility. From this perspective humanitarianism appears a limited, if vital endeavor, analogous to urgent care.

But the twin problems of scale and inequality pose other challenges. A billion dollars only goes so far, and charitable donations are not the basis of a viable, or sustainable, health care system. For this reason, the group issues regular moral exhortations and occasional denunciations aimed at those deemed responsible for the health of a given population. Governments and international agencies should do more, pharmaceutical corporations should charge less, problems should be solved at their roots. With this last point medical power reaches a limit. Violence, whether overt or structural, lies beyond a purely technical remedy. As MSF bitterly observed with regard to Rwanda in 1994, "you can't stop genocide with doctors."[2]

★ ★ ★ ★

Twenty years after the crisis in Rwanda, and a decade on from the cholera project described above, concerns about Ebola fill media headlines worldwide. Although the total number of dead has yet to reach the annual toll of cholera (let alone malaria or AIDS), the disease is extremely deadly and inspires fear in the manner of nineteenth-century outbreaks. Due to the mode of its transmission, the virus is unusually dangerous for health care workers, and ravages health care systems. Experimental treatment aside, biomedicine offers no cure for Ebola, only a reduction of mortality rates through supportive care. Amid the growing disaster in West Africa, MSF has received a new wave of attention, often cast in a heroic role as it struggles on the frontlines. Although overwhelmed and unable to offer treatment to those seeking it, the group recognized and proclaimed the severity of need earlier than most others, and its call for reinforcements have helped define a state of emergency. Its protocols for protective equipment have recently featured in discussions of shortcomings in the preparedness of US hospitals. As much as any official government body or intergovernmental agency, here an NGO defines a standard of action, however erratically and uncertainly life-saving. Yet at the same time, MSF has limited ability to actually solve the larger problem or rebuild health systems. An Ebola treatment center, just like one for cholera, is an immediate response, not a solution.

For all that emergency might offer humanitarians the allure of moral clarity—action as pure reaction—that clarity wavers when the frame widens. MSF continually opens programs in response to perceived crisis, and closes them when conditions return to a more ordinary state. In doing so, it confronts the fact that what counts as normal varies considerably from place to place, and that it cannot respond to all problems. Being 'without borders' the group continually struggles to define limits. The choices it makes factor in the work of other organizations and the larger realities of poverty and inequality, as well as its own relentless need to move on. Within the frame of a global emergency, there are always more lives to save.

### *Notes*

1. *MSF International Financial Report 2013* gives the group's overall income as just over 1,000 million Euro in 2013, up from about 938 million the year before. Of this amount, 89.5 percent came from private sources, the vast majority derived from nearly 5 million contributors worldwide.
2. http://www.doctorswithoutborders.org/news-stories/field-news/new-msf-case-study-response-rwandan-genocide

### *Reference*

Nurok, M. 2003 Elements of the medical emergency's epistemological alignment: 18th–20th century perspectives. *Social Studies of Science* 33(4):563–579.

## Unexceptional Moments

While groups such as MSF provide care to those in need during 'exceptional moments,' anthropologists often work at the other end of the spectrum of excitement—that is, they usually dwell in the most unexceptional of times and places—places like the aftermath of the Ebola epidemic. It is unexceptional in many ways: most volunteers and their attendant medical equipment have packed up and left West Africa for home; the press has moved on to other topics. It is unexceptional because the after-effects of the epidemic are chronic in nature and their cure is uncertain. At this writing (September 2015), we are just beginning to understand the long-term effects of having lived through the 2014 Ebola epidemic; survivors of the disease caused by the Ebola virus are now, six months to a year after the infection, suffering from intense joint pain, headaches, and PTSD-like symptoms of depression and anxiety. A quarter of the survivors suffer from eye problems and blindness as the virus continues to survive in the eye long after it clears from the rest of the body. Ebola virus also seems to survive in semen, increasing worries that it could be sexually transmitted (Grady 2015). The focus now shifts to treating the chronic effects of the epidemic, once more confronting the inadequacies of the medical systems of Guinea, Sierra Leone, and Liberia (see Henry and Shepler 2015). Post-aid scenarios are complex and show just how frail the immediate-response infrastructures of care are once the 'emergency' has past. It is also unexceptional in that, like most health problems, we haven't vanquished Ebola; we may have, sort of, won the battle, but the outcome of the war is far more uncertain. The Ebola virus is simply carrying out its lifecycle in other mammals, quietly waiting to re-emerge through a chance encounter with a person interacting with an infected animal in the ever-present bush. Perhaps Ebola is reminding us that it is the unexceptional times that really deserve more attention.

The case studies in this chapter provide fine-grained analyses of how biomedical understandings and practice travel to far-flung places. They illustrate how biomedicine and its attendant clinical landscapes pulse back and forth across the globe, sometimes taking hold and taking off like a plant finding welcoming soil, sometimes changing beyond recognition and sometimes retreating back to its home institutions in places more welcoming. The process is not a one-way street, with a war or a flood or a shift in fortunes; clinical assemblages can vanish as quickly as they arrived. Biomedicine is extremely resource-intensive and therefore, quite possibly, not sustainable over the *long durée*, as global resources for interventions diminish, as the world population increases, and as natural resources become increasingly limited. In a century or two, the techniques and approaches of biomedicine will have changed beyond recognition, not only on the periphery, but also at the heart of what passes currently for state-of-the-art medical care.

## References


Bornstein, Erica, and Peter Redfield, eds. 2010 *Forces of Compassion: Humanitarianism Between Ethics and Politics*. Santa Fe, NM: School for Advanced Research Press.

Cartwright, Elizabeth 1998 The logic of heartbeats: Electronic fetal monitoring and biomedically constructed birth. In *Cyborg Babies: From Techno-Sex to Techno-Tots*. Robbie Davis-Floyd and Joseph Dumit, eds. Pp. 240–254. New York: Routledge.

Ezeh, Alex C., Chimaraoke Izugbara, Caroline Kabiru, Sharon Fonn, Kathleen Kahn, Ashiwel S. Undieh, Akinyinka Omigbodun, and Margaret Thorogood 2010 Building capacity for public and population health research in Africa: Research training in Africa (CARTA) model. *Global Health Action* 3:5693.

Foucault, Michel 1973 *The Birth of the Clinic: An Archaeology of Medical Perception*. Alan M. Sheridan, transl. London: Tavistock Publications Limited.

Gammeltoft, Tine 2007 Sonography and sociality: Obstetrical ultrasound imaging in urban Vietnam. *Medical Anthropology Quarterly* 21(2):133–153.

Georges, Eugenia 2008 *Bodies of Knowledge: The Medicalization of Reproduction in Greece*. Nashville, TN: Vanderbilt University Press.

Goffman, Erving 1963 *Stigma: Notes on the Management of Spoiled Identity*. Harmondsworth, UK: Penguin.

Grady, Denise 2015 Ebola survivors face lingering pain, fatigue and depression. *New York Times*, August 8, 2015.

Henry, Doug, and Susan Shepler 2015 AAA 2014: Ebola in focus. *Anthropology Today* 31(1):20–21.

Inhorn, Marcia C. and Daphna Birenbaum-Carmeli 2008 Assisted reproductive technologies and culture change. *Annual Review of Anthropology* 37:177–196.

Karchmer, Eric I. 2010 Chinese medicine in action: On the postcoloniality of medical practice in China. *Medical Anthropology* 29(3):226–252.

Livingston, Julie 2012 *Improvising Medicine: An African Oncology Ward in an Emerging Cancer Epidemic*. Durham, NC: Duke University Press.

Lock, Margaret, and Vinh-Kim Nguyen 2010 *The Anthropology of Biomedicine*. Chichester, UK: Wiley-Blackwell.

Meara, John G., and Sarah L. Greenberg 2015 The Lancet Commission on Global Surgery. Global Surgery 2030: Evidence and solutions for achieving health, welfare and economic development. *Surgery* 157(5):834–835.

Swaroop, Mamta 2015 Association for Academic Surgery (AAS) and Global Surgery: Update on the G4 Alliance. Available at http://www.aasurg.org/blog/aas-and-global-surgery-update-on-the-g4-alliance/AAS.

van der Sijpt, Erica 2014 The unfortunate sufferer: Discursive dynamics around pregnancy loss in Cameroon. *Medical Anthropology* 33(5):395–410.

Kinshasa Women's Collective, 2011. Bumbu, Kinshasa, Democratic Republic of the Congo.

## *About the photograph*

*In the fall of 2011 and early winter 2012, I spent time in the DRC with a local NGO. I had three objectives: observing the national election, as an international election observer; tutoring children who had experienced interruptions in schooling; and helping implement AIDS and mental health group interventions with women and children affected by ongoing genocide. Maria, pictured here, was a beneficiary of the community reconciliation program in Kinshasa. Now 25, she was displaced and unaccompanied from the eastern DRC at the age of eight. Interventions must often facilitate socio-economic opportunity in addition to counseling, medicines or psychoeducation. Our NGOs collective offers training for women to learn a trade, so that they can exit livelihoods as prostitutes on the streets. Maria reflects:* J'ai plus personne dans ma vie. Cette collective est devenue comme famille *("I have no-one left anymore in my life. This collective has become my new home").*

—Athena Madan

# 16

# Vital Signs

## Medical Anthropology in the Twenty-first Century

*Lenore Manderson, Anita Hardon and Elizabeth Cartwright*

In introducing this volume, we linked the framing and maturing of medical anthropology to social and political movements that gained momentum worldwide in the 1960s and 1970s. Since then, medical anthropologists have continued to engage with and be inspired by social movements, as well as by political changes, new developments in biomedicine, and debates in public health. While taking on insistent and emerging questions in policy and practice and pursuing these in different local settings, we have increasingly extended beyond the conventional settings of villages and small towns. We have grown increasingly comfortable working with and within organizations and agencies concerned with health, troubling national and multilateral policies and programs with our own knowledge of local biologies and globalized political economies. And as we have pushed into these wider arenas of health, we have turned, too, to the microspheres of laboratories and hospital spaces, asking both what anthropology adds to an understanding of biology and medicine, and what a different lens—one that includes material as well as human actors—contributes to ideas of the body, health and disease, healing and care. The medical anthropology of the present is very different from the discipline in which we engaged three to four decades ago.

Some of the theories and ethnographic practices of twentieth-century medical anthropology remain important and compelling; other new questions and approaches have emerged to shape what we do, how and where; the combination of the two explain the vitality of the discipline today. Feminist activism and feminist anthropology in particular have had significant effect on the discipline, supporting both its directions of enquiry (in relation to pregnancy and childbirth, for instance) and its theory and analysis. We have established the ways in which disease is gendered as a result of the differential power that determines disease, access to resources and so health outcomes.

Globally, gender hierarchies continue to disadvantage women. Women still lack the autonomy to make their own decisions about their health. The photograph by Mark Nichter, the preface to Chapter 8, depicts a woman using a cell phone to gain her husband's permission to proceed with treatment for a chronic ulcer, and to explain to him the costs that might be involved. The woman has Buruli ulcer, an infectious bacterial infection which, untreated, will cause certain disfigurement and debility. Yet it is possible that the cost will be 'too high' from her husband's point of view. Women continue too, worldwide, to have less access to: food and so have compromised nutritional status; preventive health technologies such as bed nets to protect against malaria; transport to medical services; and the right to receive care if the provider is male. The majority of people subject to sexual assault and intimate partner violence are women. Women continue to receive less education and receive lower remuneration for their work, and so, more often than men, they live with their children in poverty. In these respects, the social impact of gender is clear. But we know far less of how sex shapes the onset and course of many diseases biologically, and so our capacity as anthropologists to tease out the synergies of the biological and social is necessarily fragmentary.

Our advances in exploring the intersections of class, age, race, disability, ethnicity and sexuality, among other social divisions, while avoiding tendencies to essentialize these 'variables,' is a continuing task in relation to the interconnections of social conditions and ill health, as one disease complicates, compounds and augurs in new diseases and other bodily conditions. In this volume, several case study authors (Dedding, Smith-Morris, Wainwright, Warren) write directly or indirectly of diabetes and its impact. Diabetes is a perfect example of these refractions and synergies, underpinning a constellation of conditions which, at worst, might include: obesity and associated immobility; cardiovascular disease; kidney disease; impaired vision; peripheral neuropathy and amputation; associated depression and death. Heading off this bleak outcome, people are asked—encouraged is too gentle a term for the imperative register of clinicians in this context—to lose weight, eat well and exercise regularly, without regard to the feasibility of this for people who are cash, time and resource poor.

As we have illustrated, people also take drugs to abate symptoms and head off complications. In Chapter 10, we drew attention to how medicine is a lucrative and ever-expanding market. So are other technologies of medicine, and in this context, medical anthropologists have increasingly turned to study how markets are made, products circulate and are used, and fortunes are built in global health (Erikson 2015). In 2015, there is in particular growing enthusiasm for the capacity of mobile devices to deliver 'm-health.' Anyone with a cell phone, regardless of its capacity, can be reminded by SMS of the need to renew a prescription, of the timing of dental or medical appointments, or of infant immunization schedules, or can receive support as a new mother or as a person living with HIV. Smartphones and tablets do more than this, though, as we described in Chapters 8 and 15. They provide the 'engines' for medical devices such as ultrasound sensors, blood glucose meters and blood pressure monitors for use by health providers and increasingly their patients. Digital stethoscopes are being developed (some might ask why this has taken so long); glasses prescriptions have been simplified using a NETRA (Near-Eye Tool for Refractive Assessment) device; doctors are able to receive, on their phones or tablets, the MRI and X-ray images from the machine that took them. Methods for detecting particular conditions—cancers and STDs—using such accessible technologies—are being developed as we write. Nanotechnology, solar cells, lithium batteries and the like are enabling affordable technology with two motivations: first, the imperative to provide high-quality diagnostics and care in much of the world which, until now, has had neither the services nor resources; and second, the lucrativeness of this potentially ever-expanding market.

Since the introduction of Internet and the growing use of cell phones, there has been considerable enthusiasm for e-health—medical informatics and the wider use of Internet resources for information and for communication—and more recently, m-health. Yet there are still basic shortcomings that impact on health systems and on timely and quality care. There remains a powerful digital divide, and so there is an unrealized promise of e-health to bridge inequities of health between and within countries where there remain substantial differences in health status between people who live in rural and urban areas. In poorer countries, there are still also shortcomings in the most basic data collection, analysis and transfer, inhibiting the operation of health services at local levels (such as primary health care centers) and nationally, where data are essential for strategic planning, the procurement of drugs and other materials, and the training and placement of health and medical staff. Deaths, for example, are still not always recorded, or are recorded without cause, or are attributed to proximate cause—the ever-ubiquitous 'cardiac arrest' or 'old age.' Perinatal and infant deaths, with or without cause, are even less likely to be reported for both technical and social reasons. What is counted, who are counted, and how this information is used, are critical matters that affect identities, governance and everyday experiences (Sangaramoorthy and Benton 2012).

Between gender—an old theme, no less urgent now—and new technologies of health, an old theme in new forms, there are vast areas where medical anthropology expertise is needed. Below, we reflect on areas which trouble or engage us as medical anthropologists, and as a way of guessing the directions we might take in the twenty-first century.

## Asylum Seeking and Detention

In writing on global quests for care (Chapter 12) and on war and violence (Chapter 13), we referred to the dynamics of displacement in the twenty-first century as a result of economic crises, human rights violations and war. As we write, this is a matter of urgency. Thousands of people risk their lives as they try to navigate from North Africa to Europe, most often first to Italy and Greece, then on to Germany and the Nordic countries; many thousands are being smuggled in and must find a way to live under the radar of authorities. The crisis in the EU is evolving at the moment and demonstrates how, in the presences of increasing numbers of refugees, the hospitality of receiving countries changes from welcoming to uncertain and then to hostile. Thousands more risk their lives to travel to Australia by boat too. Not all of these migrants survive the journey, but many who do are interned, imprisoned in countries that they imagined would both acknowledge their human rights and provide them with a safe haven. Many countries have adopted strict border control and migration-related detention to prevent undocumented migrants from making asylum claims (Fazel et al. 2014). And in the United States, presidential candidate Donald Trump called for the deportation of peoples living and working without papers, and for the end of automatic citizenship for children born of non-nationals in the US (Haddon 2015).

Research on people's health and wellbeing has focused on open refugee camps, where access for researchers is difficult but not impossible. Access is far harder in detention camps, and there are few opportunities for medical anthropological research. These camps provide people, usually, with health services, including potable water and sanitation, basic obstetric care, vaccination coverage, and treatment for acute infections. But many are overcrowded and insanitary, food and fuel are limited, shelter is rudimentary, and medical staff, faculties and drugs inadequate. Disease control and prevention are barely adequate and most camps lack diagnostic capacity and have problems in the supply of drugs and other materials. Infectious diseases are transmitted as people travel and as a result of the density and poor sanitation of many camps. People's physical health is compromised too from pre-existing chronic conditions and those that develop as populations in such camps age. Women and their children, people subject to sexual violence including within camps, and people who have experienced torture and trauma (Silverman and Nethery 2015), are generally poorly served. Livelihood activities may not be possible, and in this context of constraint and boredom, people's health is compromised. Pre-existing divisions continue in camps, and, as Matthew Wilhelm-Solomon (2013) describes, camp regimens can exacerbate social exclusion even as they seek to empower people. Deprivation, isolation and interpersonal violence, and the lack of clarity about the processes of review and time that might be spent in camps, further contribute to depression and anxiety (McGuire and Martin 2007). These health costs are accumulative, and their effects continue well after resettlement or safe repatriation, reminding us of the chronic nature of the toll exacted on them.

While in some camps, residents (and anthropologists and others) are able to come and go, growing numbers of people are forcibly detained by the governments of destination countries and cannot leave; this is the case in Australia. Here, 'unauthorized arrivals' are detained in remote detention facilities for between two to seven years, but in practice indefinitely (Australia 2008). Those who are apprehended and incarcerated are without exception at risk of poor mental heath. Their loss of agency, mobility and control deprives them of the possibility of a future for themselves and their families, leading to a compounding of trauma, depression and anxiety, and consequent behavioral disturbances, rioting, self-harm, drug misuse and high levels of suicidal attempts and success. The profound negative impacts on health and wellbeing of such incarcerations everywhere are reflected by intermittent reports of riots, depression, self-harm, interpersonal conflict and violence, sexual violence and drug trafficking (Nethery and Silverman 2015). Asylum seekers detained at the offshore processing facility on the small South Pacific Island of Nauru, for example, live in tents in hot, overcrowded conditions, with limited access to clean water, showers and toilets, dependent on airlifts of both food and water.

In Chapter 2, we wrote of the inattention to children as the subject of ethnographic enquiry. This inattention is true in every context: 'the refugee' or 'the asylum seeker' is often, by default, an adult man, despite that refugees include women and children. The limited evidence of children in detention points to their particular vulnerability to infectious disease, nutritional deprivation, lack of opportunities for schooling, lack of other activities, and exposure to harsh physical and social conditions as they witness adult distress including hunger strike, self-harm and suicide, leading to psychological and emotional problems (Fazel et al. 2014; Procter et al. 2013). 'Unaccompanied minors' and young adults, once 'released' into the community, show evidence of extreme trauma, having lived in regimes of terror, watched friends and kinsmen be raped, murdered and killed on voyages of escape, and watched adults self-harm in camps (Newman 2013; Newman et al. 2013). Reports of children also deliberately self-harming, and of suicidal behavior, sleep difficulties, attachment difficulties and somatic complaints after their release from camps into communities highlights the continued risks to their mental health (Australia 2014). Such public accounts draw attention to the importance of working with these populations after their release, as they try to establish autonomous lives, opening up, in the process, multiple questions of anthropologists in relation to adaptation, suffering, resilience and social inclusion.

## Rethinking Structural Violence

In this book, we have highlighted the importance of structural violence, which Paul Farmer describes as "the degree to which agency is constrained is correlated inversely, if not always neatly, with the ability to resist marginalization and other forms of oppression" (2004: 307). Structural violence causes people to be stressed and feel hopeless, thereby contributing to poor mental health (Kohrt and Mendenhall 2015). In Chapter 5, we considered the stresses in everyday life, their medicalization and management; in Chapter 14, we pointed to the developments in epigenetics, which suggest that such feelings of stress and hopelessness can influence our genes.

Epigenetics refers to heritable changes in phenotype due to processes that are independent of the primary DNA sequence of an organism. Research in this field is leading to new paradigms for genetic inheritance and evolution, beyond the 'hard heredity' models of Mendel and Darwin, and this has lead to the rehabilitation of some of their contemporaries, such as Lamarck and Lysenko (Meloni 2015). For scholars of structural violence, the importance of this field of research lies in the blurring of boundaries between natural and social inequalities. Epigenetic research suggests that living in violent communities can affect gene-expression in the people of the current generation as well as in their children, which has huge implications for our understanding of genetics. For epigeneticists, heredity is no longer the simple transmission of DNA from one generation to the next; living conditions can actually change DNA. How do these insights affect our understanding of structural violence? What if everyday living conditions that make people feel stressed and hopeless become part of their being to the extent that violent behaviors become part of our genomic makeup? What are the implications of epigenetics for our interventions in the field of global health? How do they change our thinking about nature and nurture? We need to rethink our paradigms for mental health, while at the same time examining the hypes and hopes that accompany this new field of research.

## Humans, Other Living Things and the Environment

Margaret Lock (1993), in her work on menopause in Japan and the US, came up against the ramifications of these entanglements over two decades ago—and from that encounter coined the term 'local biologies.' She argued that it was important to understand how different environmental contexts uniquely shape health risks, health outcomes, genetics and genetic potentials of individuals, and how this in turn resulted in different constellations of embodied experiences. Beyond the local biologies of humans, we also dwell in environments that are so altered from their pre-industrialized state that they are significantly changing all the other living things present in the mix. Bruno Latour talks about this, in the context of human history on this planet, as "internalized ecology": "The intense socialization, re-education, and reconfiguration of plants and animals—so intense that they change shape, function and often genetic makeup" (1999: 208). Our reconceptualization of human–animal interactions is taking account of the emergent and flexible character of the human body and its entanglement with the environment. This embeddedness produces a body which is imprinted by its social, political and material environment, and which differs considerably from context to context (Wolf 2015: 9).

In Chapter 11, "The Anthropocene," we drew attention to the continued acceleration of environmental vulnerability with increased population pressure, pollution and associated climate change, and with this, we have witnessed both loss of biodiversity and species adaptation, resulting in changes in disease risks for humans. Zoonotic infections provide a clear link between we humans and other animals. Most zoonotic infections impact particularly on the poorest and most vulnerable people in the world, and affect families, communities and nations economically, but also economic growth and business interests, as discussed in Chapter 7. Environmental control activities conducted to reduce the transmission of vector-borne diseases, such as improvements in hygiene and sanitation, water management, vector control, early diagnosis and treatment, and personal protection, were largely introduced over a century ago. Consider the sanitation drives in Freetown, the capital of Sierra Leone in 1899 to clear the city of potential breeding sites to prevent malaria; in Cuba in 1901 and in Panama the following year against yellow fever and malaria (Ross 1967 [1902]: 8, 9). In British Malaya at this same time, malaria control programs included environmental management, vector control, mass screening and treatment for anemia and enlarged spleen, and the use of quinine for presumptive treatment (Manderson 1996). The strategies have changed little for over a hundred years, despite the extensive biomedical and entomological research throughout this period, as both parasites and vectors have outpaced human inventiveness and rapidly developed resistance to chemical and pharmaceutical interventions.

Vulnerability then and now was underscored by the overcrowded and rudimentary living conditions in rural areas where breeding conditions were ideal, particularly as geographical regions were appropriated. Forest clearing and drainage have created new ecological niches ideal for breeding new and stronger zoonotic diseases; the building of mega-cities, roads, railways and airports then and now has created new habitat for vectors of viral, bacterial and parasitic diseases despite that urbanization, by reducing human–animal contact, reduces the risk of other infections. Ecohealth, ecobiopolitics and ecorisk will become increasingly important concepts as we try to make sense of new mutations and repeated epidemics of diseases like Ebola, Chagas disease, malaria and influenza, threatening all mammalian hosts. Vector-borne infectious diseases are a major category of climate-influenced risks to health in sub-Saharan Africa, Asia and South America.

## Climate Adaptation and Resilience

We see the Anthropocene as conceptually critical to our scholarship, and as global citizens, personally critical in the decades ahead. Global warming is an increasing threat to human settlement, health and wellbeing. In a hotter world, with more evaporation, overall precipitation will increase, but its geographical distribution and seasonality will change. Poleward shifts of large-scale atmospheric circulation systems have resulted in dryer conditions in semi-arid regions, with changes in 'storm tracks' affecting future agriculture and food supply, viable cultivars, local and national productivity, community structures and household incomes. Increased mean ambient temperature, and the increased duration and severity of heat waves, will impact continental areas, such as in South Africa and Australia, with increases in the heat island effects of the cities in these areas. People will be forced to confront changing environments, including environmental degradation, with the effects compounded by continued population growth, urbanization and potential civil disorder (for further discussion, see Baer and Singer 2008; 2014).

It is increasingly difficult to ignore climate change from any vantage point, but for those living on atolls and archipelagos, on river flats and the foreshores of estuaries, in major urban settings such as Dhaka, Bangkok, Freetown, Dakar and Dar es Salaam, there is a clear risk of damage to existing settlements from sea rise. Climate migration is already a reason for population movement in the Pacific. When Lenore Manderson was working in the Solomon Islands in 2011 and 2012, people recurrently described entire communities moving to large islands from small atolls, and moving inland and upland from villages built on stilts along the foreshores. Solomon Islanders understood the vulnerability of their homes, gardens and fishing grounds to cyclones, tsunamis and tidal inundation; they spoke routinely of global warming and sea rise, incredulous that some people might regard this with skepticism. The Pacific islands that we know so well through the works of Bronislaw Malinowski (1922; 1929; 1935) and Raymond Firth (1936)—the Trobriand Islands (Papua New Guinea) and Tikopia (Solomon Islands)—are so affected.

Like the distribution of infectious disease, so the influence of climate change on human health is already felt particularly in poor countries among the poorest people. Drought, changes in precipitation and floods all affect water quality and quantity for household use, impacting agricultural production, crop choice, hunting and fishing, and so food security. In Argentina, drought compounds other socioeconomic stressors as families respond to crop failure, loss of livestock and indebtedness by moving to farming areas with better current climatic conditions, or abandoning farms and relocating to cities (Wehbe et al. 2007). Global warming in the Arctic, as a very different example, has dramatically narrowed the times when people can hunt, the speed at which they must work to prevent meat spoilage, and the impact of this on other economic activities and on food reserves for the rest of the year. As one man from interior Alaska reflected, "we've got such a short window that when that window's there everybody's got to go out . . . the seasons are really putting a crimp on traditional subsistence activities of going out when it's appropriate, when the weather, you can take care of the meat and so on" (McNeeley and Shulski 2011: 469) (see Johnson's case study in Chapter 11). Lack of wild food forces dietary change, and with it, increasing dependence on nutritionally poor, expensive store-purchased food, contributing to the increased incidence of non-communicable diseases (Chapter 7). Poor nutrition and the increased risk of cardiometabolic disease is only one dimension of the impact of changing climate on human health. As Hans Baer (2008) illustrates, there is already an increase in respiratory illness from wind-borne dust, and increased diarrheal diseases, cholera, leptospirosis and *E.coli* bacterial infection as a result of chronic water pollution and flood-related contamination. There is also, already, changes in the habitat of vectors and the epidemiology of parasitic infections and arboviruses (Alley and Sommerfeld 2014; Bezirtzoglou et al. 2011).

Amber Wutich and Alexandra Brewis (2014) have argued the need for research on the impact of water and food insecurity on individual emotional wellbeing and mental health, as well as its detrimental impacts on households, including from increased indebtedness and unemployment, family disruption, increased illness and mental health problems (see also Berry et al. 2011). Wutich and Brewis (2014) also argue the need to explore 'coping' mechanisms at both household and community levels in face of such resource insecurity. A growing number of anthropologists have illustrated how people have been able to adapt to climate change, drawing on local knowledge of climate variation, and associated changes in the plant and animal life. At the same time, as we witness now, international, regional and local priorities, local histories of environmental change and action, and systems of governance all shape adaptation, mitigation and resilience (Lazrus 2015; McNeeley and Lazrus 2014; Orlove et al. 2014).

## Towards an Anthropology of Us

While infectious disease control is concerned with the detrimental effects to our bodies of specific pathogens, and climate mitigation might begin to address epidemiological shifts parallel to local ecological changes, a new line of research has emerged in human biology which focuses on the microbiome, which is defined as "the ecological community of commensal, symbiotic, and pathogenic microorganisms that literally share our body space" (Lederberg and McCray 2001: 8). The whole genome sequencing technologies that we described in Chapter 14 have made it possible to do rapid identification of these organisms, which has resulted in the frequently cited statistic that only 10 percent of the DNA in our bodies is actually human DNA. Donna Haraway suggests that it is therefore more accurate to describe our bodies as "us" rather than "me" to account for this complexity, arguing: "I am vastly outnumbered by my tiny companions; better put, I become an adult being in company with these tiny messmates. To be one is always to become with many" (2007: 4). Inspired by these developments, anthropologists and other social scientists are turning their attention to the symbiotic relations between humans, microorganisms and other living things (Wolf-Meyer and Collins 2012). This was the topic of the conference of the Association of Social Anthropologists of the United Kingdom and Commonwealth in April 2015. Increasing progress in the field of microbiome research points to the role of the intestinal microbiome and levels of inflammation in human health and mental wellbeing. Imbalances in intestinal flora have been related to diabetes, cardiovascular disorders and Alzheimer's disease, among others (Yatsunenko et al. 2012). Gut health has been found to be related to our eating patterns (Selhub et al. 2014), with overeating causing obesogenic bacterial growth; this in turn influences the so-called endocannabinoid system which regulates brain signals that it is time to eat (Romijn et al. 2008; Verdam et al. 2013). At the same time beneficial microbes—'probiotica'—can have therapeutic value for people suffering from depression and fatigue, decreasing anxiety and perceptions of stress, and contributing to healthy glucose tolerance. Not surprisingly, these products are emerging on the health market worldwide as cure-alls.

For medical anthropologists, the insights into the microbiome demand a radical rethinking of the body as the existential ground of self as defined by phenomenologists such as Csordas (1990), as well as the mindful body, proposed by Margaret Lock and Nancy Scheper-Hughes (1986) as a way to integrate the existential body with the social body and the body politic. It calls for a more ecological understanding of our being in the world, where the boundaries between us and the foods that we eat, and the soil upon which we walk, are related to health and happiness. These complex issues require us to rethink our role in global health interventions, where advocating for hygiene, use of antibiotics and eradication of microbes may not necessarily good for health. To understand the dynamics we will need to work more closely with biomedical researchers, biologists, ecologists and physical anthropologists, while at the same time being alert to the overstatements that travel with these new biological knowledges.

## Governing the Global Environment

As we take our new found biological knowledges with us and move further into the Anthropocene, we see the effects of pollution and overpopulation putting pressure on water, food and even, ultimately, air supplies. There is a risk of an increase in human-caused environmental disasters, despite that we are more aware of these than ever before. The increased incidence of environmental hazards will continue to pit citizens against governments and big industry, precipitating health litigation and strife around environmentally caused illnesses. Medical anthropology has a critical role to play in giving voice to those negatively affected by these circumstances, in documenting illnesses associated with them, and in maintaining a critique of the exercise of power involved in their resolution, as Vincanne Adams (2013) illustrated in the aftermath of Hurricane Katrina (see also her case study, Chapter 11).

The mining industry is a good example (one of many) of a rapidly expanding, environmentally disastrous, set of behaviors that humans are engaged in to the profound detriment of their health and the health of the earth (Donoghue 2004). Mining is a dangerous activity in many ways for people working in the mines, their families who live in close proximity to mines, and those in the larger watersheds and downwind from the mines. In coal mines, for instance, chemical exposures occur from breathing in the silica and the coal dust that is released into the environment, causing coal mine dust lung disease (CMDLD), a group of illnesses that includes black lung, silicosis, dust-related diffuse fibrosis and COPD. Other chemicals are commonly released in mines and environs, including asbestos, diesel particulates and arsenic. Small-scale artisanal gold mines release airborne mercury that causes neurological, respiratory and cognitive problems to those who come in contact with it (Webster 2012). These chemically induced illnesses are also spread into the environment through polluted watersheds where acid mine drainage that arises from the interactions between water and pyrite creates sulphuric acid that then leaches out heavy metals into the surrounding streams and lakes; it sickens and kills plants, animals and humans (Kirsch 2008). Psychologically, the stresses of mining include living in suboptimal conditions, risky sex and violence, and in the widespread use of addictive drugs and the copious amounts of alcohol that often make these scenes of environmental wreckage profoundly toxic at many levels (Cartwright 2016).

Workers are routinely at risk as they go about their everyday work. Adrienne Pine's case study in Chapter 13 provides a provocative example of how nursing in the context of the everyday violence of Honduras has a particular constellation of risks; the miners in the oil sands of Canada described in Chapter 11 are another occupational group whose unique ecobiological context results in a particular constellation of health problems resulting from chemical and physical hazards—unique 'eco-risks.' Workers in off-shore drilling, deep-sea fishing, nuclear power plants and other, newly created occupations are inserted in a very bodily fashion into ways of being that have profound effects on one's health and wellbeing. Conversely, the lack of one lifelong occupation—as seen in individuals moving through many types of jobs in their lifetimes—makes tracing the health effects of a particular industry difficult to prove, especially in the case of class action suits and other forms of retaliative litigation.

Issues of governmentality, and the ways that states exercise control over their populations, arise as individuals of differing ethnic and socioeconomic resources are put into the position of needing to attain recourse for the health effects coming from environmental and industrial pollutants. The NIMBY ('not in my backyard') syndrome has been long documented in anthropology (Douglas and Wildavsky 1983); while enjoying the benefits of cheap energy manufactured from coal and oil, while using our personal devices that run on the conflict mineral coltan, and while consuming beautiful things made of gold, silver and diamonds, someone else's health and the health of their communities is compromised (Strauss et al. 2013). Moving towards the future, dumping of wastes on land and sea, cross-border litigation, and the pressures that this will put on human habitat and social life, will be a critical focus of medical (and other) anthropologists.

## Revisiting Cultural Competency

In the introduction to this volume, in passing we referred to our roles in the practical application of medical anthropology, and from this, the important development of rapid assessments and focused ethnographic manuals. This was linked in part to the task of integrating anthropological methods and interests within collaborative research, but also to aiding the translation of our work into practice. This move partly reflected the interests of epidemiologists, clinicians and other health professionals of how 'culture' affected clinical consultations, hospitalization, health communication, and community participation in public health interventions. It reflected, too, ethical concerns about the value of our research to the people with whom we worked, and the expanding environments in which anthropologists were working outside of the academy: local governments and national health ministries, development aid programs and development banks, non-governmental organizations, community services and multilateral agencies. Anthropology had an important role, we demonstrated, in both local and international settings to ensure that policies, programs and protocol were appropriate to the setting (Whiteford and Manderson 2000; Whiteford and Whiteford 2005).

Interests in translating medical anthropological research into public health and clinical services resonated with growing concerns about the 'cultural appropriateness' of services to people from diverse backgrounds. Arthur Kleinman's 'explanatory model' proved helpful to explain to non-anthropological health workers cultural variations in the etiology and symptomatology of different conditions, while also contributing to anthropological research into cultural variability in symptomatology and preferred treatment for a range of mental and physical health conditions (Good and Kleinman 1985; Good et al. 2010; Hahn 1985; Kleinman 1980; Lock 1993). Although now we might argue that this approach privileged and reified 'culture' over other variables (class, ethnicity, gender, for instance), it made anthropology and its methods accessible to others, as reflected to the kinds of questions included in various manuals, with the inclusion of cultural-bound syndromes in *DSM-IV* (American Psychiatric Association 1994; Guarnaccia and Rogler 1999; Mezzich et al. 1999). This led to various studies testing the extent of cultural-specificity (Lee 2002), and its radical revision in the most recent diagnostic and statistical manual—*DSM-5*—which has replaced culture-bound syndromes with *cultural syndromes, cultural idioms of distress*, and *cultural explanations of distress and its causes* (American Psychiatric Association 2013: 758).

The attention to 'cultural difference' should not efface the importance of structural and historic factors that result in different health outcomes. Even so, there is utility to this approach in part in addressing such structural gaps, particularly as clients of health services have drawn attention to the lack of cultural appropriateness, responsiveness or sensitivity to their needs, and as we anthropologists have explained why this is so. In elaborating on this, Cristiana Giordano (2014) highlights the 'logics of difference' and the challenges of translation in the use of language, the choice of premises for consultations, and in therapeutic practice. Her focus is the ethnopsychiatric clinic in Turin, Centro Frantz Fanon, established in 1996 with psychiatrists, psychologists and anthropologists (including psychiatrist and anthropologist Roberto Beneduce, see Beneduce and Martelli 2005). The clinic operates with collaborations between cultural mediators and clients, and the judicious combination of different therapeutic approaches—faith-based practices, healing rituals from immigrants' home countries, medical psychiatric therapies and social support. In doing so, the clinicians engage in what Giordano describes as a "clinical listening" that is informed by a politics of recognition; they aim "to fight deep-seated forms of social injustice in relation to identity and difference" (2014: 8, 9). This adds to our understanding of the difficulties faced in developing effective, acceptable health and social services with and for other populations with long histories of subjugation and abuse, and contributes to an urgent discussion among Indigenous peoples, worldwide, in relation to cultural safety. At its core, the principle is one of respect, and of the significance of respect of other peoples for social repair as well as physical healing. Emerging examples of 'better than expected' health outcomes point to the importance of community engagement and support, while acknowledging the tenuousness of goodwill towards and trust in health care providers from dominant populations (Anderson and Kowal 2012; Giles and Darroch 2014; Josewski 2012). Given continued health disparities across populations and continued social exclusion, as medical anthropologists we need to revisit ideas around cultural appropriateness and 'cultural safety,' a term and implicit approach used by Maori, Aboriginal Canadians and Indigenous Australians, and to work to support communities struggling to ensure that this is standard procedure.

## How Medicine Is Practiced

As new evidence emerges in biomedicine and as current 'evidence' is shown to be incomplete and incorrect, practitioners will be faced with changing the ways they practice, or will try to continue to practice medicine consistent with old and out-dated guidelines with which they feel comfortable, such as PSA tests for prostate cancer or fecal occult blood tests for colorectal cancer. Screening policy varies worldwide, not because of the lack of equipment and cost alone, but also because of the uncertainty of what we accept as the current 'best practices.' What is the evidence and what are the recommendations in the US at the time of this writing (mid-2015) for breast cancer screening, for example? Currently, the American Cancer Association and the American Congress of Obstetricians and Gynecologists recommend that annual mammography be offered to low-risk women 40 and older. In a slightly less conservative take on the issues, the US Preventive Services Taskforce draft proposal of new guidelines, released on April 20, 2015, increases age at first mammogram and decreases frequency of screening, recommending that low-risk women 50–74 be offered a mammogram bi-annually. And finally, according to a recent statistical study published in the *JAMA Internal Medicine Journal*, "(a)mong 1000 US women aged 50 years who are screened annually for a decade, 0.3 to 3.2 will avoid a breast cancer death, 490 to 670 will have at least 1 false alarm, and 3 to 14 will be over diagnosed and treated needlessly" (Welch and Passow 2014: 448). These guidelines are difficult to interpret and implement for both clinicians and patients.

This situation brings up two issues that will continue to be critical components of clinical realities long into the future. First, both practitioners and patients have long been subjected to public health information that is targeted at them to increase their participation in screening. Fear, guilt and shame are commonly integrated into these 'health promotion' materials. Women in the US and in some other high-resource countries have internalized these messages and indeed now often demand screenings such as annual mammography, even when it may not be warranted. Physicians may allow or even encourage patients to screen for reasons that include their own fears of the disease and fears of being sued or blamed were their patient to contract breast cancer at some time in the future. Second, clinical guidelines are based on evidence and so, rightly, these change as new evidence is amassed regarding the efficacy of screening procedures, techniques and timing. As Welch and Passow (2014) show, increasingly sophisticated and reliable large databases are being established that allow standard practices to be re-evaluated. Techniques of visualization are refined, as are our understandings of human physiology. Thus, guidelines that are familiar to practitioners and patients change—this can be an uncomfortable process because inherently, it is an acknowledgment of the fact that we often know much less than we think or pretend to know about common, serious health conditions.

Health policy makers worldwide are increasingly viewing the health of their populations through interconnected sets of 'big data.' These provide them with dashboards, which illustrate progress made on key health indicators and highlight the inequities still present. Without us knowing, data from our GPs are being connected to those of our insurers, gyms and supermarkets. While big data analysis can help prioritize health interventions to those most in need, they can also be used to monitor whether we are adhering to healthy lifestyle prescriptions. We are able to acquire gadgets through which we can monitor how many steps we take per day, what we eat, and what our blood pressure is. Mobile phone companies have the capacity to track where we are, and who we are in touch with, while Facebook posts can be 'swept' to target advertising at the same time that we share what concerns us and how we feel. Anthropologists need to reflect on the ethics of these new ways of monitoring populations, at a time when there seems to be less and less concern about privacy (but see Manderson et al. 2015; Nuttall and Mbembe 2015; Sundaram 2015).

As Foucault described in *The Birth of the Clinic* (1973), the available technologies of visualization create the possibilities for the clinical gaze to penetrate into the recesses of the body; this is reflected in the depth of power that practitioners, health policy makers, insurers, phone companies and supermarket chains believe that they have over biological life itself. As technologies change, so too does the extent of that gaze. Whether we willingly strap on personal monitoring devices, or are subjected to various forms of possibly harmful radiographic tests, through a mixture of control and consent the hegemony of biomedicine will be reified in new and ever more interesting and twisted ways.

## References

Adams, Vincanne 2013 *Markets of Sorrow, Labors of Faith: New Orleans in the Wake of Katrina*. Durham, NC: Duke University Press.

Alley, Christopher, and Johannes Sommerfeld 2014 Infectious disease in times of social and ecological change. *Medical Anthropology* 33(2):85–91.

American Psychiatric Association 1994 *Diagnostic and Statistical Manual of Mental Disorders, 4th ed. (DSM-IV)*. Arlington, VA: American Psychiatric Association.

——— 2013 *Diagnostic and Statistical Manual of Mental Disorders, 5th ed. (DSM-5).* Arlington, VA: American Psychiatric Association.

Anderson, Heather, and Emma Kowal 2012 Culture, history, and health in an Australian aboriginal community: The case of Utopia. *Medical Anthropology* 31(5):438–457.

Australia, Human Rights and Equal Opportunity Commission 2008 *Immigration Detention Report: Summary of Observations Following Visits to Australia's Immigration Detention Facilities.* Sydney, Australia: Human Rights and Equal Opportunity Commission.

Australia, Human Rights Commission 2014 *The Forgotten Children: National Inquiry into Children in Immigration Detention 2014.* Sydney, Australia: Australian Human Rights Commission.

Baer, Hans A. 2008 Toward a critical anthropology on the impact of global warming on health and human societies. *Medical Anthropology* 27(1):2–8.

Baer, Hans A., and Merrill Singer 2008 *Global Warming and the Political Ecology of Health: Emerging Crises and Systemic Solutions.* Walnut Creek, CA: Left Coast Press.

——— 2014 *The Anthropology of Climate Change: An Integrated Critical Perspective.* Abingdon, UK: Routledge.

Beneduce, Roberto, and Pompeo Martelli 2005 Politics of healing and politics of culture: Ethnopsychiatry, identities and migration. *Transcultural Psychiatry* 42(3):367–393.

Berry, Helen L., Anthony Hogan, Jennifer Owen, Debra Rickwood, and Lynn Fragar 2011 Climate change and farmers' mental health: Risks and responses. *Asia-Pacific Journal of Public Health* 23:119S–132S.

Bezirtzoglou, Christos, Konstantinos Dekas, and Ekatherina Charvalos 2011 Climate changes, environment and infection: Facts, scenarios and growing awareness from the public health community within Europe. *Anaerobe* 17(6):337–340.

Cartwright, Elizabeth 2016 Mining and its health consequences: From Matewan to fracking. In *A Companion to Environmental Health: Anthropological Perspectives.* Merrill Singer, ed. Pp. 417–434. Hoboken, NJ: Wiley.

Csordas, Thomas 1990 Embodiment as a paradigm for anthropology. *Ethos* 18:5–47.

Donoghue, A. Michael 2004 Occupational health hazards in mining: An overview. *Occupational Medicine* 54(5):283–289.

Douglas, Mary, and Aaron Wildavsky 1983 *Risk and Culture: An Essay on the Selection of Technological and Environmental Dangers.* Berkeley, CA: University of California Press.

Erikson, Susan L. 2015 Secrets from whom? Following the money in global health finance. *Current Anthropology* 56(S12):306–316.

Farmer, Paul 2004 An anthropology of structural violence. *Current Anthropology* 45(3):305–325.

Fazel, Mina, Unni Karunakara, and Elizabeth A. Newnham 2014 Detention, denial, and death: Migration hazards for refugee children. *The Lancet Global Health* 2(6):e313–e314.

Firth, Raymond 1936 *We the Tikopia: A Sociological Study of Kinship in Primitive Polynesia.* London: Allen and Unwin.

Foucault, Michel 1973 *The Birth of the Clinic: An Archaeology of Medical Perception.* Alan M. Sheridan, transl. London: Tavistock Publications Limited.

Giles, Audrey R., and Francine E. Darroch 2014 The need for culturally safe physical activity promotion and programs. *Canadian Journal of Public Health-Revue Canadienne De Sante Publique* 105(4):E317–E319.

Giordano, Cristiana 2014 *Migrants in Translation: Caring and the Logics of Difference in Contemporary Italy.* Berkeley, CA: University of California Press.

Good, Byron J., Carla Raymondalexis Marchira, Nida Ul Hasanat, Muhana Sofiati Utami, and M.A.Subandi 2010 Is "chronicity" inevitable for psychotic illness?: Studying heterogeneity in the course of schizophrenia inYogyakarta, Indonesia. In *Chronic Conditions, Fluid States: Chronicity and the Anthropology of Illness.* Lenore Manderson and Carolyn Smith-Morris, eds. Pp. 54–74. New Brunswick, NJ: Rutgers University Press.

Good, Byron, and Arthur Kleinman 1985 Culture and anxiety: Cross-cultural evidence for the patterning of anxiety disorders. In *Anxiety and the Anxiety Disorders.* A. Hussain Tuma and Jack Mazur, eds. Pp. 297–323. New York: L. Earlbaum.

Guarnaccia, Peter J., and Lloyd H. Rogler 1999 Research on culture-bound syndromes: New directions. *American Journal of Psychiatry* 156(9):1322–1327.

Haddon, Heather 2015 Donald Trump says he would deport illegal immigrants. *Wall Street Journal* August 16.

Hahn, Robert 1985 Culture-bound syndromes unbound. *Social Science & Medicine* 21(2):165–171.

Haraway, Donna 2007 *When Species Meet.* Minneapolis, MN: University of Minnesota Press.

Josewski, Viviane 2012 Analysing 'cultural safety' in mental health policy reform: Lessons from British Columbia, Canada. *Critical Public Health* 22(2):223–234.

Kirsch, Stuart 2008 Social relations and the green critique of capitalism in Melanesia. *American Anthropologist* 110(3):288–298.

Kleinman, Arthur 1980 *Patients and Healers in the Context of Culture: An Exploration of the Borderland Between Anthropology, Medicine, and Psychiatry*. Berkeley, CA: University of California Press.

Kohrt, Brandon A., and Emily Mendenhall 2015 Social and structural origins of mental illness in global context. In *Global Mental Health: Anthropological Perspectives*. Brandon A. Kohrt and Emily Mendenhall, eds. Pp. 51–55. Walnut Creek, CA: Left Coast Press.

Latour, Bruno 1999 *Pandora's Hope: Essays on the Reality of Science Studies*. Cambridge, MA: Harvard University Press.

Lazrus, Heather 2015 Risk perception and climate adaptation in Tuvalu: A combined cultural theory and traditional knowledge approach. *Human Organization* 74(1):52–61.

Lederberg, Joshua, and Alexa T. McCray 2001 'Ome sweet 'omics: A genealogical treasury of words. *The Scientist* 15(7):8.

Lee, Sing 2002 Socio-cultural and global health perspectives for the development of future psychiatric diagnostic systems. *Psychopathology* 35(2–3):152–157.

Lock, Margaret 1993 *Encounters with Aging: Mythologies of Menopause in Japan and North America*. Berkeley, CA: University of California Press.

Lock, Margaret, and Nancy Scheper-Hughes 1986 The mindful body: A prolegomenon to future work in medical anthropology. *Medical Anthropology Quarterly* 1(1):6–41.

Malinowski, Bronislaw 1922 *Argonauts of the Western Pacific*. London: G. Routledge & Sons.

——— 1929 *The Sexual Life of Savages in North-Western Melanesia*. New York: Eugenics Publishing Company.

——— 1935 *Coral Gardens and Their Magic*. London: Routledge.

Manderson, Lenore 1996 *Sickness and the State: Health and Illness in Colonial Malaya, 1870–1940*. Cambridge: Cambridge University Press.

Manderson, Lenore, Mark Davis, Chip Colwell, and Tanja Ahlin 2015 On secrecy, disclosure, the public and the private in anthropology. *Current Anthropology* 56(S12):S183–S190.

McGuire, Sharon, and Kate Martin 2007 Fractured migrant families: Paradoxes of hope and devastation. *Family & Community Health* 30(3):178–188.

McNeeley, Shannon M., and Heather Lazrus 2014 The cultural theory of risk for climate change adaptation. *Weather, Climate and Society* 6(4):506–519.

McNeeley, Shannon M., and Martha D. Shulski 2011 Anatomy of a closing window: Vulnerability to changing seasonality in Interior Alaska. *Global Environmental Change: Human and Policy Dimensions* 21(2):464–473.

Meloni, Maurizio 2015 Epigenetics for the new social sciences: Justie, embodiment, and inheritance in the postgenomic age. *New Genetics and Society* 34(2):125–151.

Mezzich, Juan E., Lawrence Kirmayer, Arthur Kleinman, and Spiro Manson 1999 The place of culture in *DSM-IV*. *Journal of Nervous and Mental Disease* 187(8):457–464.

Nethery, Amy, and Stephanie Silverman, eds. 2015 *Understanding Immigration Detention: The Migration of a Policy and Its Human Impact*. London: Routledge.

Newman, Louise 2013 Seeking asylum—trauma, mental health, and human rights: An Australian perspective. *Journal of Trauma & Dissociation* 14(2):213–223.

Newman, Louise, Nicholas Procter, and Michael Dudley 2013 Seeking asylum in Australia: Immigration detention, human rights and mental health care. *Australasian Psychiatry* 21(4):315–320.

Nuttall, Sarah, and Achille Mbembe 2015 Secrecy's softwares. *Current Anthropology* 56(S12):S317–S324.

Orlove, Ben, Heather Lazrus, Grete K. Hovelsrud, and Alessandra Giannini 2014 Recognitions and responsibilities on the origins and consequences of the uneven attention to climate change around the world. *Current Anthropology* 55(3):249–275.

Procter, Nicholas G., Diego De Leo, and Louise Newman 2013 Suicide and self-harm in immigration detention. *The Medical Journal of Australia* 199(11):730–732.

Romijn, Johannes A., Eleonora P. Corssmit, Louis M. Havekes, and Hanno Pijl 2008 Gut-brain axis. *Current Opinion in Clinical Nutrition and Metabolic Care* 11(4):518–521.

Ross, Ronald 1967 [1902] *Researches on Malaria. Nobel Lecture, December 12, 1902*. In *Nobel Lectures, Physiology or Medicine 1901–1921*. Amsterdam, the Netherlands: Elsevier Publishing Company.

Sangaramoorthy, Thurka, and Adia Benton 2012 Enumeration, identity and health: Introduction. *Medical Anthropology* 31(4):287–291.

Selhub, Eva, Alan Logan, and Alison Bested 2014 Fermented foods, mirobiota and mental health: Ancient practice meets nutritional psychiatry. *Journal of Physiological Anthropology* 33:2.

Silverman, Stephanie, and Amy Nethery 2015 Introduction: Understanding immigration detention. In *Immigration Detention: The Migration of a Policy and Its Human Impact*. Amy Nethery and Stephanie Silverman, eds. Pp. 1–12. London: Routledge.
Strauss, Sarah, Stephanie Rupp, and Thomas Love, eds. 2013 *Cultures of Energy: Power, Practices and Technologies*. Walnut Creek, CA: Left Coast Press.
Sundaram, Ravi 2015 Publicity, transparency and the circulation engine: The media sting in India. *Current Anthropology* 56(S12):S297–S305.
Verdam, Froukje J., Susana Fuentes, Charlotte de Jonge, Erwin G. Zoetendal, Runi Erbil, Jan Willem Greve, Wim A. Buurman, Willem M. de Vos, and Sander S. Rensen 2013 Human intestinal microbiota composition is associated with local and systemic inflammation in obesity. *Obesity* 21:e607–615.
Webster, Paul C. 2012 Not all that glitters: Mercury poisoning in Colombia. *The Lancet* 379(9824):1379–1380.
Wehbe, Monika, Hallie Eakin, Roberto Seiler, Marta Vinocur, Cristian Avila, Cecilia Maurutto and Gerardo Sánchez Torres 2007 Local perspectives on adaptation to climate change: Lessons from Mexico and Argentina. In *Climate Change and Adaptation*. Neil Leary, James Adejuwon, Vicente Barros, Ian Burton, Jyoti Kulkarni and Rodel Lasco, eds. Pp. 315–331. London: Earthscan.
Welch, H. Gilbert, and Honor J. Passow 2014 Quantifying the benefits and harms of screening mammography. *JAMA Internal Medicine* 174(3):448–454.
Whiteford, Linda M., and Lenore Manderson, eds. 2000 *Global Health Policy, Local Realities: The Fallacy of the Level Playing Field*. Boulder, CO: Lynne Reinner Publishers.
Whiteford, Linda M., and Scott Whiteford, eds. 2005 *Globalization, Water and Health: Resource Management in Times of Scarcity*. Santa Fe, NM: School of American Research Press.
Wilhelm-Solomon, Matthew 2013 The priest's soldiers: HIV therapies, health identities, and forced encampment in Northern Uganda. *Medical Anthropology* 32(3):227–246.
Wolf, M. 2015 Is there really such a thing as "one health"? Thinking about a more than human world from the perspective of cultural anthropology. *Social Science & Medicine* 129:5–11.
Wolf-Meyer, Matthew J., and Samuel Collins 2012 Parasitic and symbiotic—The ambivalence of necessity. *Semiotic Review* 1:1–6.
Wutich, Amber, and Alexandra Brewis 2014 Food, water, and scarcity: Toward a broader anthropology of resource insecurity. *Current Anthropology* 55(4):444–468.
Yatsunenko, Tanya, Federico E. Rey, Mark J. Manary, Indi Trehan, Maria Gloria Dominguez-Bello, Monica Contreras, Magda Magris, Glida Hidalgo, Robert N. Baldassano, Andrey P. Anokhin, Andrew C. Heath, Barbara Warner, Jens Reeder, Justin Kuczynski, J. Gregory Caporaso, Catherine A. Lozupone, Christian Lauber, Jose Carlos Clemente, Dan Knights, Rob Knight, and Jeffrey I. Gordon 2012 Human gut microbiome viewed across age and geography. *Nature* 486(7402):222–227.

# Index

*Page numbers in italics refer to figures or photos.*